Human-Computer Interaction

Fundamentals

Human Factors and Ergonomics

Series Editor
Gavriel Salvendy

Human-Computer Interaction

Fundamentals

Edited by

Andrew Sears
Julie A. Jacko

CRC Press
Taylor & Francis Group
Boca Raton London New York

CRC Press is an imprint of the
Taylor & Francis Group, an **informa** business

This material was previously published in *The Human-Computer Interaction Handbook: Fundamentals, Evolving Technologies and Emerging Applications, Second Edition,* © Taylor & Francis, 2007.

CRC Press
Taylor & Francis Group
6000 Broken Sound Parkway NW, Suite 300
Boca Raton, FL 33487-2742

First issued in paperback 2017

© 2009 by Taylor & Francis Group, LLC
CRC Press is an imprint of Taylor & Francis Group, an Informa business

No claim to original U.S. Government works

ISBN-13: 978-1-4200-8881-6 (hbk)
ISBN-13: 978-1-138-11660-3 (pbk)

Library of Congress Cataloging-in-Publication Data

Human-computer interaction. Fundamentals / editors, Andrew Sears, Julie A. Jacko.
     p. cm. -- (Human factors and ergonomics)
  "Select set of chapters from the second edition of The Human computer interaction handbook"--Pref.
  Includes bibliographical references and index.
  ISBN 978-1-4200-8881-6 (hardcover : alk. paper)

  1. Human-computer interaction. I. Sears, Andrew. II. Jacko, Julie A. III. Human-computer interaction handbook. IV. Title. V. Series.
QA76.9.H85H8566 2008   004.01'9--dc22
2008049134

Visit the Taylor & Francis Web site at
http://www.taylorandfrancis.com

and the CRC Press Web site at
http://www.crcpress.com

For Beth, Nicole, Kristen, François, and Nicolas.

CONTENTS

CONTRIBUTORS

Scott Brave
Baynote Inc., USA

Stephen Brewster
Department of Computing Science, University of Glasgow, UK

Michael D. Byrne
Department of Psychology, Rice University, USA

Pascale Carayon
Department of Industrial Engineering, University of Wisconsin-Madison, USA

Romeo Chua
School of Human Kinetics, University of British Columbia, Canada

William Cohen
Exponent Failure Analysis Associates, USA

Gregory Cuellar
Communication Department, Stanford University, USA

David Danielson
Communication Department, Stanford University, USA

Alan Dix
Computing Department, Lancaster University, UK

B. J. Fogg
Persuasive Technology Lab, Center for the Study of Language and Information, Stanford University, USA

David Goodman
School of Kinesiology, Simon Fraser University, Canada

P. A. Hancock
Department of Psychology, and The Institute for Simulation and Training, University of Central Florida, USA

Ken Hinckley
Microsoft Research, USA

Hiroo Iwata
Graduate School of Systems and Information Engineering, University of Tsukuba, Japan

Holger Luczak
Institute of Industrial Engineering and Ergonomics, RWTH Aachen University, Germany

Asim S. Mailagic
College of Engineering, Carnegie-Mellon University, USA

Clifford Nass
Department of Communication, Stanford University, USA

Milda Park
Institute of Industrial Engineering and Ergonomics, RWTH Aachen University, Germany

Stephen J. Payne
University of Manchester, UK

Robert W. Proctor
Department of Psychological Sciences, Purdue University, USA

Christopher Schlick
Institute of Industrial Engineering and Ergonomics, RWTH Aachen University, Germany

Daniel P. Siewiorek
Human–Computer Interaction Institute, Carnegie-Mellon University, USA

Philip J. Smith
Institute for Ergonomics, Ohio State University, USA

Neville A. Stanton
School of Engineering and Design, Brunel University, UK

Thad Starner
College of Computing, Georgia Institute of Technology, USA

J. L. Szalma
Department of Psychology, University of Central Florida, USA

Kim-Phuong L. Vu
Department of Psychology, California State University Long Beach, USA

Daniel J. Weeks
Department of Psychology, Simon Fraser University, Canada

Timothy N. Welsh
Faculty of Kinesiology, University of Calgary, Canada

Andrew Wilson
Microsoft Research, USA

Martina Ziefle
Institute for Psychology, RWTH Aachen University, Germany

ADVISORY BOARD

PREFACE

We are pleased to offer access to a select set of chapters from the second edition of *The Human–Computer Interaction Handbook*. Each of the four books in the set comprises select chapters that focus on specific issues including fundamentals which serve as the foundation for human–computer interactions, design issues, issues involved in designing solutions for diverse users, and the development process.

While human–computer interaction (HCI) may have emerged from within computing, significant contributions have come from a variety of fields including industrial engineering, psychology, education, and graphic design. The resulting interdisciplinary research has produced important outcomes including an improved understanding of the relationship between people and technology as well as more effective processes for utilizing this knowledge in the design and development of solutions that can increase productivity, quality of life, and competitiveness. HCI now has a home in every application, environment, and device, and is routinely used as a tool for inclusion. HCI is no longer just an area of specialization within more traditional academic disciplines, but has developed such that both undergraduate and graduate degrees are available that focus explicitly on the subject.

The HCI Handbook provides practitioners, researchers, students, and academicians with access to 67 chapters and nearly 2000 pages covering a vast array of issues that are important to the HCI community. Through four smaller books, readers can access select chapters from the Handbook. The first book, *Human–Computer Interaction: Fundamentals,* comprises 16 chapters that discuss fundamental issues about the technology involved in human–computer interactions as well as the users themselves. Examples include human information processing, motivation, emotion in HCI, sensor-based input solutions, and wearable computing. The second book, *Human–Computer Interaction: Design Issues,* also includes 16 chapters that address a variety of issues involved when designing the interactions between users and computing technologies. Example topics include adaptive interfaces, tangible interfaces, information visualization, designing for the web, and computer-supported cooperative work. The third book, *Human–Computer Interaction: Designing for Diverse Users and Domains,* includes eight chapters that address issues involved in designing solutions for diverse users including children, older adults, and individuals with physical, cognitive, visual, or hearing impairments. Five additional chapters discuss HCI in the context of specific domains including health care, games, and the aerospace industry. The final book, *Human–Computer Interaction: The Development Process,* includes fifteen chapters that address requirements specification, design and development, and testing and evaluation activities. Sample chapters address task analysis, contextual design, personas, scenario-based design, participatory design, and a variety of evaluation techniques including usability testing, inspection-based techniques, and survey design.

Andrew Sears and Julie A. Jacko

March 2008

ABOUT THE EDITORS

Andrew Sears is a Professor of Information Systems and the Chair of the Information Systems Department at UMBC. He is also the director of UMBC's Interactive Systems Research Center. Dr. Sears' research explores issues related to human-centered computing with an emphasis on accessibility. His current projects focus on accessibility, broadly defined, including the needs of individuals with physical disabilities and older users of information technologies as well as mobile computing, speech recognition, and the difficulties information technology users experience as a result of the environment in which they are working or the tasks in which they are engaged. His research projects have been supported by numerous corporations (e.g., IBM Corporation, Intel Corporation, Microsoft Corporation, Motorola), foundations (e.g., the Verizon Foundation), and government agencies (e.g., NASA, the National Institute on Disability and Rehabilitation Research, the National Science Foundation, and the State of Maryland). Dr. Sears is the author or co-author of numerous research publications including journal articles, books, book chapters, and conference proceedings. He is the Founding Co-Editor-in-Chief of the *ACM Transactions on Accessible Computing,* and serves on the editorial boards of the *International Journal of Human–Computer Studies*, the *International Journal of Human–Computer Interaction,* the *International Journal of Mobil Human–Computer Interaction,* and *Universal Access in the Information Society*, and the advisory board of the upcoming *Universal Access Handbook*. He has served on a variety of conference committees including as Conference and Technical Program Co-Chair of the Association for Computing Machinery's Conference on Human Factors in Computing Systems (CHI 2001), Conference Chair of the ACM Conference on Accessible Computing (Assets 2005), and Program Chair for Asset 2004. He is currently Vice Chair of the ACM Special Interest Group on Accessible Computing. He earned his BS in Computer Science from Rensselaer Polytechnic Institute and his Ph.D. in Computer Science with an emphasis on Human–Computer Interaction from the University of Maryland—College Park.

Julie A. Jacko is Director of the Institute for Health Informatics at the University of Minnesota as well as a Professor in the School of Public Health and the School of Nursing. She is the author or co-author of over 120 research publications including journal articles, books, book chapters, and conference proceedings. Dr. Jacko's research activities focus on human–computer interaction, human aspects of computing, universal access to electronic information technologies, and health informatics. Her externally funded research has been supported by the Intel Corporation, Microsoft Corporation, the National Science Foundation, NASA, the Agency for Health Care Research and Quality (AHRQ), and the National Institute on Disability and Rehabilitation Research. Dr. Jacko received a National Science Foundation CAREER Award for her research titled, "Universal Access to the Graphical User Interface: Design For The Partially Sighted," and the National Science Foundation's Presidential Early Career Award for Scientists and Engineers, which is the highest honor bestowed on young scientists and engineers by the US government. She is Editor-in-Chief of the *International Journal of Human–Computer Interaction* and she is Associate Editor for the *International Journal of Human Computer Studies*. In 2001 she served as Conference and Technical Program Co-Chair for the ACM Conference on Human Factors in Computing Systems (CHI 2001). She also served as Program Chair for the Fifth ACM SIGCAPH Conference on Assistive Technologies (ASSETS 2002), and as General Conference Chair of ASSETS 2004. In 2006, Dr. Jacko was elected to serve a three-year term as President of SIGCHI. Dr. Jacko routinely provides expert consultancy for organizations and corporations on systems usability and accessibility, emphasizing human aspects of interactive systems design. She earned her Ph.D. in Industrial Engineering from Purdue University.

Human-Computer Interaction

Fundamentals

HUMANS IN HCI

◆ 1 ◆

PERCEPTUAL-MOTOR INTERACTION:
SOME IMPLICATIONS FOR HCI

Timothy N. Welsh
University of Calgary

Daniel J. Weeks
Simon Fraser University

Romeo Chua
University of British Columbia

David Goodman
Simon Fraser University

PERCEPTUAL-MOTOR INTERACTION: A BEHAVIORAL EMPHASIS

Many of us can still remember purchasing our first computers to be used for research purposes. The primary attributes of these new tools were their utilities in solving relatively complex mathematical problems and performing computer-based experiments. However, it was not long after that word processing brought about the demise of the typewriter, and our department secretaries no longer prepared our research manuscripts and reports. It is interesting to us that computers are not so substantively different from other tools such that we should disregard much of what the study of human factors and experimental psychology has contributed to our understanding of human behavior in simple and complex systems. Rather, it is the computer's capacity for displaying, storing, processing, and even controlling information that has led us to the point at which the manner with which we interact with such systems has become a research area in itself.

In our studies of human–computer interaction (HCI), also known as human-machine interaction, and perceptual-motor interaction in general, we have adopted two basic theoretical and analytical frameworks as part of an integrated approach. In the first framework, we view perceptual-motor interaction in the context of an information-processing model. In the second framework, we have used analytical tools that allow detailed investigations of both static and dynamic interactions. Our chapter in the previous edition of this handbook (Chua, Weeks, & Goodman, 2003) reviewed both avenues of research and their implications for HCI with a particular emphasis on our work regarding the translation of perceptual into motor space. Much of our more recent research, however, has explored the broader interplay between the processes of action and attention. Thus, in the present chapter, we turn our focus to aspects of this work that we believe to have considerable implications for those working in HCI.

Human Information Processing and Perceptual-Motor Behavior

The information-processing framework has traditionally provided a major theoretical and empirical platform for many scientists interested in perceptual-motor behavior. The study of perceptual-motor behavior within this framework has inquired into such issues as the information capacity of the motor system (e.g., Fitts, 1954), the attentional demands of movements (e.g., Posner & Keele, 1969), motor memory (e.g., Adams & Dijkstra, 1966), and processes of motor learning (e.g., Adams, 1971). The language of information processing (e.g., Broadbent, 1958) has provided the vehicle for discussions of mental and computational operations of the cognitive and perceptual-motor system (Posner, 1982). Of interest in the study of perceptual-motor behavior is the nature of the cognitive processes that underlie perception and action.

The information-processing approach describes the human as an active processor of information, in terms that are now commonly used to describe complex computing mechanisms.

An information-processing analysis describes observed behavior in terms of the encoding of perceptual information, the manner in which internal psychological subsystems utilize the encoded information, and the functional organization of these subsystems. At the heart of the human cognitive system are processes of information transmission, translation, reduction, collation, storage, and retrieval (e.g., Fitts, 1964; Marteniuk, 1976; Stelmach, 1982; Welford, 1968). Consistent with a general model of human information processing (e.g., Fitts & Posner, 1967), three basic processes have been distinguished historically. For our purposes, we refer to these processes as stimulus identification, response selection, and response programming. Briefly, stimulus identification is associated with processes responsible for the perception of information. Response selection pertains to the translation between stimuli and responses and the selection of a response. Response programming is associated with the organization of the final output (see Proctor & Vu, 2003, or the present volume).

A key feature of early models of information processing is the emphasis upon the cognitive activities that precede action (Marteniuk, 1976; Stelmach, 1982). From this perspective, action is viewed only as the end result of a complex chain of information-processing activities (Marteniuk, 1976). Thus, chronometric measures, such as reaction time and movement time, as well as other global outcome measures, are often the predominant dependent measures. However, even a cursory examination of the literature indicates that time to engage a target has been a primary measure of interest. For example, a classic assessment of perceptual-motor behavior in the context of HCI and input devices was conducted by Card, English, and Burr (1978; see also English, Engelhart, & Berman, 1967). Employing measures of error and speed, Card et al. (1978) had subjects complete a cursor positioning task using four different control devices (mouse, joystick, step keys, text keys). The data revealed the now well-known advantage for the mouse. Of interest is that the speed measure was decomposed into "homing" time, the time that it took to engage the control device and initiate cursor movement, and "positioning" time, the time to complete the cursor movement. Although the mouse was actually the poorest device in terms of the homing time measure, the advantage in positioning time produced the faster overall time. That these researchers sought to glean more information from the time measure acknowledges the importance of the movement itself in perceptual-motor interactions such as these.

The fact that various pointing devices depend on hand movement to control cursory movement has led to the emphasis that researchers in HCI have placed on Fitts' law (Fitts, 1954) as a predictive model of time to engage a target. The law predicts pointing (movement) time as a function of the distance to and width of the target—where, in order to maintain a given level of accuracy, movement time must increase as the distance of the movement increases and/or the width of the target decreases. The impact of Fitts' law is most evident by its inclusion in the battery of tests to evaluate computer-pointing devices in ISO 9241-9. We argue that there are a number of important limitations to an exclusive reliance on Fitts' law in this context.

First, although the law predicts movement time, it does this based on distance and target size. Consequently, it does not allow for determining what other factors may influence movement time. Specifically, Fitts' law is often based on a movement to a single target at any given time (although it was originally developed using reciprocal movements between two targets). However, in most HCI and graphical user interface (GUI) contexts, there is an array of potential targets that can be engaged by an operator. As we will discuss later in this chapter, the influence of these distracting nontarget stimuli on both the temporal and physical characteristics of the movements to the imperative target can be significant.

Second, we suggest that the emphasis on Fitts' law has diverted attention from the fact that cognitive processes involving the selection of a potential target from an array are an important, and time consuming, information processing activity that must precede movement to that target. For example, the Hick-Hyman law (Hick, 1952; Hyman, 1953) predicts the decision time required to select a target response from a set of potential responses—where the amount of time required to choose the correct response increases with the number of possible alternative responses. What is important to understand is that the two laws work independently to determine the total time it takes for an operator to acquire the desired location. In one instance, an operator may choose to complete the decision-making and movement components sequentially. Under these conditions, the total time to complete the task will be the sum of the times predicted by the Hick-Hyman and Fitts' laws. Alternatively, an operator may opt to make a general movement that is an approximate average of the possible responses and then select the final target destination while the movement is being completed. Under such conditions, Hoffman and Lim (1997) reported interference between the decision and movement component that was dependent on their respective difficulties (see also Meegan & Tipper, 1998).

Finally, although Fitts' law predicts movement time given a set of movement parameters, it does not actually reveal much about the underlying movement itself. Indeed, considerable research effort has been directed toward revealing the movement processes that give rise to Fitts' law. For example, theoretical models of limb control have been forwarded that propose that Fitts' law emerges as a result of multiple submovements (e.g., Crossman & Goodeve, 1963/1983), or as a function of both initial movement impulse variability and subsequent corrective processes late in the movement (Meyer, Abrams, Kornblum, Wright, & Smith, 1988). These models highlight the importance of conducting detailed examinations of movements themselves as a necessary complement to chronometric explorations.

For these reasons, HCI situations that involve dynamic perceptual-motor interactions may not be best indexed merely by chronometric methods (cf., Card et al., 1978). Indeed, as HCI moves beyond the simple key press interfaces that are characteristic of early systems to include virtual and augmented reality, teleoperation, gestural, and haptic interfaces, among others, the dynamic nature of perceptual-motor interactions are even more evident. Consequently, assessment of the actual movement required to engage such interfaces would be more revealing.

To supplement chronometric explorations of basic perceptual-motor interactions, motor behaviour researchers have also advocated a movement-process approach (Kelso, 1982). The

argument is that, in order to understand the nature of movement organization and control, analyses should also encompass the movement itself, and not just the activities preceding it (e.g., Kelso, 1982; 1995; Marteniuk, MacKenzie, & Leavitt, 1988). Thus, investigators have examined the kinematics of movements in attempts to further understand the underlying organization involved (e.g., Brooks, 1974; Chua & Elliott, 1993; Elliott, Carson, Goodman, & Chua, 1991; Kelso, Southard, & Goodman, 1979; MacKenzie, Marteniuk, Dugas, Liske, & Eickmeier, 1987; Marteniuk, MacKenzie, Jeannerod, Athenes, & Dugas, 1987). The relevance of this approach will become apparent in later sections.

Translation, Coding, and Mapping

As outlined above, the general model of human information processing (e.g., Fitts & Posner, 1967) distinguishes three basic processes: stimulus identification, response selection, and response programming. While stimulus identification and response programming are functions of stimulus and response properties, respectively, response selection is associated with the translation between stimuli and responses (Welford 1968).

Translation is the seat of the human "interface" between perception and action. Moreover, the effectiveness of translation processes at this interface is influenced to a large extent by the relation between perceptual inputs (e.g., stimuli) and motor outputs (e.g., responses). Since the seminal work of Fitts and colleagues (Fitts & Seeger, 1953; Fitts & Deninger, 1954), it has been repeatedly demonstrated that errors and choice reaction times to stimuli in a spatial array decrease when the stimuli are mapped onto responses in a spatially compatible manner. Fitts and Seeger (1953) referred to this finding as stimulus-response (S-R) compatibility and ascribed it to cognitive codes associated with the spatial locations of elements in the stimulus and response arrays. Presumably, it is the degree of coding and recoding required to map the locations of stimulus and response elements that determine the speed and accuracy of translation and thus response selection (e.g., Wallace, 1971).

The relevance of studies of S-R compatibility to the domain of human factors engineering is paramount. It is now well understood that the design of an optimal human-machine interface in which effective S-R translation facilitates fast and accurate responses is largely determined by the manner in which stimulus and response arrays are arranged and mapped onto each other (e.g., Bayerl, Millen, & Lewis, 1988; Chapanis & Lindenbaum, 1959; Proctor & Van Zandt, 1994). As a user, we experience the recalibrating of perceptual-motor space when we take hold of the mouse and move it in a fairly random pattern when we interact with a computer for the first time. Presumably, what we are doing here is attempting to calibrate our actual movements to the resulting virtual movements of the cursor on the screen. Thus, for optimal efficiency of functioning, it seems imperative that the system is designed to require as little recalibration as possible. Again, our contribution to the previous edition of this handbook reviews our work in the area of stimulus-response translation and the implications of this work for HCI (Chua et al., 2003). We encourage those who are more interested in these issues to read that chapter.

PERCEPTUAL-MOTOR INTERACTION: ATTENTION AND PERFORMANCE

The vast literature on selective attention and its role in the filtering of target from nontarget information (e.g. Cherry, 1953; Treisman, 1964a, 1964b, 1986; Deutsch & Deutch, 1963; Treisman & Gelade, 1980) has no doubt been informative in the resolution of issues in HCI pertaining to stimulus displays and inputs (e.g., the use of color and sound). However, attention should not be thought of as a unitary function, but rather as a set of information processing activities that are important for perceptual, cognitive, and motor skills. Indeed, the evolution of HCI into the realm of augmented reality, teleoperation, gestural interfaces, and other areas that highlight the importance of dynamic perceptual-motor interactions, necessitates a greater consideration of the role of attention in the selection and execution of action. Recent developments in the study of how selective attention mediates perception and action and, in turn, how intended actions influence attentional processes, are poised to make just such a contribution to HCI. We will now turn to a review of these developments and some thoughts on their potential relevance to HCI.

Attention

We are all familiar with the concept of attention on a phenomenological basis. Even our parents, who likely never formally studied cognition, demonstrated their understanding of the essential characteristics of attention when they directed us to pay attention when we were daydreaming or otherwise not doing what was asked. They knew that humans, like computers, have a limited capacity to process information in that we can only receive, interpret, and act upon a fixed amount of information at any given moment. As such, they knew that any additional, nontask processing would disrupt the performance of our goal task, be it homework, cleaning, or listening to their lectures. But what is attention? What does it mean to pay attention? What influences the direction of our attention? The answers to these questions are fundamental to understanding how we interact with our environment. Thus, it is paramount for those who are involved in the design of HCI to consider the characteristics of attention and its interactive relationship with action planning.

Characteristics of Attention

Attention is the collection of processes that allow us to dedicate our limited information processing capacity to the purposeful (cognitive) manipulation of a subset of available information. Stated another way, attention is the process through which information enters into working memory and achieves the level of consciousness. There are three important characteristics of attention: (a) attention is selective and allows only a specific subset of information to enter the limited processing system; (b) the focus of attention can be shifted from one source of information to another; and, (c) attention can be divided such that, within certain limitations, one may selectively attend to

more that one source of information at a time. The well-known "cocktail party" phenomenon (Cherry, 1953) effectively demonstrates these characteristics.

Picture yourself at the last busy party or poster session you attended where there was any number of conversations continuing simultaneously. You know from your own experience that you are able to filter out other conversations and selectively attend to the single conversation in which you are primarily engaged. You also know that there are times when your attention is drawn to a secondary conversation that is continuing nearby. These shifts of attention can occur automatically, especially if you hear your name dropped in the second conversation, or voluntarily, especially when your primary conversation is boring. Finally, you know that you are able to divide your attention and follow both conversations simultaneously. However, although you are able to keep track of each discussion simultaneously, you will note that your understanding and contributions to your primary conversation diminish as you dedicate more and more of your attentional resources to the secondary conversation. The diminishing performance in your primary conversation is, of course, an indication that the desired amount of information processing has exceeded your limited capacity.

What does the "cocktail party" phenomenon tell us about designing HCI environments? The obvious implication is that, in order to facilitate the success of the performer, the HCI designer must be concerned about limiting the stress on the individuals' information processing systems by (a) creating interfaces that assist in the selection of the most appropriate information; (b) being knowledgeable about the types of attention shifts and about when (or when not) to use them; and (c) understanding that, when attention must be divided amongst a series of tasks, that each of these tasks should be designed to facilitate automatic performance so as to avoid conflicts in the division of our limited capacity and preserve task performance. While these suggestions seem like statements of the obvious, the remainder of the chapter will delve deeper into these general characteristics and highlight situations in which some aspects of design might not be as intuitive as it seems. Because vision is the dominant modality of information transfer in HCI, we will concentrate our discussion on visual selective attention. It should be noted, however, that there is a growing literature on cross-modal influences on attention, especially visual-auditory system interactions (e.g., Spence, Lloyd, McGlone, Nichols, & Driver, 2000), that will be relevant in the near future.

Shifts and Coordinate Systems of Attention

Structural analyses of the retinal (photo sensitive) surface of the eye has revealed two distinct receiving areas—the fovea and the perifoveal (peripheral) areas. The fovea is a relatively small area (about two to three degrees of visual angle) near the center of the retina, which has the highest concentration of color-sensitive cone cells. It is this high concentration of color-sensitive cells that provides the rich, detailed information that we typically use to identify objects. There are several important consequences of this structural and functional arrangement. First, because of the foveas' pivotal role in object identification and the importance of object identification for the planning of action and many other cognitive processes, visual attention is typically dedicated to the information received by the fovea. Second, because the fovea is such a small portion of the eye, we are unable to derive a detailed representation of the environment from a single fixation. As a result, it is necessary to constantly move information from objects in the environment onto the fovea by rotating the eye rapidly and accurately. These rapid eye movements are known as saccadic eye movements. Because of the tight link between the location of visual attention and saccadic eye movements, these rapid eye movements are referred to as overt shifts of attention.

Although visual attention is typically dedicated to foveal information, it must be remembered that the perifoveal retinal surface also contains color-sensitive cells and, as such, is able to provide details about objects. A covert shift of attention refers to any situation in which attention is being dedicated to a nonfoveated area of space. Covert shifts of attention are employed when the individual wants or needs to maintain the fovea on a particular object while continuing to scan the remaining environment for other stimuli. Covert shifts of attention also occur immediately prior to the onset of an overt shift of attention or other type of action (e.g., Shepherd, Findlay, & Hockey, 1986). For this reason, people are often able to identify stimuli at the location of covert attention prior to the acquisition of that location by foveal vision (e.g., overt attention) (Deubel & Schneider, 1996).

Because attention is typically dedicated to the small foveal subdivision of the retinal surface, attention is often considered to work as a spotlight or zoom lens that constantly scans the environment (e.g., Eriksen & Eriksen, 1974). More often, however, the objects that we attend to are larger than the two to three degrees of visual angle covered by the fovea. Does this mean that the components of objects that are outside of foveal vision do not receive attentional processing? No, in fact it has been repeatedly shown that attention can work in an object-based coordinate system where attention actually spreads along the full surface of an object when attention is dedicated to a small section of the object (Davis & Driver, 1997; Egly, Driver, & Rafal, 1994; see also Duncan, 1984). These object-centered attentional biases occur even when other objects block connecting sections of the continuous object (e.g., Pratt & Sekuler, 2001). Finally, it should be noted that, although entire objects receive attentional processing, particular sections of the object often receive preferential attentional treatment often based on the action potential of the object (see Attention and Stimulus-Response Compatibility below). Thus, attention should be seen as a flexible resource allocation instead of a fixed commodity with a rigid structure. The spotlight coding system is typically employed during detailed discrimination tasks, for example when reading the text on this page, whereas object-based coding might be more effective when gaining an appreciation for the context of an object in the scene or the most interactive surface of the object. The relevance of the flexible, action-dependent nature of attentional coding systems will be readdressed in latter sections.

Stimulus Characteristics and Shifts of Attention

Both overt and covert shifts of attention can be driven by stimuli in the environment or by the will of the performer. Shifts

of attention that are driven by stimuli are known as exogenous, or bottom-up, shifts of attention. They are considered to be automatic in nature and thus, for the most part, are outside of cognitive influences. Exogenous shifts of attention are typically caused by a dynamic change in the environment such as the sudden, abrupt appearance (onset) or disappearance (offset) of a stimulus (e.g., Pratt & McAuliffe, 2001), a change in the luminance or color of a stimulus (e.g., Folk, Remington, & Johnston, 1992; Posner, Nissen, & Ogden, 1978; Posner & Cohen, 1984), or the abrupt onset of object motion (e.g., Abrams & Chirst, 2003; Folk, Remington, & Wright, 1994). The effects of exogenous shifts have a relatively rapid onset, but are fairly specific to the location of the dynamic change and are transient, typically reaching their peak influence around 100 milliseconds after the onset of the stimulus (Cheal & Lyon, 1991; Müller & Rabbitt, 1989). From an evolutionary perspective, it could be suggested that these automatic shifts of attention developed because such dynamic changes would provide important survival information such as the sudden, unexpected appearance of a predator or prey. However, in modern times, these types of stimuli can be used to quickly draw one's attention to the location of important information.

In contrast, performer-driven, or endogenous, shifts of attention are under complete voluntary control. The effects of endogenous shifts of attention take longer to develop, but can be sustained over a much longer period of time (Cheal & Lyon, 1991; Müller & Rabbitt, 1989). From an HCI perspective, there advantages and disadvantages to the fact that shifts of attention can be under cognitive control. The main benefit of cognitive control is that shifts of attention can result from a wider variety of stimuli such as symbolic cues like arrows, numbers, or words. In this way, performers can be cued to locations or objects in the scene with more subtle or permanent information than the dynamic changes that are required for exogenous shifts. The main problem with endogenous shifts of attention is that the act of interpreting the cue requires a portion of the limited information processing capacity and thus can interfere with, or be interfered by, concurrent cognitive activity (Jonides, 1981).

Although it was originally believed that top-down processes could not influence exogenous shifts of attention (e.g., that dynamic changes reflexively capture attention regardless of intention), Folk et al. (1992) demonstrated that this is not always the case. The task in the Folk et al. (1992) study was to identify a stimulus that was presented in one of four possible locations. For some participants, the target stimulus was a single abrupt onset stimulus (the target appeared in one location and nothing appeared in the other three locations), whereas for the remaining participants the target stimulus was a color singleton (a red stimulus that was presented at the same time as white stimuli that appeared in the other three possible locations). One-hundred fifty milliseconds prior to the onset of the target, participants received cue information at one of the possible target locations. The cue information was either abrupt onset stimuli at a single location or color singleton information. Across a series of experiments, Folk et al. (1992) found that the cue tended to slow reaction times to the target stimulus when the cue information was presented at a location that was different from where the target subsequently appeared indicating that attention had initially been exogenously drawn to the cue.

Importantly, the cue stimuli only interfered with the identification of the target stimulus when the characteristics of cue stimuli matched the characteristics of the target stimulus (e.g., onset cue-onset target and color cue-color target conditions). When the characteristics of the cue did not match the target stimulus (e.g., onset cue-color target and color cue-onset target conditions), the location of the cue did not influence reaction times. Thus, these results reveal that dynamic changes only capture attention when the performer is searching for a dynamic change stimulus. Stated another way, it seems that automatic attentional capture is dependent upon the expectations of the performer. Folk et al. (1992) suggested that people create an attention set in which they establish their expectations for the characteristics of the target stimulus. Stimuli meeting the established set will automatically capture attention, whereas stimuli that do not meet the established set will not (see also Folk et al., 1994). The obvious implication of these results is that the most efficient HCIs will be those for which the designer has considered perceptual expectations of the person controlling the system. As we will discuss in the Action-Centered Attention section, however, consideration of the perceptual components alone is, at best, incomplete.

Facilitation and inhibition of return. While it is the case that our attention can be endogenously and exogenously shifted to any location or object in our environment, it seems there are unconscious mechanisms that work to hinder the movement of our attention to previously investigated locations and objects. The existence of these reflexive mechanisms was first revealed through a series of studies by Posner and Cohen (1984) and has been reliably demonstrated many times since. The basic task in these studies is to respond as quickly as possible following the onset of a target that randomly appears at one of any number of possible locations. The presentation of the target is preceded by a briefly presented cue that is not predictive of the target location (e.g., if there are 2 possible target locations, the target will appear at the location of the cue on 50% of the trials and at the uncued location of the remaining 50% of the trials). The key findings of these studies are that (a) when there is a short time interval between the onset of the cue and the onset of the target (less than 200 ms), participants are faster at responding to targets presented at the cued location than at the uncued location; whereas, (b) when there is a longer time interval between the onset of the cue and the onset of the target (greater than 400–500 ms), participants are faster at responding to the target presented at the uncued location than at the cued location (see Fig. 1.1 for some representative data). It is important to remember that the reaction times to cued targets are facilitated at short intervals and inhibited at longer intervals even though the cue has no predictive relation to the location of the subsequent target. The earlier facilitation effect is thought to arise because attention has been exogenously drawn from a central fixation point to the location of the cue and is still there when the target subsequently appears—with attention already at the target location, subsequent target processing is efficient. As the time elapses after the onset of the cue, however, the performer knows that the target is equally likely to appear at any location and so endogenously returns attention back to the central

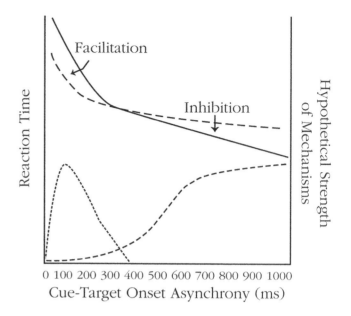

FIGURE 1.1. Exemplar reaction time data as a function of Target Location (cued [large dashed black line] vs. uncued [solid black line]) and Cue-Target Onset Asynchrony (CTOA) and hypothetical activation levels of the facilitatory [dotted line] and inhibitory [small dashed line] mechanisms that cause Facilitation and Inhibition of Return.

point. It is suggested that in moving attention back to a central point, the performer places an inhibitory code on the location of the cue. This inhibitory code subsequently interferes with the processing of target when it appears at the location of the cue and increases reactions for the cued target relative to any uncued (e.g., uninhibited) target leading to the inhibition of return (IOR) effect.

The twenty plus years of research on these phenomena has revealed many characteristics of the mechanisms of attention system that are relevant for HCI. First is the time course of the development of the mechanisms underlying these facilitation and IOR effects. It is thought that the mechanism of facilitation has a very short onset and a small life, whereas the mechanism of inhibition has a longer latency but has a much longer lasting influence (up to three to four seconds; see Fig. 1.1). Thus, if a designer intends on using a dynamic environment in which irrelevant, but perhaps aesthetic, stimuli constantly appear, disappear, or move, then it is important to realize that the users' performance might be negatively influenced because of the inadvertent facilitation of the identification of nontarget information or the inhibition of the identification of target information depending on the spatiotemporal relationship between the relevant and irrelevant information. Alternatively, video game designers could exploit these effects to suit their purpose when they want to facilitate or hinder the performance of a gamer in a certain situation as it has been shown that even experienced video game players demonstrate facilitation and IOR effects of similar magnitude to novice gamers (Castel, Drummond, & Pratt, 2005).

The second important feature is that facilitation and IOR only occur when attention has been shifted to the location of the cue. Thus, if the designer is using cues that typically only produce endogenous shifts of attention (e.g., by using a symbolic cue such as an arrow or word that indicates a particular location), and then reaction times will be similar across cued and uncued targets (Posner & Cohen, 1984). The lack of facilitation and IOR following symbolic cues is thought to occur because the participant can easily ignore the cue and prevent the shift of attention to the cued location that activates the facilitatory or inhibitory mechanisms. It is important to note, however, the mechanisms of IOR are activated each time attention has been shifted. Thus, symbolic information can result in IOR if the participant does shift their attention to the cued location (even if they were asked not to) or if the participant is required to respond to the cue information (Rafal, Calabresi, Brennan, & Sciolto, 1989; Taylor & Klein, 2000). Although symbolic information presented at locations that are outside of the target locations do not typically activate the possibly detrimental mechanisms of facilitation and IOR, one must still use caution when presenting such information and be sensitive to context (e.g., how similar symbolic cues have been used in previous interactions) and the response expectations of the user.

Finally, as briefly mentioned in the previous paragraph, it is important to realize that the inhibitory mechanisms of IOR are also activated by responding to a location. That is, when people are required to make a series of responses to targets that are randomly presented at one of a number of possible locations, they are slower at initiating response *n* when it is the same as response *n-1* than when it is different from response *n-1* (Maylor & Hockey, 1985; Tremblay, Welsh, & Elliott, 2005). This target-target IOR effect has important implications for two reasons. First, if the designer intends on requiring the user to complete a series of choice responses to targets that appear at the same locations each time and the user is uncertain as to where each target will appear (as in typical interactions with automated banking machines), then the user will be slower at responding to targets that appear successively at the same location than to targets that appear at new locations. When it is also considered that it has been shown that, in a cue-target IOR experiments, inhibitory codes can be maintained at up to four locations simultaneously (Tipper, Weaver, & Watson, 1996; Wright & Richard, 1996; cf., Abrams & Pratt, 1996), the designer must be cautious about designing displays that use similar locations on successive interactions. The second reason that an understanding of target-target IOR is important for HCI is that we have recently shown that target-target IOR effects transfer across people (Welsh, Elliott, Anson, Dhillon, Weeks, Lyons, et al., 2005; Welsh, Lyons, Weeks, Anson, Chua, Mendoza, et al., in press). That is, if two people are performing a task in which they must respond to a series of successive targets, then an individual will be slower at returning to the previously responded-to location regardless if that person or their partner completed the initial response. Although the research on social attention is in its infancy, the results of our work indicate that those involved in the emerging fields of virtual reality or immersive collaborative and multiuser environments must be aware of and consider how attention is influenced by a variety of stimuli including the actions of other participants.

Action-Centered Attention

The majority of the literature reviewed thus far has involved experiments that investigated attentional processes through tasks that employed simple or choice key press actions. Cognitive scientists typically use these arbitrary responses (a) because key press responses are relatively uncomplicated and provide simple measures of performance, namely reaction time and error; and (b) because, by using a simple response, the researcher assumes that they have isolated the perceptual and attentional processes of interest from additional, complex motor programming and control processes. While there are certainly numerous examples of HCIs in which the desired response is an individual key press or series of key presses, there are perhaps as many situations in which movements that are more complicated are required. As HCIs move increasingly into virtual reality, touch screen, and other more complex environments, it will become more and more important to consider the ways in which attention and motor processes interact. Thus, it will become more and more critical to determine if the same principles of attention apply when more involved motor responses are required. In addition, some cognitive scientists have suggested that, because human attention systems have developed through evolution to acquire the information required to plan and control complex actions, studying attention under such constrained response conditions may actually provide an incomplete or biased view of attention (Allport, 1987; 1993). The tight link between attention and action is apparent when one recognizes that covert shifts of attention occur prior to saccadic eye movements (Deubel & Schneider, 1996) and that overt shifts of attention are tightly coupled to manual aiming movements (Helsen, Elliott, Starkes, & Ricker, 1998; 2000). Such considerations, in combination with neuroanatomical studies revealing tight links between the attention and motor centers (Rizzolatti, Riggio, & Sheliga, 1994), have led to the development of action-centered models of attention (Rizzolatti, Riggio, Dascola, & Umilta, 1987; Tipper, Howard, & Houghton, 1999; Welsh & Elliott, 2004a).

Arguably the most influential work in the development of the action-centered models was the paper by Tipper, Lortie, and Baylis (1992). Participants in these studies were presented with nine possible target locations, arranged in a three-by-three matrix, and were asked to identify the location of a target stimulus appearing at one of these locations while ignoring any non-target stimuli presented at one of the remaining eight locations. The key innovation of this work was that Tipper and colleagues (1992) asked participants to complete a rapid aiming movement to the target location instead of identifying it with a key press. Consistent with traditional key press studies, the presence of a distractor was found to increase response times to the target. Although the finding of distractor interference in this selective reaching task was an important contribution to the field in and of itself, the key discovery was that the magnitude of the interference effects caused by a particular distractor location was dependent on the aiming movement being completed. Specifically, it was found that distractors (a) along the path of the movement caused more interference than those that were outside the path of the movement (the proximity-to-hand effect);

and, (b) ipsilateral to the moving hand caused more interference than those in the contralateral side of space (the ipsilateral effect). Based on this pattern of interference, Tipper et al. (1992) concluded that attention and action are tightly linked such that the distribution of attention is dependent on the action that was being performed (e.g., attention was distributed in an action-centered coordinate system) and that the dedication of attention to a stimulus evokes a response to that stimulus.

While the Tipper et al. (1992) paper was critical in initiating investigations into action-centered attention, more recent research has demonstrated that the behavioral consequences of selecting and executing target-directed actions in the presence of action-relevant nontarget stimuli extend beyond the time taken to prepare and execute the movement (e.g., Meegan & Tipper, 1998; Pratt & Abrams, 1994). Investigations in our labs and others have revealed that the actual execution of the movement changes in the presence of distractors. For example, there are reports that movements will deviate towards (Welsh, Elliott, & Weeks, 1999; Welsh & Elliott, 2004a; 2005) or away from (Howard & Tipper, 1997; Tipper, Howard, & Jackson, 1997; Welsh & Elliott, 2004a) the nontarget stimulus. Although these effects seem paradoxical, Welsh and Elliott (2004a) have formulated and tested (e.g., Welsh & Elliott, 2004a; 2004b; 2005) a conceptual model that can account for these movement execution effects.

The model of response activation. Consistent with the conclusions of Tipper et al. (1992), Welsh and Elliott (2004a) based the model of response activation on the premise that attention and action processes are so tightly linked that the dedication of attention to a particular stimulus automatically initiates response-producing processes that are designed to interact with that stimulus. Responses are activated to attended stimuli regardless of the nature of attentional dedication (e.g., reflexive or voluntary). It is proposed that each time a performer approaches a known scene, a response set is established in working memory in which the performer identifies and maintains the characteristics of the expected target stimulus and the characteristics of the expected response to that stimulus. Thus, the response set in the model of response activation is an extension of the attentional set of Folk et al. (1992) in that the response set includes the performer's expectations of the target stimulus as well as preexcited (preprogrammed) and/or preinhibited response codes. Each stimulus that matches the physical characteristics established in the response set captures attention and, as a result, activates an independent response process. Stimuli that do not possess at least some of the expected characteristics do not capture attention and thus do not activate responses. Thus, if only one stimulus in the environment matches the response set, then that response process is completed unopposed and the movement emerges rapidly and in an uncontaminated form. On the other hand, under conditions in which more than one stimulus matches the response set, multiple response representations are triggered and subsequently race one another to surpass the threshold level of neural activation required to initiate a response. It is important to note that this is not a winner-take-all race where only the characteristics of the winning response influence the characteristics of actual movement alone. Instead, the characteristics of the observed movement

are determined by the activation level of each of the competing responses at the moment of movement initiation. In this way, if more than one neural representation is active (or if one is active and one is inhibited) at response initiation, then the emerging response will have characteristics of both responses (or characteristics that are opposite to the inhibited response).

The final relevant element of the model is that the activation level of each response is determined by at least three interactive factors: the salience of the stimulus and associated response, an independent inhibitory process, and the time course of each independent process. The first factor, the salience or action-relevancy of the stimulus, is in fact the summation of a number of separate components including the degree attentional capture (based on the similarity between the actual and anticipated stimulus within the response set), the complexity of the response afforded by the stimulus, and the S-R compatibility. When attentional capture and stimulus-response compatibility are maximized and response complexity minimized, the salience of an individual response is maximized and the response to that stimulus will be activated rapidly. The second factor that influences the activation level of a response is an independent inhibitory process that works to eliminate nontarget responses. When this inhibitory mechanism has completed its task, the neuronal activity associated with the inhibited response will have been reduced to below baseline. Thus, in effect, an inhibited response will add characteristics that are opposite to it to the formation of the target response. The final factor that contributes to the activation level of each independent response process is the time course of the development of the representation. It is assumed that the neural representations of each response do not develop instantaneously and that the inhibitory mechanism that eliminates nontarget responses does not instantly remove the undesired neural activity. Instead, each of these processes requires time to reach an effective level and the time course of each responses' development will be independent of one another. For example, a response to a very salient stimulus will achieve a higher level of activation sooner than a response to a stimulus with a lower saliency. If this stimulus of high salience evokes the desired response, then responses to other, less salient stimuli will cause little interference because they simply do not reach as high an activation level than the target response. In contrast, when nontarget stimuli are more salient, then the interference is much more severe (Welsh & Elliott, 2005). In sum, to extend the analogy of the race model, responses that run faster (the responses were evoked by more salient stimuli), have a head start relative to another (one stimulus was presented prior to another), or have a shorter distance to go to the finish line (the response was partially preprogrammed in the response set) will achieve a higher level of activation and will, as a result, contribute more to the characteristics of the final observable movement than ones that are further behind.

So, what implications does the model of response activation have for the design of HCI? In short, because the model of response activation provides a fairly comprehensive account of movement organization in complex environments, it could be used as the basis for the design of interfaces that consider the cognitive system as an interactive whole as opposed to separate units of attention and movement organization. One of the more obvious implications is that a designer should consider the time intervals between the presentation of each stimulus in a multiple stimuli set as this can have dramatic effects on the performer's ability to quickly respond to each stimulus (e.g., psychological refractory period; Telford, 1931; Pashler, 1994) and the physical characteristics of each response (Welsh & Elliott, 2004a).

Perhaps more importantly, the model highlights the importance for the designer to consider the interactions between attention and motor processing because there are some situations in which the transfer from simple to complex movements is straightforward, whereas there are others that do not transfer at all. The study by Bekkering and Pratt (2004) provided a good demonstration of a situation in which the interaction between attention and action provides the same results in key press and aiming responses. Specifically, they showed that a dynamic change on one location of an object facilitates reaction times for aiming movements to any other portion of the object (e.g., attention can move in object-centered coordinate systems in aiming responses as it does for key press response). However, Welsh & Pratt (2005) recently found that the degree of attentional capture by some dynamic changes is different when key press and spatially-directed responses are required. In this study, participants were asked to identify the location of an onset or offset target stimulus while ignoring a distractor stimulus of the opposite characteristics (e.g., onset targets were paired with offset distractors and vice versa). In separate experiments, participants responded to the target stimulus with a choice key press response or an aiming movement to the target location. The results indicated that an onset distractor slowed responding to an offset target in both tasks. Offset distractor, on the other hand, only interfered with task performance when a key press was required. When participants were asked to perform an aiming movement, an offset distractor actually caused a nonsignificant facilitation effect. Similar action-specific interference effects have been shown across pointing and grasping actions (Bekkering & Neggers, 2002), pointing and verbal responses (Meegan & Tipper, 1999), and different types of pointing responses (Meegan & Tipper, 1999; Tipper, Meegan, & Howard, 2002). In sum, now that HCI is moving into virtual reality and other types of assisted response devices (voice activated, head mounted, roller ball, and eye-gaze mouse systems), it will become increasingly important to consider the required and/or anticipated action when designing HCI environments. Given our emphasis on the importance of considering the interaction of attention and motor processes, we will explore this issue in greater detail in the following sections.

Attention and action requirements. Initial investigations into action-centered attention were focused primarily on the influence that the spatial location of distractors with respect to the target had on the planning and execution of action (e.g., Meegan & Tipper, 1998; Pratt & Abrams, 1994; Tipper et al., 1992). In that context, an action-centered framework could offer a useful perspective for the spatial organization of perceptual information presented in an HCI context. However, often the reason for engaging a target in an HCI task is that the target symbolically represents an outcome or operation to be achieved. Indeed, this is what defines an icon as a target—target features symbolically carry a meaning that defines it as the appropriate target. An interest in the application of the action-centered model to human factors and HCI led Weir, Weeks, Welsh, Elliott, Chua, Roy, and Lyons (2003) to consider whether or not distractor effects could be elicited based upon the specific actions required to engage a target and distractor object. The question was whether the engagement properties of target and distractor objects (e.g., turn or pull) in a control array would mediate the influence of the distractor on the control of movement. In that study, participants executed their movements on a control panel that was located directly in front of them. On some trials, the control panel consisted of a single pull-knob or right-turn dial located at the midline either near or far from a starting position located proximal to the participant. On other trials, a second control device (pull knob or dial) was placed into the other position on the display. If this second device was present, it served as a distractor object and was to be ignored. Weir et al. (2003) found that the distractor object only interfered with the programming of the response to the target stimulus when the distractor afforded a different response from the target response (e.g., when the pull knob was presented with the turn dial). These results suggest that when moving in an environment with distracting stimuli or objects, competing responses may be programmed in parallel and that these parallel processes will only interfere with one another when they are incompatible. The implication is that the terminal action required to engage the objects in the environment is also important to the distribution of attention and movement planning and execution.

In addition to considering the terminal action requirements, other work in our labs suggests that the manner in which the actions are completed is an equally important concern. For example, the "cluttered" environment of response buttons employed by researchers interested in selective reaching struck us as being analogous to the array of icons present in a typical GUI. In a study, Lyons, Elliott, Ricker, Weeks, and Chua (1999) sought to determine whether these paradigms could be imported into a "virtual" environment and ultimately serve as a test bed for investigations of perceptual-motor interactions in an HCI context. The task space in the Lyons et al. (1999) study utilized a three-by-three matrix similar to that used by Tipper et al. (1992). The matrix, made up of nine blue circles, was displayed on a monitor placed vertically in front of the participant. Participants were required to move the mouse on the graphics tablet, which would in turn move a cursor on the monitor in the desired direction toward the target (red) circle while ignoring any distractor (yellow) circles. The participants were unable to view their hand; the only visual feedback of their progress was from the cursor moving on the monitor. The graphics tablet allowed the researchers to record displacement and time data of the mouse throughout the trial. In contrast to previous experiments (e.g., Meegan & Tipper, 1998; Tipper et al., 1992), the presence of a distractor had relatively little influence on performance. Lyons et al. (1999) postulated that, in a task environment in which perceptual-motor interaction is less direct (e.g., using a mouse to move a cursor on a remote display) perceptual and motor workspaces are misaligned, and the increased translation processing owing to the misalignment serves to limit the impact of distractor items.

To test this idea, Lyons et al. (1999) modified the task environment so as to align the perceptual and motor workspaces. The monitor was turned and held screen down inside a support frame. The same three-by-three matrix was displayed on the monitor and reflected into a half-silvered mirror positioned above the graphics tablet allowing for sufficient space for the participant to manipulate the mouse and move the cursor to the target without vision of the hand. With this configuration, the stimulus display was presented and superimposed on the same plane as the motor workspace (e.g., the graphics tablet). Under this setup, distractor effects emerged and were consistent with an action-centered framework of attention. Taken together, these findings underscore the influence of translation requirements demanded by relative alignment of perceptual and motor workspaces. More importantly, these findings suggest that even relatively innocuous changes to the layout of the task environment may have significant impact on processes associated with selective attention in the mediation of action in an HCI context.

Attention and stimulus-response compatibility. Thus far we have only lightly touched on the issue of attention and S-R compatibility. However, the action-centered model of selective attention clearly advocates the view that attention and action are intimately linked. The fundamental premise is that attention mediates perceptual-motor interactions, and these, in turn, influence attention. Consistent with this perspective, the role of attention in the translation between perceptual inputs and motor outputs has also received considerable interest over the past decade. As discussed above, a key element in the selection of an action is the translation between stimuli and responses, the effectiveness of which is influenced to a large extent by the spatial relation between the stimuli and responses. The degree of coding and recoding required to map the locations of stimulus and response elements has been proposed to be a primary determinant of the speed and accuracy of translation (e.g. Wallace, 1971). Attentional processes have been implicated in the issue of how relative spatial stimulus information is coded. Specifically, the orienting of attention to the location of a stimulus has been proposed to result in the generation of the spatial stimulus code.

Initial interest in the link between attention orienting and spatial translation have emerged as a result of attempts to explain the Simon effect. The Simon effect (Simon, 1968; Simon & Rudell, 1969), often considered a variant of spatial S-R compatibility, occurs in a situation in which a nonspatial stimulus attribute indicates the correct response and the spatial attribute is irrelevant to the task. Thus, the spatial dimension of the stimulus is an irrelevant attribute, and a symbolic stimulus feature constitutes the relevant attribute. Although the spatial stimulus attribute is irrelevant to the task, faster responding is found when the position of the stimulus and the position of the response happen to correspond. A number of researchers (e.g., Umiltà & Nicoletti, 1992) have suggested that attentional processes may be a unifying link between the Simon effect and the spatial compatibility effect proper. Specifically, the link between attention and action in these cases is that a shift in attention is postulated to be the mechanism that underlies the generation

of the spatial stimulus code (e.g., Nicoletti & Umiltà, 1989, 1994; Proctor & Lu, 1994; Rubichi, Nicoletti, Iani, & Umiltà, 1997; Stoffer, 1991; Stoffer & Umiltà, 1997; Umiltà & Nicoletti, 1992). According to an attention-shift account, when a stimulus is presented to the left or right of the current focus of attention, a reorienting of attention occurs toward the location of the stimulus. This attention shift is associated with the generation of a spatial code that specifies the position of the stimulus with respect to the last attended location. If this spatial stimulus code is congruent with the spatial code of the response, then S-R translation, and therefore response selection, is facilitated. If the two codes are incongruent, response selection is hindered.

Recent work in our lab has also implicated a role for attention shifts in compatibility effects and object recognition. In these studies, Lyons, Weeks, and Chua (2000a; 2000b) sought to examine the influence of spatial orientation on the speed of object identification. Participants were presented with video images of common objects that possessed a graspable surface (e.g., a tea cup, frying pan), and were instructed to make a left or right key press under two distinct mapping rules depending on whether the object was in an upright or inverted vertical orientation. The first mapping rule required participants to respond with a left key press when the object was inverted and a right key press when the object was upright. The opposite was true for the second mapping rule. The orientation of the object's graspable surface was irrelevant to the task. The results showed that identification of object orientation was facilitated when the graspable surface of the object was also oriented to the same side of space as the response (see also Tucker & Ellis, 1998). In contrast, when participants were presented with objects that possessed symmetrical graspable surfaces on both sides (e.g., a sugar bowl with two handles), identification of object orientation was not facilitated. Lyons et al. (2000b) also showed that response facilitation was evident when the stimuli consisted simply of objects that, though may not inherently be graspable, possessed a left-right asymmetry. Taken together, these results were interpreted in terms of an attentional mechanism. Specifically, Lyons et al. (2000a; 2000b) proposed that a left-right object asymmetry (e.g., a protruding handle) might serve to capture spatial attention (cf., Tucker & Ellis, 1998). If attention is thus oriented toward the same side of space as the ensuing action, the spatial code associated with the attention shift (e.g., see discussion above) would lead to facilitation of the response. In situations in which no such object asymmetry exists, attentional capture and orienting may be hindered, and as a result, there is no facilitation of the response.

Taken into the realm of HCI, it is our position that the interplay between shifts of attention, spatial compatibility, and object recognition will be a central human performance factor as technological developments continue to enhance the directness of direct-manipulation systems (cf., Shneiderman, 1983; 1992). Specifically, as interactive environments become better abstractions of reality with greater transparency (Rutkowski, 1982), the potential influence of these features of human information-processing will likely increase. Thus, it is somewhat ironic that the view toward virtual reality, as the solution to the problem of creating the optimal display representation, may bring with it an "unintended consequence" (Tenner, 1996). Indeed, the operator in such an HCI environment will be subject to the same constraints that are present in everyday life.

The primary goal of human factors research is to guide technological design in order to optimize perceptual-motor interactions between human operators and the systems they use within the constraints of maximizing efficiency and minimizing errors. Thus, the design of machines, tools, interfaces, and other sorts of devices utilizes knowledge about the characteristics, capabilities, as well as limitations, of the human perceptual-motor system. In computing, the development of input devices such as the mouse and graphical user interfaces was intended to improve human–computer interaction. As technology has continued to advance, the relatively simple mouse and graphical displays have begun to give way to exploration of complex gestural interfaces and virtual environments. This development may be, perhaps in part, a desire to move beyond the artificial nature of such devices as the mouse, to ones that provide a better mimic of reality. Why move an arrow on a monitor using a hand-held device to point to a displayed object, when instead, you can reach and interact with the object? Perhaps such an interface would provide a closer reflection of real-world interactions—and the seeming ease with which we interact with our environments, but also subject to the constraints of the human system.

PERCEPTUAL-MOTOR INTERACTION IN APPLIED TASKS: A FEW EXAMPLES

As we mentioned at the outset of this chapter, the evolution of computers and computer-related technology has brought us to the point at which the manner with which we interact with such systems has become a research area in itself. Current research in motor behavior and experimental psychology pertaining to attention, perception, action, and spatial cognition is poised to make significant contributions to the area of HCI. In addition to the continued development of a knowledge base of fundamental information pertaining to the perceptual-motor capabilities of the human user, these contributions will include new theoretical and analytical frameworks that can guide the study of HCI in various settings. In this final section, we highlight just a few specific examples of HCI situations that offer a potential arena for the application of the basic research that we have outlined in this chapter.

Remote Operation and Endoscopic Surgery

Work by Hanna, Shimi, and Cuschieri (1998) examined task performance of surgeons as a function of the location of the image display used during endoscopic surgical procedures. In their study, the display was located in front, to the left, or to the right of the surgeon. In addition, the display was placed either at eye level or at the level of the surgeon's hands. The surgeons' task performance was observed with the image display

positioned at each of these locations. Hanna et al. (1998) showed that the surgeons' performance was affected by the location of the display. Performance was facilitated when the surgeons were allowed to view their actions with the monitor positioned in front and at the level of the immediate workspace (the hands). Similar findings have also been demonstrated by Mandryk and MacKenzie (1999). In addition to the frontal image display location employed by Hanna et al. (1998), Mandryk and MacKenzie (1999) also investigated the benefits of projecting and superimposing the image from the endoscopic camera directly over the workspace. Their results showed that performance was superior when participants were initially exposed to the superimposed viewing conditions. This finding was attributed to the superimposed view allowing the participants to better calibrate the display space with the workspace. These findings are consistent with our own investigations of action-centered attention in virtual environments (Lyons et al., 1999). We would suggest that the alignment of perceptual and motor workspaces in the superimposed viewing condition facilitated performance due to the decreased translation requirements demanded by such a situation. However, the findings of Lyons et al. (1999) would also lead us to suspect that this alignment may have additional implications with respect to processes associated with selective attention in the mediation of action. Although the demands on perceptual-motor translation may be reduced, the potential intrusion of processes related to selective attention and action selection may now surface. Thus, we would cautiously suggest that the optimal display location in this task environment placed less translation demands on the surgeon during task performance.

In addition to considering the orientation of the workspace of the surgeon, it must be recognized that the surgeons work with a team of support personnel performing a variety of actions all designed to achieve a common goal. Thus, the group of people can be considered to be a synergistic unit much like an individual consists of a synergistic group of independent processes. This similarity in functional structure led us (Welsh et al., 2005, in press) and others to contemplate whether the behavior of a group of people in perceptual-motor and attention tasks follow the same principles as those that govern the behavior of an individual. While these initial experiments provide converging evidence that the same processes are involved in group and individual behavior, this line of research is still in its infancy and additional research is required to determine possible conditions in which different rules apply. For example, an interesting qualification is that the interactions between attention and action that occur when two humans interact do not emerge when a human interacts with a robotic arm (Castiello, 2003). Thus, given the different vantage points and goals of each member of the surgical team, an important empirical and practical question will be the manner in which perceptual-motor workspace can be effectively optimized to maximize the success of the team.

Personal Digital Assistants

Hand-held computer devices (PDAs and other similar communication devices like the Blackberry) are becoming increasingly sophisticated as they become more and more popular as mobile information processing and communication systems. One of the very relevant aspects of the evolving design of PDAs is the incorporation of a stylus touch screen system. This design allows for a tremendous flexibility in possible functions while at the same time maintains a consistent coding between real and virtual space. This latter advantage should allow for the straightforward application of the principles of action and attention discussed in this chapter to this virtual environment (e.g., Tipper et al., 1992 vs. Lyons et al., 1999). However, caution should once more be urged when implementing design change without due consideration of issues such as S-R compatibility as it has been shown that novices are fairly poor judges of configurations that facilitate the most efficient performance (Payne, 1995; Vu & Proctor, 2003). In addition to the principles of action-centered attention for lower-level interactions such as program navigation, the design of higher-order language-based functions must also take into account the experience and expectations of the user. For example, Patterson and Lee (2005) found that participants had greater difficulty in learning to use characters from the new written language developed for some PDAs when those characters did not resemble the characters of traditional English. Thus, as been highlighted many times in this chapter, consideration of perceptual and motor expectations of the user is necessary for the efficient use of these devices.

Eye-Gaze vs. Manual Mouse Interactions

Based on our review of attention and action, it is obvious that the spatiotemporal arrangement of relevant and nonrelevant stimulus information will affect the manner in which actions are completed. Although the work conducted in our labs (e.g., Lyons et al., 1999; Welsh et al., 1999; Welsh & Elliott, 2004a; 2004b) revealed that manual responses, such as those controlling conventional mouse or touch screen systems, are affected by nontarget distracting information, research from other labs has revealed that eye movements are similarly affected by the presentation of nontarget information (e.g., Godjin & Theewues, 2004). Given the intimate link between eye movements and attention, this similarity between the action-centered effects on eye and hand movements should not be surprising. Although the similarities in the interactions between attention and the manual and eye movement systems would initially lead one to believe that environments designed for manual responses can be immediately imported into eye-gaze controlled systems, there are important differences between the systems that must be considered. For example, it has been reported that one of the fundamental determinants of manual reaction times, the number of possible target locations (Hick-Hyman law; Hick, 1952; Hyman, 1953) does not influence eye movement reaction times (Kveraga, Boucher, & Hughes, 2002). Thus, while the programming of eye-gaze controlled HCI systems may be much more complicated than those involved in manual systems, the productivity of the user of eye-gaze systems may be more efficient than those of traditional manual (mouse) systems (Sibert & Jacob, 2000). Although eye movements still seem to be susceptible to the goals of the manual system (Bekkering & Neggers, 2002), additional investigation into the viability of eye-gaze and/or interactive eye-gaze and manual response systems is certainly warranted.

SUMMARY

The field of HCI offers a rich environment for the study of perceptual-motor interactions. The design of effective human-computer interfaces has been, and continues to be, a significant challenge that demands an appreciation of the entire human perceptual-motor system. The information-processing approach has provided a dominant theoretical and empirical framework for the study of perceptual-motor behavior in general, and for consideration of issues in HCI and human factors in particular. Texts in the area of human factors and HCI (including the present volume) are united in their inclusion of chapters or sections that pertain to the topic of human information processing. Moreover, the design of effective interfaces reflects our knowledge of the perceptual (e.g., visual displays, use of sound, graphics), cognitive (e.g., conceptual models, desktop metaphors), and motoric constraints (e.g., physical design of input devices, ergonomic keyboards) of the human perceptual-motor system.

Technological advances have undoubtedly served to improve the HCI experience. For example, we have progressed beyond the use of computer punch cards and command-line interfaces to more complex tools such as graphical user interfaces, speech recognition, and eye-gaze control systems. As HCI has become not only more effective, but by the same token more elaborate, the importance of the interaction between the various perceptual, cognitive, and motor constraints of the human system has come to the forefront. In our previous chapter, we presented an overview of some topics of research in stimulus-response compatibility in perceptual-motor interactions that we believed were relevant to HCI. In this chapter, we have shifted the focus to current issues in the interaction between attention and action-planning. While action-centered models will require additional development to describe full-body behavior in truly complex environments (e.g., negotiating a busy sidewalk or skating rink), the settings in which these models have been tested are, in fact, very similar to modern HCI environments. Thus, we believe that the relevance of this work for HCI cannot be underestimated. Clearly, considerable research will be necessary to evaluate the applicability of both of these potentially relevant lines of investigation to specific HCI design problems. Nevertheless, the experimental work to date leads us to conclude that the allocation of attention carries an action-centered component. For this reason, an effective interface must be sensitive to the perceptual and action expectations of the user, the specific action associated with a particular response location, the action relationship between that response and those around it, and the degree of translation required to map the perceptual-motor workspaces.

ACKNOWLEDGMENTS

We would like to recognize the financial support of the Natural Sciences and Engineering Research Council of Canada and the Alberta Ingenuity Fund.

References

Abrams, R. A., & Chirst, S. E. (2003). Motion onset captures attention. *Psychological Science, 14*, 427–432.

Abrams, R. A., & Pratt, J. (1996). Spatially-diffuse inhibition affects multiple locations: A reply to Tipper, Weaver, and Watson. *Journal of Experimental Psychology: Human Perception and Performance, 22*, 1294–1298.

Adams, J. A. (1971). A closed-loop theory of motor learning. *Journal of Motor Behavior, 3*, 111–150.

Adams, J. A., & Dijkstra, S. (1966). Short-term memory for motor responses. *Journal of Experimental Psychology, 71*, 314–318.

Allport, A. (1987). Selection for action: Some behavioral and neurophysiological considerations of attention and action. In H. Heuer & A. F. Sanders (Eds.), *Perspectives on perception and action* (pp. 395–419). Hillsdale, NJ: Erlbaum.

Allport, A. (1993). Attention and control: Have we been asking the wrong questions? A critical review of twenty-five years. In D.E. Meyer & S. Kornblum (Eds.), *Attention and performance 14: Synergies in experimental psychology, artificial intelligence, and cognitive neuroscience* (pp. 183–218). Cambridge, MA: MIT Press.

Bayerl, J., Millen, D., & Lewis, S. (1988). Consistent layout of function keys and screen labels speeds user responses. *Proceedings of the Human Factors Society 32nd Annual Meeting* (pp. 334–346). Santa Monica, CA: Human Factors Society.

Bekkering, H. & Neggers, S. F. W. (2002). Visual search is modulated by action intentions. *Psychological Science, 13*, 370–374.

Bekkering, H. & Pratt, J. (2004). Object-based processes in the planning of goal-directed hand movements. *The Quarterly Journal of Experimental Psychology, 57A*, 1345–1368.

Broadbent, D. E. (1958). *Perception and Communication.* New York, NY: Pergamon.

Brooks, V. B. (1974). Some examples of programmed limb movements. *Brain Research, 71*, 299–308.

Card, S. K., English, W. K., & Burr, B. J. (1978). Evaluation of mouse, rate-controlled isometric joystick, step keys, and text keys for text selection on a CRT. *Ergonomics, 21*, 601–613.

Castel, A. D., Drummond, E., & Pratt, J. (2005). The effects of action video game experience on the time course of inhibition of return and the efficiency of visual search. *Acta Psychologica, 119*, 217–230.

Castiello, U. (2003). Understanding other people's actions: Intention and attention. *Journal of Experimental Psychology: Human Perception and Performance, 29*, 416–430.

Chapanis, A., & Lindenbaum, L. E. (1959). A reaction time study of four control-display linkages, *Human Factors, 1*, 1–7.

Cheal, M. L., & Lyon, D. R. (1991). Central and peripheral precuing of forced-choice discrimination. *The Quarterly Journal of Experimental Psychology, 43A*, 859–880.

Cherry, E. C. (1953). Some experiments on the recognition speech, with one and with two ears. *Journal of the Acoustical Society of America, 25*, 975–979.

Chua, R., & Elliott, D. (1993). Visual regulation of manual aiming. *Human Movement Science, 12*, 365–401.

Chua, R., Weeks, D. J., & Goodman, D. (2003). Perceptual-motor interaction: Some implications for HCI. In J. A. Jacko & A. Sears (Eds.), *The Human–Computer Interaction Handbook: Fundamentals, evolving technologies, and emerging applications* (pp. 23–34). Hillsdale, NJ: Lawrence Earlbaum.

Crossman, E. R. F. W., & Goodeve, P. J. (1983). Feedback control of hand movement and Fitts' Law. Paper presented at the meeting of the Experimental Psychology Society, Oxford, July 1963. Published in *Quarterly Journal of Experimental Psychology, 35A*, 251–278.

Davis, G., & Driver, J. (1997). Spreading of visual attention to modally versus amodally completed regions. *Psychological Science, 8*, 275–281.

Deubel, H., & Schneider, W. X. (1996). Saccade target selection and object recognition: Evidence for a common attentional mechanism. *Vision Research, 36*, 1827–1837.

Deutsch, J. A., & Deutsch, D. (1963). Attention: Some theoretical considerations. *Psychological Review, 70*, 80–90.

Duncan, J. (1984). Selective attention and the organization of visual information. *Journal of Experimental Psychology: General, 113*, 501–517.

Egly, R., Driver, J., & Rafal, R. D. (1994). Shifting visual attention between objects and locations: Evidence from normal and parietal lesion subjects. *Journal of Experimental Psychology: General, 123*, 161–177.

Elliott, D., Carson, R. G., Goodman, D., & Chua, R. (1991). Discrete vs. continuous visual control of manual aiming. *Human Movement Science, 10*, 393–418.

English, W. K., Engelhart, D. C., & Berman, M. L. (1967). Display-selection techniques for text manipulation. *IEEE Transactions on Human Factors in Electronics, 8*, 5–15.

Eriksen, B. A., & Eriksen, C. W. (1974). Effects of noise letters upon the identification of a target letter in a non-search task. *Perception and Psychophysics, 16*, 143–146.

Fitts, P. M. (1954). The information capacity of the human motor system in controlling the amplitude of movement. *Journal of Experimental Psychology, 47*, 381–391.

Fitts, P. M. (1964). Perceptual-motor skills learning. In A. W. Melton (Ed.), *Categories of human learning* (pp. 243–285). New York: Academic Press.

Fitts, P. M., & Deninger, R. I. (1954). S-R compatibility: Correspondence among paired elements within stimulus and response codes. *Journal of Experimental Psychology, 48*, 483–491.

Fitts, P. M., & Posner, M. I. (1967). *Human performance.* Belmont, CA: Brooks-Cole.

Fitts, P. M., & Seeger, C. M. (1953). S-R compatibility: Spatial characteristics of stimulus and response codes. *Journal of Experimental Psychology, 46*, 199–210.

Folk, C. L., Remington, R. W., & Johnson, J. C. (1992). Involuntary covert orienting is contingent on attentional control settings. *Journal of Experimental Psychology: Human Perception and Performance, 18*, 1030–1044.

Folk, C. L., Remington, R. W., & Wright, J. H. (1994). The structure of attentional control: Contingent attentional capture by apparent motion, abrupt onset, and color. *Journal of Experimental Psychology: Human Perception and Performance, 20*, 317–329.

Godijn, R., & Theeuwes, J. (2004). The relationship between inhibition of return and saccade trajectory deviations. *Journal of Experimental Psychology: Human Perception and Performance, 30*, 538–554.

Hanna, G. B., Shimi, S. M., & Cuschieri, A. (1998). Task performance in endoscopic surgery is influenced by location of the image display. *Annals of Surgery, 4*, 481–484.

Helsen, W. F., Elliot, D., Starkes, J. L., & Ricker, K. L. (1998). Temporal and spatial coupling of point of gaze and hand movements in aiming. *Journal of Motor Behavior, 30*, 249–259.

Helsen, W. F., Elliott, D., Starkes, J. L., & Ricker, K. L. (2000). Coupling of eye, finger, elbow, and shoulder movements during manual aiming. *Journal of Motor Behavior, 32*, 241–248.

Hick, W. E. (1952). On the rate of gain of information. *The Quarterly Journal of Experimental Psychology, 4*, 11–26.

Hoffman, E. R., & Lim, J. T. A. (1997). Concurrent manual-decision tasks. *Ergonomics, 40*, 293–318.

Howard, L. A., & Tipper, S. P. (1997). Hand deviations away from visual cues: indirect evidence for inhibition. *Experimental Brain Research, 113*, 144–152.

Hyman, R. (1953). Stimulus information as a determinant of reaction time. *Journal of Experimental Psychology, 45*, 188–196.

Jonides, J. (1981). Voluntary versus automatic control over the mind's eye's movement. In J. Long & A. Baddley (Eds.), *Attention and performance IX* (pp. 187–203). Hillsdale, NJ: Lawrence Erlbaum.

Kelso, J. A. S. (1982). The process approach to understanding human motor behavior: An introduction. In J. A. S. Kelso (Ed.), *Human Motor Behavior: An Introduction* (pp. 3–19). Hillsdale, NJ: Lawrence Erlbaum.

Kelso, J. A. S. (1995). *Dynamic Patterns: The self-organization of brain and behavior.* Cambridge, MA: MIT Press.

Kelso, J. A. S., Southard, D. L., & Goodman, D. (1979). On the coordination of two-handed movements. *Journal of Experimental Psychology: Human Perception and Performance, 5*, 229–238.

Kveraga, K., Boucher, L., & Hughes, H. C. (2002). Saccades operate in violation of Hick's law. *Experimental Brain Research, 146*, 307–314.

Lyons, J., Elliott, D., Ricker, K. L., Weeks, D. J., & Chua, R. (1999). Action-centered attention in virtual environments. *Canadian Journal of Experimental Psychology, 53*(2), 176–178.

Lyons, J., Weeks, D. J., & Chua, R. (2000a). The influence of object orientation on speed of object identification: Affordance facilitation or cognitive coding? *Journal of Sport & Exercise Psychology, 22*, Supplement, S72.

Lyons, J., Weeks, D. J., & Chua, R. (2000b). Affordance and coding mechanisms in the facilitation of object identification. Paper presented at the Canadian Society for Psychomotor Learning and Sport Psychology Conference, Waterloo, Ontario, Canada.

MacKenzie, C. L., Marteniuk, R. G., Dugas, C., Liske, D., & Eickmeier, B. (1987). Three dimensional movement trajectories in Fitts' task: Implications for control. *The Quarterly Journal of Experimental Psychology, 39A*, 629–647.

Mandryk, R. L., & MacKenzie, C. L. (1999). Superimposing display space on workspace in the context of endoscopic surgery. *ACM CHI Companion,* 284–285.

Marteniuk, R. G. (1976). *Information processing in motor skills.* New York: Holt, Rinehart and Winston.

Marteniuk, R. G., MacKenzie, C. L., & Leavitt, J. L. (1988). Representational and physical accounts of motor control and learning: Can they account for the data? In A. M. Colley & J. R. Beech (Eds.), *Cognition and action in skilled behavior* (pp. 173–190). Amsterdam: Elsevier Science Publishers.

Marteniuk, R. G., MacKenzie, C. L., Jeannerod, M., Athenes, S., & Dugas, C. (1987). Constraints on human arm movement trajectories. *Canadian Journal of Psychology, 41*, 365–378.

Maylor, E. A., & Hockey, R. (1985). Inhibitory component of externally controlled covert orienting in visual space. *Journal of Experimental Psychology: Human Perception and Performance, 11*, 777–787.

Meegan, D. V., & Tipper, S. P. (1998). Reaching into cluttered visual environments: Spatial and temporal influences of distracting objects. *The Quarterly Journal of Experimental Psychology, 51A*, 225–249.

Meegan, D. V., & Tipper, S. P. (1999). Visual search and target-directed action. *Journal of Experimental Psychology: Human Perception and Performance, 25*, 1347–1362.

Meyer, D. E., Abrams, R. A., Kornblum, S., Wright, C. E., & Smith, J. E. K. (1988). Optimality in human motor performance: Ideal control of rapid aimed movements. *Psychological Review, 95*, 340–370.

Müller, H. J., & Rabbitt, P. M. A. (1989). Reflexive and voluntary orienting of visual attention: Time course of activation and resistance to interruption. *Journal of Experimental Psychology: Human Perception and Performance, 15*, 315–330.

Nicoletti, R., & Umiltà, C. (1989). Splitting visual space with attention. *Journal of Experimental Psychology: Human Perception and Performance, 15*, 164–169.

Nicoletti, R., & Umiltà, C. (1994). Attention shifts produce spatial stimulus codes. *Psychological Research, 56*, 144–150.

Pashler, H. (1994). Dual-task interference in simple tasks: Data and theory. *Psychological Bulletin, 116*, 220–244.

Patterson, J. T., & Lee, T. D. (2005, in press). Learning a new human-computer alphabet: The role of similarity and practice. *Acta Psychologica.*

Payne, S. J. (1995). Naïve judgments of stimulus-response compatibility. *Human Factors, 37*, 495–506.

Posner, M. I. (1982). Cumulative development of attentional theory. *American Psychologist, 37*, 168–179.

Posner, M. I., & Cohen, Y. (1984). Components of visual orienting. In H. Bouma & D. G. Bouwhuis (Eds.), *Attention and Performance X* (pp. 531–556). Hillsdale, NJ: Lawrence Earlbaum.

Posner, M. I., & Keele, S. W. (1969). Attentional demands of movement. *Proceedings of the 16th Congress of Applied Psychology.* Amsterdam: Swets and Zeitlinger.

Posner, M. I., Nissen, M. J., & Ogden, W. C. (1978). Attended and unattended processing modes: The role of set for spatial location. In H. Pick & E. Saltzman (Eds.), *Modes of perceiving and processing information* (pp. 137–157). Hillsdale, NJ: Lawrence Earlbaum.

Pratt, J., & Sekuler, A. B. (2001). The effects of occlusion and past experience on the allocation of object-based attention. *Psychonomic Bulletin & Review, 8*, 721–727.

Pratt, J., & Abrams, R. A. (1994). Action-centered inhibition: Effects of distractors in movement planning and execution. *Human Movement Science, 13*, 245–254.

Pratt, J., & McAuliffe, J. (2001). The effects of onsets and offsets on visual attention. *Psychological Research, 65*(3), 185–191.

Pratt, J., & Sekuler, A. B. (2001). The effects of occlusion and past experience on the allocation of object-based attention. *Psychonomic Bulletin and Review, 8*(4), 721–727.

Proctor, R. W., & Vu, K. L. (2003). Human information processing. In J. A. Jacko & A. Sears (Eds.), *The Human–Computer Interaction Handbook: Fundamentals, evolving technologies, and emerging applications* (pp. 35–50). Hillsdale, NJ: Lawrence Earlbaum.

Proctor, R. W., & Lu, C. H. (1994). Referential coding and attention-shifting accounts of the Simon effect. *Psychological Research, 56*, 185–195.

Proctor, R. W., & Van Zandt, T. (1994). *Human factors in simple and complex systems.* Boston: Allyn and Bacon.

Rafal, R. D., Calabresi, P. A., Brennan, C. W., & Sciolto, T. K. (1989). Saccade preparation inhibits reorienting to recently attended locations. *Journal of Experimental Psychology: Human Perception and Performance, 15*, 673–685.

Rizzolatti, G., Riggio, L., & Sheliga, B. M. (1994). Space and selective attention. In C. Umiltà & M. Moscovitch (Eds.), *Attention and performance XV* (pp. 231–265). Cambridge: MIT Press.

Rizzolatti, G., Riggio, L., Dascola, J., & Umilta, C. (1987). Reorienting attention across the horizontal and vertical meridians: Evidence in favor of a premotor theory of attention. *Neuropsychologia, 25*, 31–40.

Rubichi, S., Nicoletti, R, Iani, C., & Umiltà, C. (1997). The Simon effect occurs relative to the direction of an attention shift. *Journal of Experimental Psychology: Human Perception and Performance, 23*, 1353–1364.

Rutkowski, C. (1982). An introduction to the human applications standard computer interface, part I: Theory and principles. *Byte, 7*, 291–310.

Shepherd, M., Findlay, J. M., & Hockey, R. J. (1986) The relationship between eye-movements and spatial attention. *The Quarterly Journal of Experimental Psychology, 38A*, 475–491.

Shneiderman, B. (1983). Direct manipulation: A step beyond programming languages. *IEEE Computer, 16*, 57–69.

Shneiderman, B. (1992). *Designing the user interface: Strategies for effective human–computer interaction.* Reading, MA: Addison-Wesley Publishing Company.

Sibert, L. E., & Jacob, R. J. K. (2000). Evaluation of eye gaze interaction. *CHI Letters, 2*, 281–288.

Simon, J. R. (1968). Effect of ear stimulated on reaction time and movement time. *Journal of Experimental Psychology, 78,* 344–346.

Simon, J. R., & Rudell, A. P. (1967). Auditory S-R compatibility: The effect of an irrelevant cue on information processing. *Journal of Applied Psychology, 51,* 300–304.

Spence, C., Lloyd, D., McGlone, F., Nichols, M. E. R., & Driver, J. (2000). Inhibition of return is supramodal: A demonstration between all possible pairings of vision, touch, and audition. *Experimental Brain Research, 134,* 42–48.

Stelmach, G. E. (1982). Information-processing framework for understanding human motor behavior. In J. A. S. Kelso (Ed.), *Human motor behavior: An introduction* (pp. 63–91). Hillsdale, NJ: Lawrence Erlbaum.

Stoffer, T. H. (1991). Attentional focusing and spatial stimulus response compatibility. *Psychological Research, 53,* 127–135.

Stoffer, T. H., & Umiltà, C. (1997). Spatial stimulus coding and the focus of attention in S-R compatibility and the Simon effect. In B. Hommel & W. Prinz (eds.), *Theoretical issues in stimulus-response compatibility* (pp. 373–398). Amsterdam: Elsevier Science B.V.

Taylor, T. L., & Klein, R. M. (2000). Visual and motor effects in inhibition of return. *Journal of Experimental Psychology: Human Perception and Performance, 26,* 1639–1656.

Telford, C. W. (1931). The refractory phase of voluntary and associative responses. *Journal of Experimental Psychology, 14,* 1–36.

Tenner, E. (1996). *Why things bite back: Technology and the revenge of unintended consequences.* NY: Alfred A. Knopf.

Tipper, S. P., Howard, L. A., & Houghton, G. (1999). Behavioral consequences of selection form neural population codes. In S. Monsell & J. Driver (Eds.), *Attention and performance XVIII* (pp. 223–245). Cambridge, MA: MIT Press.

Tipper, S. P., Howard, L. A., & Jackson, S. R. (1997). Selective reaching to grasp: Evidence for distractor interference effects. *Visual Cognition, 4,* 1–38.

Tipper, S. P., Lortie, C., & Baylis, G. C. (1992). Selective reaching evidence for action-centered attention. *Journal of Experimental Psychology: Human Perception and Performance, 18,* 891–905.

Tipper, S. P., Meegan, D., & Howard, L. A. (2002). Action-centered negative priming: Evidence for reactive inhibition. *Visual Cognition, 9,* 591–614.

Tipper, S. P., Weaver, B., & Watson, F. L. (1996). Inhibition of return to successively cued spatial locations: Commentary on Pratt and Abrams (1995). *Journal of Experimental Psychology: Human Perception and Performance, 22,* 1289–1293.

Treisman, A. M. (1964a). The effect of irrelevant material on the efficiency of selective listening. *American Journal of Psychology, 77,* 533–546.

Treisman, A. M. (1964b). Verbal cues, language, and meaning in selective attention. *American Journal of Psychology, 77,* 206–219.

Treisman, A. M. (1986). Features and objects in visual processing. *Scientific American, 255,* 114–125.

Treisman, A. M., & Gelade, G. (1980). A feature-integration theory of attention. *Cognitive Psychology, 12,* 97–136.

Tremblay, L., Welsh, T. N., & Elliott, D. (2005). Between-trial inhibition and facilitation in goal-directed aiming: Manual and spatial asymmetries. *Experimental Brain Research, 160,* 79–88.

Tucker, M., & Ellis, R. (1998). On the relations between seen objects and components of potential actions. *Journal of Experimental Psychology: Human Perception and Performance, 24,* 830–846.

Umiltà, C., & Nicoletti, R. (1992). An integrated model of the Simon effect. In J. Alegria, D. Holender, J. Junca de Morais, & M. Radeau (Eds.), *Analytic approaches to human cognition* (pp. 331–350). Amsterdam: North-Holland.

Vu, K-P. L., & Proctor, R. W. (2003). Naïve and experienced judgments of stimulus-response compatibility: Implications for interface design. *Ergonomics, 46,* 169–187.

Wallace, R. J. (1971). S-R compatibility and the idea of a response code. *Journal of Experimental Psychology, 88,* 354–360.

Weir, P. L., Weeks, D. J., Welsh, T. N., Elliott, D., Chua, R., Roy, E. A., & Lyons, J. (2003). Action-centered distractor effects in discrete control selection. *Experimental Brain Research, 149,* 207–213.

Welford, A. T. (1968). *Fundamentals of skill.* London: Methuen.

Welsh, T. N., & Elliott, D. (2004a). Movement trajectories in the presence of a distracting stimulus: Evidence for a response activation model of selective reaching. *Quarterly Journal of Experimental Psychology–Section A, 57,* 1031–1057.

Welsh, T. N., & Elliott, D. (2004b). The effects of response priming and inhibition on movement planning and execution. *Journal of Motor Behavior, 36,* 200–211.

Welsh, T. N., & Elliott, D. (2005). The effects of response priming on the planning and execution of goal-directed movements in the presence of a distracting stimulus. *Acta Psychologica, 119,* 123–142.

Welsh, T. N., Elliott, D., Anson, J. G., Dhillon, V., Weeks, D. J., Lyons, J. L., & Chua, R. (2005). Does Joe influence Fred's action? Inhibition of return across different nervous systems. *Neuroscience Letters, 385,* 99–104.

Welsh, T. N., Elliott, D., & Weeks, D. J. (1999). Hand deviations towards distractors: Evidence for response competition. *Experimental Brain Research, 127,* 207–212.

Welsh, T. N., Lyons, J., Weeks, D. J., Anson, J. G., Chua, R., Mendoza, J. E., & Elliott, D. (in press). Within- and between-nervous system inhibition of return: Observation is as good as performance. *Psychonomic Bulletin and Review.*

Welsh, T. N., & Pratt, J. (2005, November). The salience of offset and onset stimuli to attention and motor systems: Evidence through distractor interference. Paper presented at the annual meeting of the Canadian Society for Psychomotor Learning and Sports Psychology, Niagara Falls, ON, Canada.

Wright, R. D., & Richard, C. M. (1996). Inhibition of return at multiple locations in visual space. *Canadian Journal of Experimental Psychology, 50,* 324–327.

•2•

HUMAN INFORMATION PROCESSING: AN OVERVIEW FOR HUMAN–COMPUTER INTERACTION

Robert W. Proctor
Purdue University

Kim-Phuong L. Vu
California State University Long Beach

*It is natural for an applied psychology of human-
computer interaction to be based theoretically
on information-processing psychology.*
—Card, Moran, & Newell, 1983

Human-computer interaction (HCI) is fundamentally an information-processing task. In interacting with a computer, a user has specific goals and subgoals in mind. The user initiates the interaction by giving the computer commands that are directed toward accomplishing those goals. The commands may activate software programs designed to allow specific types of tasks, such as word processing or statistical analysis to be performed. The resulting computer output, typically displayed on a screen, must provide adequate information for the user to complete the next step, or the user must enter another command to obtain the desired output from the computer. The sequence of interactions to accomplish the goals may be long and complex, and several alternative sequences, differing in efficiency, may be used to achieve these goals. During the interaction, the user is required to identify displayed information, select responses based on the displayed information, and execute those responses. The user must search the displayed information and attend to the appropriate aspects of it. She or he must also recall the commands and resulting consequences of those commands for different programs, remember information specific to the task that is being performed, and make decisions and solve problems during the process. For the interaction between the computer and user to be efficient, the interface must be designed in accordance with the user's information processing capabilities.

HUMAN INFORMATION PROCESSING APPROACH

The rise of the human information-processing approach in psychology is closely coupled with the growth of the fields of cognitive psychology, human factors, and human engineering. Although research that can be classified as falling within these fields has been conducted since the last half of the 19th century, their formalization dates back to World War II (see Hoffman & Deffenbacher, 1992). As part of the war efforts, experimental psychologists worked along with engineers on applications associated with using the sophisticated equipment being developed. As a consequence, the psychologists were exposed not only to applied problems but also to the techniques and views being developed in areas such as communications engineering (see Roscoe, 2005). Many of the concepts from engineering (for instance, the notion of transmission of information through a limited capacity communications channel) were seen as applicable to analyses of human performance.

The human information-processing approach is based on the idea that human performance, from displayed information to response, is a function of several processing stages. The nature of these stages, how they are arranged, and the factors that influence how quickly and accurately a particular stage operates can be discovered through appropriate research methods. It is often said that the central metaphor of the information-processing approach is that a human is like a computer (e.g., Lachman, Lachman, & Butterfield, 1979). However, even more fundamental than the computer metaphor is the assumption that the human is a complex system that can be analyzed in terms of subsystems and their interrelation. This point is evident in the work of researchers on attention and performance, such as Paul Fitts (1951) and Donald Broadbent (1958), who were among the first to adopt the information-processing approach in the 1950s.

The systems perspective underlies not only human information-processing but also human factors and HCI, providing a direct link between the basic and applied fields (Proctor & Van Zandt, 1994). Human factors in general, and HCI in particular, begin with the fundamental assumption that a human-machine system can be decomposed into machine and human subsystems, each of which can be analyzed further. The human information-processing approach provides the concepts, methods, and theories for analyzing the processes involved in the human subsystem. Posner (1986) stated, "Indeed, much of the impetus for the development of this kind of empirical study stemmed from the desire to integrate description of the human within overall systems" (p. V-6). Young, Clegg, and Smith (2004) emphasized that the most basic distinction between three processing stages (perception, cognition, and action), as captured in a block diagram model of human information processing, is important even for understanding the dynamic interactions of an operator with a vehicle for purposes of computer-aided augmented cognition. They noted:

This block diagram model of the human is important because it not only models the flow of information and commands between the vehicle and the human, it also enables access to the internal state of the human at various parts of the process. This allows the modeling of what a cognitive measurement system might have access to (internal to the human), and how that measurement might then be used as part of a closed-loop human-machine interface system. (pp. 261–262)

In the first half of the 20th century, the behaviorist approach predominated in psychology, particularly in the United States. Within this approach, many sophisticated theories of learning and behavior were developed that differed in various details (Bower & Hilgard, 1981). However, the research and theories of the behaviorist approach tended to minimize the role of cognitive processes and were of limited value to the applied problems encountered in World War II. The information-processing approach was adopted because it provided a way to examine topics of basic and applied concern, such as attention, that were relatively neglected during the behaviorist period. It continues to be the main approach in psychology, although contributions have been made from other approaches, some of which we will consider in the last section of the chapter.

Within HCI, human information-processing analyses are used in two ways. First, empirical studies evaluate the information-processing requirements of various tasks in which a human uses a computer. Second, computational models are developed which are intended to characterize human information processing when interacting with computers, and predict human performance with alternative interfaces. In this chapter, we survey methods used to study human information processing and summarize the major findings and the theoretical frameworks developed to explain them. We also tie the methods, findings, and theories to HCI issues to illustrate their use.

INFORMATION-PROCESSING METHODS

Any theoretical approach makes certain presuppositions and tends to favor some methods and techniques over others. Information-processing researchers have used behavioral and, to an increasing extent, psychophysiological measures, with an emphasis on chronometric (time-based) methods. There also has been a reliance on flow models that are often quantified through computer simulation or mathematical modeling.

Signal Detection Methods and Theory

One of the most useful methods for studying human information processing is that of signal detection (Macmillan & Creelman, 2005). In a signal-detection task, some event is classified as a signal and the subject's task is to detect whether the signal is present. Trials in which it is not present are called "noise trials." The proportion of trials in which the signal is correctly identified as present is called the "hit rate," and the proportion of trials in which the signal is incorrectly identified as present is called the "false alarm rate." By using the hit and false alarm rates, whether the effect of a variable is on detectability or response bias can be evaluated.

Signal-detection theory is often used as the basis for analyzing data from such tasks. This theory assumes that the response on each trial is a function of two discrete operations: encoding and decision. In a trial, the subject samples the information presented and decides whether this information is sufficient to warrant a *signal present* response. The sample of information is assumed to provide a value along a continuum of evidence states regarding the likelihood that the signal was present. The noise trials form a probability distribution of states, as do the signal trials. The decision that must be made on each trial can be characterized as to whether the event is from the signal or noise distribution. The subject is presumed to adopt a criterion value of evidence above which he or she responds "signal present" and below which he or she responds "signal absent."

In the simplest form, the distributions are assumed to be normal and equal variance. In this case, a measure of detectability (d') can be derived. This measure represents the difference in the means for the signal and noise distributions in standard deviation units. A measure of response bias (β), which represents the relative heights of the signal and noise distributions at the criterion, can also be calculated. This measure reflects the subject's overall willingness to say "signal present," regardless of whether it actually is present. There are numerous alternative measures of detectability and bias based on different assumptions and theories, and many task variations to which they can be applied (see Macmillan & Creelman, 2005).

Signal-detection analyses have been particularly useful because they can be applied to any task that can be depicted in terms of binary discriminations. For example, the proportion of words in a memory task correctly classified as old can be treated as a hit rate and the proportion of new lures classified as old can be treated as a false alarm rate (Lockhart & Murdock, 1970). In cases such as these, the resulting analysis helps researchers determine whether variables are affecting detectability of an item as old or response bias.

An area of research in which signal-detection methods have been widely used is that of vigilance (Parasuraman & Davies, 1977). In a typical vigilance task, a display is monitored for certain changes in it (e.g., the occurrence of an infrequent stimulus). Vigilance tasks are common in the military, but many aspects also can be found in computer-related tasks such as monitoring computer network operations (Percival & Noonan, 1987). A customary finding for vigilance tasks is the vigilance decrement, in which the hit rate decreases as time on the task increases. The classic example of this vigilance decrement is that, during World War II, British radar observers detected fewer of the enemy's radar signals after 30 minutes in a radar observation shift (Mackworth, 1948). Parasuraman and Davies concluded that, for many situations, the primary cause of the vigilance decrement is an increasingly strict response criterion. That is, the false alarm rate as well as the hit rate decreases as a function of time on task. They provided evidence that perceptual sensitivity also seems to decrease when the task requires comparison of each event to a standard held in memory and the event rate is high. Subsequently, See, Howe, Warm, and Dember (1995) concluded that a decrease in perceptual sensitivity occurs for a broad range of tasks. Although signal-detection theory can be used to help determine whether a variable affects encoding quality or decision, as in the vigilance example, it is important to keep in mind that the measures of detectability and bias are based on certain theoretical assumptions. Balakrishnan (1998) argued, on the basis of an analysis that does not require the assumptions of signal-detection theory, that the vigilance decrement is not a result of a biased placement of the response criterion, even when the signal occurs infrequently and time on task increases.

Chronometric Methods

Chronometric methods, for which time is a factor, have been the most widely used for studying human information processing. Indeed, Lachman et al. (1979) portrayed reaction time (RT) as the main dependent measure of the information-processing approach. Although many other measures are used, RT still predominates in part because of its sensitivity and in part because of the sophisticated techniques that have been developed for analyzing RT data.

A technique called the "subtractive method," introduced by F. C. Donders (1868/1969) in the 1860s, was revived in the 1950s and 60s. This method provides a way to estimate the duration of a particular processing stage. The assumption of the subtractive method is that a series of discrete processing stages intervenes between stimulus presentation and response execution. Through careful selection of tasks that differ by a single stage, the RT for the easier task can be subtracted from that for the more difficult task to yield the time for the additional process. The subtractive method has been used to estimate the durations of a variety of processes, including rates of mental rotation (approximately 12 to 20 ms per degree of rotation; Shepard & Metzler, 1971) and memory search (approximately 40 ms per item; Sternberg, 1969). An application of the subtractive method to HCI would be, for example, to compare the time to find a target link on two web pages that are identical except for the number of links displayed, and to attribute the extra time to the additional visual search required for the more complex web page.

The subtractive method has several limitations (Pachella, 1974). First, it is only applicable when discrete, serial processing stages can be assumed. Second, the processing for the two tasks being compared must be equivalent except for the additional process that is being evaluated. This requires an assumption of pure insertion, which is that the additional process for the more complex of two tasks can be inserted without affecting the processes held in common by the two tasks. However, this assumption often is not justified.

Sternberg (1969) developed the additive factors method to allow determination of the processes involved in performing a task. The additive factors method avoids the problem of pure insertion because the crucial data are whether two variables affect RT for the same task in an additive or interactive manner. Sternberg assumed, as did Donders, that information processing occurs in a sequence of discrete stages, each of which produces a constant output that serves as input to the next stage in the sequence. With these assumptions, he showed that two variables that affect different stages should have additive effects on RT. In contrast, two variables that affect the same stage should have interactive effects on RT. Sternberg performed detailed analyses of memory search tasks in which a person holds a set of letters or digits in memory and responds to a target stimulus by indicating whether it is in the memory set. Based on the patterns of additive and interactive effects that he observed, Sternberg concluded that the processing in such tasks involves four stages: target identification, memory search, response selection, and response execution. Grobelny, Karwowski, and Drury (2005) provided an application of additive-factors logic to usability of graphical icons in the design of HCI interfaces. Mode of icon array (menu or dialog box), number of icons, and difficulty of movement had additive effects on response times, implying that these variables affect different processing stages.

Both the subtractive and additive factors methods have been challenged on several grounds (Pachella, 1974). First, the assumption of discrete serial stages with constant output is difficult to justify in many situations. Second, both methods rely on analyses of RT, without consideration of error rates. This can be problematic because performance is typically not error free, and, as described below, speed can be traded for accuracy. Despite these limitations, the methods have proved to be robust and useful (Sanders, 1998). For example, Salthouse (2005) noted that the process analysis approach employed in contemporary research into aging effects on cognitive abilities "has used a variety of analytical methods such as subtraction, additive factors . . . to partition the variance in the target variable into theoretically distinct processes" (p. 288).

Speed-Accuracy Methods

The function relating response speed to accuracy is called the "speed-accuracy tradeoff" (Pachella, 1974). The function, illustrated in Fig. 2.1, shows that very fast responses can be performed with chance accuracy, and accuracy will increase as responding slows down. Of importance is the fact that when accuracy is high, as in most RT studies, a small increase in errors

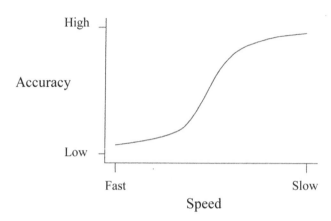

FIGURE 2.1. Speed-accuracy operating characteristic curve. Faster responding occurs at the cost of lower accuracy.

can result in a large decrease in RT. With respect to text entry on computing devices, MacKenzie and Soukoreff (2002) stated, "Clearly, both speed and accuracy must be measured and analyzed. . . . Participants can enter text more quickly if they are willing to sacrifice accuracy" (pp. 159–160).

In speed-accuracy tradeoff studies, the speed-accuracy criterion is varied between blocks of trials or between subjects by using different instructions regarding the relative importance of speed versus accuracy, varying payoffs such that speed or accuracy is weighted more heavily, or imposing different response deadlines (Wickelgren, 1977). These studies have the potential to be more informative than RT studies because they can provide information about whether variables affect the intercept (time at which accuracy exceeds chance), asymptote (the maximal accuracy), and rate of ascension from the intercept to the asymptote, each of which may reflect different processes. For example, Boldini, Russo, and Avons (2004) obtained evidence favoring dual-process models of recognition memory over single-process models by varying the delay between a visually presented test word and a signal to respond. Recognition accuracy benefited from a modality match at study and test (better performance when the study words were also visual rather than auditory) at short response-signal delays, but it benefited from deep processing during study (judging pleasantness) over shallow processing (repeating aloud each word) at long response-signal delays. Boldini et al. interpreted these results as consistent with the view that recognition judgments are based on a fast familiarity process or a slower recollection process.

Because the speed-accuracy criterion is manipulated in addition to any other variables of interest, much more data must be collected in a speed-accuracy study than in a RT study. Consequently, use of speed-accuracy methods has been restricted to situations in which the speed-accuracy relation is of major concern, rather than being widely adopted as the method of choice.

Psychophysiological and Neuroimaging Methods

In recent years, psychophysiological methods have been used to evaluate implications of information-processing models and to relate the models to brain processes. Such methods can provide details regarding the nature of processing by examining physiological activity as a task is being performed. The most widely used method involves measurement of electroencephalograms (EEGs), which are recordings of changes in brain activity as a function of time measured by electrodes placed on the scalp (Rugg & Coles, 1995). Of most concern for information-processing research are event-related potentials (ERPs), which are the changes in brain activity that are elicited by an event such as stimulus presentation or response initiation. ERPs are obtained by averaging across many trials of a task to remove background EEG noise and are thought to reflect postsynaptic potentials in the brain.

There are several features of the ERP that represent different aspects of processing. These features are labeled according to their polarity, positive (P) or negative (N), and their sequence or latency. The first positive (P1) and first negative (N1) components are associated with early perceptual processes. They are called "exogenous components" because they occur in close temporal proximity to the stimulus event and have a stable latency with respect to it. Later components reflect cognitive processes and are called "endogenous" because they are a function of the task demands and have a more variable latency than the exogenous components. One such component that has been studied extensively is the P3 (or P300), which represents post-perceptual processes. When an occasional target stimulus is interspersed in a stream of standards, the P3 is observed in response to targets, but not to standards. By comparing the effects of task manipulations on various ERP components such as P3, their onset latencies, and their scalp distributions, relatively detailed inferences about the cognitive processes can be made.

One example of applying a P3 analysis to HCI is a study by Trimmel and Huber (1998). In their study, subjects performed three HCI tasks (text editing, programming, and playing the game *Tetris*) for seven minutes each. They also performed comparable paper/pencil tasks in three other conditions. The P3 was measured after each experimental task by having subjects monitor a stream of high- and low-pitched tones, keeping count of each separately. The P3 varied as a function of type of task, as well as medium (computer vs. paper/pencil). The amplitude of the P3 was smaller following the HCI tasks than following the paper/pencil tasks, suggesting that the HCI tasks caused more fatigue or depletion of cognitive resources than the paper/pencil task. The P3 latency was shorter after the programming task than after the others, which the authors interpreted as an after-effect of highly focused attention.

Another measure that has been used in studies of human information processing is the lateralized readiness potential (LRP; Eimer, 1998). The LRP can be recorded in choice-reaction tasks that require a response with the left or right hand. It is a measure of differential activation of the lateral motor areas of the visual cortex that occurs shortly before and during execution of a response. The asymmetric activation favors the motor area contralateral to the hand making the response, because this is the area that controls the hand. The LRP has been obtained in situations in which no overt response is ever executed, allowing it to be used as an index of covert, partial-response activation. The LRP is thus a measure of the difference in activity from the two sides of the brain that can be used as an indicator of covert reaction tendencies, to determine whether a response has been prepared even when it is not actually executed. It can also be used to determine whether the effects of a variable are prior or subsequent to response preparation.

Electrophysiological measurements do not have the spatial resolution needed to provide precise information about the brain structures that produce the recorded activity. Recently developed neuroimaging methods, including positron-emission tomography (PET) and functional magnetic resonance imaging (fMRI), measure changes in blood flow associated with neuronal activity in different regions of the brain (Huettel, Song, & McCarthy, 2004). These methods have poor temporal resolution but much higher spatial resolution than the electrophysiological methods. In an imaging study, both control and experimental tasks are performed, and the functional neuroanatomy of the cognitive processes is derived by subtracting the image during the control task from that during the experimental task.

Application of cognitive neuroscience to HCI has been advocated under the heading of neuroergonomics. According to Parasuraman (2003), "Neuroergonomics focuses on investigations of the neural bases of mental functions and physical performance in relation to technology, work, leisure, transportation, health care and other settings in the real world" (p. 5). Neuroergonomics has the goal of using knowledge of the relation between brain function and human performance to design interfaces and computerized systems that are sensitive to brain function with the intent of increasing the efficiency and safety of human-machine systems.

INFORMATION-PROCESSING MODELS

Discrete and Continuous Stage Models

It is common to assume that the processing between stimuli and responses consists of a series of discrete stages for which the output for one stage serves as the input for the next, as Donders and Sternberg assumed. This assumption is made for the Model Human Processor (Card et al., 1983) and the Executive-Process Interactive Control (EPIC; Meyer & Kieras, 1997) architectures, both of which have been applied to HCI. However, models can be developed that allow for successive processing stages to operate concurrently. A well-known model of this type is McClelland's (1979) cascade model, in which partial information at one subprocess, or stage, is transferred to the next. Each stage is continuously active, and its output is a continuous value that is always available to the next stage. The final stage results in selection of which of the possible alternative responses to execute. Many parallel distributed processing, or neural network, models are of a continuous nature.

According to J. Miller (1988), models of human information processing can be classified as discrete or continuous along three dimensions: (a) representation, (b) transformation, and (c) transmission. "Representation" refers to whether the input and output codes for the processing stage are continuous or discrete. "Transformation" refers to whether the operation performed by the processing stage (e.g., spatial transformation) is continuous or discrete. "Transmission" is classified as discrete if the processing of successive stages does not overlap temporally.

The discrete stage model proposed by Sternberg (1969) has discrete representation and transmission, whereas the cascade model proposed by McClelland (1979) has continuous representation, transmission, and transformation. Models can be intermediate to these two extremes. For example, J. Miller's (1988) asynchronous discrete coding model assumes that most stimuli are composed of features, and these features are identified separately. Discrete processing occurs for feature identification, but once a feature is identified, this information can be passed to response selection while the other features are still being identified.

Sequential Sampling Models

Sequential sampling models are able to account for both RT and accuracy, and consequently, the tradeoff between them (Ratcliff & Smith, 2004; Van Zandt, Colonius, & Proctor, 2000). According to such models, information from the stimulus is sequentially sampled, resulting in a gradual accumulation of information on which selection of one of the alternative responses is based. A response is selected when the accumulated information exceeds a threshold amount required for that response. Factors that influence the quality of information processing have their effects on the rate at which the information accumulates, whereas factors that bias speed-versus-accuracy or specific responses have their effects on the response thresholds.

Balakrishnan (1998) argued that sequential sampling may be a factor even when the experiment does not stress speed of responding. As mentioned before, he showed that an analysis of vigilance data that does not make the assumptions of signal-detection theory suggests that attribution of the vigilance decrement to a change toward a more conservative response bias is incorrect. One reason why signal detection theory may lead to an incorrect conclusion is that the model assumes that the decision is based on a fixed sample of information, rather than information that is accumulating across time. Balakrishnan argued that even though there are no incentives to respond quickly in the typical vigilance task, subjects may choose not to wait until all of the stimulus information has been processed before responding. He proposed that a sequential sampling model, in which the subject continues to process the information until a stopping rule condition is satisfied, provides a better depiction. In this model, there are two potential sources of bias: the stopping rule and decision rule. Based on this model, Balakrishnan concluded that there is a response bias initially when the signal rate is low, and that the vigilance decrement is due to a gradual reduction of this response bias toward a more optimal decision during the time-course of the vigil.

INFORMATION PROCESSING IN CHOICE REACTION TASKS

In a typical choice reaction task in which each stimulus is assigned to a unique response, it is customary to distinguish between three stages of processing: stimulus identification, response selection, and response execution (Proctor & Van Zandt, 1994). The stimulus identification stage involves processes that are entirely dependent on stimulus properties. The response-selection stage concerns those processes involved in determining which response to make to each stimulus. Response execution refers to programming and execution of motor responses. Based on additive factors logic, Sanders (1998) decomposed the stimulus identification stage into three subcategories and the response execution stage into two subcategories, resulting in six stages (see Fig. 2.2).

Stimulus Identification

The preprocessing stage of stimulus identification refers to peripheral sensory processes involved in the conduction of the sensory signal along the afferent pathways to the sensory projection areas of the cerebral cortex. These processes are affected by variables such as stimulus contrast and retinal location. As stimulus contrast, or intensity, increases, RT decreases until reaching asymptote. For example, Bonin-Guillaume, Possamäi, Blin, and Hasbroucq (2000) had younger and older adults perform a two-choice reaction task, in which a left or right keypress was made to a bright or dim light positioned to the left or right. Stimulus intensity interacted with age, with RTs being about 25 ms shorter to a bright stimulus than to a dim stimulus for younger adults compared to 50 ms for older adults. The effect of stimulus intensity did not interact with variables that affect response selection or motor adjustment, suggesting that the age-related deficit in sensory preprocessing did not affect the later processing stages.

Feature extraction involves lower-level perceptual processing based in area V1 (the visual cortex) and other early visual cortical areas. Stimulus discriminability, word priming, and stimulus quality affect the feature extraction process. For example, manipulations of stimulus quality (such as superimposing a grid) slow RT, presumably by creating difficulty for the extraction of features. Identification itself is influenced by word frequency and mental rotation. The latter refers to the idea that, when a stimulus is rotated from the upright position, the time it takes to identify the stimulus increases as an approximately linear function of angular deviation from upright (Shepard & Metzler, 1971). This increase in identification time is presumed to reflect a normalization process by which the image is mentally rotated in a continuous manner to the upright position.

Response Selection

Response selection refers to those processes involved in determining what response to make to a particular stimulus. It is affected by the number of alternatives, stimulus-response compatibility, and precuing. RT increases as a logarithmic function of the number of stimulus-response alternatives (Hick, 1952; Hyman 1953). This relation is known as the Hick-Hyman law, which for N equally likely alternatives is:

$$RT = a + b \log_2 N$$

where a is the base processing time and b is the amount that RT increases with increases in N. The slope of the Hick-Hyman

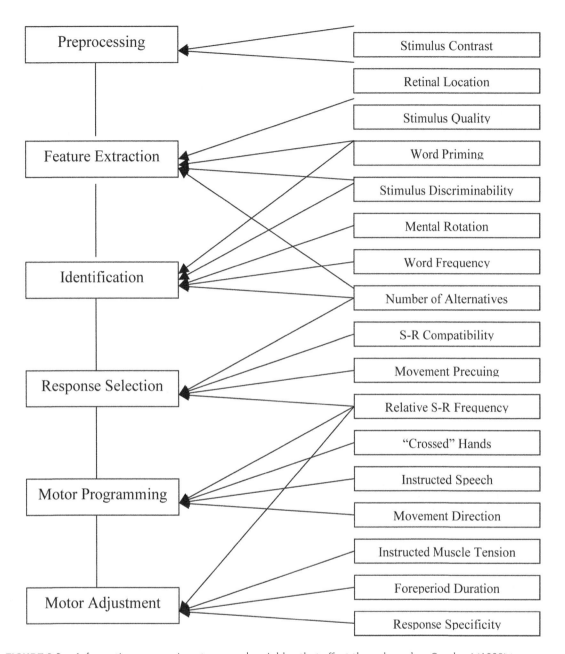

FIGURE 2.2.   Information-processing stages and variables that affect them, based on Sanders' (1998) taxonomy.

function is influenced by many factors. For example, the slope decreases as subjects become practiced at a task (Teichner & Krebs, 1974). Usher, Olami, and McClelland (2002) provided evidence from fits of a sequential sampling model that the Hick-Hyman law is due to subjects' adjusting their response criteria upward as the number of alternatives increases, in an attempt to maintain a constant high level of accuracy.

One variable that influences the slope of the Hick-Hyman function is stimulus-response compatibility, which has considerable impact on response-selection efficiency (see Proctor & Vu, 2006, for a review of compatibility principles). Compatibil-

ity effects are differences in speed and accuracy of responding as a function of how natural, or compatible, the relation between stimuli and responses is. Two types of compatibility effects can be distinguished (Kornblum, Hasbroucq, & Osman, 1990). For one type, certain sets of stimuli are more compatible with certain sets of responses than with others. For example, the combinations of verbal-vocal and spatial-manual sets yield better performance than the combinations of verbal-manual and spatial-vocal sets (Wang & Proctor, 1996). For the other type, within a specific stimulus-response set, some mappings of individual stimuli to responses produce better performance than others. If one stimulus has the meaning "left" and the other "right," performance is better if the left stimulus is mapped to the left response and the right stimulus to the right response, for all stimulus and response modes.

Fitts and Seeger (1953) and Fitts and Deininger (1954) demonstrated both types of compatibility effects for spatially arranged display and response panels. However, compatibility effects occur for a much wider variety of other stimulus-response sets. According to Kornblum et al. (1990), dimensional overlap (similarity) between the stimulus and response sets is the critical factor. When the sets have dimensional overlap, a stimulus will activate its corresponding response automatically. If this response is correct (compatible mapping), responding will be facilitated, but if it is not correct (incompatible mapping), responding will be inhibited. A second factor contributing to the advantage for the compatible mapping is that intentional translation of the stimulus into a response will occur quicker when the mapping is compatible than when it is not. Most contemporary models of stimulus-response compatibility incorporate both automatic and intentional response-selection routes (Hommel & Prinz, 1997), although they differ regarding the exact conditions under which each plays a role and the way in which they interact.

One reason why automatic activation is considered to contribute to compatibility effects is that such effects occur when irrelevant stimulus information overlaps with the response set (Lu & Proctor, 1995). The Stroop color-naming effect, for which an incongruent color word produces interference in naming a relevant stimulus color, is the most well known example. An irrelevant stimulus location also produces interference when it is incongruent with the location of a key-press to a relevant stimulus dimension, a phenomenon known as the Simon effect (Simon, 1990). Psychophysiological studies in which the LRP has been measured have provided evidence that the Simon effect is due at least in part to activation of the response corresponding to stimulus location (Valle-Inclán, de Labra, & Redondo, 2000).

For completely unrelated stimulus and response sets that are structured, performance is better when structural correspondence is maintained (Reeve & Proctor, 1990). For instance, when stimuli and responses are ordered (e.g., a row of four stimulus locations and a row of four response locations), RT is faster when the stimulus-response mapping can be characterized by a rule (e.g., pressing the key at the mirror opposite location) than when the mapping is random (Duncan, 1977). Spatial compatibility effects also occur when display and response elements refer to orthogonal spatial dimensions (Cho & Proctor, 2003). However, stimulus-response compatibility effects sometimes do not occur under conditions in which one would expect them to. For example, when compatible and incompatible mappings are mixed within a single block, the typical compatibility effect is eliminated (Shaffer, 1965; Vu & Proctor, 2004). Moreover, the same display and response elements can be coded along multiple dimensions in certain situations (e.g., vertical position versus horizontal position). The relative importance of maintaining compatibility on each dimension is a function of how salient the dimensions are made by the task environment (Rubichi, Vu, Nicoletti, & Proctor, 2006).

Because when and where compatibility effects are going to occur is not always obvious, interface designers are likely to make poor decisions if they rely only on their intuitions. Payne (1995), Vu and Proctor (2003), and Tlauka (2004) have shown that naïve subjects can predict simple compatibility effects, such as that performance will be better with a mapping that is spatially compatible than with one that is not. However, they are not able to accurately predict many other compatibility effects that occur, such as the benefit of maintaining a consistent stimulus-response mapping rule. One encouraging finding is that estimates of relative compatibility can be improved by a small amount of experience performing with the different stimulus-response mappings (Vu & Proctor, 2003). The important point for HCI is that designers need to be aware of the potential problems created by various types of incompatibility between display and response elements because their effects are not always obvious. A designer can get a better feel for the relative compatibility of alternative arrangements by performing tasks that use them. However, after the designer selects a few arrangements that would seem to yield good performance, these remaining arrangements need to be tested more thoroughly on groups of users.

Response Execution

"Motor programming" refers to specification of the physical response that is to be made. This process is affected by variables such as relative stimulus-response frequency and movement direction. One factor that influences this stage is movement complexity. The longer the sequence of movements that is to be made upon occurrence of a stimulus in a choice-reaction task, the longer the RT to initiate the sequence (Sternberg, Monsell, Knoll, & Wright, 1978). This effect is thought to be due to the time required to load the movement sequence into a buffer before initiating the movements. Time to initiate the movement sequence decreases with practice, and recent fMRI evidence suggests that this decrease in RT involves distinct neural systems that support visuomotor learning of finger sequences and spatial learning of the locations of the finger movements on a keypad (Parsons, Harrington, & Rao, 2005).

One of the most widely known relations attributed to response execution is Fitts' Law, which describes the time to make aimed movements to a target location (Fitts, 1954). This law, as originally specified by Fitts, is:

$$\text{Movement Time} = a + b \log_2 (2D/W)$$

where a and b are constants, D is distance to the target, and W is target width. However, there are slightly different versions of the law. According to Fitts' Law, movement time is a direct function of distance and an inverse function of target width. Fitts' Law has been found to provide an accurate description of movement time in many situations, although alternatives have been proposed for certain situations. One of the factors that contribute to the increase in movement time as the index of difficulty increases is the need to make a corrective submovement based on feedback in order to hit the target location (Meyer, Abrams, Kornblum, Wright, & Smith, 1988).

The importance of Fitts' Law for HCI is illustrated by the fact that the December 2004 issue of *International Journal of Human-Computer Studies* was devoted to the 50th anniversary of Fitts' original study. In the preface to the issue, the editors, Guiard and Beudouin-Lafon (2004), stated: "What has come to be known as Fitts' law has proven highly applicable in Human–Computer Interaction (HCI), making it possible to predict reliably the minimum time for a person in a pointing task to reach a specified target" (p. 747). Several illustrations of this point follow.

One implication of the law for interface design is that the slope of the function (*b*) may vary across different control devices, in which case movement times will be faster for the devices that yield lower slopes. Card, English, and Burr (1978) conducted a study that evaluated how efficient text keys, step keys, a mouse, and a joystick are at a text-selection task in which users selected text by positioning the cursor on the desired area and pressing a button or key. They showed that the mouse was the most efficient device for this task; positioning time for the mouse and joystick could be accounted for by Fitts' Law, with the slope of the function being less steep for the mouse; positioning time with the keys was proportional to the number of keystrokes that had to be executed.

Another implication of Fitts' Law is that any reduction in the index of difficulty should decrease the time for movements. Walker, Smelcer, and Nilsen (1991) evaluated movement time and accuracy of menu selection for the mouse. Their results showed that reducing the distance to be traveled (which reduces the index of difficulty) by placing the initial cursor in the middle of the menu, rather than the top, improved movement time. Placing a border around the menu item in which a click would still activate that item, and increasing the width of the border as the travel distance increases, also improved performance. The reduction in movement time by use of borders is predicted by Fitts' Law because borders increase the size of the target area.

Gillan, Holden, Adam, Rudisill, and Magee (1992) noted that designers must be cautious when applying Fitts' Law to HCI because factors other than distance and target size play a role when using a mouse. Specifically, they proposed that the critical factors in pointing and dragging are different than those in pointing and clicking (which was the main task in Card et al.'s (1978) study). Gillan et al. showed that, for a text-selection task, both point-click and point-drag movement times can be accounted for by Fitts' Law. For point-click sequences, the diagonal distance across the text object, rather than the horizontal distance, provided the best fit for pointing time. For point-drag, the vertical distance of the text provided the best fit. The reason why the horizontal distance is irrelevant is that the cursor must be positioned at the beginning of the string for the point-drag sequence. Thus, task requirements should be taken into account before applying Fitts' Law to the interface design.

Motor adjustment deals with the transition from a central motor program to peripheral motor activity. Studies of motor adjustment have focused on the influence of foreperiod duration on motor preparation. In a typical study, a neutral warning signal is presented at various intervals prior to the onset of the imperative stimulus. Bertelson (1967) varied the duration of the warning foreperiod and found that RT reached a minimum at a foreperiod of 150 ms and then increased slightly at the 200- and 300-ms foreperiods. However, error rate increased to a maximum at the 150-ms foreperiod and decreased slightly at the longer foreperiods. This relatively typical pattern suggests that it takes time to attain a state of high motor preparation, and that this state reflects an increased readiness to respond quickly at the expense of accuracy. Courtire, Hardouin, Vidal, Possamai, and Hasbroucq (2003) recently concluded that nitrous oxide impaired motor adjustment in rats because inhalation of nitrous oxide interacted with foreperiod duration but not with stimulus luminance.

MEMORY IN INFORMATION PROCESSING

Memory refers to explicit recollection of information in the absence of the original stimulus and to persisting effects of that information on information processing that may be implicit. Memory may involve recall of an immediately preceding event or one many years in the past, knowledge derived from everyday life experiences and education, or procedures learned to accomplish complex perceptual-motor tasks. Memory can be classified into several categories. Episodic memory refers to memory for a specific event such as going to the movie last night, whereas semantic memory refers to general knowledge such as what a movie is. Declarative memory is verbalizable knowledge, and procedural memory is knowledge that can be expressed nonverbally. In other words, declarative memory is knowledge that something is the case, whereas procedural memory is knowledge of how to do something. For example, telling your friend your new phone number involves declarative memory, whereas riding a bicycle involves procedural knowledge. A memory test is regarded as explicit if a person is asked to judge whether a specific item or event has occurred before in a particular context; the test is implicit if the person is to make a judgment, such as whether a string of letters is a word or nonword, that can be made without reference to earlier "priming" events. In this section, we focus primarily on explicit episodic memory.

Three types of memory systems are customarily distinguished: sensory stores, short-term memory (STM; or working memory), and long-term memory (LTM). Sensory stores, which we will not cover in detail, refer to brief modality-specific persistence of a sensory stimulus from which information can be retrieved for one or two seconds (see Nairne, 2003). STM and LTM are the main categories by which investigations of episodic memory are classified, and as the terms imply, the distinction is based primarily on duration. The dominant view is that these are distinct systems that operate according to different principles, but there has been debate over whether the processes involved in these two types of memories are the same or different. A recent fMRI study by Talmi, Grady, Goshen-Gottstein, and Moscovitch (2005) found that recognition of early items in the list was accompanied by activation of areas in the brain associated with LTM whereas recognition of recent items did not, supporting a distinction between STM and LTM stores.

Short-Term (Working) Memory

STM refers to representations that are currently being used or have recently been used and last for a short duration. A distinguishing characteristic is that STM is of limited capacity. This point was emphasized in G. A. Miller's (1956) classic article,

"The Magical Number Seven Plus or Minus Two," in which he indicated that capacity is not simply a function of the number of items, but rather the number of "chunks." For example, "i," "b," and "m" are three letters, but most people can combine them to form one meaningful chunk: "IBM." Consequently, memory span is similar for strings of unrelated letters and strings of meaningful acronyms or words. Researchers refer to the number of items that can be recalled correctly, in order, as "memory span."

As most people are aware from personal experience, if distracted by another activity, information in STM can be forgotten quickly. With respect to HCI, Oulasvirta and Saariluoma (2004) noted that diversion of attention from the current task to a competing task is a common occurrence, for example, when an unrequested pop-up dialogue box requiring an action appears on a computer screen. Laboratory experiments have shown that recall of a string of letters that is within the memory span decreases to close to chance levels over a retention interval of 18 s when rehearsal is prevented by an unrelated distractor task (Brown, 1958; Peterson & Peterson, 1959). This short-term forgetting was thought initially to be a consequence of decay of the memory trace due to prevention of rehearsal. However, Keppel and Underwood (1962) showed that proactive interference from items on previous lists is a significant contributor to forgetting. They found no forgetting at long retention intervals when only the first list in a series was examined, with the amount of forgetting being much larger for the second and third lists as proactive interference built up. Consistent with this interpretation, "release" from proactive inhibition—that is, improved recall—occurs when the category of the to-be-remembered items on the current list differs from that of previous lists (D. D. Wickens, 1970).

The capacity limitation of STM noted by G. A. Miller (1956) is closely related to the need to rehearse the items. Research has shown that the memory span, the number of words that can be recalled correctly in order, varies as a function of word length. That is, the number of items that can be retained decreases as word length increases. Evidence has indicated that the capacity is the number of syllables that can be said in about two seconds (Baddeley, Thomson, & Buchanan, 1975; Schweickert & Boruff, 1986). That pronunciation rate is critical suggests a time-based property of STM, which is consistent with a decay account. Consequently, the most widely accepted view is that both interference and decay contribute to short-term forgetting, with decay acting over the first few seconds and interference accounting for the largest part of the forgetting.

As the complexity of an HCI task increases, one consequence is the overloading of STM. Jacko and Ward (1996) varied four different determinants of task complexity (multiple paths, multiple outcomes, conflicting interdependence among paths, or uncertain or probabilistic linkages) in a task requiring use of a hierarchical menu to acquire specified information. When one determinant was present, performance was slowed by approximately 50%, and when two determinants were present in combination, performance was slowed further. That is, as the number of complexity determinants in the interface increased, performance decreased. Jacko and Ward attributed the decrease in performance for all four determinants to the increased STM load they imposed.

The best-known model of STM is Baddeley and Hitch's (1974) working memory model, which partitions STM into three main parts: central executive, phonological loop, and visuospatial sketchpad. The central executive controls and coordinates the actions of the phonological loop and visuospatial sketchpad. The phonological loop is composed of a phonological store that is responsible for storage of the to-be-remembered items, and an articulatory control process that is responsible for recoding verbal items into a phonological form and rehearsal of those items. The items stored in the phonological store decay over a short interval and can be refreshed through rehearsal from the articulatory control process. The visuospatial sketchpad retains information regarding visual and spatial information, and it is involved in mental imagery.

The working memory model has been successful in explaining several phenomena of STM (Baddeley, 2003)—for example, that the number of words that can be recalled is affected by word length. However, the model cannot explain why memory span for visually presented material is only slightly reduced when subjects engage in concurrent articulatory suppression (such as saying the word "the" aloud repeatedly). Articulatory suppression should monopolize the phonological loop, preventing any visual items from entering it. To account for such findings, Baddeley revised the working memory model to include an episodic buffer (see Fig. 2.3). The buffer is a limited-capacity temporary store that can integrate information from the phonological loop, visuospatial sketchpad, and long-term memory. By attending to a given source of information in the episodic buffer, the central executive can create new cognitive representations that might be useful in problem solving.

Long-Term Memory

LTM refers to representations that can be remembered for durations longer than can be attributed to STM. LTM can involve information presented from minutes to years ago. Initially, it was

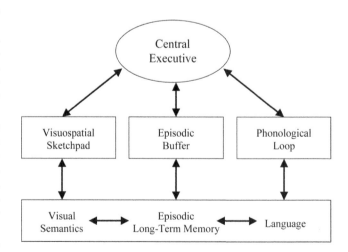

FIGURE 2.3. Baddeley's (2000) revised working memory model. From "The episodic buffer: A new component of working memory?" by A. D. Baddeley, 2000, *Trends in Cognitive Sciences*, 4, p. 421. Copyright 2000 by Elsevier Science Ltd. with permission.

thought that the probability of an item being encoded into LTM was a direct function of the amount of time that it was in STM, or how much it was rehearsed. However, Craik and Watkins (1973) showed that rehearsal in itself is not sufficient, but rather that deep-level processing of the meaning of the material is the important factor in transferring items to LTM. They presented subjects with a list of words and instructed them that when the experimenter stopped the presentation, they were to recall the last word starting with the letter "a." The number of other words between instances of "a" words was varied, with the idea that the amount of time a word was rehearsed would depend on the number of words before the next "a" word. At the end of the session, subjects were given a surprise test in which they were to recall all "a" words. There was no effect of number of intervening words on recall, suggesting that although subjects rehearsed the words longer, their recall did not improve because the words were not processed deeply.

Craik and Watkins' (1973) results were consistent with the levels of processing framework proposed by Craik and Lockhart (1972). According to this view, encoding proceeds in a series of analyses, from shallow perceptual features to deeper, semantic levels. The deeper the level of processing, the more strongly the item is encoded in memory. A key study supporting the levels of processing view is that of Hyde and Jenkins (1973). In their study, groups of subjects were presented a list of words for which they engaged in shallow processing (e.g., deciding whether each word contained a capital letter) or deep processing (e.g., identifying whether each word was a verb or a noun). Subjects were not told in advance that they would be asked to recall the words, but were given a surprise recall test at the end of the session. The results showed that the deep processing group recalled more words than the shallow processing group. Of direct relevance to HCI, Oulasvirta, Kärkkäinen, and Laarni (2005) recently found that participants who viewed the content area of a web page had no better memory for the locations of content objects than did a control group who guessed where those objects would be placed, because the participants' task was to locate navigational objects on the page and not to process the content information.

Another well-known principle for LTM is encoding specificity, which states that the probability that a retrieval cue results in recollection of an earlier event is an increasing function of the match between the features encoded initially and those provided by the retrieval cue (Tulving & Thomson, 1973). An implication of this principle is that memory will be context dependent. Godden and Baddeley (1975) demonstrated a context-dependent memory effect by having divers learn a list of words on land or under water, and recall the words on land or under water. Recall was higher for the group who learned on land when the test took place on land than under water, and vice versa for the group who learned under water. A related principle is that of transfer-appropriate processing (Morris, Bransford, & Franks, 1977). Morris et al. showed that deep-level semantic judgments during study produced better performance than shallow rhyme judgments on a standard recognition memory test. However, when the memory test required decisions about whether the test words rhymed with studied words, the rhyme judgments led to better performance than the semantic judgments. Healy, Wohldmann, and Bourne (2005) have proposed that encoding

specificity and transfer-appropriate processing can be incorporated within the single principle of procedural reinstatement: Retention will be evident to the extent that the procedures engaged in during study or training are reinstated during the retention test.

Research has confirmed that the levels-of-processing framework must accommodate the effects of the retention context, as captured by the above principles, to explain the effects of processing performed during encoding. Although levels-of-processing has a strong effect on accuracy of explicit recall and recognition, Jacoby and Dallas (1981) found no effect on an implicit memory test. Later studies have shown a robust effect of levels-of-processing on implicit tests similar to that obtained for recall and recognition if the test is based on conceptual cues, rather than on perceptual cues (see Challis, Velichkovsky, & Craik, 1996). Challis et al. constructed direct recognition tests, in which the words were graphemically, phonologically, or semantically similar to the studied words, that showed no levels of processing effect. They emphasized that, to account for levels-of-processing results, it is necessary to specify the types of information produced by the levels-of-processing, the types of information required for the specific test, and how task instructions modify encoding and retrieval processes.

Other Factors Affecting Retrieval of Earlier Events

Memory researchers have studied many factors that influence long-term retention. Not surprisingly, episodic memory improves with repetition of items or events. Also, massed repetition (repeating the same item in a row) is less effective than spaced repetition (repeating the same item with one or more intervening items). This benefit for spaced repetition, called the "spacing effect" or "lag effect," is often attributed to two main factors. First, study time for the same items appearing in succession is less than study time for the same items appearing further apart. Second, when the items are studied over a longer period of time, there is an opportunity for the items to be associated with different cues that can aid recall later. The spacing or lag effect is widespread and occurs for both recall and recognition (Hintzman, 1974). Bahrick and Hall (2005) noted that a similar spacing benefit is found for learning lists of items when practice sessions, each with test and learning phases, are separated by several days. They presented evidence that a large part of the spacing benefit in this case arises from individuals determining which study strategies are more effective at promoting long-term retention and then using those strategies more.

Another widely studied phenomenon is the generation effect, in which recall is better when subjects have to generate the to-be-remembered words rather than just studying the words as they are presented (Slamecka & Graf, 1978). In a generation effect experiment, subjects are divided into two groups, read and generate. Each group receives a series of words, with each word spelled out completely for the read group and missing letters for the generate group. An example is as follows:

Read group: CAT; ELEPHANT; GRAPE; CAKE
Generate group: C _ T; E_E_H _ _ NT; G _ APE; CAK_

The typical results show that subjects in the generate group can recall more words than those in the read group. One application of the generation effect to HCI is that when a computer user needs a password for an account, the system should allow the user to generate the password rather than providing her with one because the user would be more likely to recall the generated password. The common method of proactive password generation, in which users are asked to generate a password that meets certain restrictions (e.g., contain an uppercase letter, a lowercase letter, a digit, etc.), is intended to result in more memorable and secure passwords (e.g., Vu, Bhargav, & Proctor, 2003).

Events that precede or follow an event of interest can interfere with recall of that event. The former is referred to as "proactive interference," and was discussed in the section on STM; the latter is referred to as "retroactive interference." One area of research in which retroactive interference is of central concern is that of eyewitness testimony. Loftus and Palmer (1974) showed that subsequent events could distort a person's memory of an event that the person witnessed. Subjects were shown a sequence of events depicting a car accident. Subsequently, they were asked the question, "How fast were the cars going when they _____ each other." When the verb "contacted" was used, subjects estimated the speed to be 32 miles per hour, and only one-tenth of them reported seeing broken glass. However, when the verb "smashed" was used, the estimated speed increased to 41 miles per hour, and almost one third of the subjects reported seeing broken glass. Demonstrations like these indicate not only that retroactive interference can cause forgetting of events, but that it also can cause the memory of events to be changed. More recent research has shown that completely false memories can be implanted (see Roediger & McDermott, 1995).

Mnemonic techniques can also be used to improve recall. The basic idea behind mnemonics is to connect the to-be-remembered material with an established organizational structure that can be easily accessible later on. Two widely used mnemonic techniques are the pegword method (Wood & Pratt, 1987) and the method of loci (Verhaeghen & Marcoen, 1996). In the pegword method, a familiar rhyme provides the organizational structure. A visual image is formed between each pegword in the rhyme and the associated target item. At recall, the rhyme is generated, and the associated items come to mind. For the method of loci, locations from a well-known place, such as your house, are associated with the to-be-remembered items. Although specific mnemonic techniques are limited in their usefulness, the basic ideas behind them (utilizing imagery, forming meaningful associations, and using consistent encoding and retrieval strategies) are of broad value for improving memory performance.

Vu, Tai, Bhargav, Schultz, and Proctor (2004) examined the effectiveness of a "first-letter" mnemonic technique to help users relate individual characters of a password to a structured sentence in order to aid recall at a later time. In one condition, Vu et al. had users generate a sentence and take the first letter of each word in the sentence to form a password; in another condition, users generated a sentence that also included numbers and special characters embedded into the sentence and resulting password. Passwords generated using the first-letter tech-

nique were more memorable when users did not have to embed a digit and special character into the sentence, but were more secure (i.e., more resistant to cracking) when the sentence and resulting password included the digit and special character. Thus, when it comes to memory and security of computer passwords, there seems to be a tradeoff between memorability and security.

ATTENTION IN INFORMATION PROCESSING

Attention is increased awareness directed at a particular event or action to select it for increased processing. This processing may result in enhanced understanding of the event, improved performance of an action, or better memory for the event. Attention allows us to filter out unnecessary information so that we can focus on a particular aspect that is relevant to our goals. Several significant information-processing models of attention have been proposed.

Models of Attention

In an influential study, Cherry (1953) presented subjects with different messages to each ear through headphones. Subjects were to repeat aloud one of the two messages while ignoring the other. When subsequently asked questions about the two messages, subjects were able to accurately describe the message to which they were attending but could not describe anything except physical characteristics, such as gender of the speaker, about the unattended message.

To account for such findings, Broadbent (1958) developed the filter theory, which assumes that the nervous system acts as a single-channel processor. According to filter theory, information is received in a preattentive temporary store and then is selectively filtered, based on physical features such as spatial location, to allow only one input to access the channel. Broadbent's filter theory implies that the meaning of unattended messages is not identified, but later studies showed that the unattended message could be processed beyond the physical level, in at least some cases (Treisman, 1964).

To accommodate the finding that meaning of an unattended message can influence performance, Treisman (1964) reformulated filter theory into what is called the "filter-attenuation theory." According to attenuation theory, early selection by filtering still precedes stimulus identification, but the filter only attenuates the information on unattended channels. This attenuated signal may be sufficient to allow identification if the stimulus is one with a low identification threshold, such as a person's name or an expected event. Deutsch and Deutsch (1963) proposed that unattended stimuli are always identified and the bottleneck occurs in later processing, a view called "late-selection theory." The difference between attenuation theory and late-selection theory and is that latter assumes that meaning is fully analyzed, but the former does not.

Lavie, Hirst, de Fockert, and Viding (2004) have proposed a load theory of attention, which they claim "resolves the long-standing early versus late selection debate" (p. 339). Specifically, the load theory includes two selective attention mechanisms: a perceptual selection mechanism and a cognitive control mechanism. When perceptual load is high (i.e., great demands are placed on the perceptual system), the perceptual mechanism excludes irrelevant stimuli from being processed. When memory load is high, it is not possible to suppress irrelevant information at a cognitive level. In support of load theory, Lavie et al. showed that interference from distracting stimuli is reduced under conditions of high perceptual load but increased under conditions of high working memory load.

In divided attention tasks, a person must attend to multiple sources of information simultaneously. Kahneman (1973) proposed a unitary resource model that views attention as a single resource that can be divided up among different tasks in different amounts, based on task demands and voluntary allocation strategies. Unitary resource models provided the impetus for dual-task methodologies (such as performance operating characteristics) and mental workload analyses that are used widely in HCI (Eberts, 1994). The expectation is that multiple tasks should produce interference when their resource demands exceed the supply that is available.

Many studies have shown that it is easier to perform two tasks together when they use different stimulus or response modalities than when they use the same modalities. Performance is also better when one task is verbal and the other visuospatial than when they are the same type. These result patterns provide the basis for multiple resource models of attention such as that of C. D. Wickens (1984). According to multiple resource models, different attentional resources exist for different sensory-motor modalities and coding domains. Multiple resource theory captures the fact that multiple task performance typically is better when the tasks use different input-output modes than when they use the same modes. However, it is often criticized as being too flexible because new resources can be proposed arbitrarily to fit any finding of specificity of interference (Navon, 1984).

A widely used metaphor for visual attention is that of a spotlight that is presumed to direct attention to everything in its field (Posner & Cohen, 1984). Direction of attention is not necessarily the same as the direction of gaze because the attentional spotlight can be directed independently of fixation. Studies show that when a location is cued as likely to contain a target stimulus, but then a probe stimulus is presented at another location, a spatial gradient surrounds the attended location such that items nearer to the focus of attention are processed more efficiently than those farther away from it (Yantis, 2000). The movement of the attentional spotlight to a location can be triggered by two types of cues, exogenous and endogenous. An exogenous cue is an external event such as the abrupt onset of a stimulus at a peripheral location that involuntarily draws the attentional spotlight to its location. Exogenous cues produce rapid performance benefits, which dissipate quickly, for stimuli presented at the cued location. This is followed by a period in which performance is worse for stimuli at the cued location than for ones presented at the uncued location, a phenomenon called "inhibition of return" (Posner & Cohen, 1984). An endogenous cue is typically a symbol such as a central arrowhead that must be identified before a voluntary shift in attention to the designated location can be made. The performance benefits for endogenous cues take longer to develop and are sustained

for a longer period of time when the cues are relevant, indicating that their benefits are due to conscious control of the attentional spotlight (Klein & Shore, 2000).

In a visual search task, subjects are to detect whether a target is present among distractors. Treisman and Gelade (1980) developed Feature Integration Theory to explain the results from visual search studies. When the target is distinguished from the distractors by a basic feature such as color (feature search), RT and error rate often show little increase as the number of distractors increases. However, when two or more features must be combined to distinguish the target from distractors (conjunctive search), RT and error rate typically increase sharply as the number of distractors increases. To account for these results, feature integration theory assumes that basic features of stimuli are encoded into feature maps in parallel across the visual field at a preattentive stage. Feature search can be based on this preattentive stage because a "target-present" response requires only detection of the feature. The second stage involves focusing attention on a specific location and combining features that occupy the location into objects. Attention is required for conjunctive search because responses cannot be based on detection of a single feature. According to feature integration theory, performance in conjunctive search tasks decreases as the number of distractors increases because attention must be moved sequentially across the search field until a target is detected or all items present have been searched. Feature integration theory served to generate a large amount of research on visual search that showed, as is typically the case, that the situation is not as simple as depicted by the theory. This has resulted in modifications of the theory, as well as alternative theories. For example, Wolfe's (1994) Guided Search Theory maintains the distinction between an initial stage of feature maps and a second stage of attentional binding, but assumes that the second stage is guided by the initial feature analysis.

In HCI, a common visual search task involves locating menu items. When users know exactly what option to search for, they can use identity matching, in which they search the display for the menu name that they want to find. Perlman (1984) suggested that when identity search is used, the menu options should be displayed in alphabetical order to facilitate search. When users do not know where an option is included within a main list of menus, inclusion matching is used. The users must decide in which group the specific option would be categorized and then search the list of items within that group. With inclusion matching, search times may be longer for items that can be classified in more than one of the main groupings or when the items are less well-known examples of a main grouping (Somberg & Picardi, 1983). Equivalence search occurs when the user knows what option to select, but does not know how that option is labeled. McDonald, Stone, and Liebelt (1983) showed that alphabetical and categorical organizations yield shorter search times than randomized organization for equivalence search. Search can also be affected by the breadth versus depth of the menu design. Lee and MacGregor (1985) showed that deep hierarchies are preferred over broad ones. However, more recently, Tullis, Catani, Chadwick-Dias, and Cianchette (2005) suggested that, for complex or ambiguous situations, there is a benefit for broad menu designs because they facilitate comparison between categories. The main point is that, when structuring menus, designers must take into account the type of search in which the user would most likely be engaged.

The role of attention in response selection has been investigated extensively using the psychological refractory period (PRP) paradigm (Pashler, 1998). In the PRP paradigm, a pair of choice-reaction tasks must be performed, and the stimulus onset asynchrony (SOA) of the second stimulus is presented at different intervals. RT for Task 2 is slowed at short SOAs, and this phenomenon is called the "PRP effect." The experimental results have been interpreted with what is called "locus of slack logic" (Schweickert, 1978), which is an extension of additive factors logic to dual-task performance. The basic idea is that if a Task 2 variable has its effect prior to a bottleneck, that variable will have an underadditive interaction with SOA. This underadditivity occurs because, at short SOAs, the slack period during which post-bottleneck processing cannot begin can be used for continued processing for the more difficult condition. If a Task 2 variable has its effect after the bottleneck, the effect will be additive with SOA.

The most widely accepted account of the PRP effect is the response-selection bottleneck model (Pashler, 1998). The primary evidence for this model is that perceptual variables typically have underadditive interactions with SOA, implying that their effects are prior to the bottleneck. In contrast, post-perceptual variables typically have additive effects with SOA, implying that their effects are after the bottleneck. There has been dispute as to whether there is also a bottleneck at the later stage of response initiation (De Jong, 1993); whether the response-selection bottleneck is better characterized as a parallel processor of limited capacity that divides resources among to-be-performed tasks (Tombu & Jolicœur, 2005); and whether the apparent response-selection bottleneck is structural, or simply a strategy adopted by subjects to comply with task instructions (Meyer & Kieras, 1997). This latter approach is consistent with a recent emphasis on the executive functions of attention in the coordination and control of cognitive processes (Monsell & Driver, 2000).

Automaticity and Practice

Attention demands are high when a person first performs a new task. However, these demands decrease and performance improves as the task is practiced. Because the quality of performance and attentional requirements change substantially as a function of practice, it is customary to describe performance as progressing from an initial cognitively demanding phase to a phase in which processing is automatic (Anderson, 1982; Fitts & Posner, 1967).

The time to perform virtually any task from choice RT to solving geometry problems decreases with practice, with the largest benefits occurring early in practice. Newell and Rosenbloom (1981) proposed a power function to describe the changes in RT with practice:

$$RT = BN^{-\alpha}$$

where N is the number of practice trials, B is RT on the first trial, and α is the learning rate. Although the power function has be-

come widely accepted as a law that describes the changes in RT, Heathcote, Brown, and Mewhort (2000) indicated that it does not fit the functions for individual performers adequately. They showed that exponential functions provided better fits than power functions to 40 individual datasets, and proposed a new exponential law of practice. The defining characteristic of the exponential function is that the relative learning rate is a constant at all levels of practice, whereas, for the power function, the relative learning rate is a hyperbolically decreasing function of practice trials.

PROBLEM SOLVING AND DECISION MAKING

Beginning with the work of Newell and Simon (1972), it has been customary to analyze problem solving in terms of a problem space. The problem space consists of the following: (a) an initial state, (b) a goal state that is to be achieved, (c) operators for transforming the problem from the initial state to the goal state in a sequence of steps, and (d) constraints on application of the operators that must be satisfied. The problem-solving process itself is conceived of as a search for a path that connects the initial and goal states.

Because the size of a problem space increases exponentially with the complexity of the problem, most problem spaces are well beyond the capacity of short-term memory. Consequently, for problem solving to be effective, search must be constrained to a limited number of possible solutions. A common way to constrain search is to use heuristics. For example, people often use a means-ends heuristic for which, at each step, an operator is chosen that will move the current state closer to the goal state (Atwood & Polson, 1976). Such heuristics are called "weak methods" because they do not require much knowledge about the exact problem domain. Strong methods, such as those used by experts, rely on prior domain-specific knowledge and do not require much search because they are based on established principles applicable only to certain tasks.

The problem space must be an appropriate representation of the problem if the problem is to be solved. One important method for obtaining an appropriate problem space is to use analogy or metaphor. Analogy enables a shift from a problem space that is inadequate to one that may allow the goal state to be reached. There are several steps in using analogies (Holland, Holyoak, Nisbett, & Thagard, 1986), including detecting similarity between source and target problems, and mapping the corresponding elements of the problems. Humans are good at mapping the problems, but poor at detecting that one problem is an analog of another. An implication for HCI is that potential analogs should be provided to users for situations in which they are confronted by novel problems.

The concept of mental model, which is closely related to that of the problem space, has become widely used in recent years (see chapter 3, this volume). The general idea of mental models with respect to HCI is that as the user interacts with the computer, he or she receives feedback from the system that allows him or her to develop a representation of how the system is functioning for a given task. The mental model incorporates the goals of the user, the actions taken to complete the goals, and expectations of the system's output in response to the actions. A designer can increase the usability of an interface by using metaphors that allow transfer of an appropriate mental model (e.g., the desktop metaphor), designing the interface to be consistent with other interfaces with which the user is familiar (e.g., the standard web interface), and conveying the system's functions to the user in a clear and accurate manner. Feedback to the user is perhaps the most effective way to communicate information to the user and can be used to guide the user's mental model about the system.

Humans often have to make choices regarding situations in which the outcome depends on events that are outside of those humans' control. According to expected utility theory, a normative theory of decision making under uncertainty, the decision maker should determine the expected utility of a choice by multiplying the subjective utility of each outcome by the outcome's probability and summing the resulting values (see Proctor & Van Zandt, 1994). The expected utility should be computed for each choice, and the optimal decision is the choice with the highest expected utility. It should be clear from this description that for all but the simplest of problems, a human decision maker cannot operate in this manner. To do so would require attending to multiple cues that exceed attentional capacity, accurate estimates of probabilities of various events, and maintenance of, and operation on, large amounts of information that exceed short-term memory capacity.

Research of Kahneman and Tversky (2000) and others has shown that what people do when the outcome associated with a choice is uncertain is to rely heavily on decision-making heuristics. These heuristics include representativeness, availability, and anchoring. The representativeness heuristic is that the probability of an instance being a member of a particular category is judged on the basis of how representative the instance is of the category. The major limitation of the representativeness heuristic is that it ignores base rate probabilities for the respective categories. The availability heuristic involves determining the probability of an event based on the ease with which instances of the event can be retrieved. The limitation is that availability is affected not only by relative frequency, but also by other factors. The anchoring heuristic involves making a judgment regarding probabilities of alternative states based on initial information, and then adjusting these probabilities from this initial "anchor" as additional information is received. The limitation of anchoring is that the initial judgment can produce a bias for the probabilities. Although heuristics are useful, they may not always lead to the most favorable decision. Consequently, designers need to make sure that the choice desired for the user in a particular situation is one that is consistent with the user's heuristic biases.

NEW DEVELOPMENTS AND ALTERNATIVE APPROACHES

It should not be surprising that since its ascendance, the human information-processing approach has been challenged on many grounds, and alternative approaches have been championed. We describe three of those approaches below. In advocating

one, Smith and Henning (2005) recently stated, "The basic scientific problem with information processing models that have come to dominate cognitive science in the past four decades is that they are non-refutable" (p. 641). Such an argument misses the point that although the theoretical framework itself is non-refutable, as with any other such framework, the specific models and theories developed within the framework are refutable. The issue is the relative success of the information-processing approach at providing solutions to issues of theoretical and applied importance. Pashler (1998), who took an information-processing approach to the study of attention, noted, "The success or failure of information-processing psychology can be assessed only on the basis of insights that do or do not emerge from the research it spawns" (p. 7). It is the assessment of many researchers that the information-processing approach has generally fared well on many counts in comparison to alternative approaches.

Cognitive Neuroscience

A quick scan of journals and job announcements would reveal that cognitive neuroscience is a rapidly emerging field. As noted earlier, the major goal of research in this area is to analyze the brain mechanisms responsible for cognitive functioning. The knowledge gained from behavioral studies of human information processing provides the foundation for much of this research. This point is acknowledged explicitly in the instructions for authors for the *Journal of Cognitive Neuroscience*, which state, "The Journal publishes papers that bridge the gap between descriptions of information processing and specifications of brain activity" (MIT Press, 2007). Note that there would not be a gap to bridge were it not for the success of research on human information processing. If this gap is bridged successfully, the advances in neuroergonomics and augmented cognition envisioned by Parasuraman (2003) and Schmorrow (2005) will become reality.

The Ecological Approach

Gibson (1979) developed an ecological approach to perception and action that is typically presented as an alternative to the information-processing approach (e.g., Schvaneveldt, 2005). The ecological approach places emphasis on analyzing the information that is available in the optic array and the dynamics of this information as individuals and objects move in the environment. It also takes the position that perception is direct and that accounts relying on mental representations and cognitive processes are unnecessary and incorrect (see, for example, Michaels & Stins, 1997). Much of HCI involves performing information-intensive tasks in artificial, human-made environments that do not resemble carrying out actions in the natural environment. For example, Aziz and Macredie (2005) wrote, "The use of web-based information system[s] mainly involves processing of information" (p. 1). Because such environments are similar to those studied in basic human information-processing experiments, the approach has much to offer in analyzing and understanding the tasks performed in them.

For HCI tasks of a more ecological nature, such as navigating in a virtual world, understanding the information available to an observer/actor in the environment being simulated is clearly essential. As an example, Schvaneveldt (2005) noted that he and his colleagues have focused their recent studies of aviation on analyses of the information in the flight environment that allows pilots to perform effectively during different stages of flight. From an information-processing perspective, although such ecological analyses are valuable, and indeed necessary, they must be incorporated within information-processing models (e.g., Ullman, 1980).

Cybernetic Approach

The cybernetic view of cognition is that cognition emerges as a consequence of motor control over sensory feedback. The cybernetic approach is closely related to the ecological approach, but places greater emphasis on self-regulated control of perception and cognition. Smith and Henning (2005), who advocated a cybernetic approach, took a strong position against the information-processing approach, stated:

The cybernetic perspective presented here differs in a number of fundamental scientific respects from an information processing model. Behavioral cybernetics . . . emphasizes active control of information as sensory feedback via motor-sensory behavior, with motor-sensory behavior mediating both perception and cognition. In contrast, information processing models treat information as a fixed commodity presented on the input side of the processing system, ignoring the need for specific design factors to promote human motor control over this information as a source of sensory feedback. This failure to include motor-sensory control is a direct consequence of using an information processing model, where overt/motor behavior is viewed as a system output with no central role in the organization and control over information as feedback, nor over the control of subsequent behavior through the reciprocal effects of motor control on psychophysiological state. (p. 641)

Smith and Henning's (2005) statement stands in contrast to Young et al.'s (2004) assessment, described earlier in the chapter, that an information-processing analysis is essential to understanding the dynamic interactions of an operator with a vehicle for purposes of computer-aided augmented cognition. Although Smith and Henning may be correct that greater emphasis should be placed on the role of sensory feedback produced by actions, they have mistakenly attributed a simplifying tactic used by researchers to fundamental assumptions underlying the human information-processing approach. Recollect that human information processing has its roots in control theory.

Situated Cognition

Another approach that has been advocated recently is that of situated cognition, or situated action. Kiekel and Cooke (2005) stated that, according to situated cognition theories, "[m]uch of what symbolic [information-processing] theorists assign to the individual's head takes place outside of the confines of the individual and is directed by a world that is a much richer place than symbolic theories tend to suggest" (p. 92). They go on to say, "Information processing research tends to isolate psychological principles from a generic context, such as a laboratory . . ., while SA [situated action] research focuses on understanding the specific contextual constraints of the environment" (pp. 92–93). The extent to which the principles and theories developed from information-processing research conducted in the laboratory generalize to other contexts is an empirical issue. Although the evidence is not all in, the widespread application of information-processing principles and theories to human factors and HCI (e.g., Wickens, Lee, Liu, & Becker, 2004) suggests that what has been learned about human information processing is applicable to a variety of domains.

SUMMARY AND CONCLUSION

The methods, theories, and models in human information processing are currently well developed. The knowledge in this area, which we are only able to describe at a surface level in this chapter, is relevant to a wide range of concerns in HCI, from visual display design to representation and communication of knowledge. For HCI to be effective, the interaction must be made compatible with the human information-processing capabilities. Cognitive architectures that incorporate many of the facts about human information processing have been developed that can be applied to HCI. The Model Human Processor of Card et al. (1983) is the most widely known, but applications of other more recent architectures, including the ACT model of Anderson and colleagues (Anderson, Matessa, & Lebiere, 1997), the SOAR Model of Newell and colleagues (Howes & Young, 1997), and the EPIC Model of Kieras and Meyer (1997), hold considerable promise for the field, as demonstrated in chapter 5 of this volume.

References

Anderson, J. R. (1982). Acquisition of cognitive skill. *Psychological Review, 89,* 369–406.

Anderson, J. R., Matessa, M., & Lebiere, C. (1997). ACT-R: A theory of higher level cognition and its relation to visual attention. *Human–Computer Interaction, 12,* 439–462.

Atwood, M. E., & Polson, P. G. (1976). A process model for water jug problems. *Cognitive Psychology, 8,* 191–216.

Aziz, M., & Macredie, R. (2005). Specifying a model to explore the role of perceived usefulness in web-based information system use. In G. Salvendy (Ed.), *HCI International 2005.* Mahwah, NJ: Erlbaum.

Baddeley, A. (2003). Working memory and language: An overview. *Journal of Communication Disorders, 36*, 189–208.

Baddeley, A. D., & Hitch, G. J. (1974). Working memory. In G. H. Bower (Ed.), *The psychology of learning and motivation* (Vol. 8, pp. 47–89). New York: Academic Press.

Baddeley, A. D., Thomson, N., & Buchanan, M. (1975). Word length and the structure of shortterm memory. *Journal of Verbal Learning and Behavior, 14*, 575–589.

Bahrick, H. P., & Hall, L. K. (2005). The importance of retrieval failures to long-term retention: A metacognitive explanation of the spacing effect. *Journal of Memory and Language, 52*, 566–577.

Balakrishnan, J. D. (1998). Measures and interpretations of vigilance performance: Evidence against the detection criterion. *Human Factors, 40*, 601–623.

Bertelson, P. (1967). The time course of preparation. *Quarterly Journal of Experimental Psychology, 19*, 272–279.

Boldini, A., Russo, R., & Avons, S. E. (2004). One process is not enough! A speed-accuracy tradeoff study of recognition memory. *Psychonomic Bulletin & Review, 11*, 353–361.

Bonin-Guillaume, S., Possamäi, C.-A., Blin, O., & Hasbroucq, T. (2000). Stimulus preprocessing, response selection, and motor adjustment in the elderly: An additive factor analysis. *Cahiers de Psychologie Cognitive/Current Psychology of Cognition, 19*, 245–255.

Bower, G. H., & Hilgard, E. R. (1981). *Theories of learning* (5th ed.). Englewood Cliffs, NJ: Prentice-Hall.

Broadbent, D. E. (1958). *Perception and communication*. Oxford, UK: Pergamon Press.

Brown, J. (1958). Some tests of the decay theory of immediate memory. *Quarterly Journal of Experimental Psychology, 10*, 12–21.

Card, S. K., English, W. K., & Burr, B. J. (1978). Evaluation of the mouse, rate-controlled isometrick joystick, step keys, and text keys for text selection on a CRT. *Ergonomics, 21*, 601–613.

Card, S. K., Moran, T. P., & Newell, A. (1983). *The psychology of human–computer interaction*. Hillsdale, NJ: Erlbaum.

Challis, B. H., Velichkovsky, B. M., & Craik, F. I. M. (1996). Levels-of-processing effects on a variety of memory tasks: New findings and theoretical implications. *Consciousness and Cognition, 5*, 142–164.

Cherry, E. C. (1953). Some experiments on the recognition of speech, with one and with two ears. *Journal of the Acoustical Society of America, 25*, 975–979.

Cho, Y. S., & Proctor, R. W. (2003). Stimulus and response representations underlying orthogonal stimulus-response compatibility effects. *Psychonomic Bulletin & Review, 10*, 45–73.

Courtire, A., Hardouin, J., Vidal, F., Possamai, C.-A., & Hasbroucq, T. (2003). An additive factor analysis of the effect of sub-anaesthetic doses of nitrous oxide on information processing: Evidence for an impairment of the motor adjustment stage. *Psychopharmacology, 165*, 321–328.

Craik, F. I. M., & Lockhart, R. S. (1972). Levels of processing: A framework for memory research. *Journal of Verbal Learning and Verbal Behavior, 11*, 671–684.

Craik, F. I. M., & Watkins, M. J. (1973). The role of rehearsal in short-term memory. *Journal of Verbal Learning and Verbal Behavior, 12*, 599–607.

De Jong, R. (1993). Multiple bottlenecks in overlapping task performance. *Journal of Experimental Psychology: Human Perception and Performance, 19*, 965–980.

Deutsch, J. A., & Deutsch, D. (1963). Attention: Some theoretical considerations. *Psychological Review, 70*, 80–90.

Donders, F. C. (1969). On the speed of mental processes. In W. G. Koster (Ed.), *Acta Psychologica, 30, Attention and Performance II* (pp. 412–431). Amsterdam: North-Holland. (Original work published in 1868.)

Duncan, J. (1977). Response selection rules in spatial choice reaction tasks. In S. Dornic (Ed.). *Attention and performance VI* (pp. 49–71). Hillsdale, NJ: Erlbaum.

Eberts, R. E. (1994). *User interface design*. Englewood Cliffs, NJ: Prentice-Hall.

Eimer, M. (1998). The lateralized readiness potential as an on-line measure of central response activation processes. *Behavior Research Methods, Instruments, & Computers, 30*, 146–156.

Fitts, P. M. (1951). Engineering psychology and equipment design. In S. S. Stevens (Ed.) *Handbook of experimental psychology* (pp. 1287–1340). New York: Wiley.

Fitts, P. M. (1954). The information capacity of the human motor system in controlling the amplitude of movement. *Journal of Experimental Psychology, 47*, 381–391.

Fitts, P. M., & Deininger, R. L. (1954). S-R compatibility: Correspondence among paired elements within stimulus and response codes. *Journal of Experimental Psychology, 48*, 483–492.

Fitts, P. M., & Posner, M. I. (1967). *Human performance*. Belmont, CA: Brooks/Cole.

Fitts, P. M., & Seeger, C. M. (1953). S-R compatibility: Spatial characteristics of stimulus and response codes. *Journal of Experimental Psychology, 46*, 199–210.

Gibson, J. J. (1979). *The ecological approach to visual perception*. Boston: Houghton Mifflin.

Gillan, D. J., Holden, K., Adam, S. Rudisill, M., & Magee, L. (1992). How should Fitts' law be applied to human–computer interaction? *Interacting with Computers, 4*, 291–313.

Godden, D. R., & Baddeley, A. D. (1975). Context-dependent memory in two natural environments: On land and underwater. *British Journal of Psychology, 66*, 325–331.

Grobelny, J., Karwowski, W., & Drury, C. (2005). Usability of graphical icons in the design of human-computer interfaces. *International Journal of Human–Computer Interaction, 18*, 167–182.

Guiard, Y., & Beaudouin-Lafon, M. (2004). Fitts' law 50 years later: Applications and contributions from human–computer interaction. *International Journal of Human-Computer Studies, 61*, 747–750.

Healy, A. F., Wohldmann, E. L., & Bourne, L. E., Jr. (2005). The procedural reinstatement principle: Studies on training, retention, and transfer. In A. F. Healy (Ed.), *Experimental cognitive psychology and its applications* (pp. 59–71). Washington, DC: American Psychological Association.

Heathcote, A., Brown, S., & Mewhort, D. J. K. (2000). The power law repealed: The case for an exponential law of practice. *Psychonomic Bulletin & Review, 7*, 185–207.

Hick, W. E. (1952). On the rate of gain of information. *Quarterly Journal of Experimental Psychology, 4*, 11–26.

Hintzman, D. L. (1974). Theoretical implications of the spacing effect. In R. L. Solso (Ed.), *Theories of cognitive psychology: The Loyola symposium* (pp. 77–99). Hillsdale, NJ: Erlbaum.

Hoffman, R. R., & Deffenbacher, K. A. (1992). A brief history of applied cognitive psychology. *Applied Cognitive Psychology, 6*, 1–48.

Holland, J. H., Holyoak, K. J., Nisbett, R. E., & Thagard, P. R. (1986). *Induction*. Cambridge, MA: MIT Press.

Hommel, B., & Prinz, W. (Eds.) (1997). *Theoretical issues in stimulus-response compatibility*. Amsterdam: North-Holland.

Howes, A., & Young, R. M. (1997). The role of cognitive architecture in modeling the user: Soar's learning mechanism. *Human–Computer Interaction, 12*, 311–343.

Huettel, S. A., Song, A. W., & McCarthy, G. (2004). *Functional magnetic resonance imaging*. Sunderland, MA: Sinauer Associates.

Hyde, T. S., & Jenkins, J. J. (1973). Recall of words as a function of semantic, graphic, and syntactic orienting tasks. *Journal of Verbal Learning and Verbal Behavior, 12*, 471–480.

Hyman, R. (1953). Stimulus information as a determinant of reaction time. *Journal of Experimental Psychology, 45*, 188–196.

Jacko, J. A., & Ward, K. G. (1996). Toward establishing a link between psychomotor task complexity and human information processing. *19th International Conference on Computers and Industrial Engineering, 31*, 533–536.

Jacoby, L. L., & Dallas, M. (1981). On the relationship between autobiographical memory and perceptual learning. *Journal of Experimental Psychology: General, 110,* 306–340.

Kahneman, D. (1973). *Attention and effort.* Englewood Cliffs, NJ: Prentice Hall.

Kahneman, D., & Tversky, A. (Eds.) (2000). *Choices, values, and frames.* New York: Cambridge University Press.

Keppel, G., & Underwood, B. J. (1962). Proactive inhibition in short-term retention of single items. *Journal of Verbal Learning and Verbal Behavior, 1,* 153–161.

Kiekel, P. A., & Cooke, N. J. (2005). Human factors aspects of team cognition. In R. W. Proctor & K.-P. L. Vu (Eds.), *Handbook of human factors in web design* (pp. 90–103). Mahwah, NJ: Erlbaum.

Kieras, D. E., & Meyer, D. E. (1997). An overview of the EPIC architecture for cognition and performance with application to human–computer interaction. *Human–Computer Interaction, 12,* 391–438.

Klein, R. M., & Shore, D. I. (2000). Relation among modes of visual orienting. In S. Monsell, & J. Driver (Eds.), *Control of cognitive processes: Attention and performance XVIII* (pp. 195–208). Cambridge, MA: MIT Press.

Kornblum, S., Hasbroucq, T., & Osman, A. (1990). Dimensional overlap: Cognitive basis for stimulus-response compatibility—A model and taxonomy. *Psychological Review, 97,* 253–270.

Lachman, R., Lachman, J. L., & Butterfield, E. C. (1979). *Cognitive psychology and information processing: An introduction.* Hillsdale, NJ: Erlbaum.

Lavie, N., Hirst, A., de Fockert, J. W., & Viding, E. (2004). Load theory of selective attention and cognitive control. *Journal of Experimental Psychology: General, 133,* 339–354.

Lee, E., & MacGregor, J. (1985). Minimizing user search time in menu retrieval systems. *Human Factors, 27,* 157–162.

Lockhart, R. S., & Murdock, B. B., Jr. (1970). Memory and the theory of signal detection. *Psychological Bulletin, 74,* 100–109.

Loftus, E. F., & Palmer, J. C. (1974). Reconstruction of automobile destruction: An example of the interaction between language and memory. *Journal of Experimental Psychology: Human Learning and Memory, 4,* 19–41.

Lu, C.-H., & Proctor, R. W. (1995). The influence of irrelevant location information on performance: A review of the Simon effect and spatial Stroop effects. *Psychonomic Bulletin & Review, 2,* 174–207.

MacKenzie, I. S., & Soukoreff, R. W. (2002). Text entry for mobile computing: Models and methods, theory and practice. *Human–Computer Interaction, 17,* 147–198.

Mackworth, N. H. (1948). The breakdown of vigilance during prolonged visual search. *Quarterly Journal of Experimental Psychology, 1,* 6–21.

Macmillan, N. A., & Creelman, C. D. (2005). *Detection theory: A user's guide* (2nd ed.). Mahwah, NJ: Erlbaum.

McClelland, J. L. (1979). On the time relations of mental processes: A framework for analyzing processes in cascade. *Psychological Review, 88,* 375–407.

McDonald, J. E., Stone, J. D., & Liebelt, L. S. (1983). Searching for items in menus: The effects of organization and type of target. *Proceedings of the Human Factors Society 27th Annual Meeting* (pp. 289–338). Hillsdale, NJ: Erlbaum.

Meyer, D. E., Abrams, R. A., Kornblum, S., Wright, C. E., & Smith, J. E. K. (1988). Optimality in human motor performance: Ideal control of rapid aimed movements. *Psychological Review, 86,* 340–370.

Meyer, D. E., & Kieras, D. E. (1997). A computational theory of executive cognitive processes and multiple-task performance: Part 2. Accounts of psychological refractory-period phenomena. *Psychological Review, 104,* 749–791.

Michaels, C. F., & Stins, J. F. (1997). An ecological approach to stimulus-response compatibility. In B. Hommel & W. Prinz (Eds.), *Theoretical issues in stimulus-response compatibility* (pp. 333–360). Amsterdam: North-Holland.

Miller, G. A. (1956). The magical number seven plus or minus two: Some limits on our capacity for processing information. *Psychological Review, 63,* 81–97.

Miller, J. (1988). Discrete and continuous models of human information processing: Theoretical distinctions and empirical results. *Acta Psychologica, 67,* 191–257.

MIT Press (2007). The Journal of Cognitive Neuroscience Instructions for Authors. Retrieved February 20, 2007, from http://jocn.mit press.org/misc/ifora.shtml.

Monsell, S., & Driver, J. (Eds.) (2000). *Control of cognitive processes: Attention and performance XVIII.* Cambridge, MA: MIT Press.

Morris, C. D., Bransford, J. D., & Franks, J. J. (1977). Levels of processing versus transfer appropriate processing. *Journal of Verbal Learning and Verbal Behavior, 16,* 519–533.

Nairne, J. S. (2003). Sensory and working memory. In A. F. Healy & R. W. Proctor (Eds.), *Experimental psychology.* Volume 4 of *Handbook of psychology,* Editor-in-Chief: I. B. Weiner. Hoboken, NJ: Wiley.

Navon, D. (1984). Resources—a theoretical soup stone? *Psychological Review, 91,* 216–234.

Newell, A., & Rosenbloom, P. S. (1981). Mechanisms of skill acquisition and the law of practice. In J. R. Anderson (Ed.), *Cognitive skills and their acquisition* (pp. 1–55). Hillsdale, NJ: Erlbaum.

Newell, A., & Simon, H. A. (1972). *Human problem solving.* Englewood Cliffs, NJ: Prentice-Hall.

Oulasvirta, A., Kärkkäinen, L., & Laarni, J. (2005). Expectations and memory in link search. *Computers in Human Behavior, 21,* 773–789.

Oulasvirta, A., & Saariluoma, P. (2004). Long-term working memory and interrupting messages in human–computer interaction. *Behaviour & Information Technology, 23,* 53–64.

Pachella, R. G. (1974). The interpretation of reaction time in information-processing research. In B. H. Kantowitz (Ed), *Human information processing: Tutorials in performance and cognition* (pp. 41–82). Hillsdale, NJ: Erlbaum.

Parasuraman, R. (2003). Neuroergonomics: Research and practice. *Theoretical Issues in Ergonomics Science, 4,* 5–20.

Parasuraman, R., & Davies, D. R. (1977). A taxonomic analysis of vigilance performance. In R. R. Mackie (Ed.), *Vigilance: Theory, operational performance, and physiological correlates* (pp. 559–574). New York: Plenum.

Parsons, M. W., Harrington, D. L., & Rao, S. M. (2005). Distinct neural systems underlie learning visuomotor and spatial representations of motor skills. *Human Brain Mapping, 24,* 229–247.

Pashler, H. (1998). *The psychology of attention.* Cambridge, MA: MIT Press.

Payne, S. J. (1995). Naïve judgments of stimulus-response compatibility. *Human Factors, 37,* 495–506.

Percival, L. C., & Noonan, T. K. (1987). Computer network operation: Applicability of the vigilance paradigm to key tasks. *Human Factors, 29,* 685–694.

Perlman, G. (1984). Making the right choices with menus. *Proceedings of INTERACT '84* (pp. 291–295). London: IFIP.

Peterson, L. R., & Peterson, M. J. (1959). Short-term retention of individual verbal items. *Journal of Experimental Psychology, 58,* 193–198.

Posner, M. I. (1986). Overview. In K. R. Boff, L. Kaufman, & J. P. Thomas (Eds.). *Handbook of perception and human performance vol. II: Cognitive processes and performance* (pp. V3–V10). New York: Wiley.

Posner, M. I., & Cohen, Y. (1984). Components of visual orienting. In H. Bouma & D. G. Bouwhuis (Eds.) *Attention and performance X* (pp. 531–556). Hillsdale, NJ: Erlbaum.

Proctor, R. W., & Van Zandt, T. (1994). *Human factors in simple and complex systems.* Boston: Allyn & Bacon.

Proctor, R. W., & Vu, K.-P. L. (2006). *Stimulus-response compatibility: Data, theory and application.* Boca Raton, FL: CRC Press.

Ratcliff, R., & Smith, P. L. (2004). A comparison of sequential sampling models for two-choice reaction time. *Psychological Review, 111,* 333–367.

Reeve, T. G., & Proctor, R. W. (1990). The salient features coding principle for spatial- and symbolic-compatibility effects. In R. W. Proctor & T. G. Reeve (Eds.) *Stimulus-response compatibility: An integrated perspective* (pp. 163–180). Amsterdam: North-Holland.

Roediger, H. L., III, & McDermott, K. B. (1995). Creating false memories: Remembering words not presented in lists. *Journal of Experimental Psychology: Learning, Memory, and Cognition, 21,* 803–814.

Roscoe, S. (2005). Historical overview of human factors and ergonomics. In R. W. Proctor & K.-P. L. Vu (Eds.), *Handbook of human factors in web design* (pp. 3–12). Mahwah, NJ: Erlbaum.

Rubichi, S., Vu, K.-P. L., Nicoletti, R., & Proctor, R. W. (2006). Spatial coding in two dimensions. *Psychonomic Bulletin & Review, 13,* 201–216..

Rugg, M. D., & Coles, M. G. H. (Eds.) (1995). *Electrophysiology of mind: Event-related brain potentials and cognition.* Oxford, England: Oxford University Press.

Salthouse, T. A. (2005). From description to explanation in cognitive aging. In R. J. Sternberg & J. E. Pretz (Eds.), *Cognition & intelligence: Identifying the mechanisms of the mind* (pp. 288–305). Cambridge, England: Cambridge University Press.

Sanders, A. F. (1998). *Elements of human performance.* Mahwah, NJ: Erlbaum.

Schmorrow, D. D. (Ed.) (2005). *Foundations of augmented cognition.* Mahwah, NJ: Erlbaum.

Schvaneveldt, R. W. (2005). Finding meaning in psychology. In A. F. Healy (Ed.), *Experimental cognitive psychology and its applications* (pp. 211–235). Washington, DC: American Psychological Association.

Schweickert, R. (1978). A critical path generalization of the additive factor method: Analysis of a Stroop task. *Journal of Mathematical Psychology, 18,* 105–139.

Schweickert, R., & Boruff, B. (1986). Short-term memory capacity: Magic number or magic spell? *Journal of Experimental Psychology: Learning, Memory, and Cognition, 12,* 419–425.

See, J. E., Howe, S. R., Warm, J. S., & Dember, W. N. (1995). Meta-analysis of the sensitivity decrement in vigilance. *Psychological Bulletin, 117,* 230–249.

Shaffer, L. H. (1965). Choice reaction with variable S-R mapping. *Journal of Experimental Psychology, 70,* 284–288.

Shepard, R. N., & Metzler, J. (1971). Mental rotation of three-dimensional objects. *Science, 171,* 701–703.

Simon, J. R. (1990). The effects of an irrelevant directional cue on human information processing. In R. W. Proctor & T. G. Reeve (Eds.), *Stimulus-response compatibility: An integrated perspective* (pp. 31–86). Amsterdam: North-Holland.

Slamecka, N. J. (1968). An examination of trace storage in free recall. *Journal of Experimental Psychology, 76,* 504–513.

Slamecka, N. J., & Graf, P. (1978). The generation effect: Delineation of a phenomenon. *Journal of Experimental Psychology: Human Learning and Memory, 4,* 592–604.

Smith, T. J., & Henning, R. A. (2005). Cybernetics of augmented cognition as an alternative to information processing. In D. D. Schmorrow (Ed.), *Foundations of augmented cognition* (pp. 641–650). Mahwah, NJ: Erlbaum.

Somberg, B. L., & Picardi, M. C. (1983). Locus of information familiarity effect in search of computer menus. *Proceedings of the Human Factors Society 27th Annual Meeting* (pp. 826–830). Santa Monica, CA: Human Factors Society.

Sternberg, S. (1969). The discovery of processing stages: Extensions of Donders' method. In W. G. Koster (Ed), *Attention and Performance II. Acta Psychologica, 30,* 276–315.

Sternberg, S., Monsell, S, Knoll, R. L., & Wright, C. E. (1978). The latency and duration of rapid movement sequences. In G. E. Stelmach (Ed.), *Information processing in motor control and learning.* New York: Academic Press.

Talmi, D., Grady, C. L., Goshen-Gottstein, Y., & Moscovitch, M. (2005). Neuroimaging the serial position curve: A test of single-store versus dual-store models. *Psychological Science, 16,* 716–723.

Teichner, W. H., & Krebs, M. J. (1974). Laws of visual choice reaction time. *Psychological Review, 81,* 75–98.

Tlauka, M. (2004). Display-control compatibility: The relationship between performance and judgments of performance. *Ergonomics, 47,* 281–295.

Tombu, M., & Jolicœur, P. (2005). Testing the predictions of the central capacity sharing model. *Journal of Experimental Psychology: Human Perception and Performance, 31,* 790–802.

Treisman, A. M. (1964). Selective attention in man. *British Medical Bulletin, 20,* 12–16.

Treisman, A. M., & Gelade, G. (1980). A feature-integration theory of attention. *Cognitive Psychology, 12,* 97–136.

Trimmel, M., & Huber, R. (1998). After-effects of human–computer interaction indicated by P300 of the event-related brain potential. *Ergonomics, 41,* 649–655.

Tulving, E., & Thomson, D. M. (1973). Encoding specificity and retrieval processes in episodic memory. *Psychological Review, 80,* 359–380.

Tullis, T. S., Catani, M., Chadwick-Dias, A., & Cianchette, C. (2005). Presentation of information. In R. W. Proctor & K.-P. L. Vu (Eds.), *Handbook of human factors in web design* (pp. 107–133). Mahwah, NJ: Erlbaum.

Usher, M., Olami, Z., & McClelland, J. L. (2002). Hick's law in a stochastic race model with speed-accuracy tradeoff. *Journal of Mathematical Psychology, 46,* 704–715.

Valle-Inclán, F., de Labra, C., & Redondo, M. (2000). Psychophysiological studies of unattended information processing. *The Spanish Journal of Psychology, 3,* 76–85.

Van Zandt, T., Colonius, H., & Proctor, R. W. (2000). A comparison of two response time models applied to perceptual matching. *Psychonomic Bulletin & Review, 7,* 208–256.

Verhaeghen, P., & Marcoen, A. (1996). On the mechanisms of plasticity in young and older adults after instruction in the method of loci: Evidence for an amplification model. *Psychology & Aging, 11,* 164–178.

Vu, K.-P. L., Bhargav, A., & Proctor, R. W. (2003). Imposing password restrictions for multiple accounts: Impact on generation and recall of passwords. *Proceedings of the 47th Annual Meeting of the Human Factors and Ergonomics Society* (pp. 1331–1335). Santa Monica, CA: HFES.

Vu, K.-P. L, & Proctor, R. W. (2003). Naïve and experienced judgments of stimulus-response compatibility: Implications for interface design. *Ergonomics, 46,* 169–187.

Vu, K.-P. L, & Proctor, R. W. (2004). Mixing compatible and incompatible mappings: Elimination, reduction, and enhancement of spatial compatibility effects. *Quarterly Journal of Experimental Psychology, 57A,* 539–556.

Vu, K.-P. L., Tai, B.-L., Bhargav, A., Schultz, E. E., & Proctor, R. W. (2004). Promoting memorability and security of passwords through sentence generation. *Proceedings of the Human Factors and Ergonomics Society 48th Annual Meeting* (pp. 1478–1482). Santa Monica, CA: HFES.

Walker, N., Smelcer, J. B., & Nilsen, E. (1991). Optimizing speed and accuracy of menu selection: A comparison of walking and pulldown menus. *International Journal of Man-Machine Studies, 35,* 871–890.

Wang, H., & Proctor, R. W. (1996). Stimulus-response compatibility as a function of stimulus code and response modality. *Journal of Experimental Psychology: Human Perception and Performance, 22,* 1201–1217.

Wickelgren, W. A. (1977). Speed-accuracy tradeoff and information processing dynamics. *Acta Psychologica, 41*, 67–85.

Wickens, C. D. (1984). Processing resources in attention. In R. Parasuraman and D. R. Davies (Eds.), *Varieties of attention* (pp. 63–102). San Diego, CA: Academic Press.

Wickens, C. D., Lee, J., Liu, Y., & Becker, S. (2004). *An introduction to human factors engineering* (2nd ed.). New York: Prentice Hall.

Wickens, D. D. (1970). Encoding categories of words: An empirical approach to meaning. *Psychological Review, 77*, 1–15.

Wolfe, J. M. (1994). Guided Search 2.0: A revised model of visual search. *Psychonomic Bulletin & Review, 1*, 202–238.

Wood, L. E., & Pratt, J. D. (1987). Pegword mnemonic as an aid to memory in the elderly: A comparison of four age groups. *Educational Gerontology, 13*, 325–339.

Yantis, S. (2000). Goal-directed and stimulus-driven determinants of attentional control. In S. Monsell, & J. Driver (Eds.), *Control of cognitive processes: Attention and performance XVIII* (pp. 195–208). Cambridge, MA: MIT Press.

Young, P. M., Clegg, B. A., & Smith, C. A. P. (2004). Dynamic models of augmented cognition. *International Journal of Human–Computer Interaction, 17*, 259–273.

·3·

MENTAL MODELS IN
HUMAN–COMPUTER INTERACTION

Stephen J. Payne
University of Manchester

The plan for this chapter is as follows. It begins by reviewing and discussing the term "mental models" as it has been used in the literature on human–computer interaction (HCI), and in the neighboring disciplines of cognitive psychology where it was first coined. There is little consensus on what exactly is and is not a mental model, and yet it is too widely used for any posthoc attempt at a narrower definition to somehow cleanse the field. In consequence, I characterize several layers of theoretical commitment that the term may embrace, following an earlier discussion (Payne, 2003). To illustrate the argument, several classic and more recent studies from the HCI literature will be reviewed, with pointers to others. This first part of the chapter is based on material published in Payne (2003).

In cognitive psychology, mental models have major currency in two sub-disciplines—text comprehension and reasoning, although in the former they more often currently go by the name "situation models." Discussion in the latter focuses on quite refined theoretical disputes that currently have little relevance for HCI. The work on text comprehension, however, is germane. With the advent of the Web, the comprehension of text of various kinds has become a dominant mode of HCI, with important design issues for Websites, digital libraries, and so on. Interaction with text is in some ways a paradigm for interaction with information. With these points in mind, the concept of mental models in text comprehension will be discussed, with a particular eye to the issues that HCI accentuates, such as understanding multiple texts.

Two of the major practical questions raised by mental models are (a) how are they acquired and how can their acquisition be supported by instruction? The third section of this chapter will discuss two angles on these questions in HCI: first, the use of interactive computation and multimedia as an instructional method; second, the important tension between exploration and instruction, first systematically discussed in the HCI literature by Carroll's (1990) work on minimalism.

Finally, the paper will review some recent work on the importance of mental models for understanding aspects of collaborative teamwork. This area suggests that a relatively expansive view of human knowledge representations may be necessary for progress in HCI.

Throughout the chapter, a particular approach is taken to review: to choose one or two key studies and report them in some detail. I hope that this will allow some of the empirical methodologies and the rich variation in these to be conveyed. The chosen studies will be accompanied by some further references to the literature, but there are too many subtopics reviewed to aim for completeness.

WHAT IS A MENTAL MODEL?

The user's mental model of the device is one of the more widely discussed theoretical constructs in HCI. Alongside wide-ranging research literature, even commercial style guides have appealed to mental models for guidance (i.e., Mayhew, 1992; Tognazzini, 1992; Apple Human Interface Guidelines Apple Computer Inc., 1987).

Yet a casual inspection of the HCI literature reveals that mental models are used to label many different aspects of users'

knowledge about the systems they use. Nevertheless, I propose that even this simple core construct—what users know and believe about the systems they use—is worth highlighting and promoting. It is more distinctive than it might first seem, especially in comparison with other cognitive-science approaches. Further, beyond the core idea there is a progression of stronger theoretical commitments that have been mobilized by the mental models label, each of which speaks to important issues in HCI research, if not yet in practice.

The fundamental idea is that the *contents* of people's knowledge, including their theories and beliefs, can be an important explanatory concept for understanding users' behavior in relation to systems. This idea may seem obvious and straightforward, but in fact it suggests research questions that go against the grain of most contemporary cognitive psychology, which has concerned itself much more with the general limits of the human-information-processing system, such as the constraints on attention, retrieval, and processing. Thus, cognitive psychology tends to focus on the *structure* of the mind, rather than its contents. (The major exception to the rule that cognitive psychology has been obsessed with architecture over content is the work on expertise, and even here, recent work has focused on explanations of extreme performance in terms of general independent variables such as "motivated practice," i.e., Ericsson, Krampe, and Tesch-Romer (1993), rather than epistemological analysis.)

Refocusing attention on mental content about particular domains is what made mental models a popular idea in the early 1980s, such as the papers in Gentner and Stevens (1983). For example, work on naïve physics (i.e., McClosky, 1983) attempts to explain people's reasoning about the physical world, not in terms of working memory limits or particular representations, but in terms of their beliefs about the world, such as the nature of their theories of mechanics or electricity, for example. This focus on people's knowledge, theories, and beliefs about particular domains transfers naturally to questions in HCI, where practical interest may focus on how users conceive the workings of a particular device, how their beliefs shape their interactive behavior, and what lessons may be drawn for design.

In this mold, consider a very simple study of my own (Payne, 1991). Students were interviewed about ATMs. Following Collins and Gentner (1987) among others, "what if" questions were posed to uncover student's theories about the design and function of ATMs. For example, students were asked whether machines sometimes took longer to process their interactions; what information was stored on the plastic card; and what would happen if they "typed ahead" without waiting for the next machine prompt.

The interviews uncovered a wide variety in students' beliefs about the design of ATMs. For example, some assumed that the plastic card was written to as well as read from during transactions, and thus could encode the current balance of their account. Others assumed that the only information on the card was the user's personal identification number, allowing the machine to check the identity of the user (as it turns out, both these beliefs are incorrect). A conclusion from this simple observation is that users of machines are eager to form explanatory models and will readily go beyond available data to infer models that are consistent with their experiences. (One might wonder whether such explanations were not merely ad hoc, prompted during the interview: in fact some were, but explicit linguistic cues—such as "I've always thought"—strongly suggested that many were not.)

Another observation concerning students' "models" of ATMs was that they were fragmentary, perhaps more fragmentary than the term "model" might ordinarily connote: they were collections of beliefs about parts of the system, processes, or behaviors, rather than unified models of the whole design. Students would happily recruit an analogy to explain one part of the machine's operation that bore no relation to the rest of the system. This fragmentary character of mental models of complex systems may be an important aspect (see i.e., Norman, 1983), allowing partial understandings to be maintained. One implication is that users' mental models of single processes or operations might be a worthwhile topic for study and practical intervention (in design or instruction).

One widely held belief about a particular process affected the students' behavior as users. Almost all respondents believed that it was not possible to type ahead during machine pauses. At the time the study was conducted this was true for some, but not all, designs in use. Consequently, in some cases transactions were presumably being needlessly slowed because of an aspect of users' mental models.

A more recent study of a similar kind is an investigation of users' models of the navigation facilities provided by Internet browsers (Cockburn & Jones, 1996). Internet browsers, like Internet Explorer, maintain history lists of recently visited pages, providing direct access to these pages without needing to enter the URL or follow a hyperlink. The "back" and "forward" buttons provide a very frequently used mechanism for browsing history lists, but do users have good mental models for how they work? Cockburn and Jones (1996) showed that many do not.

The history list of visited pages can be thought of as a stack: a simple last-in-first-out data structure to which elements can be added (pushed) or taken out (popped) only from the top (consider a stack of trays in a canteen). When a new web page is visited by following a hyperlink, or by entering a URL, its address is pushed onto the top of the stack. This is true even if the page is already in the history list, so that the history list may contain more than one copy of the same page. However, when a page is visited by using the Back button (or, at least typically, by choosing from the history list), the page is not pushed onto the stack. So, what happens when the currently displayed page is not at the top of the stack (because it has been visited via the history list) and a new link is followed (or a new URL entered)? The answer is that all the pages in the history list that were above the current page are popped from the stack, and the newly visited page is pushed onto the stack in their place. For this reason the history list does *not* represent a complete record, or time-line of visited pages, and not all pages in the current browsing episode can be backed-up to. In Cockburn and Jones' study, few users appreciated this aspect of the device.

This then, has been the major thrust of work on mental models in HCI: what do people know and believe to be true about the way the systems they interact with are structured? How do their beliefs affect their behavior? In this literature a "mental model" is little more than a pointer to the relevant parts of the user's knowledge, yet this is not to deny its usefulness. One approach that it has engendered is a typology of knowledge—making groupings and distinctions about types of knowledge that are relevant in certain circumstances. It is in exactly this way that a literature on "shared mental models" as an explanatory concept in teamwork has been developed. This topic is perhaps the most rapidly growing area of mental models research in HCI and will be reviewed in the final section of this chapter.

However, as argued at length in Payne (2003), there are approaches to mental models in HCI that go beyond a concern with user knowledge and beliefs to ask more nuanced theoretical questions. The first of these is to investigate the form of mental models by inspecting the processes through which mental models might have their effects on behavior.

A powerful idea here is that mental models of machines provide a problem space that allows more elaborate encoding of remembered methods, and in which novice or expert problem solvers can search for new methods to achieve tasks.

The classic example of this approach is the work of Halasz and Moran (1983) on Reverse Polish Notation (RPN) calculators. RPN is a post-fix notation for arithmetic, so that to express 3 + 4, one would write 3 4 +. RPN does away with the need for parentheses to disambiguate composed operations. For example (1 + 2) * 3 can be expressed 1 2 + 3 * with no ambiguity. RPN calculators need a key to act as a separator between operands, which is conventionally labeled ENTER, but they do not need an = key, as the current total can be computed and displayed whenever an operator is entered.

Halasz and Moran taught one group of students how to use an RPN calculator using instructions, like a more elaborate version of the introduction above, which simply described the appropriate syntax for arithmetic expressions. A second group of subjects was instructed, using a diagram, about the stack model that underlies RPN calculation. Briefly, when a number is keyed in, it is "pushed" on top of a stack-data structure (and the top slot is displayed). The ENTER key copies the contents of the top slot down to the next slot. Any binary arithmetic operation is always performed on the contents of the top two slots and leads to the result being in the top slot, with the contents of slots 3 and below moving up the stack.

Halasz and Moran discovered that the stack-model instructions made no difference to participants' ability to solve routine arithmetic tasks: the syntactic "method-based" instructions sufficed to allow participants to transform the tasks into RPN notation. However, for more creative problems (such as calculating (6 + 4) and (6 + 3) and (6 + 2) and only keying the number 6 once) the stack group was substantially better. Verbal protocols showed that these subjects reasoned about such problems by mentally stepping through the transformations to the stack at each keystroke.

This kind of reasoning, stepping through a sequence of states in some mental model of a machine, is often called "mental simulation" in the mental models literature, and the kind of model that allows simulation is often called a "surrogate" (Young, 1983; Carroll & Olson, 1988). From a practical standpoint, the key property of this kind of reasoning is that it results in behavior that is richer and more flexible than the mere rote following of learned methods. The idea that the same method may be encoded more richly, so that it is more flexible and less prone to forgetting will be returned to later in the chapter when a theory of mental models of interactive artifacts is considered, and when ideas about instruction for mental models are reviewed.

A second example of mental models providing a problem space elaboration of rote methods comes in the work of Kieras and Bovair (1984). This research was similar to that of Halasz and Moran (1983) in that it compared the learning performance of two groups: (a) one instructed with rote procedures, (b) the other additionally with a diagrammatic model of the device on which the procedures were enacted. In this case, the device was a simple control panel, in which each rote procedure specified a sequence of button-pushes and knob-positions leading to a sequence of light-illuminations. The model was a circuit diagram showing the connections between power-source switches and display-lights.

Kieras and Bovair (1984) found that the participants instructed with the model learned the procedures faster, retained the procedures more accurately, executed the procedures faster, and could simplify inefficient procedures that contained redundant switch settings. They argued that this was because the model (circuit diagram) explained the contingencies in the rote-action sequences (i.e., if a switch is set to MA, so that the main accumulator circuit is selected, then the FM, fire main, button must be used).

A related theoretical idea is that mental models are a special kind of representation, sometimes called an *analog* representation: one that shares the structure of the world it represents. This was taken as the definitional property of mental models by the modern originator of the term, the British psychologist Kenneth Craik (1943). It is this intuition that encourages the use of terms like "mental simulation"—the intuition that a mental model is like a physical model, approximating the structure of what it represents, just as a model train incorporates (aspects of) the physical structure of a train.

The idea that mental models are analog in this sense is a definitional property in the work on reasoning and comprehension by Johnson-Laird (Johnson-Laird, 1983, 1989; this will be further discussed in part 2, concerning representations of text) and also in the theory of Holland, Holyoak, Nisbett, and Thagard (1986) and Moray (1999). However, there are different nuances to the claim, which must be considered. And, in addition, there is a vexed question to be asked; namely, what is the explanatory or predictive force of a commitment to analog representational form? Is there any reason for HCI researchers to pay attention to theoretical questions at this level?

Certainly, this is the view of Moray (1999) who is concerned with mental models of complex dynamic systems, such as industrial plants. He proposes that models of such systems are structure-sharing *homomorphisms* rather than isomorphisms, i.e. they are many to one rather than one-to-one mappings of objects, properties, and relations. (In this he follows Holland, Holyoak, Nisbett, & Thagard, 1986).

Homomorphic models of dynamic systems may not share structure with the system at the level of static relations, but only at the level of state-changes. Thus, such models have the character of state-transition diagrams, making the empirical consequences of structure sharing somewhat unclear, because any problem space can be represented in this way.

In my view, a clearer view of the explanatory force of analog mental models can be derived by carefully considering the ideas of computational and informational equivalence first introduced by Simon (1978).

It is obviously possible to have two or more distinct representations of the same information. Call such representations "informationally equivalent" if all the information in one is inferable from the other, and vice versa. Two informationally equivalent representations may or may not additionally be "computationally equivalent," meaning that the cost structure of accessing and processing the information is equivalent in both cases, or, as Larkin and Simon (1987) put it: "information given explicitly in the one can also be drawn easily and quickly from the information given explicitly in the other, and vice versa." As Larkin and Simon point out, "easily" and "quickly" are not precise terms, and so this definition of computational equivalence is inherently somewhat vague; nevertheless it points to empirical consequences of a representation (together with the processes that operate upon it) that depend on form, and therefore go beyond mere informational content.

In Payne (2003), I propose adopting *task-relative* versions of the concepts of informational and computational equivalence. Thus, representations are informationally equivalent, *with respect to a set of tasks*, if they allow the same tasks to be performed (i.e. contain the requisite information for those tasks). The representations are, additionally, computationally equivalent with respect to the tasks they allow to be performed, if the *relative difficulty* of the tasks is the same, whichever representation is being used. (Note that according to these definitions, two representations might be computationally equivalent with regard to a subset of the tasks they support but not with regard to the total set, so that in Larkin and Simon's sense they would merely be informationally equivalent. The task-relative versions of the constructs thus allow more finely graded comparisons between representations.)

This idea can express what is behaviorally important about the idea of analog models, or structure-sharing mental representations of a state of affairs of a dynamic system. An analog representation is computationally equivalent (with respect to some tasks) to external perception and manipulation of the state of affairs it represents.

Bibby and Payne (1993; 1996) exploited this distinction between computational and informational equivalence in the domain of HCI, using a computer simulation of a device derived from that studied by Kieras and Bovair (1984). The device was a multiroute circuit, in which setting switches into one of several configurations would make a laser fire; various indicator lights showed which components of the circuit were receiving power. What concerned Bibby and Payne (1993) was the idea of computational equivalence between a mental model and a diagram of the device, rather than the device itself.

Bibby and Payne asked participants to repeatedly perform two types of tasks: a switch task, in which all but one switch was already in position to make a laser fire (the participant had to key the final switch) and a fault task, in which the pattern of indicator lights was such that one of the components must be broken (the participant had to key the name of the broken component).

Participants were instructed about the device with either a table, which showed the conditions under which each indicator light would be illuminated, or with procedures, sequences of switch positions enabling the laser to be fired. Both instructions were sufficient for both switch and fault tasks; they were informationally equivalent with respect to those tasks. However, the table made the fault task easier than the switch task, whereas the procedures made the switch task easier.

During practice, when participants consulted the instructions, this pattern of relative difficulty was confirmed by a crossover interaction in response times. Furthermore, when the instructions were removed from the participants, so that they had to rely on their mental representation of the device, the crossover interaction persevered, demonstrating that the mental representations were computationally equivalent to the external instructions.

In subsequent experiments, Bibby and Payne (1996) demonstrated that this pattern persevered even after considerable interaction with the device that might have been expected to provide an opportunity to overcome the representational constraints of the initial instruction. The crossover interaction eventually disappeared only after extended practice on the particular fault-and-switch task (80 examples of each: perhaps because of asymptotic performance having been reached). At this point, Bibby and Payne introduced two similar but new types of tasks designed so that once again, the table favored one task whereas procedures favored the other. (However, the device instructions were *not* re-presented.) At this point the crossover re-appeared, demonstrating that participants were consulting their instructionally derived mental model of the device, and that this was still in a form computationally equivalent to the original external representation of the instructions.

Practically, this research shows that the exact form of instructions may exert long-lasting effects on the strategies that are used to perform tasks, so that designers of such instructions must be sensitive not only to their informational content but also to their computational properties. In this light, they also suggest that one instructional representation of a device is very unlikely to be an optimal vehicle for supporting all user tasks: it may well be better to provide different representations of the same information, each tailored to particular tasks. In this sense, perhaps instructions should mirror and exploit the natural tendency, noted above, for users to form fragmentary mental models, with different fragments for different purposes.

In terms of theory, Bibby and Payne's findings lend support to the suggestion developed above that mental models of a device that are formed from instructions may be computationally equivalent to the external representations of the device. This idea gives a rather new reading, and one with more ready empirical consequences to the theoretically strong position that mental models are essentially analog, homomorphic representations.

MENTAL MODELS OF TEXT AND OTHER ARTIFACTS

The psychological literature on text comprehension has been transformed by the idea of a situation model, first put forward as part of a general theory of text comprehension by van Dijk and Kintsch (1983), and developed over the years by Kintsch (e.g.,

1998) and followers. The central idea of the general theory is that readers construct mental representations of what they read at several different levels. First, they encode the surface form of the text: the words and syntax. Second, they go beyond this to a representation of the propositional content of the text. Finally, they go beyond the propositional context of the text itself to represent what the text is about, incorporating their world knowledge to construct a situation model or mental model of the described situation.

(Under this view, it is the content that distinguishes a situation model from a text base, rather than a representational format. However, some researchers, notably Johnson-Laird (1983), and followers have pursued the idea of mental models derived from text as analog representations of the described situation. Thus, in text comprehension, there is a version of the issue discussed in part one.)

It is instructive to consider some of the evidence for situation models, and what important issues in text comprehension the theory of situation models allows us to address.

A classic early study was conducted by Bransford, Barclay, and Franks (1972). They asked participants to read simple sentences such as,

Three turtles rested beside/on a floating log, and a fish swam beneath them.

(The slash indicates that some subjects read the sentence with the word "beside", and others read the same sentence with the word "on").

In a later recognition test, interest centered on how likely readers were to falsely accept minor rewordings of the original sentences. In the above case, the foil sentence was

Three turtles rested beside/on a floating log, and a fish swam beneath it.

The key finding was that people who had read the "on" versions of the sentences were much more likely to accept the changed version of the sentence, despite the fact that that at the level of the sentences the difference between original and foil sentences in the two conditions is identical, limited in each case to the last word of the sentence. The reason for false recognition in one case is because, in this case, but not when "on" is replaced by "beside," the original and foil sentences describe the same situation.

A related series of experiments was reported by Fletcher and Chrysler (1990). In a series of carefully controlled experiments, they varied the overlap between sentences in a recognition test and sentences from 10 different texts read by the participants. Each text described a state of affairs (i.e., the relative cost of antiques) consistent with a linear ordering among a set of five objects. They found that participants were influenced by overlap between sentences at study and test corresponding to the three levels of discourse representation proposed by van Dijk and Kintsch (1983): surface form, text base, and situation model. Recognition performance was best when distracter items were inconsistent with all three levels of representation. Recognition was above chance when distracters violated merely the surface form of the original sentences (i.e. substituting rug for carpet). It improved further when propositional information from the text base, but not the linear ordering of the situation, was violated. Recognition was best of all when the distracters were inconsistent with the situation described by the text. This suggests that the some aspects of the structure of the situation (in this case a set of linear orderings) were retained.

Next, consider work by Radvansky and Zacks (Radvansky & Zacks, 1991; Radvansky, Spieler, & Zacks, 1993). In these experiments, participants read sentences such as, "The cola machine is in the hotel," each of which specified the location of an object. In one condition sentences shared a common object (i.e. cola machine) but different locations. In a second condition, different objects share a common location (i.e. the city hall). Later in the experiment participants were given a speeded-recognition test. Radvansky and Zacks found a significant fan effect (Anderson, 1974) for the common object condition; times to verify sentences increased as the number of different locations rose. For the common location sentences no significant fan effect emerged. This was interpreted as evidence that participants formed mental models around the common location (a representation of such a location containing all the specified objects) and retrieval from long-term memory was organized around these mental models. It is impossible, or much harder, to form such a representation of the same object in multiple locations.

What all these studies, and many like them, reveal is that when understanding text, readers spontaneously construct a mental representation that goes beyond the text itself and what it means, and use inferences to construct a richer model of what the text is about—a situation model.

Beyond these refined and clever, but undeniably rather narrow experimental contexts, the construct of situation models has been put to work to illuminate some practical issues concerning text comprehension, and exactly this issue will be returned to later, where we will see how it can inform attempts to understand instructional strategies for engendering useful mental models.

There are two principal ways in which the literature on text comprehension is relevant to HCI. First, it provides support for the idea that a mental model is a representation of what a representational artifact represents. The layered model of text comprehension previously outlined can be generalized to the claim that the user of any representational artifact must construct a representation of the artifact itself, and of what the artifact represents, and of the mapping between the two (how the artifact represents). This is the basis of the Yoked State Space (YSS) hypothesis (Payne, Squibb, & Howes, 1990).

If a reader's goal is just to understand a text, as it was in the experiments just reviewed, then the text-representation can be discarded once a model has been constructed. However, there are many tasks of text *use,* in which it is necessary to maintain a representation of the text, alongside a mental model of the meaning of the text. Consider, for example, the tasks of writing and editing, or of searching for particular content in a text. In such tasks, it is necessary to keep in mind the relation between the surface form of the text—wording, spatial layout, etc.—and its meaning. Text is a representational artifact, and to *use* it in this sense one needs a mental representation of the structure of the text, and of the "situation" described by the text and of the mapping between the two.

According to the Yoked State Space hypothesis (Payne, Squibb, & Howes, 1990), this requirement is general to all representational artifacts, including computer systems. To use such artifacts requires some representation of the domain of application of the artifact—the concepts the artifact allows you to represent and process. The user's goals are states in this domain, which is therefore called the goal space. However, states in the goal space cannot be manipulated directly. Instead, the user interacts with the artifact, and therefore needs knowledge of the artifact, and of the operations that allow states of the artifact to be transformed. Call this problem space the device space. In order to solve problems in the goal space by searching in the device space, the user must know how the device space represents the goal space. In this sense the two spaces need to be yoked. The minimal device space for a certain set of tasks must be capable of representing all the states in the corresponding goal space. More elaborate device spaces may incorporate device states that do not directly represent goal states, but which allow more efficient performance of tasks, just as the stack model of an RPN calculator allows an elaboration of methods for simple arithmetic.

The work of Halasz and Moran (1983) can readily be assimilated into the YSS framework. The no-model condition was provided with enough information to translate algebraic expressions into their Reverse Polish equivalent. However, in this understanding of RP expressions, the ENTER key was given merely an operational account, serving simply as a separator of operands, and did not transform the device state. The stack model, however, provides a figurative account of the ENTER key.

This discussion illustrates a practical lesson for the design of interfaces and instructions. In the case of the copy buffer and the calculator stack, the standard interface does not allow the appropriate device space readily to be induced, so that conceptual instructions must fill the gap. The obvious alternative, which has been developed to some extent in both cases, is to redesign the user interface so as to make the appropriate device space visible. These examples suggest a simple heuristic for the provision of conceptual instructions that may help overcome the considerable controversy over whether or not such instructions (as opposed to simple procedural instructions) are useful (see, i.e., Wright, 1988). According to this heuristic, conceptual instructions will be useful if they support construction of a YSS that the user would otherwise have difficulty inducing (Payne, Howes, & Hill, 1992).

A more direct way in which text comprehension research is relevant to HCI is that so much HCI is reading text. Beyond the standard issues, the widespread availability of electronic texts raises some new concerns that have not yet seen much work, yet are perhaps the most directly relevant to HCI design. Two issues stand out: (a) the usability of documents that incorporate multiple media alongside text, and (b) the exploitation by readers of multiple texts on the same topic.

How are multimedia "texts" that incorporate graphics comprehended? There is only a small literature on this within the mainstream field of text comprehension, but this literature exploits the idea of a mental model.

Glenberg and Langston (1992) argued that the widespread idea that diagrams can assist the comprehension of technical text had, at the time, been little tested or understood and that mental models were an important explanatory construct. In their analysis, diagrams are useful in concert with texts precisely because they assist the construction of mental models. This idea has been pursued in a very active program of work on multimedia instruction by Mayer and colleagues, which will be reviewed in the next section.

What about when the multiple sources of information are not presented as part of a single text, but rather independently, covering overlapping ground, so that the reader has to perform all the integration and mapping? This is the issue of multiple texts, and it has become commonplace in the age of the Internet. It is now rarely the case that a student struggles to find relevant source documents on a topic. Instead, students are typically faced with an overabundance of relevant materials and must somehow allocate their time across them, and integrate the knowledge they derive from different sources.

Perfetti (1997) has suggested that learning from multiple texts is one of the most important new challenges for text researchers. Research has shown, for example, that integrating information across multiple texts is a skill that does not come readily but can be acquired and taught (Stahl, Hind, Britton, McNish, & Bosquet, 1996; Rouet, Favart, Britt, & Perfetti, 1997).

The YSS theory raises important issues here. As previously noted, everyday reading of text can be seen as engendering a progression of mental representations moving from the surface form through the propositional content to a situation model. When reading, earlier representations can be discarded as later ones are formed, but for other tasks of text use, the reader needs to maintain a representation of the form of the multitext, and map this form onto the content. Payne and Reader (in press) refer to such a representation as a structure map.

The usefulness of a structure map becomes even more apparent when multiple texts are considered. In this case, structure maps could play a role in encoding *source* information, which might be important not only for locating information, but also for integrating diverse and potentially contradictory information and for making judgments of trust or confidence in the information. Source information might additionally encode temporal properties of information sources, and thus be useful for memory updating—revising knowledge in the light of new information, making distinctions between current and superseded propositions.

The widespread availability of the Web not only means that multiple texts are more widely encountered, but also encourages a situation where multiple texts are read in an interleaved fashion, in a single sitting, or at least temporally close, raising the importance of the above challenges, and meaning that recency in autobiographical memory is unlikely to accomplish source identification, so further stressing the importance of a structure map.

Payne and Reader (in press) studied readers' ability to search for specific ideas in multiple texts that they had just read. They found evidence that readers spontaneously constructed structure maps, as just described, in that they showed some memory of which documents contained which ideas, even when they did not expect to need such knowledge when reading the texts.

INSTRUCTIONS FOR MENTAL MODELS

Multimedia Instruction

If mental models are important for operating devices, how should they best be taught? We have seen that models are constructed automatically by readers of text, but can modern computational media, such as animations, be used to improve the acquisition of mental models from instructional texts, just as Glenberg and Langston (1992) suggested in the case of simple diagrams? Just such a question has been addressed in a long-standing programme of work by Richard Mayer and colleagues, which will be reviewed in this section.

Mayer and Moreno (2002) present a cognitive theory of multimedia learning, which builds on three main ideas:

1. From dual coding theory the authors suppose that humans have separate visual and verbal information processing systems (Clark & Paivio, 1991; Paivio, 1986);
2. From cognitive load theory the authors assume that the processing capacity of both the visual and the verbal memory system is strictly limited (Baddeley, 1992; Chandler & Sweller, 1991) and that cognitive load during instruction can interfere with learning;
3. From constructivist learning theory the authors take the idea that meaningful learning requires learners actively to select relevant information, to structure it into coherent representations, and make connections with other relevant knowledge (Mayer, 1996; Mayer, 1999a).

This latter process, of building coherent representations that connect information from different modalities with pre-existing knowledge, bears clear relation to Johnson-Laird's construct of mental models, and indeed Mayer and colleagues use the term in this context. In the case of the physical systems that many of their studies have addressed, mental models may take the form of cause-effect chains. According to Mayer and Moreno (2002) a key design principle for instructional materials is that they should maximise the opportunity for these model-construction processes to be completed.

Mayer and colleagues have conducted a large number of experiments comparing learning from multimedia source materials with learning from components of these materials (words, pictures, etc) successively or in other kinds of combination. Based on this research, Mayer (1999b) and Mayer and Moreno (2002) have identified some principles of instructional design that foster multimedia learning.

The *multiple presentation principle* states that explanations in words and pictures will be more effective than explanations that use only words (Mayer & Moreno, 2002, p. 107). When words only are presented, learners may find it difficult to construct an appropriate mental image, and this difficulty may block effective learning. Mayer and Anderson (1991; Experiment 2b) compared four treatment groups: words with pictures, words only, pictures only, and control, on tests of creative problem solving involving reasoning how a bicycle pump works. Results demonstrated that participants in the words with pictures group generated a greater number of creative problem solutions than did participants in the other groups. Interestingly, animation without narration was equivalent to no instruction at all. Other studies have offered support for the general idea that learners will acquire richer knowledge from narration and animation than from narration alone (Mayer & Anderson, 1991, Experiment 2a; Mayer & Anderson, 1992, Experiments 1 and 2).

The *contiguity principle* is the claim that simultaneous as opposed to successive presentation of visual and verbal materials is preferred (Mayer & Moreno, 2002), because this will enable learners to build referential connections more readily (Mayer & Sims, 1994). Mayer and Anderson (1991, Experiments 1 and 2) studied a computer-based animation of how a bicycle pump works. They compared a version that presented words with pictures against the same content presenting words before pictures, and tested acquisition with tests of creative problem solving. Those in the words-with-pictures group generated about 50% more solutions to the test problems than did subjects in the words-before-pictures group.

The *individual differences principle* predicts that factors such as prior knowledge or spatial ability will influence transfer of learning from multimedia materials, moderating the effects of other principles (Mayer, 1999). With regard to domain specific knowledge, Mayer proposed that experienced learners may suffer little decrease in problem solving transfer when receiving narration and animation successively because their background knowledge will allow a mental model to be constructed from the words alone, then linked to the visual information. Low-experience learners, on the other hand, will have no means to over-ride the effects underlying the contiguity principle, and their problem solving transfer will suffer (Mayer & Sims, 1994). In support of this suggestion, experimental work by Mayer and Gallini (1990) demonstrated across three studies that the synchronization of words and pictures served to improve transfer for low- but not high-experience learners.

The *chunking principle* refers to a situation in which visual and verbal information must be presented successively, or alternately (against the contiguity principle). It states that learners will demonstrate better learning when such alternation takes place in short rather than long segments. The reasoning is straightforward, given the assumptions of the framework: working memory may become overloaded by having to hold large chunks before connections can be formed (Mayer, 1999b). An experiment by Mayer and Moreno (1998) investigated this *chunking principle* using explanations of how lightning storms develop. The ability to solve novel, transfer problems about lightning exhibited by a 'large chunk' group (who received all the visual information before or after all the verbal information) was compared with that of a 'small chunk' group (alternating presentations of a short portion of visual followed by a short portion of narration). The gain in performance of the small chunk group over the large chunk group was circa 100% (Mayer & Moreno, 1998).

The debt of Mayer's work to Sweller's programme of research on Cognitive Load Theory is obvious. Mayer's design principles reflect the premise that students will learn more deeply when their visual and/or verbal memories are not overloaded. Students are better able to make sense of information when they receive both verbal and visual representations rather than only verbal; when they can hold relevant visual and verbal representations in working memory at the same time; when they have domain specific knowledge and/or high spatial ability; and when they receive small bits of information at a time from each mode of presentation.

Despite incredibly positive research results, at this stage Mayer's work should be viewed with a little caution. Almost all of the experiments utilise very short instructional presentations, with some of the animations lasting only 30 seconds. Subjects are then required to answer problem-solving questions that seem ambiguous, requiring students to be fairly creative in order to generate solutions. Mayer's work also typically neglects to include any tests of long-term retention. It may conceivably be falling into the instructional trap of maximising performance during learning at the expense of longer-term performance. This issue is the focus of the next section.

The Theory of Learning by Not Doing

Mayer's theory of multimedia instruction adheres to the common assumption that the optimal design of instructional material involves minimizing the cognitive burden on the learner due to the limits of the working memory.

Yet minimizing the mental effort of learners is not necessarily or always a good instructional strategy. According to Schmidt and Bjork (1992), instructional conditions that achieve the training goals of generalizability and long-term retention are not necessarily those that maximize performance during the acquisition phase.

They argue that the goal of instruction and training in real-world settings should first be to support a level of performance in the long term, and second to support the capability to transfer that training to novel-tasks environments. Methodologically, in order to measure a genuine *learning effect,* some form of long-term assessment of retention must take place; skill acquisition is not a reliable indicator of learning.

Schmidt and Bjork (1992) discussed three situations in which introducing difficulties for the learner can enhance long-term learning. First, studies that vary the scheduling of tasks during practice were reported. Random practice is more difficult than blocked schedules of practice, as a given task is never practiced on the successive trial. Using a complex motor task involving picking up a tennis ball and using it to knock over a particular set of barriers, Shea and Morgan (1979) reported a clear advantage for subjects who practiced under blocked conditions (subsets of barriers to knock), in terms of performance during practice. However, the amount of learning as demonstrated by the retention phase favored the random condition. Similar results have been reported by Baddeley and Longman (1978), Lee and Magill (1983), and (with verbal tasks) Landauer and Bjork, (1978).

Schmidt and Bjork offer an explanation for this paradigm, in which retrieval practice may play a key role. They suggest that there may be a benefit, in terms of long-term retention, for activities that actually cause forgetting of the information to be recalled, forcing the learner to practice retrieving this information (Bjork & Allen, 1970).

Experiments that vary the feedback the learner receives have demonstrated a similar phenomenon. A study by Schmidt, Young, Swinnen, and Shapiro (1989) demonstrated that delaying the feedback that subjects received during motor tasks interfered with performance. However, on a delayed-retention test, those who had received the feedback least often demonstrated the most effective performance. This seems to contradict the established opinion that feedback is vital for effective learning. Schmidt and Bjork (1992) suggested that frequent feedback may actually serve to block information-processing activities that are important during the skill-acquisition phase.

A final area reviewed by Schmidt and Bjork concerns the introduction of variability during practice, such as when practicing tossing a beanbag at a target at a particular distance. Practicing at variable distances is more effective than practicing at a fixed distance (Kerr & Booth, 1978).

Does the Schmidt and Bjork approach extend to HCI tasks, and in particular to instruction for mental models?

One impressive example of an instructional effect in the Schimdt and Bjork (1992) paradigm is informed by the idea of mental models or situation models derived from text, as discussed in part 2 of this chapter. Informed by the distinction between a text base and a situation model, work by McNamara, Kintsch, Songer, and Kintsch (1996) has shown how expository text can be designed to introduce difficulties for readers in exactly the productive manner advocated by the Schmidt and Bjork conception of training. These authors created two versions of target texts, one more coherent than the other (one experiment used a text about traits of mammals, a second used a text about heart disease). Coherence cues were provided by linking clauses with appropriate connectives and by inserting topic headings. The level of readers' background knowledge on the topic of the text was also assessed with a pretest. After reading a text, participants were given tests of the text base (free recall of the text propositions and specific factual questions about the contents of the text) and tests of the situation model (problem-solving-based questions, questions requiring inferences from the text, and a concept-sorting task).

McNamara et al. (1996) reported that for measures that tested the text base, the high coherence texts produced better performance. However, for situation-model measures, test performance for high-knowledge readers was better when they read the low-coherence text. McNamara et al. argued that limiting the coherence of a text forced readers to engage in compensatory processing to infer unstated relations in the text. This compensatory processing supported a deeper understanding of the text, in that the information in the text became more integrated with background knowledge. Thus, for high-knowledge readers, the texts that were more difficult to read improved the situation model by encouraging more transfer-appropriate processing. Low-knowledge readers were, presumably, unable to achieve the compensatory inferences, and therefore did better with more coherent texts. Because the text base does not incorporate background knowledge, it was not enhanced by any compensatory processing. (This finding is related to the work of Mayer and Sims (1994) reviewed above.)

One very successful practical approach to the design of instructions for interactive devices which is well known in the HCI community, is perhaps quite strongly related to this more theo-retically oriented work. The concept of a "minimal manual" was outlined by Carroll (1990). It sought to minimize the extent to which instructional materials obstruct learning. Crucially, a well-designed Minimal Manual does not necessarily optimize the speed at which users can perform procedures as they read. Carroll's manuals avoided explicit descriptions that encouraged rapid but mindless rote performance. Instead, the emphasis was on active learning whereby learners were encouraged to generate their own solutions to meaningful tasks. This process was facilitated in part by reducing the amount of text provided and including information about error recovery.

O'Hara and Payne (1998, 1999) argued that learning from a problem-solving experience might be enhanced to the extent that problem solvers planned their moves through the problem space. Many puzzles with an interactive user interface, and indeed many user interfaces to commercial systems, encourage a one-step-at-a-time approach to problem solving, in which a move is chosen from the currently available set. This may be quick and relatively effortless, yet lead to little learning and inefficient solutions. For example, in an HCI task, participants had to copy names and addresses from a database to a set of letters. Each item category from the database had to be copied to several letters, so that the most obvious and perhaps least effortful strategy of preparing letters one at a time was inefficient in terms of database access. O'Hara and Payne's manipulation was to increase the cost of making each move (in the copying experiment by adding a system lock-out time). This resulted in more planning, more think-time per move, meaning slower solutions in the first instance, but more efficient behavior in the long term, and the discovery of strategies that required fewer database accesses and fewer user inputs.

Recent work by Duggan and Payne (2001) combined several of the insights in the work just reviewed to explore acquisition of interactive procedures during instruction following. Good procedural instructions for interactive devices must satisfy two criteria. First, they must support performance. Like all procedural instructions they should effectively communicate the procedure they describe, so as to allow users who don't know the procedure to enact it successfully and efficiently. Second, they must support learning. In common with instructions for all procedures that will be used repeatedly, they should facilitate subsequent memory for the procedure, so that it might later be performed without consulting the instructions.

How might procedural instructions be designed so as to follow the Schmidt and Bjork paradigm and provide transfer-appropriate practice opportunities for the learner? Of course, not all manipulations that introduce difficulties during learning are beneficial for the learner. Simply making the instructions unclear is unlikely to be effective. However, much this idea may have informed the design of some commercial user manuals. The criterion that quality instructions must communicate the procedure that they describe cannot be ignored.

The work of Diehl and Mills (1995) further illustrated the relevance of the theory of text comprehension to the design of instruction for interactive procedures. They argued that in the case of procedural instructions the distinction between situation model and text base maps directly onto a distinction between memory for the procedure (as tested by later task performance) and memory for the instructions themselves.

Texts describing how to complete a task using a device (setting an alarm clock or constructing a child's toy) were provided. While reading a text, participants were required to either perform the task (read and do), or do nothing (read only). (In addition, Diehl and Mills studied some intermediate conditions, such as read and watch experimenter. These conditions produced intermediate results and are not relevant to the current argument.) The effect of these training methods was then examined by asking participants to recall the text and then complete the task.

Diehl and Mills reported that the increased exposure to the device in the read-and-do condition resulted in improved task performance times relative to the read-only condition. However, text recall was better in the read-only condition, supporting the conceptual separation of text base and situation model.

Inspired by this work, Duggan and Payne (2001) introduced a particular technique to exploit the principle of Schmidt and Bjork (1992) and the methods of McNamara and colleagues (1996). Like the manipulations of Diehl and Mills (1995), their innovation centered not on the design of the instructions per se, but rather on the way the instructions are read and used. Diehl and Mills' reported advantage for reading and doing over reading alone has no real practical implication, as it is difficult to imagine anyone advocating isolated reading as a preferred method. However, Duggan and Payne suggested that the way learners manage the interleaving of reading and doing will affect their later retention, and thus offers an important lever for improving instruction.

Many procedural instructions have a natural step-wise structure, and in these cases it is possible to execute the procedure while reading with minimal load on memory. Learners can read a single step, and then execute it before reading the next step. Such an approach is low in effort (and therefore attractive to the learner), but also low in transfer-appropriate practice and therefore, one would argue on the basis of the reviewed work, poor at encouraging retention. If learners could instead be prompted to read several procedural steps before enacting them, performance would be made more effortful, but learning might benefit. Readers would be encouraged to integrate the information across the chunk of procedural steps, and the increased memory load would provide transfer-appropriate practice.

Duggan and Payne (2001) developed this idea as follows. First, by implementing an online help system in the context of experimental tasks (programming a VCR) they forced participants into either a step-wise or a chunk-based strategy for interleaving reading and acting. These experiments demonstrated that reading by chunks did tax performance during training, but improved learning, in particular retention of the procedure. Next, they developed a more subtle, indirect manipulation of chunking. By adding a simple cost to the access of online instructions (c.f., O'Hara & Payne, 1998), they encouraged readers to chunk steps so as to minimize the number of times the instructions were accessed. Just as with enforced chunking, this led to improved retention of the procedures.

SHARED MENTAL MODELS

In the last 10 years or so there has been a rapid surge of interest in the concept of shared mental models in the domain of teamwork and collaboration. The use of mental models in this literature, to date, is somewhat inexact, with little theoretical force, except to denote a concern with what the team members know, believe, and want. As the name suggests, shared mental models refers to the overlap in individuals' knowledge and beliefs.

The central thesis and motive force of the literature is that team performance will improve when team members share relevant knowledge and beliefs about their situation, task, equipment, and team. Different investigations and different authors have stressed different aspects of knowledge, and indeed proposed different partitions into knowledge domains. (And recently, as we shall see, some investigators have questioned the extent to which overlapping knowledge is a good thing. There are some situations in which explicit distribution or division of knowledge may serve the team goals better.)

At first glance, the idea that teams need to agree about or share important knowledge seems intuitively plain. Models of communication (i.e., Clark, 1992) stress the construction of a common ground of assumptions about each partner's background and intentions. The idea of shared mental models develops this idea in a plausible practical direction.

A recent study by Mathieu, Heffner, Goodwin, Salas, and Cannon-Bowers (2000) was one of the most compelling demonstrations of the basic phenomenon under investigation, as well as being centered on an HCI paradigm. For these reasons, this study will be described and used as a framework to introduce the space of theoretical and empirical choices that characterize the mainstream of the shared mental models literature.

Mathieu, Heffner, Goodwin, Salas, and Cannon-Bowers (2000) considered team members' mental models as comprising knowledge of four separate domains: (a) technology (essentially the mental models described in part one of this chapter); (b) job or task; (c) team interaction (such as roles, communication channels and information flow) and (d) other teammates' knowledge and attitudes. Knowledge of the last three types would rarely be called a mental model outside this literature, and so straight away we can see a broader and more practical orientation than in individually oriented mental models literatures.

For the purposes of operationalization, the authors suggested that these four categories of knowledge may be treated as two: task related and team related. This binary distinction mirrors a distinction that has been made in terms of team behaviors and communications, which have been considered in terms of a task track and a teamwork track (McIntyre & Salas, 1995).

Mathieu and colleagues studied dyads using a PC-based flight simulator. One member of each dyad was assigned to the joystick to control aircraft position. The other was assigned to keyboard, speed, weapon systems, and information gathering. Both members could fire weapons. The experimental procedure incorporated a training phase, including the task and basics of teamwork, and then the flying of six missions, divided into three equally difficult blocks of two, each mission lasting around 10 minutes. Performance on a mission was scored in terms of survival, route following, and shooting enemy planes. Team processes were scored by two independent raters viewing videotapes to assign scores, for, example, how well the dyad communicated with each other.

Mental models were measured after each pair of missions. At each measurement point, each individual's task or team mental model was elicited by the completion of a relatedness matrix (one for task, one for team), in which the team member rated the degree to which each pair from a set of dimensions was related. For the task model there were eight dimensions, including diving versus climbing; banking or turning; and choosing airspeed. For the team model there were seven dimensions, including amount of information and roles and team spirit.

Thus, at each measurement point, participants had to assign numbers between -4 (negatively related, a high degree of one requires a low degree of the other) and $+4$ (positively related, a high degree of one requires a high degree of the other) to each pair of dimensions in each domain. For example, they had to rate the relatedness of diving versus climbing to choosing airspeed, and the relatedness of roles to team spirit. For each team at each time for each model-type a convergence index was calculated by computing a correlation co-efficient (QAP correlation) between the two matrices. The co-efficient could vary from -1 (complete disagreement) to $+1$ (completely shared mental models).

The main findings of this investigation were as follows. Contrary to hypothesis, convergence of mental models did not increase over time; rather it was stable across missions 1 to 3. This runs counter to a major and plausible assumption of the shared mental models program, which is that agreement between team members should increase with extent of communication and collaboration.

Nevertheless, convergence of both task and team models predicted the quality of team process and the quality of performance. Further, the relationship between convergence and performance was fully mediated by quality of team process.

The most natural interpretation of these findings is that team process is supported by shared mental models. In turn, good team processes lead to good performance. According to its authors, this study provided the first clear empirical support for the oft-supposed positive relationship between shared mental models and team effectiveness (Mathieu et al., 2000, p. 280).

As well as being paradigmatic in illustrating the key ideas in the shared mental models literature, this study has several aspects that highlight the range of approaches and the controversy in the field.

First, it is worth considering what particular properties of the task and teams may have contributed to the positive relation between shared mental models and team process and performance. Compared with most situations in which coordination and collaboration are of prime interest, including most situations addressed by CSCW researchers, the teams studied by Matheiu et al. were minimal (two members) and the tasks were very short term and relatively circumscribed. Beyond these obvious remarks, I would add that the division of labor in the task was very "close," and the workers' performance was extremely interdependent. Of course, interdependence is the signature of collaborative tasks; nevertheless, a situation in which one person controls airspeed and another controls altitude may make this interdependence more immediate than is the norm.

It is also possible that the relatively circumscribed nature of the task and collaboration contributed to the failure of this study to find evidence for the sharing of mental models increasing across the duration of collaboration.

As just mentioned, although the literature contains many proposals that shared mental models will positively influence process and performance, there has been much less empirical evidence. Another study of particular relevance to HCI is concerned with the workings of software development teams.

Software development is an ideal scenario for the study of team coordination for several reasons. First, much modern software development is quintessentially team based (Crowston & Kammerer, 1998; Curtis, Krasner, & Iscoe, 1998; Kraut & Streeter, 1995), and relies heavily on the complex coordinations of team members. Secondly, this effort is often geographically dispersed, further stressing collaboration and putting an emphasis on communications technologies. Finally, software development takes place in technologically advanced settings with technologically savvy participants, so that it provides something of a test bed for collaboration and communication technologies.

One study of complex geographically distributed software teams has been reported that partially supports the findings of the Mathieu et al. (2000) study and provided complementary evidence for positive effects of shared mental models on team performance. Espinosa, Kraut, Slaughter, Lerch, Herbsleb, and Mockus (2002) reported a multimethod investigation of software teams in two divisions of a multinational telecommunications company. The most relevant aspect of their study was a survey of 97 engineers engaged in team projects of various sizes ranging from 2 to 7. Team coordination and shared mental models (SMM) were both measured by simple survey items, followed by posthoc correlational analysis to uncover the relation between shared mental models and team process. As in the Mathieu et al. (2000) study, shared mental models were considered in two categories: task and team. A positive relation between team SMM and coordination was discovered, but the effect of task SMM was not significant.

It is worth being clear about the positive relation and how it was computed. Team SMM was computed for each team by assessing the correlations between each team member's responses to the team SMM survey items. This index was entered as an independent variable in a multiple regression to predict average reported levels of team coordination. It is, of course, hard to infer any causal relation from such correlational analyses, and one might also wonder about the validity of purely questionnaire-based measures of some of the constructs, yet nevertheless the study is highly suggestive that SMM can have a positive influence in group-work situations far removed from pairs of students interactive with a flight simulator. Additionally, Espinosa, Kraut, Slaughter, Lerch, Herbsleb, and Mockus (2002) reported an interview study in which respondents confirmed their own belief that SMM contributed positively to project communications and outcomes.

Nevertheless, Espinosa et al. (2002) failed to find any relation between task SMM and team process. It seems to me that, in view of the survey methodology, this would have been the more compelling evidence in favor of the SMM construct. It seems less surprising and perhaps less interesting that there should be a correlation between participants' survey responses concerning how well they communicated on their team, and, for example, their agreement about which teammates had high knowledge about the project.

Levesque, Wilson, and Wholey (2001) reported a different study of software development teams, using ad hoc student-project groupings to study whether sharing of Team SMM increased over time. They only measured Team SMM, using Likert scale items on which participants signaled amount of agreement or disagreement with statements like, "Most of our team's communication is about technical issues," "Voicing disagreement on this team is risky," or "Lines of authority on this team are clear.". Team SMM was measured by computing correlations among team members of these responses after 1, 2, and 3 months of working on a joint project.

Levesque, Wilson, and Wholey (2001) found that, contrary to their hypothesis, team SMM decreased over time. They argue that this is because projects were managed by a division of labor thath required much initial collaboration but meant that later activity was more individual.

There are surely many teamwork situations in which role differentiation is critical for success, and this observation suggested that the most straightforward interpretation of shared mental models is overly simple. Indeed, even in teams that continue to meet, communicate, and collaborate, it may be that role differentiation means that task mental models should not so much be "shared" as "distributed" to allow for effective team performance. (Studies of intimate couples have explored a similar process of specialization of memory functions, under the name "transactive memory," i.e., Wegner, 1987, 1995).

When roles are differentiated, it is no longer important that task knowledge is shared, but rather that individuals' knowledge about who knows what is accurate. Thus, one would expect team SMMs to support communication and collaboration even in teams with highly differentiated roles. This may explain the findings reviewed above. In the Mathieu et al. study, the team members' technical roles remained tightly interdependent, so that both task and team models had to be shared for successful performance. In the Espinosa et al. (2002) study, the technical roles may have been differentiated but the level of communication remained high, so that team SMM affected performance but task SMM did not. In the Levesque et al. study, the teams divided their labor to the extent that communication and collaboration ceased to be necessary (apart, perhaps for some final pooling of results). In this case, we would predict that neither task nor team SMMs would affect performance once the division of labor had been accomplished. No data on performance were reported, but team models became less shared over the course of the projects.

Although there has been quite a sudden flurry of interest in shared mental models, this brief review makes clear that much empirical and conceptual work remains to be done. Of particular relevance to this chapter is the question of what exactly is meant by a mental model in this context.

To date throughout the field, mental models have been considered as semantic knowledge, using traditional associative networks as a representation. Thus, mental models have typically been tapped using simple likert scales or direct questions about the relations (i.e. similarity) between constructs, analyzed with multidimensional techniques such as pathfinder (for a review of measurement techniques in this field, see Mohammed, Kilmoski, & Rentsch, 2000). Because interest has focused on the extent to which knowledge and beliefs are common among team members, these approaches have been useful, allowing quantitative measures of similarity and difference. Nevertheless, compared with the literature on individual mental models, they tend to reduce participants' understanding to mere associations, and yet the thrust of the individual work shows that this may not be appropriate, because the particular conceptualizations of the domain, the analogies drawn, the computational as well as informational relations between internal and external representations, etc., can have real effects on performance. It seems that an important path of development may be to adopt this more refined cognitive orientation and investigate the impact of shared models—as opposed to shared networks of facts and associations—on collaboration.

References

Baddeley, A. (1992). Working memory. *Science, 255,* 556–559.

Baddeley, A. D., & Longman, D. J. A. (1978). The influence of length and frequency of training session on the rate of learning to type. *Ergonomics, 21,* 627–635.

Bibby, P. A., & Payne, S. J. (1993). Internalization and use–specificity of device knowledge. *Human–Computer Interaction, 8,* 25–56.

Bibby, P. A., & Payne, S. J. (1996). Instruction and practice in learning about a device. *Cognitive Science, 20,* 539–578.

Bjork, R. A., & Allen, T. W. (1970). The spacing effect: Consolidation or differential encoding? *Journal of Verbal Learning and Verbal Behavior, 9,* 567–572.

Bransford, J. D., Barclay, J. R., & Franks, J. J. (1972). Sentence memory: a constructive versus interpretive approach. *Cognitive Psychology, 3,* 193–209.

Carroll, J. M. (1990). *The Nurnberg funnel: Designing minimalist instruction for practical computer skill,* Cambridge, MA: MIT Press.

Carroll, J. M., & Olson, J. R. (1988). Mental models in human–computer interaction. In M. Helander (Ed.), *Handbook of human–computer interaction* (pp. 45–65). New York: Elsevier.

Chandler, P., & Sweller, J. (1991). Cognitive load theory and the format of instruction. *Cognition and Instruction, 8,* 293–332.

Clark, J. M., & Paivio, A. (1991). Dual coding theory and education. *Educational Psychology Review, 3*(3), 149–170.

Clark, H. H. (1992). *Arenas of language use.* Chicago: Chicago University Press.

Cockburn, A., & Jones, S. (1996).Which way now? Analysing and easing inadequacies in WWW navigation. *International Journal of Human-Computer Studies, 45,* 195–130.

Collins, A., & Gentner, D. (1987). How people construct mental models. In D. Holland, & N. Quinn (Eds.), *Cultural models in language and thought.* Cambridge, UK: Cambridge University Press.

Craik, K. J. W. (1943). *The nature of explanation.* Cambridge: Cambridge University Press.

Crowston, K., & Kammerer, E. E. (1998). Coordination and collective mind in software requirements development. *IBM Systems Journal, 37*(2), 227–245.

Curtis, B., Krasner, H., & Iscoe, N. (1988). A field study of the software design process for large systems. *Communications of the ACM 31*(11), 1268–1286.

Diehl, V. A., & Mills, C. B. (1995). The effects of interaction with the device described by procedural text on recall, true/false, and task performance. *Memory and Cognition, 23*(6), 675–688.

Duggan, G. B., & Payne, S. J. (2001). Interleaving reading and acting while following procedural instructions. *Journal of Experimental Psychology: Applied, 7*(4), 297–307.

Ericsson, K. A., Krampe, R. T., & Tesch-Romer, C. (1993). The role of deliberate practice in the acquisition of expert performance. *Psychological Review, 100*(3), 363–406.

Espinosa, J. A., Kraut, R. E., Slaughter, S. A., Lerch, J. F., Herbsleb, J. D., & Mockus, A. (2002). Shared mental models, familiarity and coordination: A multi-method study of distributed software teams. *Proceedings of the 23rd International Conference in Information Systems (ICIS)*, Barcelona, Spain, 425–433.

Fletcher, C. R., & Chrysler, S. T. (1990). Surface forms, textbases and situation models: recognition memory for three types of textual information. *Discourse Processes, 13*, 175–190.

Gentner, D., & Stevens, A. L. (1983). *Mental models*. Hillsdale, NJ: Erlbaum.

Glenberg, A. M., & Langston, W. E. (1992). Comprehension of illustrated text: pictures help to build mental models. *Journal of Memory and Language, 31*, 129–151.

Halasz, F. G., & Moran, T. P. (1983). Mental models and problem-solving in using a calculator. *Proceedings of CHI 83 Human Factors in Computing Systems*, New York: ACM.

Holland, J. H., Holyoak, K. J., Nisbett, R. E., & Thagard, P. R. (1986). *Induction*. Cambridge, MA: MIT Press.

Johnson-Laird, P. N. (1983). *Mental models*. Cambridge, UK: Cambridge University Press.

Johnson-Laird, P. N. (1989). Mental models. In M. I. Posner (Ed.), *Foundations of cognitive science*. Cambridge, MA: MIT Press.

Kerr, R., & Booth, B. (1978). Specific and varied practice of a motor skill. *Perceptual and Motor Skills, 46*, 395–401.

Kieras, D. E., & Bovair, S. (1984). The role of a mental model in learning to use a device. *Cognitive Science, 8*, 255–273.

Kintsch, W. (1998). *Comprehension*. Cambridge, UK: Cambridge University Press.

Kraut, R. E., & Streeter, L. A. (1995). Coordination in software development, *Communications of the ACM, 38*(3), 69–81.

Landauer, T. K., & Bjork, R. A. (1978). Optimum rehearsal patterns and name learning. In M. M. Gnineberg, P. E. Morris, & R. N. Sykes (Eds.), *Practical aspects of memory* (pp. 625–6321). London: Academic Press.

Larkin, J. H., & Simon, H. A. (1987). Why a diagram is (sometimes) worth ten thousand words. *Cognitive Science, 11*, 65–100.

Lee, T. D., & Magill, R. A. (1983). The locus of contextual interference in motorskill acquisition. *Journal of Experimental Psychology: Learning, Memory, and Cognition, 9*, 730–746.

Levesque, L. L., Wilson, J. M., & Wholey, D. R. (2001). Cognitive divergence and shared mental models in software development project teams, *Journal of Organizational Behavior, 22*, 135–144.

Mathieu, J. E., Heffner, T. S., Goodwin, G. F., Salas, E., & Cannon-Bowers, J. A. (2000). The influence of shared mental models on team process and performance. *Journal of Applied Psychology, 85*, 273–283.

Mayer, R. E. (1996). Learning strategies for making sense out of expository text: The SOI model for guiding three cognitive processes in knowledge construction. *Educational Psychology Review, 8*, 357–371.

Mayer, R. E. (1999a). Research-based principles for the design of instructional messages: The case of multimedia explanations. *Document Design, 1*, 7–20.

Mayer, R. E. (1999b). Multimedia aids to problem solving transfer. *International Journal of Educational Research, 31*, 611–623.

Mayer, R. E. (1999). Multimedia aids to problem-solving transfer. *International Journal of Educational Research, 31*, 661–624.

Mayer, R. E., & Anderson, R. B. (1992). The instructive animation: Helping students build connections between words and pictures in multimedia learning. *Journal of Educational Psychology, 84*, 444–452.

Mayer, R. E., & Anderson, R. B. (1991). Animations need narrations: An experimental test of a dual-coding hypothesis. *Journal of Educational Psychology, 83*, 484–490.

Mayer, R. E., & Gallini, J. K. (1990). When is an illustration worth ten thousand words? *Journal of Educational Psychology, 82*, 715–726.

Mayer, R. E., & Moreno, R. (1998). A split-attention affect in multimedia learning: Evidence for dual processing systems in working memory. *Journal of Educational Psychology, 90*, 312–320.

Mayer, R. E., & Moreno, R. (2002). Aids to computer-based multimedia learning. *Learning and Instruction, 12*, 107–119.

Mayer, R. E., & Sims, V. K. (1994). For whom is a picture worth a thousand words? Extensions of a dual-coding theory of multimedia learning. *Journal of Educational Psychology, 86*, 389–401.

Mayhew, D. J. (1992). *Principles and guidelines in software user interface design*. Englewood Cliffs, NJ: Prentice Hall.

McCloskey, M. (1983). Naïve theories of motion. In D. Gentner, & A. L. Stevens (Eds.), *Mental models* (pp. 299–323). Hillsdale, NJ: Erlbaum.

McIntyre, R. M., & Salas, E. (1995). Measuring and managing for team performance: Emerging principles from complex environments. In R. A. Guzzo & E. Salas (Eds.), *Team effectiveness and decision making in organizations* (pp. 9–45). San Francisco: Jossey-Bass.

McNamara, D. S., Kintsch, E., Songer, N. B., & Kintsch, W. (1996). Are good texts always better? Text coherence, background knowledge, and levels of understanding in learning from text. *Cognition and Instruction, 14*, 1–43.

Moheammet, S., & Dunville, B. C. (2001). Team mental models in a team knowledge framework: Expanding theory and measurement across disciplinary boundaries. *Journal of Organizational Behavior, 22*(2), 89.

Mohammed, S., Klimoski, R., & Rentsch, J. R. (2000). The measurement of team mental models: We have no shared schema. *Organizational Research Methods, 3*(2), 123–165.

Moray, N. (1999). Mental models in theory and practice. In D. Gopher, & A. Koriat (Eds.), *Attention and performance XVII* (pp. 223–258). Cambridge, MA: MIT Press.

Norman, D. A. (1983). Some observations on mental models. In D. Gentner & A. L. Stevens (Eds.), *Mental models* (pp. 7–14). Hillsdale, NJ: Erlbaum.

Newell, A., & Simon, H. A. (1972). *Human problem solving*. Englewood Cliffs, NJ: Prentice Hall.

O'Hara, K. P., & Payne, S. J. (1998). The effects of operator implementation cost on planfulness of problem solving and learning. *Cognitive Psychology, 35*, 34–70.

O'Hara, K. P., & Payne, S. J. (1999). Planning and user interface: The effects of lockout time and error recovery cost. *International Journal of Human-Computer Studies, 50*, 41–59.

Schmidt, R. A., & Bjork, R. A. (1992). New conceptualizations of practice: Common principles in three paradigms suggest new concepts for training. *Psychological Science, 3*, 207–217.

Paivio, A. (1986). *Mental representations*. New York: Oxford University Press.

Payne, S. J. (1991). A descriptive study of mental models. *Behavior and Information Technology, 10*, 3–21.

Payne, S. J. (2003). Users' mental models of devices: The very ideas. In. J.M. Carroll (Ed.), *HCI models, theories and frameworks: Towards a multi-disciplinary science* (pp. 135–156). San Francisco: Morgan Kaufmann.

Payne, S. J., Howes, A., & Hill, E. (1992). Conceptual instructions derived from an analysis of device models. *International Journal of Human–Computer Interaction, 4*, 35–58.

Payne, S. J., & Reader, W. R. (in press). Constructing structure maps of multiple on-line texts. *International Journal of Human-Computer Studies*.

Payne, S. J., Squibb, H. R., & Howes, A. (1990). The nature of device models: The yoked state space hypothesis and some experiments with text editors. *Human–Computer Interaction, 5*, 415–444.

Perfetti, C. A. (1997). Sentences, individual differences, and multiple texts. Three issues in text comprehension. *Discourse Processes, 23*, 337–355.

Radvansky, G. A., Spieler, D. H., & Zacks, R. T. (1993). Mental model organization. *Journal of Experimental Psychology: Learning, Memory, and Cognition, 19*, 95–114.

Radvansky, G. A., & Zacks, R. T. (1991). Mental models and fact retrieval. *Journal of Experimental Psychology: Learning, Memory, and Cognition, 17*, 940–953.

Rouet, J. F., Favart, M., Britt, M. A., & Perfetti, C. A. (1997). Studying and using multiple documents in history: Effects of discipline expertise. *Cognition and Instruction, 75*(1), 85–106.

Schmidt, R. A., Young, D. E., Swinnen, S., & Shapiro, D. C. (1989). Summary knowledge of results for skill acquisition: Support for the guidance hypothesis. *Journal of Experimental Psychology: Learning, Memory, and Cognition, 15*, 352–359.

Shea, J. B., & Morgan, R. L. (1979). Contextual interference effects on the acquisition, retention, and transfer of a motor skill. *Journal of Experimental Psychology: Human Learning and Memory, 5*, 179–187.

Simon, H. A. (1955). A behavioral model of rational choice. *Quarterly Journal of Economics, 69*, 99–118.

Simon, H. A. (1978). On the forms of mental representation. In C. W. Savage (Ed.), *Minnesota studies in the philosophy of science* (pp. 3–18) (Vol. 9). Minneapolis: University of Minnesota Press.

Simon, H. A. (1992). What is an "explanation" of behavior? *Psychological Science, 3*, 150–161.

Stahl, S. A., Hind, C. R., Britton, B. K., McNish, M. M., & Bosquet, D. (1996). What happens when students read multiple source documents in history. *Reading Research Quarterly, 31*(4), 430–456.

Tognazzini, B. (1992). *Tog on interface*. Reading, MA: Addison-Wesley.

van Dijk, T. A., & Kintsch, W. (1983). *Strategies of discourse comprehension*. New York: Academic Press.

Wegner, D. M. (1987). Transactive memory: A contemporary analysis of the group mind. In I. B. Mullen, & G. R. Goethals (Eds.), *Theories of group behaviour* (pp. 185–208). New York: Springer-Verlag.

Wegner, D. M. (1995). A computer network model of human transactive memory. *Social Cognition, 13*, 319–339.

Young, R. M. (1983). Surrogates and mappings. Two kinds of conceptual models for interactive devices. In D. Gentner, & A. L. Stevens (Eds.), *Mental models* (pp. 35–52). Hillsdale, NJ: Erlbaum.

·4·

EMOTION IN HUMAN–COMPUTER INTERACTION

Scott Brave and Clifford Nass
Stanford University

Emotion is a fundamental component of being human. Joy, hate, anger, and pride, among the plethora of other emotions, motivate action and add meaning and richness to virtually all human experience. Traditionally, human–computer interaction (HCI) has been viewed as the "ultimate" exception; users must discard their emotional selves to work efficiently and rationality with computers, the quintessentially unemotional artifact. Emotion seemed at best marginally relevant to HCI and at worst oxymoronic.

Recent research in psychology and technology suggests a different view of the relationship between humans, computers, and emotion. After a long period of dormancy and confusion, there has been an explosion of research on the psychology of emotion (Gross, 1999). Emotion is no longer seen as limited to the occasional outburst of fury when a computer crashes inexplicably, excitement when a video game character leaps past an obstacle, or frustration at an incomprehensible error message. It is now understood that a wide range of emotions plays a critical role in *every* computer-related, goal-directed activity, from developing a three-dimensional (3D) CAD model and running calculations on a spreadsheet, to searching the Web and sending an e-mail, to making an online purchase and playing solitaire. Indeed, many psychologists now argue that it is impossible for a person to have a thought or perform an action without engaging, at least unconsciously, his or her emotional systems (Picard, 1997b).

The literature on emotions and computers has also grown dramatically in the past few years, driven primarily by advances in technology. Inexpensive and effective technologies that enable computers to assess the physiological correlates of emotion, combined with dramatic improvements in the speed and quality of signal processing, now allow even personal computers to make judgments about the user's emotional state in real time (Picard, 1997a). Multimodal interfaces that include voices, faces, and bodies can now manifest a much wider and more nuanced range of emotions than was possible in purely textual interfaces (Cassell, Sullivan, Prevost, & Churchill, 2000). Indeed, any interface that ignores a user's emotional state or fails to manifest the appropriate emotion can dramatically impede performance and risks being perceived as cold, socially inept, untrustworthy, and incompetent.

This chapter reviews the psychology and technology of emotion, with an eye toward identifying those discoveries and concepts that are most relevant to the design and assessment of interactive systems. The goal is to provide the reader with a more critical understanding of the role and influence of emotion, as well as the basic tools needed to create emotion-conscious and consciously emotional interface designs.

The "seat" of emotion is the brain; hence, we begin with a description of the psychophysiological systems that lie at the core of how emotion emerges from interaction with the environment. By understanding the fundamental basis of emotional responses, we can identify those emotions that are most readily manipulable and measurable. We then distinguish emotions from moods (longer-term affective states that bias users'

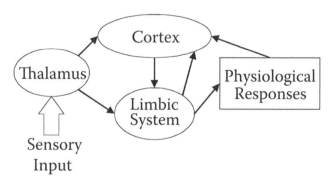

FIGURE 4.1. Neurological structure of emotion.

responses to any interface) and other related constructs. The following section discusses the cognitive, behavioral, and attitudinal effects of emotion and mood, focusing on attention and memory, performance, and user assessments of the interface. Designing interfaces that elicit desired affective states requires knowledge of the causes of emotions and mood; we turn to that issue in the following section. Finally, we discuss methods for measuring affect, ranging from neurological correlates to questionnaires, and describe how these indicators can be used both to assess users and to manifest emotion in interfaces.

UNDERSTANDING EMOTION

What is emotion? Although the research literature offers a plethora of definitions (Kleinginna & Kleinginna, 1981), two generally agreed-upon aspects of emotion stand out: (a) emotion is a reaction to events deemed relevant to the needs, goals, or concerns of an individual; and, (b) emotion encompasses physiological, affective, behavioral, and cognitive components. Fear, for example, is a reaction to a situation that threatens (or seems to threaten, as in a frightening picture) an individual's physical well-being, resulting in a strong negative affective state, as well as physiological and cognitive preparation for action. Joy, on the other hand, is a reaction to goals being fulfilled and gives rise to a more positive, approach-oriented state.

A useful model for understanding emotion, based on a simplified view of LeDoux's (1996) work in neuropsychology, is shown in Fig. 4.1. There are three key regions of the brain in this model: (a) the thalamus, (b) the limbic system, and (c) the cortex. All sensory input from the external environment is first received by the thalamus, which functions as a basic signal processor. The thalamus then sends information simultaneously both to the cortex, for "higher-level" processing, and directly to the limbic system (LeDoux, 1995). The limbic system,[1] often called the "seat of emotion," constantly evaluates the need/goal relevance of its inputs. If relevance is determined, the limbic system sends appropriate signals both to the body, coordinating the physiological response, and also to the cortex, biasing attention and other cognitive processes.

---

[1]The limbic system is often considered to include the hypothalamus, the hippocampus, and the amygdala. According to LeDoux, the amygdala is the only critical area (LeDoux & Phelps, 2000).

The direct thalamic-limbic pathway is the mechanism that accounts for the more primitive emotions, such as startle-based fear, as well as innate aversions and attractions. Damasio (1994) called these the "primary" emotions. In an HCI context, on-screen objects and events have the potential to activate such primitive emotional responses (Reeves & Nass, 1996). For example, objects that appear or move unexpectedly (i.e. pop-up windows, sudden animations) and loud or sharp noises are likely to trigger startle-based fear. Visual stimuli that tend to be particularly arousing include images that fill a large fraction of the visual field (either because the image or screen is large or because the eyes are close to the screen; (Detenber & Reeves, 1996; Voelker, 1994)), images that seem to approach the user (i.e., a rapidly expanding image on the screen, an image that appears to be flying out from the screen, or a character that walks toward the user), and images that move in peripheral vision (i.e., on the side of the screen; (Reeves & Nass, 1996)). Finally, certain images and sounds may be innately disturbing or pleasing due to their evolutionary significance (i.e., screeching or crying noises or explicit sexual or violent imagery; (see, i.e., Lang, 1995; Malamuth, 1996)).

Most of the emotions that we are concerned with in the design of HCI—and the ones we will focus on in the remainder of this chapter—require more extensive cognitive (i.e., knowledge-based) processing. These "secondary" emotions, such as frustration, pride, and satisfaction, result from activation of the limbic system by processing in the cortex. Such cortical processing can occur at various levels of complexity, from simple object recognition (i.e., seeing the Microsoft Office Paperclip) to intricate rational deliberation (i.e., evaluating the consequences of erasing a seldom-used file), and may or may not be conscious. The cortex can even trigger emotion in reaction to internally generated stimuli (i.e., thinking about how difficult it will be to configure a newly purchased application).

Finally, an emotion can result from a combination of both the thalamic-limbic and the cortico-limbic mechanisms. For example, an event causing an initial startle/fear reaction can be later recognized as harmless by more extensive, rational evaluation (i.e., when you realize that the flash of your screen suddenly going blank is just the initiation of the screen saver). In other situations, higher-level processing can reinforce an initial evaluation. Whatever the activation mechanism—thalamic or cortical, conscious or nonconscious—the cortex receives input from an activated limbic system, as well as feedback from the body, both contributing to the conscious "experience" of emotion.

The previous discussion provides a useful framework for considering one of the classic debates in emotion theory: are emotions innate or learned? At one extreme, evolutionary theorists argue that all emotions (including complex emotions such as regret and relief) are innate, each evolved to address a specific environmental concern of our ancestors (Darwin, 1872/1998; Neese, 1990; Tooby & Cosmides, 1990; see also Ekman, 1994; Izard, 1992). These theories are consistent with a hypothesis of high differentiation within the limbic system, corresponding to each of the biologically determined emotions. From this perspective, it is also reasonable to speculate that each emotion is associated with a unique set of physiological and cognition-biasing responses.

At the other extreme, many emotion theorists argue that, with the exception of startle and innate affinity/disgust (which they would consider pre-emotional), emotions are almost entirely learned social constructions (Averill, 1980; Ortony & Turner, 1990; Shweder, 1994; Wierzbicka, 1992). Such theories emphasize the role of higher cortical processes in differentiating emotions and concede minimal, if any, specificity within the limbic system (and consequently, within physiological responses). For example, the limbic system may operate in simply an on/off manner, or at most be differentiated along the dimensions of valence (positive/negative or approach/avoidance) and arousal (low/high) (Barrett & Russell, 1999; Lang, 1995). From this perspective, emotions are likely to vary considerably across cultures, with any consistency being based in common social structure, not biology.

Between these two extremes lie those who believe that there are "basic emotions." Citing both cross-cultural universals and primate studies, these theorists contend that there is a small set of innate, basic emotions shared by all humans (Ekman, 1992; Oatley & Johnson-Laird, 1987; Panksepp, 1992). Which emotions qualify as basic is yet another debate, but the list typically includes fear, anger, sadness, joy, disgust, and sometimes also interest and surprise. Other emotions are seen either as combinations of these basic emotions or as socially learned differentiations within the basic categories (i.e., agony, grief, guilt, and loneliness are various constructions of sadness; Bower, 1992). In this view, the limbic system is prewired to recognize the basic categories of emotion, but social learning and higher cortical processes still play a significant role in differentiation.

If the "basic emotions" view is correct, a number of implications for interaction design and evaluation emerge. First, the basic categories would likely be the most distinguishable, and therefore measurable, emotional states (both in emotion recognition systems as well as in postinteraction evaluations). Further, the basic emotions would be less likely to vary significantly from culture to culture; facilitating the accurate translation and generalizability of questionnaires intended to assess such emotions. Lower variability also enables more reliable prediction of emotional reactions to interface content, both across cultures and across individuals. Finally, for users interacting with on-screen characters, depictions of the basic emotions would presumably be most immediately recognizable. If the social construction view of emotions is valid, then emotion measurement and assessment, prediction, and depictions are more challenging and nuanced.

DISTINGUISHING EMOTION FROM RELATED CONSTRUCTS

Mood

It is useful to distinguish among several terms often used ambiguously: emotion, mood, and sentiment. Emotion can be distinguished from mood by its object-directedness. As Frijda (1994) explained, emotions are *intentional*: They "imply and involve relationships with a particular object." We get scared *of* something, angry *with* someone, and excited *about* some event.

Moods, on the other hand, though they may be indirectly caused by a particular object, are "nonintentional"; they are not *directed* at any object in particular and are thus experienced as more diffuse, global, and general. A person can be sad about something (an emotion) or generally depressed (a mood). Unfortunately, the English language often allows the same term to describe both emotion and mood (i.e., "happy").

Another distinction between emotion and mood emerges from a functional perspective. As a reaction to a particular situation, emotions bias action—they prepare the body and the mind for an appropriate, immediate response. As such, emotions also tend to be relatively short lived. Moods, in contrast, tend to bias cognitive strategies and processing over a longer term (Davidson, 1994). More generally, moods can be seen to serve as a background affective filter through which both internal and external events are appraised. A person in a good mood tends to view everything in a positive light, while a person in a bad mood does the opposite. The interaction between emotions and moods is also important. Moods tend to bias which emotions are experienced, lowering the activation thresholds for mood-related emotions. Emotions, on the other hand, often cause or contribute to moods.

When assessing user response to an interface, it is important to consider the biasing effects of user mood. Users entering a usability or experimental study in a good mood, for instance, are more likely to experience positive emotion during an interaction than users in a bad mood. Pretesting for mood and including it as a variable in analysis can, therefore, reduce noise and increase interpretive power. If pretesting users immediately prior to an interaction is inappropriate, there is a second noise-reducing option: assessment of temperament. Temperament reflects the tendency of certain individuals to exhibit particular moods with great frequency. Participants can be pretested for temperament at any point prior to the study, enabling the exclusion of extreme cases of depressive or excitable individuals (i.e., Bishop, Jacks, & Tandy, 1993). Finally, if user testing involves multiple stimuli, order of presentation can also influence the results. For example, earlier stimuli may establish a mood that biases emotional reactions to subsequent stimuli. To combat this problem, the order of stimuli should be varied from participant to participant, when feasible.

Sentiment

Sentiment is also often confused with emotion. Unlike emotions (and moods), sentiments are not states of an individual, but assigned properties of an object. When people say that they "like" an interface or find an interface to be "frustrating," what they really mean is that that they associate the interface with a positive or frustrating emotional state; in other words, they expect interaction with the interface to lead to positive or frustrating emotions. The basis for this judgment often comes from direct experience and subsequent generalization, but may also arise from social learning (Frijda, 1994).

One reason for the confusion between emotions and sentiment is that many languages use the same words for both. For example, the word "like" can be used both to indicate prediction

or opinion (sentiment) as well as a current emotional state (i.e. "I like receiving e-mail" vs. "I like the e-mail that just arrived"). Clore (1994, p. 108) offers an interesting explanation for this ambiguity, theorizing that sentiments are judged by bringing the object to mind and observing the affective reaction. But, while emotions and moods are fleeting—emotions last only seconds and moods last for hours or even days—sentiments can persist indefinitely and are thus responsible for guiding our propensities to seek out or avoid particular objects and situations. In this sense, sentiments are of critical importance for HCI because they motivate users to return to particular software products or Websites.

Although direct interaction with an object is the most accurate way for a user to create a sentiment (consider the colloquial phrase, "how do you know you don't like it unless you try it"), sentiments can also be caused by assumptions based on the communicated properties of an object. People may, for example, base a sentiment on someone else's description of their interaction with the object, or even immediately adopt the sentiment of someone they know or respect (i.e., consider the presumed influence of celebrities in software advertisements).

As a predictive construct, sentiments are often generalizations about a class of objects with a given recognizable property, i.e., stereotypes. Although some of these generalizations may be logical and accurate, others may not—in fact, they may not even be conscious. Negative experiences with a particular computer character, for example, may lead users to conclude that they dislike all character-based interfaces. However, using a character that people know and like already—Mickey Mouse, for example—may be able to leverage sentiment to an interface's advantage. Similarly, many people have well-established sentiments regarding certain types of applications (i.e. "I hate spreadsheet applications"). For such users, interfaces that avoid triggering their negative stereotypes have the advantage. Positive stereotypes, on the other hand, should be encouraged whenever possible, such as when learning applications are framed as entertainment.

EFFECTS OF AFFECT

Attention

One of the most important effects of emotion lies in its ability to capture attention. Emotions have a way of being completely absorbing. Functionally, they direct and focus our attention on those objects and situations that have been appraised as important to our needs and goals so that we can deal with them appropriately. Emotion-relevant thoughts then tend to dominate conscious processing—the more important the situation, the higher the arousal, and the more forceful the focus (Clore & Gasper, 2000). In an HCI context, this attention-getting function can be used advantageously, as when a sudden beep is used to alert the user, or can be distracting, as when a struggling user is frustrated and can only think about his or her inability.

Emotion can further influence attention through a secondary process of emotion regulation (Gross, 1998). Once an emotion is triggered, higher cognitive processes may determine that the emotion is undesirable. In such cases, attention is often directed away from the emotion-eliciting stimulus for the purpose of distraction. For example, becoming angry with an onscreen agent may be seen as ineffectual (i.e., because it doesn't recognize your anger) or simply unreasonable. An angered user may then actively try to ignore the agent, focusing instead on other onscreen or off-screen stimuli, or even take the next step and completely remove the agent from the interaction (which could mean leaving an application or Website entirely). Positive emotions may likewise require regulation at times, such as when amusing stimuli lead to inappropriate laughter in a work environment. If the emotionally relevant stimulus is too arousing, however, regulation through selective attention is bound to fail (Wegner, 1994), because users will be unable to ignore the stimulus.

Mood can have a less profound but more enduring effect on attention. At the most basic level, people tend to pay more attention to thoughts and stimuli that have some relevance to their current mood state (Bower & Forgas, 2000). However, people also often consciously regulate mood, selecting and attending to stimuli that sustain desired moods or, alternatively, counteract undesired moods. An interface capable of detecting—or at least predicting—a user's emotional or mood state could similarly assume an affect-regulation role, helping to guide attention away from negative and toward more positive stimuli. For example, a frustrated user could be encouraged to work on a different task, focus on a different aspect of the problem at hand, or simply take a break (perhaps by visiting a suggested online entertainment site).

Memory

Emotion's effect on attention also has implications for memory. Because emotion focuses thought on the evoking stimulus, emotional stimuli are generally remembered better than unemotional events (Thorson & Friestad, 1985). Negative events, which tend to be highly arousing, are typically remembered better than positive events (Newhagen & Reeves, 1991, 1992; Reeves & Nass, 1996, Chapter 10; Reeves, Newhagen, Maibach, Basil, & Kurz, 1991). In addition, emotionality "improves memory for central details while undermining memory for background details" (see Heuer & Reisberg, 1992; Parrott & Spackman, 2000).

Mood also comes into play in both memory encoding and retrieval. Research has shown that people will remember "mood-congruent" emotional stimuli better than incongruent stimuli. Bower, Gilligan, and Monteiro (1981), for example, hypnotized subjects into either a happy or sad mood before having them read stories about various characters. The next day, subjects were found to remember more facts about characters whose mood had agreed with their own than about other characters. Similarly, on the retrieval end, people tend to better recall memories consistent with their current mood (Ellis & Moore, 1999). However, the reverse effect has also been shown to occur in certain situations; people will sometimes better recall mood-incongruent memories (i.e., happy memories while in a sad mood). Parrott and Spackman (2000) hypothesized that mood regulation is responsible for this inverse effect: When a given mood is seen as inappropriate or distracting, people will often actively try to evoke memories or thoughts to modify that mood (see Forgas, 1995) Affect Infusion Model (AIM) for insight into these contradictory findings (also see Erber & Erber, 2001)). Finally, there is some evidence for mood-dependent recall: Memories encoded while in a particular mood are better recalled when in that same mood. This effect is independent of the emotional content of the memory itself (Ucros, 1989). It should be noted, however, that the effects of mood on memory are often unreliable and therefore remain controversial.

Performance

Mood has also been found to affect cognitive style and performance. The most striking finding is that even mildly positive affective states profoundly affect the flexibility and efficiency of thinking and problem solving (Hirt, Melton, McDonald, & Harackiewicz, 1996; Isen, 2000; Murray, Sujan, Hirt, & Sujan, 1990). In one of the best-known experiments, subjects were induced into a good or bad mood and then asked to solve Duncker's (1945) candle task. Given only a box of thumbtacks, the goal of this problem was to attach a lighted candle to the wall, such that no wax drips on the floor. The solution required the creative insight to thumbtack the box itself to the wall and then tack the candle to the box. Subjects who were first put into a good mood were significantly more successful at solving this problem (Isen, Daubman, & Nowicki, 1987). In another study, medical students were asked to diagnose patients based on X-rays after first being put into a positive, negative, or neutral mood. Subjects in the positive-affect condition reached the correct conclusion faster than did subjects in other conditions (Isen, Rosenzweig, & Young, 1991). Positive affect has also been shown to increase heuristic processing, such as reliance on scripts and stereotypes. Though some have argued that such reliance is at the expense of systematic processing (Schwartz & Bless, 1991), more recent evidence suggests that heuristic processing and systematic processing are not mutually exclusive (Isen, 2000). Keeping a user happy may, therefore, not only affect satisfaction, but may also lead to efficiency and creativity.

Assessment

Mood has also been shown to influence judgment and decision making. As mentioned earlier, mood tends to bias thoughts in a mood-consistent direction, while also lowering the thresholds of mood-consistent emotions. One important consequence of this is that stimuli—even those unrelated to the current affective state—are judged through the filter of mood (Clore et al., 2001; Erber & Erber, 2001; Niedenthal, Setterlund, & Jones, 1994). This suggests that users in a good mood will likely judge both the interface and their work more positively, regardless of any direct emotional effects. It also suggests that a happy user at an e-commerce site would be more likely to evaluate the products or services positively.

Positive mood also decreases risk-taking, likely in an effort to preserve the positive mood. That is, although people in a positive mood are more risk-prone when making hypothetical decisions, when presented with an actual risk situation, they tend to be more cautious (Isen, 2000). In an e-commerce purchasing situation, then, one can predict that a low-risk purchase is more likely during a good mood, due to a biased judgment in favor of the product, while a high-risk purchase may be more likely in a less cautious, neutral, or negative mood (consistent with the adage that desperate people resort to desperate measures).

A mood's effect on judgment, combined with its effect on memory, can also influence the formation of sentiments. Sentiments are not necessarily determined during interaction with an object; they often are grounded in reflection. This is important to consider when conducting user tests, as the mood set by the interaction immediately prior to a questionnaire may bias like/dislike assessments of earlier interactions. Thus, varying order of presentation ensures both that later stimuli do not influence the *assessment* of earlier stimuli, and that earlier stimuli do not influence the *experience* of later stimuli (as discussed earlier).

CAUSES OF EMOTION

What causes emotions? The answer to this question is critical for HCI because an understanding of emotions' antecedents will better enable us to design interfaces that encourage desired emotional states and understand interfaces that do not.

Needs and Goals

As we discussed in the first section, emotions are reactions to situations deemed relevant to the needs and goals of the individual. Clearly, a user comes to a computer hoping to achieve certain application-specific goals—composing a document, sending an e-mail, finding a piece of information, etc. The degree to which an interface facilitates or hampers those goals has a direct effect on the emotional state of the user. An interface capable of detecting emotion could, therefore, use such information as feedback regarding whether the user's goals are being met, modifying its behavior as necessary. In an information-seeking context, for example, emotional reactions to displayed content could be used to improve the goal-relevance of future retrievals. Similarly, if an interface detects frustration, desperation, or anger in a user, goals may be facilitated by trying a new approach or offering assistance (Klein, Moon, & Picard, 1999; Picard, 1997a). (If the particular goals implicated by an emotion are not clear, there can be advantages to an interface that empathizes with the user; (Klein et al., 1999)). More generally, user preferences can be automatically determined based on a user's emotional reactions to interface elements (Picard, 1997a).

There are also a host of more abstract needs underlying, and often adjacent to, application-specific goals. A user may have a strong need to feel capable and competent, maintain control, learn, or be entertained. A new user typically needs to feel com-

fortable and supported, while an expert is more focused on aesthetic concerns of efficiency and elegance. Acknowledging these more abstract goals in interface design can be as instrumental in determining a user's affective state as meeting or obstructing application-specific goals. Maslow's hierarchy (Maslow, 1968) presents a useful starting place for considering the structure of these more abstract user needs. In his later work, Maslow (1968) grouped an individual's basic needs into eight categories:

- **Physiological:** hunger, thirst, bodily comforts, etc.
- **Safety/security:** being out of danger
- **Social:** affiliate with others, be accepted
- **Esteem:** to achieve, be competent, gain approval and recognition
- **Cognitive:** to know, to understand, and explore
- **Aesthetic:** symmetry, order, and beauty
- **Self-actualization:** to find self-fulfillment and realize one's potential
- **Transcendence:** to help others find self-fulfillment and realize their potential.

When a particular situation or event is deemed as promoting these needs, positive emotion results. When someone or something hampers these needs, negative emotion results. The specific emotion experienced is due in part to the category of need implicated by the event. Fright, for example, is typically associated with threatened safety/security needs; love and embarrassment with social needs; pride with esteem needs; and curiosity with cognitive needs.

Within Maslow's (1968) framework, application-specific goals of a user can be seen as instruments ultimately serving these more basic needs. For example, a user who successfully enhances a digital family photograph may simultaneously be contributing to the fulfillment of social, esteem, cognitive, and aesthetic needs. However, interfaces can also *directly* address a user's basic needs. For example, a spell-checker interface that praises a user on his or her spelling ability, regardless of the user's actual performance, is a somewhat humorous, though illustrative, approach to acknowledging a user's esteem needs. Such interfaces, by enhancing the user's affective state, have been shown also to be viewed as more intelligent and likable (Reeves & Nass, 1996, Chapter 4). As another example, an interface that takes care to establish a trusting and safe relationship with users may ultimately lead to more effective and cooperative interactions (Fogg, 1998). Educational software should address users' emotional needs, not only teaching the relevant content, but also ensuring users *believe* that they are learning. Optimized learning further requires a careful balance of esteem and self-actualization needs, offering appropriate levels of encouragement and challenge, as well as praise and criticism. Finally, one of the key arguments for social interfaces is that they meet the social needs of users (Reeves & Nass, 1996).

Although the type of need relevant in a situation offers some insight into emotional reaction, need category alone is not sufficient to differentiate fully among all emotions. Distinguishing frustration and anger, for example, cannot be achieved based solely on knowing the users' need; it also requires some notion of agency.

Appraisal Theories

"Appraisal" theories provide much greater predictive power than category or hierarchy-based schemes by specifying the critical properties of antecedent events that lead to particular emotions (Lazarus, 1991; Ortony, Clore, & Collins, 1988; Roseman, Antoniou, & Jose, 1996; Scherer, 1988). Ellsworth (1994), for example, described a set of "abstract elicitors" of emotion. In addition to *novelty* and *valence*, Ellsworth contended that the level of *certainty/uncertainty* in an event has a significant impact on the emotion experienced. For instance, "uncertainty about probably positive events leads to interest and curiosity, or to hope," while, "uncertainty about probably negative events leads to anxiety and fear" (Ellsworth, 1994, p. 152). Certainty, on the other hand, can lead to relief in the positive case and despair in the negative case.

Because slow, unclear, or unusual responses from an interface generally reflect a problem, one of the most common interface design mistakes—from an affective standpoint—is to leave the user in a state of uncertainty. Users tend to fear the worst when, for example, an application is at a standstill, the hourglass remains up longer than usual, or the hard drive simply starts grinding away unexpectedly. Such uncertainty leads to a state of anxiety that can be easily avoided with a well-placed, informative message or state indicator. Providing users with immediate feedback on their actions reduces uncertainty, promoting a more positive affective state (see Norman, 1990, on visibility and feedback). When an error has *actually* occurred, the best approach is to make the user aware of the problem and its possible consequences, but frame the uncertainty in as positive a light as possible (i.e., "this application has experienced a problem, but the document should be recoverable").

According to Ellsworth (1994), *obstacles and control* also play an important role in eliciting emotion. High control can lead to a sense of challenge in positive situations, but stress in negative situations. Lack of control, on the other hand, often results in frustration, which if sustained can lead to desperation and resignation. In an HCI context, providing an appropriate level of controllability, given a user's abilities and the task at hand, is thus critical for avoiding negative affective consequences. Control need not only be perceived to exist (Skinner, 1995; Wegner, Bargh, Gilbert, Fiske, et al., 1998), but must be understandable and visible, otherwise the interface itself is an obstacle (Norman, 1990).

Agency is yet another crucial factor determining emotional response (Ellsworth, 1994; Friedman & Kahn, 1997). When oneself is the cause of the situation, shame (negative) and pride (positive) are likely emotions. When another person or entity is the cause, anger (negative) and love (positive) are more likely. However, if fate is the agent, one is more likely to experience sorrow (negative) and joy (positive). An interface often has the opportunity to direct a user's perception of agency. In any anomalous situation, for example—be it an error in reading a file, inability to recognize speech input, or simply a crash—if the user is put in a position encouraging blame of oneself or fate, the negative emotional repercussions may be more difficult to diffuse than if the computer explicitly assumes blame (and is apologetic). For example, a voice interface encountering a recognition error can say, "This system failed to understand your command" (blaming itself), "The command was not understood" (blaming no one), or "You did not speak clearly enough for your command to be understood" (blaming the user).

Appraisal theories of emotion, such as Ellsworth's (1994), are useful not only in understanding the potential affective impacts of design decisions, but also in creating computer agents that exhibit emotion. Although in some cases scripted emotional responses are sufficient, in more dynamic or interactive contexts, an agent's affective state must be simulated to be believable. Ortony, Clore, and Collins' (1988) cognitive theory of emotion is currently the most commonly applied appraisal theory for such purposes (Bates, Loyall, & Reilly, 1994; Elliott & Brzezinski, 1998; for alternate approaches, see Ball & Breese, 2000; Bozinovski & Bozinovska, 2001; Scheutz, Sloman, & Logan, 2000). Appraisal theories can also be used to help model and predict a user's emotional state in real time (Elliott & Brzezinski, 1998).

Contagion

Another cause of emotion that does not fit cleanly into the structure just described is contagion (Hatfield, Cacioppo, & Rapson, 1994). People often "catch" other's emotions. Sometimes this social phenomenon seems logical, such as when a person becomes afraid upon seeing another experience fear. At other times, contagion seems illogical, such as when another person's laughter induces immediate, "unexplainable" amusement. Anticipatory excitement is another emotion that transfers readily from person to person.

Emotions *in* interfaces can also be contagious. For example, a character that exhibits excitement when an online product appears can make users feel more excited. Similarly, an attempt at light humor in a textual interface, even if unsuccessful, may increase positive affect (Morkes, Kernal, & Nass, 2000).

Moods and Sentiments

Mood and sentiment can also bias emotion. One of the fundamental properties of mood is that it lowers the activation threshold for mood-consistent emotions. Sentiment can act in a similar way. For example, interaction with an object, to which a sentiment is already attached, can evoke emotion either in memory of past interaction or in anticipation of the current interaction. Thus, an interface that proved frustrating in the past may elicit frustration before the user even begins working. In addition, sentiment can bias perception of an object, increasing the probability of eliciting sentiment-consistent emotions. For example, an application that users *like* can do no wrong, while one that users *dislike* does everything to anger them, regardless of the application's actual behavior. Of critical importance here is that sentiments need not derive from direct experience; they may also be inferred from stereotypes or other generalizations.

Previous Emotional State

Finally, a user's previous emotional state can affect the experience of subsequent emotions. This occurs not only through the mechanism of mood—emotions can cause moods and moods then bias the activation thresholds of emotions—but also through the mechanisms of excitation transfer and habituation. Excitation transfer (Zillmann, 1991) is based on the fact that after an emotion-causing stimulus has come and gone, an activated autonomic nervous system takes some time to return to its deactivated state. If another emotion is triggered before that decay is complete, the residual activation ("excitement") will be added to the current activation and be perceived as part of the current emotion. As Zillmann (1991) explained, "residues of excitation from a previous affective reaction will *combine* with excitation produced by subsequent affective stimulation and thereby cause an *overly intense* affective reaction to subsequent stimuli. . . . Residual arousal from anger, then, may intensify fear; residues from fear may intensify sexual behaviors; residual sexual arousal may intensify aggressive responses; and so forth" (p. 116). Thus, people who have just hit the "purchase" button associated with their web shopping cart can become particularly angry when they are presented with multiple pages before they can complete their transaction: The arousal of buying increases the intensity of their frustration with the post-purchase process. Similarly, Reeves and Nass (1996) have argued that pictorial characters "raise the volume knob" on both positive and negative feelings about an interaction, because explicitly social interactions are more arousing than their non-social counterparts are.

Habituation is, in some sense, the converse of excitation transfer. It posits that the intensity of an emotion *decreases* over time if the emotion is experienced repeatedly. One explanation for this effect relates back to appraisal theory: "Emotions are elicited not so much by the presence of favorable or unfavorable conditions, but by actual or expected changes in favorable or unfavorable conditions" (Frijda, 1988, p. 39). Repeated pleasurable affective states, therefore, become expected and thus gradually lose intensity. The same is true for negative affective states; however, particularly extreme negative emotional states may never habituate (Frijda, 1988). This may be why negative experiences with frequently used interfaces (i.e., operating systems) are remembered more vividly than positive experiences.

CAUSES OF MOOD

Mood has a number of potential causes. The most obvious is emotion itself. Intense or repetitive emotional experiences tend to prolong themselves into moods. A user who is continually frustrated will likely be put in a frustrated mood, while a user who is repeatedly made happy will likely be put in a positive mood. Mood can also be influenced, however, by anticipated emotion, based on sentiment. For example, if users know that they must interact with an application that they dislike (i.e., they associate with negative emotion), they may be in a bad mood from the start.

Contagion

Similar to emotion, moods also exhibit a contagion effect (Neumann & Strack, 2000). For example, a depressed person will often make others feel depressed and a happy person will often make others feel happy. Murphy and Zajonc (1993) have shown that even a mere smiling or frowning face, shown so quickly that the subject is not conscious of seeing the image, can affect a person's mood and subsequently bias judgment. From an interface standpoint, the implications for character-based agents are clear: Moods exhibited by onscreen characters may directly transfer to the user's mood. Onscreen mood can also lead to "perceived contagion" effects: One smiling or frowning face on the screen can influence users' perceptions of other faces that they subsequently see on the screen, perhaps because of priming (Reeves, Biocca, Pan, Oshagan, & Richards, 1989; Reeves & Nass, 1996, Chapter 22).

Color

Color can clearly be designed into an interface with its mood influencing properties in mind. Warm colors, for example, generally provoke "active feelings," while cool colors are "much less likely to cause extreme reactions" (Levy, 1984). Gerard (1957; 1958), for example, found that red light projected onto a diffusing screen produces increased arousal in subjects, using a number of physiological measures (including cortical activation, blood pressure, and respiration), while blue light has essentially the opposite "calming" effect (see Walters, Apter, & Svebak, 1982). Subjective ratings of the correlations between specific colors and moods can be more complicated. As Gardano (1986) summarized, "yellow (a warm color) has been found to be associated with both sadness (Peretti, 1974) and with cheerfulness (Wexner, 1954). Similarly, red (another warm color) is related to anger and violence (Schachtel, 1943) as well as to passionate love (Henry & Jacobs, 1978; Pecjak, 1970); and blue (a cool color), to tenderness (Schachtel, 1943) and sadness (Peretti, 1974). . . ." Nevertheless, as any artist will attest, carefully designed color schemes (combined with other design elements) can produce reliable and specific influences on mood.

Other Effects

A number of other factors can affect mood. For example, in music, minor scales are typically associated with negative emotion and mood, while major scales have more positive/happy connotations (Gregory, Worrall, & Sarge, 1996). Other possible influences on mood include weather, temperature, hormonal cycles, genetic temperament, sleep, food, medication, and lighting (Thayer, 1989).

MEASURING AFFECT

Measuring user affect can be valuable both as a component of usability testing and as an interface technique. When evaluating interfaces, affective information provides insight into what a user is feeling—the fundamental basis of liking and other sentiments. Within an interface, knowledge of a user's affect provides useful feedback regarding the degree to which a user's goals are being met, enabling dynamic and intelligent adaptation. In particular, social interfaces (including character-based interfaces) must have the ability to recognize and respond to emotion in users to execute effectively real-world interpersonal interaction strategies (Picard, 1997a).

Neurological Responses

The brain is the most fundamental source of emotion. The most common way to measure neurological changes is the electro-encephalogram (EEG). In a relaxed state, the human brain exhibits an alpha rhythm, which can be detected by EEG recordings taken through sensors attached to the scalp. Disruption of this signal (alpha blocking) occurs in response to novelty, complexity, and unexpectedness, as well as during emotional excitement and anxiety (Frijda, 1986). EEG studies have further shown that positive/approach-related emotions lead to greater activation of the left anterior region of the brain, while negative/avoidance-related emotions lead to greater activation of the right anterior region (Davidson, 1992; see also Heller, 1990). Indeed, when one flashes a picture to either the left or the right of where a person is looking, the viewer can identify a smiling face more quickly when it is flashed to the left hemisphere and a frowning face more quickly when it is flashed to the right hemisphere (Reuter-Lorenz & Davidson, 1981). Current EEG devices, however, are fairly clumsy and obstructive, rendering them impractical for most HCI applications. Recent advances in magneto resonance imaging (MRI) offer great promise for emotion monitoring, but are currently unrealistic for HCI because of their expense, complexity, and form factor.

Autonomic Activity

Autonomic activity has received considerable attention in studies of emotion, in part due to the relative ease in measuring certain components of the autonomic nervous system (ANS), including heart rate, blood pressure, blood-pulse volume, respiration, temperature, pupil dilation, skin conductivity, and more recently, muscle tension (as measured by electromyography (EMG)). However, the extent to which emotions can be distinguished on the basis of autonomic activity alone remains a hotly debated issue (see Ekman & Davidson, 1994, ch. 6; Levenson, 1988). On the one end are those, following in the Jamesian tradition (James, 1884), who believe that each emotion has a unique autonomic signature—technology is simply not advanced enough yet to fully detect these differentiators. On the other extreme, there are those, following Cannon (1927), who contended that all emotions are accompanied by the same state of nonspecific autonomic (sympathetic) arousal, which varies only in magnitude—most commonly measured by galvanic skin response (GSR), a measure of skin conductivity (Schachter & Singer, 1962). This controversy has clear connections to the nature-nurture debate in emotion, described earlier, because autonomic specificity seems more probable if each emotion has a distinct biological basis, while nonspecific autonomic (sympathetic) arousal seems more likely if differentiation among emotions is based mostly on cognition and social learning.

Though the debate is far from resolved, certain measures have proven reliable at distinguishing among "basic emotions." Heart rate, for example, increases most during fear, followed by anger, sadness, happiness, surprise, and finally disgust, which shows almost no change in heart rate (Cacioppo, Bernston, Klein, & Poehlmann, 1997; Ekman, Levenson, & Friesen, 1983; Levenson, Ekman, & Friesen, 1990). Heart rate also generally increases during excitement, mental concentration, and "upon the presentation of intense sensory stimuli" (Frijda, 1986). Decreases in heart rate typically accompany relaxation, attentive visual and audio observation, and the processing of pleasant stimuli (Frijda, 1986). As is now common knowledge, blood pressure increases during stress and decreases during relaxation. Cacioppo et al. (2000) further observed that anger increases diastolic blood pressure to the greatest degree, followed by fear, sadness, and happiness. Anger is further distinguished from fear by larger increases in blood pulse volume, more nonspecific skin conductance responses, smaller increases in cardiac output, and other measures indicating that "anger appears to act more on the vasculature and less on the heart than does fear" (Cacioppo et al., 1997). Results using other autonomic measures are less reliable.

Combined measures of multiple autonomic signals show promise as components of an emotion recognition system. Picard, Vyzas, and Healey (in press), for example, achieved 81% percent recognition accuracy on eight emotions through combined measures of respiration, blood pressure volume, and skin conductance, as well as facial muscle tension (to be discussed in the next subsection). Many autonomic signals can also be measured in reasonably nonobstructive ways (i.e., through user contact with mice and keyboards; Picard, 1997a).

However, even assuming that we could distinguish among all emotions through autonomic measures, it is not clear that we should. In real-world social interactions, humans have at least partial control over what others can observe of their emotions. If another person, or a computer, is given direct access to users' internal states, users may feel overly vulnerable, leading to stress and distraction. Such personal access could also be seen as invasive, compromising trust. It may, therefore, be more appropriate to rely on measurement of the external signals of emotion (discussed next).

Facial Expression

Facial expression provides a fundamental means by which humans detect emotion. Table 4.1 describes characteristic facial features of six basic emotions (Ekman & Friesen, 1975; Rosenfeld, 1997). Endowing computers with the ability to recognize facial expressions, through pattern recognition of captured images, have proven to be a fertile area of research (Essa & Pentland, 1997; Lyons, Akamatsu, Kamachi, & Gyoba, 1998; Martinez, 2000; Yacoob & Davis, 1996); for recent reviews, see Cowie et al., 2001; Lisetti & Schiano, 2000; Tian, Kanade, & Cohn, 2001). Ekman and Friesen's (1977) Facial Action Coding System (FACS), which identifies a highly specific set of muscular movements for each emotion, is one of the most widely accepted foundations for facial-recognition systems (Tian et al., 2001). In many systems, recognition accuracy can reach as high as 90%–98% on a small set of basic emotions. However, current

TABLE 4.1. Facial Cues and Emotion

Emotion	Observed Facial Cues
Surprise	Brows raised (curved and high)
	Skin below brow stretched
	Horizontal wrinkles across forehead
	Eyelids opened and more of the white of the eye is visible
	Jaw drops open without tension or stretching of the mouth
Fear	Brows raised and drawn together
	Forehead wrinkles drawn to the center
	Upper eyelid is raised and lower eyelid is drawn up
	Mouth is open
	Lips are slightly tense or stretched and drawn back
Disgust	Upper lip is raised
	Lower lip is raised and pushed up to upper lip or it is lowered
	Nose is wrinkled
	Cheeks are raised
	Lines below the lower lid, lid is pushed up but not tense
	Brows are lowered
Anger	Brows lowered and drawn together
	Vertical lines appear between brows
	Lower lid is tensed and may or may not be raised
	Upper lid is tense and may or may not be lowered due to brows' action
	Eyes have a hard stare and may have a bulging appearance
	Lips are either pressed firmly together with corners straight or down or open, tensed in a squarish shape
	Nostrils may be dilated (could occur in sadness too) unambiguous only if registered in all three facial areas
Happiness	Corners of lips are drawn back and up
	Mouth may or may not be parted with teeth exposed or not
	A wrinkle runs down from the nose to the outer edge beyond lip corners
	Cheeks are raised
	Lower eyelid shows wrinkles below it, and may be raised but not tense
	Crow's-feet wrinkles go outward from the outer corners of the eyes
Sadness	Inner corners of eyebrows are drawn up
	Skin below the eyebrow is triangulated, with inner corner up
	Upper lid inner corner is raised
	Corners of the lips are drawn or lip is trembling

recognition systems are tested almost exclusively on "produced" expressions (i.e., subjects are asked to make specific facial movements or emotional expressions), rather than natural expressions resulting from actual emotions. The degree of accuracy that can be achieved on more natural expressions of emotion

remains unclear. Further, "not all . . . emotions are accompanied by visually perceptible facial action" (Cacioppo et al., 1997).

An alternate method for facial expression recognition, capable of picking up both visible and extremely subtle movements of facial muscles, is facial electromyography (EMG). EMG signals, recorded through small electrodes attached to the skin, have proven most successful at detecting positive versus negative emotions and show promise in distinguishing among basic emotions (Cacioppo et al., 2000). Though the universality (and biological basis) of facial expression is also debated, common experience tells us that, at least within a culture, facial expressions are reasonably consistent. Nonetheless, individual differences may also be important, requiring recognition systems to adapt to a specific user for greatest accuracy. Gestures can also be recognized with technologies similar to those for facial-expression recognition, but the connection between gesture and emotional state is less distinct, in part due to the greater influence of personality (Cassell & Thorisson, in press; Collier, 1985).

Voice

Voice presents yet another opportunity for emotion recognition (see Cowie et al., 2001 for an extensive review). Emotional arousal is the most readily discernible aspect of vocal communication, but voice can also provide indications of valence and specific emotions through acoustic properties such as pitch range, rhythm, and amplitude or duration changes (Ball & Breese, 2000; Scherer, 1989). A bored or sad user, for example, will typically exhibit slower, lower-pitched speech, with little high frequency energy, while a user experiencing fear, anger, or joy will speak faster and louder, with strong high-frequency energy and more explicit enunciation (Picard, 1997a). Murray and Arnott (1993) provided a detailed account of the vocal effects associated with several basic emotions (see Table 4.2). Though few systems have been built for automatic emotion recognition through speech, Banse and Scherer (1996) have demonstrated the feasibility of such systems. Cowie and Douglas-Cowie's ACCESS system (Cowie & Douglas-Cowie, 1996) also presents promise (Cowie et al., 2001).

Self-Report Measures

A final method for measuring a user's affective state is to ask questions. Post-interaction questionnaires, in fact, currently

TABLE 4.2. Voice and Emotion

	Fear	Anger	Sadness	Happiness	Disgust
Speech rate	Much faster	Slightly faster	Slightly slower	Faster or slower	Very much slower
Pitch average	Very much higher	Very much higher	Slightly lower	Much higher	Very much lower
Pitch range	Much wider	Much wider	Slightly narrower	Much wider	Slightly wider
Intensity	Normal	Higher	Lower	Higher	Lower
Voice quality	Irregular voicing	Breathy chest tone	Resonant	Breathy blaring	Grumbled chest tone
Pitch changes	Normal	Abrupt on stressed syllables	Downward inflections	Smooth upward inflections	Wide downward terminal inflections
Articulation	Precise	Tense	Slurring	Normal	Normal

serve as the primary method for ascertaining emotion, mood, and sentiment during an interaction. However, in addition to the standard complexities associated with self-report measures (such as the range of social desirability effects), measuring affect in this way presents added challenges. To begin with, questionnaires are capable of measuring only the *conscious experience* of emotion and mood. Much of affective processing, however, resides in the limbic system and in nonconscious processes. Although it is debatable whether an emotion can exist without any conscious component at all, a mood surely can. Further, questions about emotion, and often those about mood, refer to past affective states and thus rely on imperfect and potentially biased memory. Alternatively, asking a user to report on an emotion as it occurs requires interruption of the experience. In addition, emotions and moods are often difficult to describe in words. Finally, questions about sentiment, although the most straightforward given their predictive nature, are potentially affected by when they are asked (both because of current mood and memory degradation). Nevertheless, self-report measures are the most direct way to measure sentiment and a reasonable alternative to direct measures of emotion and mood (which currently remain in the early stages of development).

Several standard questionnaires exist for measuring affect (Plutchik & Kellerman, 1989, Chapter 1–3). The most common approach presents participants with a list of emotional adjectives and asks how well each describes their affective state. Izard's (1972) Differential Emotion Scale (DES), for example, includes 24 emotional terms (such as delighted, scared, happy, and astonished) that participants rate on seven-point scales indicating the degree to which they are feeling that emotion (from "not at all" to "extremely"). McNair, Lorr, and Droppleman's (1981) Profile of Mood States (POMS) is a popular adjective-based measure of mood. Researchers have created numerous modifications of these standard scales (Desmet, Hekkert, & Jacobs, 2000, presented a unique nonverbal adaptation), and many current usability questionnaires include at least some adjective-based affect assessment items (i.e., the Questionnaire for User Interface Satisfaction (QUIS) (Chin, Diehl, & Norman, 1988)).

A second approach to questionnaire measurement of affect derives from dimensional theories of emotion and mood. Many researchers argue that two dimensions—arousal (activation) and valence (pleasant/unpleasant)—are nearly sufficient to describe the entire space of conscious emotional experience (Feldman, Barrett, & Russell, 1999). Lang (1995), for example, presented an interesting measurement scheme where subjects rate the arousal and valence of their current affective state by selecting among pictorial representations (rather than the standard number/word representation of degree). Watson, Clark, and Tellegen's (1988) Positive and Negative Affect Schedule (PANAS) is a popular dimensional measure of mood. Finally, to measure emotion as it occurs, with minimum interruption, some researchers have asked subjects to push one of a small number of buttons indicating their current emotional reaction during presentation of a stimulus (i.e., one button each for positive, negative, and neutral response (Breckler & Berman, 1991)).

Affect Recognition by Users

Computers are not the only (potential) affect recognizers in human–computer interactions. When confronted with an interface—particularly a social or character-based interface—users constantly monitor cues to the affective state of their interaction partner, the computer (though often nonconsciously; see Reeves & Nass, 1996). Creating natural and efficient interfaces requires not only recognizing emotion in users, but also expressing emotion. Traditional media creators have known for a long time that portrayal of emotion is a fundamental key to creating the "illusion of life" (Jones, 1990; Thomas & Johnson, 1981; for discussions of believable agents and emotion, see, i.e., Bates, 1994; Maldonado, Picard, & Hayes-Roth, 1998).

Facial expression and gesture are the two most common ways to manifest emotion in screen-based characters (Cassell et al., 2000; Kurlander, Skelly, & Salesin, 1996). Though animated expressions lack much of the intricacy found in human expressions, users are nonetheless capable of distinguishing emotions in animated characters (Cassell et al., 2000; Schiano, Ehrlich, Rahardja, & Sheridan, 2000). As with emotion recognition, Ekman and Friesen's (1977) Facial Action Coding System (FACS) is a commonly used and well-developed method for constructing affective expressions. One common strategy for improving accurate communication with animated characters is to exaggerate expressions, but whether this leads to corresponding exaggerated assumptions about the underlying emotion has not been studied.

Characters that talk can also use voice to communicate emotion (Nass & Gong, 2000). Prerecorded utterances are easily infused with affective tone, but are fixed and inflexible. Cahn (1990) has successfully synthesized affect-laden speech using a text-to-speech (TTS) system coupled with content-sensitive rules regarding appropriate acoustic qualities (including pitch, timing, and voice quality; see also Nass, Foehr, & Somoza, 2000). Users were able to distinguish among six different emotions with about 50% accuracy, which is impressive considering that people are generally only 60% accurate in recognizing affect in *human* speech (Scherer, 1981).

Finally, characters can indicate affective state verbally through word and topic choice, as well as explicit statements of affect (i.e., "I'm happy"). Characters, whose nonverbal and verbal expressions are distinctly mismatched, however, may be seen as awkward or even untrustworthy. In less extreme mismatched cases, recent evidence suggests that users will give precedence to nonverbal cues in judgments about affect (Nass et al., 2000). This finding is critical for applications in which characters/agents mediate interpersonal communication (i.e., in virtual worlds or when characters read email to a user), because the affective tone of a message may be inappropriately masked by the character's affective state. Ideally, in such computer-mediated communication contexts, emotion would be encoded into the message itself, either through explicit tagging of the message with affect, through natural language processing of the message, or through direct recognition of the sender's affective state during message composition (i.e., using autonomic nervous system or facial expression measures). Mediator characters could then display the appropriate nonverbal cues to match the verbal content of the message.

OPEN QUESTIONS

Beyond the obvious need for advancements in affect recognition and manifestation technology, it is our opinion that there are five important and remarkably unexplored areas for research in emotion and HCI:

1. *With which emotion should HCI designers be most concerned?*

Which emotion(s) should interface designers address first? The basic emotions, to the extent that they exist and can be identified, have the advantage of similarity across cultures and easy discriminability. Thus, designs that attempt to act upon or manipulate these dimensions may be the simplest to implement. However, within these basic emotions, little is known about their relative manipulability or manifestability—particularly within the HCI context—or their relative impact on individuals' attitudes and behaviors. Once one moves beyond the basic emotions, cultural and individual differences introduce further problems and opportunities.

2. *When and how should interfaces attempt to directly address users' emotions and basic needs (vs. application-specific goals)?*

If one views a computer or an interface merely as a tool, then interface design should solely focus on application-specific goals, assessed by such metrics as efficiency, learnability, and accuracy. However, if computers and interfaces are understood as a *medium*, then it becomes important to think about both uses and *gratifications* (Katz, Blumler, & Gurevitch, 1974; Rosengren, 1974; Rubin, 1986); that is, the more general emotional and basic needs that users bring to any interaction. Notions of "infotainment" or "edutainment" indicate one category of attempts to balance task and affect. However, there is little understanding of how aspects of interfaces that directly manipulate users' emotions compliment, undermine, or are orthogonal to aspects of interfaces that specifically address users' task needs.

3. *How accurate must emotion recognition be to be useful as an interface technique?*

Although humans are not highly accurate emotion detectors—the problem of "receiving accuracy" (Picard, 1997a, p. 120)—they nonetheless benefit from deducing other's emotions and acting on those deductions (Goleman, 1995). Clearly, however, a minimum threshold of accuracy is required before behavior based on emotion induction is appropriate. Very little is known about the level of confidence necessary before an interface can effectively act on a user's emotional state.

4. *When and how should users be informed that their affective states are being monitored and adapted to?*

When two people interact, there is an *implicit* assumption that each person is monitoring the other's emotional state and responding based on that emotional state. However, an explicit statement of this fact would be highly disturbing: "To facilitate our interaction, I will carefully and constantly monitor everything you say and do to discern your emotional state and respond based on that emotional state" or "I have determined that you are sad; I will now perform actions that will make you happier." However, when machines acquire and act upon information about users without making that acquisition and adaptation explicit, there is often a feeling of "surreptitiousness" or "manipulation." Furthermore, if emotion monitoring and adapting software are desired by consumers, there are clearly incentives for announcing and marketing these abilities. Because normal humans only exhibit implicit monitoring, the psychological literature is silent on the psychological and performance implications for awareness of emotional monitoring and adaptation.

5. *How does emotion play out in computer-mediated communication (CMC)?*

This chapter has focused on the direct relationship between the user and the interface. However, computers also are used to *mediate* interactions between people. In face-to-face encounters, affect not only creates richer interaction, but also helps to disambiguate meaning, allowing for more effective communication. Little is known, however, about the psychological effects of mediated affect, or the optimal strategies for encoding and displaying affective messages (see Maldonado & Picard, 1999; Rivera, Cooke, & Bauhs, 1996).

CONCLUSION

Though much progress has been made in the domain of "affective computing" (Picard, 1997a), more work is clearly necessary before interfaces that incorporate emotion recognition and manifestation can reach their full potential. Nevertheless, a careful consideration of affect in interaction design and testing can be instrumental in creating interfaces that are both efficient and effective, as well as enjoyable and satisfying. Designers and theorists, for even the simplest interfaces, are well advised to thoughtfully address the intimate and far-reaching linkages between emotion and HCI.

ACKNOWLEDGMENTS

James Gross, Heidi Maldonado, Roz Picard, and Don Roberts provided extremely detailed and valuable insights. Sanjoy Banerjee, Daniel Bochner, Dolores Canamero, Pieter Desmet, Ken Fabian, Michael J. Lyons, George Marcus, Laurel Margulis, Byron Reeves, and Aaron Sloman provided useful suggestions.

References

Averill, J. R. (1980). A contructionist view of emotion. In R. Plutchik & H. Kellerman (Eds.), *Emotion: Theory, research, and experience* (Vol. 1, pp. 305–339). New York: Academic Press.

Ball, G., & Breese, J. (2000). Emotion and personality in conversational agents. In J. Cassell, J. Sullivan, S. Prevost, & E. Churchill (Eds.), *Embodied conversational agents* (pp. 189–219). Cambridge, MA: The MIT Press.

Banse, R., & Scherer, K. (1996). Acoustic profiles in vocal emotion expression. *Journal of Personality and Social Psychology, 70*(3), 614–636.

Barrett, L. F., & Russell, J. A. (1999). The structure of current affect: Controversies and emerging consensus. *Current Directions in Psychological Science, 8*(1), 10–14.

Bates, J. (1994). The role of emotions in believable agents. *Communications of the ACM, 37*(7), 122–125.

Bates, J., Loyall, A. B., & Reilly, W. S. (1994). An architecture for action, emotion, and social behavior, *Artificial social systems: Fourth European workshop on modeling autonomous agents in a multi-agent world.* Berlin, Germany: Springer-Verlag.

Bishop, D., Jacks, H., & Tandy, S. B. (1993). The Structure of Temperament Questionnaire (STQ): Results from a U.S. sample. *Personality & Individual Differences, 14*(3), 485–487.

Bower, G. H. (1992). How might emotions affect learning? In C. Svenåke (Ed.), *The handbook of emotion and memory: Research and theory* (pp. 3–31). Hillsdale, NJ: Lawrence Erlbaum Associates.

Bower, G. H., & Forgas, J. P. (2000). Affect, memory, and social cognition. In E. Eich, J. F. Kihlstrom, G. H. Bower, J. P. Forgas, & P. M. Niedenthal (Eds.), *Cognition and emotion* (pp. 87–168). Oxford, U.K.: Oxford University Press.

Bower, G. H., Gilligan, S. G., & Monteiro, K. P. (1981). Selectivity of learning caused by affective states. *Journal of Experimental Psychology: General, 110*, 451–473.

Bozinovski, S., & Bozinovska, L. (2001). Self-learning agents: A connectionist theory of emotion based on crossbar value judgment. *Cybernetics and Systems, 32*, 5–6.

Breckler, S. T., & Berman, J. S. (1991). Affective responses to attitude objects: Measurement and validation. *Journal of Social Behavior and Personality, 6*(3), 529–544.

Cacioppo, J. T., Bernston, G. G., Klein, D. J., & Poehlmann, K. M. (1997). Psychophysiology of emotion across the life span. *Annual Review of Gerontology and Geriatrics, 17*, 27–74.

Cacioppo, J. T., Bernston, G. G., Larsen, J. T., Poehlmann, K. M., & Ito, T. A. (2000). The psychophysiology of emotion. In M. Lewis & J. M. Haviland-Jones (Eds.), *Handbook of emotions* (2nd ed., pp. 173–191). New York: The Guilford Press.

Cahn, J. E. (1990). The generation of affect in sythesized speech. *Journal of the American Voice I/O Society, 8*, 1–19.

Cannon, W. B. (1927). The James-Lange theory of emotions: A critical examination and an alternate theory. *American Journal of Psychology, 39*, 106–124.

Cassell, J., Sullivan, J., Prevost, S., & Churchill, E. (Eds.). (2000). *Embodied conversational agents.* Cambridge, MA: MIT Press.

Cassell, J., & Thorisson, K. (in press). The power of a nod and a glance: Envelope vs. emotional feedback in animated conversational agents. *Journal of Applied Artificial Intelligence.*

Chin, J. P., Diehl, V. A., & Norman, K. L. (1988). Development of an instrument measuring user satisfaction of the human-computer interface, *Proceedings of CHI '88 human factors in computing systems* (pp. 213–218). New York: ACM Press.

Clore, G. C. (1994). Why emotions are felt. In P. Ekman & R. J. Davidson (Eds.), *The nature of emotion: Fundamental questions* (pp. 103–111). New York: Oxford University Press.

Clore, G. C., & Gasper, K. (2000). Feeling is believing: Some affective influences on belief. In N. H. Frijda, A. S. R. Manstead, & S. Bem (Eds.), *Emotions and beliefs: How feelings influence thoughts* (pp. 10–44). Cambridge, UK: Cambridge University Press.

Clore, G. C., Wyer, R. S., Jr., Diened, B., Gasper, K., Gohm, C., & Isbell, L. (2001). Affective feelings as feedback: Some cognitive consequences. In L. L. Martin & G. C. Clore (Eds.), *Theories of mood and cognition: A user's handbook* (pp. 63–84). Mahwah, NJ: Lawrence Erlbaum Associates.

Collier, G. (1985). *Emotional expression.* Hillsdale, NJ: Lawrence Erlbaum Associates.

Cowie, R., & Douglas-Cowie, E. (1996). *Automatic statistical analysis of the signal and posidic signs of emotion in speech.* Paper presented at the Proceedings of the 4th International Conference on Spoken Language Processing (ICSLP-96), Oct 3–6, New Castle, DE.

Cowie, R., Douglas-Cowie, E., Tsapatsoulis, N., Votsis, G., Kollias, S., Fellenz, W., & Taylor, J. G. (2001). Emotion recognition in human–computer interaction. *IEEE Signal Processing Magazine, 18*(1), 32–80.

Damasio, A. R. (1994). *Descartes' error: Emotion, reason, and the human brain.* New York: Putnam Publishing Group.

Darwin, C. (1872/1998). *The expression of the emotions in man and animals.* London: HarperCollins.

Davidson, R. J. (1992). Anterior cerebral asymmetry and the nature of emotion. *Brain and Cognition, 20*, 125–151.

Davidson, R. J. (1994). On emotion, mood, and related affective constructs. In P. Ekman & R. J. Davidson (Eds.), *The nature of emotion* (pp. 51–55). New York: Oxford University Press.

Detenber, B. H., & Reeves, B. (1996). A bio-informational theory of emotion: Motion and image size effects on viewers. *Journal of Communication, 46*(3), 66–84.

Duncker, K. (1945). On problem-solving. *Psychological Monographs, 58*(Whole No. 5).

Ekman, P. (1992). An argument for basic emotions. *Cognition and Emotion, 6*(3/4), 169–200.

Ekman, P. (1994). All emotions are basic. In P. Ekman & R. J. Davidson (Eds.), *The nature of emotion: Fundamental questions* (pp. 7–19). New York: Oxford University Press.

Ekman, P., & Davidson, R. J. (Eds.). (1994). *The nature of emotion.* New York: Oxford University Press.

Ekman, P., & Friesen, W. V. (1975). *Unmasking the face.* Englewood Cliffs, NJ: Prentice-Hall.

Ekman, P., & Friesen, W. V. (1977). *Manual for the Facial Action Coding System.* Palo Alto: Consulting Psychologist Press.

Ekman, P., Levenson, R. W., & Friesen, W. V. (1983). Autonomic nervous system activity distinguishes among emotions. *Science, 221*, 1208–1210.

Elliott, C., & Brzezinski, J. (1998, Summer). Autonomous agents and synthetic characters. *AI magazine*, 13–30.

Ellis, H. C., & Moore, B. A. (1999). Mood and memory. In T. Dalgleish & M. J. Power (Eds.), *Handbook of cognition and emotion* (pp. 193–210). New York: John Wiley & Sons.

Ellsworth, P. C. (1994). Some reasons to expect universal antecedents of emotion. In P. Ekman & R. J. Davidson (Eds.), *The nature of emotion: Fundamental questions* (pp. 150–154). New York: Oxford University Press.

Erber, R., & Erber, M. W. (2001). Mood and processing: A view from a self-regulation perspective. In L. L. Martin & G. C. Clore (Eds.), *Theories of mood and cognition: A user's handbook* (pp. 63–84). Mahwah, NJ: Lawrence Erlbaum Associates.

Essa, I. A., & Pentland, A. P. (1997). Coding, analysis, interpretation, and recognition of facial expressions. *IEEE Transactions on Pattern Analysis and Machine Intelligence, 19*(7), 757–763.

Feldman Barrett, L., & Russell, J. A. (1999). The structure of current affect: Controversies and emerging consensus. *Current Directions in Psychological Science, 8*(1), 10–14.

Fogg, B. J. (1998). Charismatic computers: Creating more likable and persuasive interactive technologies by leveraging principles from social psychology. *Dissertation Abstracts International Section A: Humanities & Social Sciences, 58*(7-A), 2436.

Forgas, J. P. (1995). Mood and judgment: The Affect Infusion Model (AIM). *Psychological Bulletin, 117*, 39–66.

Friedman, B., & Kahn, P. H., Jr. (1997). Human agency and responsible computing: Implications for computer system design. In B. Friedman (Ed.), *Human values and the design of computer technology* (pp. 221–235). Stanford, CA: CSLI Publications.

Frijda, N. H. (1986). *The emotions*. Cambridge, UK: Cambridge University Press.

Frijda, N. H. (1988). The laws of emotion. *American Psychologist, 43*(5), 349–358.

Frijda, N. H. (1994). Varieties of affect: Emotions and episodes, moods, and sentiments. In P. Ekman & R. J. Davidson (Eds.), *The nature of emotion* (pp. 59–67). New York: Oxford University Press.

Gardano, A. C. (1986). Cultural influence on emotional response to color: A research study comparing Hispanics and non-Hispanics. *American Journal of Art Therapy, 23*, 119–124.

Gerard, R. (1957). *Differential effects of colored lights on psychophysiological functions*. Unpublished doctoral dissertation, University of California, Los Angeles, CA.

Gerard, R. (1958, July 13). Color and emotional arousal [Abstract]. *American Psychologist, 13*, 340.

Goleman, D. (1995). *Emotional intelligence*. New York: Bantam Books.

Gregory, A. H., Worrall, L., & Sarge, A. (1996). The development of emotional responses to music in young children. *Motivation and Emotion, 20*(4), 341–348.

Gross, J. J. (1998). Antecedent- and response-focused emotion regulation: Divergent consequences for experience, expression, and physiology. *Journal of Personality and Social Psychology, 74*, 224–237.

Gross, J. J. (1999). Emotion and emotion regulation. In L. A. Pervin & O. P. John (Eds.), *Handbook of personality: Theory and research* (2nd ed., pp. 525–552). New York: Guildford.

Hatfield, E., Cacioppo, J. T., & Rapson, R. L. (1994). *Emotional contagion*. Paris/New York: Editions de la Maison des Sciences de l'Homme and Cambridge University Press (jointly published).

Heller, W. (1990). The neuropsychology of emotion: Developmental patterns and implications for psychpathology. In N. L. Stein, B. Leventhal, & T. Trabasso (Eds.), *Psychological and biological approaches to emotion* (pp. 167–211). Hillsdale, NJ: Lawrence Erlbaum Associates.

Henry, D. L., & Jacobs, K. W. (1978). Color eroticism and color preference. *Perceptual and Motor Skills, 47*, 106.

Heuer, F., & Reisberg, D. (1992). Emotion, arousal, and memory for detail. In C. Sven-Åke (Ed.), *The handbook of emotion and memory: Research and theory* (pp. 3–31). Hillsdale, NJ: Lawrence Erlbaum Associates.

Hirt, E. R., Melton, R. J., McDonald, H. E., & Harackiewicz, J. M. (1996). Processing goals, task interest, and the mood-performance relationship: A mediational analysis. *Journal of Personality and Social Psychology, 71*, 245–261.

Isen, A. M. (2000). Positive affect and decision making. In M. Lewis & J. M. Haviland-Jones (Eds.), *Handbook of emotions* (2nd ed., pp. 417–435). New York: The Guilford Press.

Isen, A. M., Daubman, K. A., & Nowicki, G. P. (1987). Positive affect facilitates creative problem solving. *Journal of Personality and Social Psychology, 52*(6), 1122–1131.

Isen, A. M., Rosenzweig, A. S., & Young, M. J. (1991). The influence of positive affect on clinical problem solving. *Medical Decision Making, 11*(3), 221–227.

Izard, C. E. (1972). *Patterns of emotions*. New York: Academic Press.

Izard, C. E. (1992). Basic emotions, relations among emotions, and emotion-cognition relations. *Psychological Review, 99*(3), 561–565.

James, W. (1884). What is an emotion? *Mind, 9*, 188–205.

Jones, C. (1990). *Chuck amuck: The life and times of an animated cartoonist*. New York: Avon Books.

Katz, E., Blumler, J. G., & Gurevitch, M. (1974). Utilization of mass communication by the individual. In J. G. Blumler & E. Katz (Eds.), *The uses of mass communcations: Current perspectives on gratifications research* (pp. 19–32). Beverly Hills, CA: Sage.

Klein, J., Moon, Y., & Picard, R. W. (1999). *This computer responds to user frustration* (pp. 242–243). Paper presented at the Human factors in computing systems: CHI'99 extended abstracts, New York.

Kleinginna, P. R., Jr., & Kleinginna, A. M. (1981). A categorized list of emotion definitions, with suggestions for a consensual definition. *Motivation and Emotion, 5*(4), 345–379.

Kurlander, D., Skelly, T., & Salesin, D. (1996). *Comic chat*. Paper presented at the Proceedings of SIGGRAPH'96: International conference on computer graphics and interactive techniques, Aug 4–9, New York.

Lang, P. J. (1995). The emotion probe. *American Psychologist, 50*(5), 372–385.

Lazarus, R. S. (1991). *Emotion and adaptation*. New York: Oxford University Press.

LeDoux, J. E. (1995). Emotion: Clues from the brain. *Annual Review of Psychology, 46*, 209–235.

LeDoux, J. E. (1996). *The emotional brain*. New York: Simon & Schuster.

LeDoux, J. E., & Phelps, E. A. (2000). Emotional networks in the brain. In M. Lewis & J. M. Haviland-Jones (Eds.), *Handbook of emotions* (pp. 157–172). New York: The Guilford Press.

Levenson, R. W. (1988). Emotion and the autonomic nervous system: A prospectus for research on autonomic specificity. In H. Wagner (Ed.), *Social psychophysiology: Perspectives on theory and clinical applications* (pp. 17–42). London: Wiley.

Levenson, R. W., Ekman, P., & Friesen, W. V. (1990). Voluntary facial action generates emotion-specific autonomic nervous system activity. *Psychophysiology, 27*, 363–384.

Levy, B. I. (1984). Research into the psychological meaning of color. *American Journal of Art Therapy, 23*, 58–62.

Lisetti, C. L., & Schiano, D. J. (2000). Automatic facial expression interpretation: Where human–computer interaction, artificial intelligence and cognitive science intersect [Special issue on Facial Information Processing: A Multidisciplinary Perspective]. *Pragmatics and Cognition, 8*(1), 185–235.

Lyons, M., Akamatsu, S., Kamachi, M., & Gyoba, J. (1998). Coding facial expressions with gabor wavelets, *Proceedings of the Third IEEE international conference on automatic face and gesture recognition* (pp. 200–205). New York: Avon Books.

Malamuth, N. (1996). Sexually explicit media, gender differences, and evolutionary theory. *Journal of Communication, 46*, 8–31.

Maldonado, H., & Picard, A. (1999). The Funki Buniz playground: Facilitating multi-cultural affective collaborative play [Abstract]. *CHI99 extended abstracts* (pp. 328–329).

Maldonado, H., Picard, A., & Hayes-Roth, B. (1998). Tigrito: A high affect virtual toy. *CHI98 Summary* (pp. 367–368).

Martinez, A. M. (2000). *Recognition of partially occluded and/or imprecisely localized faces using a probabilistic approach*. Paper presented at the Proceedings of IEEE Computer Vision and Pattern Recognition (CVPR '2000). June 13–15, Hilton Head, SC, USA.

Maslow, A. H. (1968). *Toward a psychology of being*. New York: D. Van Nostrand Company.

McNair, D. M., Lorr, M., & Droppleman, L. F. (1981). *Manual of the Profile of Mood States*. San Diego, CA: Educational and Industrial Testing Services.

Morkes, J., Kernal, H. K., & Nass, C. (2000). Effects of humor in task-oriented human–computer interaction and computer-mediated communication: A direct test of SRCT theory. *Human–Computer Interaction, 14*(4), 395–435.

Murphy, S. T., & Zajonc, R. B. (1993). Affect, cognition, and awareness: Affective priming with suboptimal and optimal stimulus. *Journal of Personality and Social Psychology, 64*, 723–739.

Murray, I. R., & Arnott, J. L. (1993). Toward the simulation of emotion in synthetic speech: A review of the literature on human vocal emotion. *Journal Acoustical Society of America, 93*(2), 1097–1108.

Murray, N., Sujan, H., Hirt, E. R., & Sujan, M. (1990). The influence of mood on categorization: A cognitive flexibility interpretation. *Journal of Personality and Social Psychology, 59*, 411–425.

Nass, C., Foehr, U., & Somoza, M. (2000). *The effects of emotion of voice in synthesized and recorded speech*. Unpublished manuscript.

Nass, C., & Gong, L. (2000). Speech interfaces from an evolutionary perspective. *Communications of the ACM, 43*(9), 36–43

Neese, R. M. (1990). Evolutionary explanations of emotions. *Human Nature, 1*(3), 261–289.

Neumann, R., & Strack, F. (2000). "Mood contagion": The automatic transfer of mood between persons. *Journal of Personality and Social Psychology, 79*(2), 211–223.

Newhagen, J., & Reeves, B. (1991). Emotion and memory responses to negative political advertising. In F. Biocca (Ed.), *Televions and political advertising: Psychological processes* (pp. 197–220). Hillsdale, NJ: Lawrence Erlbaum.

Newhagen, J., & Reeves, B. (1992). This evening's bad news: Effects of compelling negative television news images on memory. *Journal of Communication, 42*, 25–41.

Niedenthal, P. M., Setterlund, M. B., & Jones, D. E. (1994). Emotional organization of perceptual memory. In P. M. Niedenthal & S. Kitayama (Eds.), *The heart's eye* (pp. 87–113). San Diego, CA: Academic Press, Inc.

Norman, D. (1990). *The design of everyday things*. Garden City, NJ: Doubleday.

Oatley, K., & Johnson-Laird, P. N. (1987). Towards a cognitive theory of emotions. *Cognition and Emotion, 1*(1), 29–50.

Ortony, A., Clore, G. C., & Collins, A. (1988). *The cognitive structure of emotions*. Cambridge, MA: Cambridge University Press.

Ortony, A., & Turner, T. J. (1990). What's basic about emotions. *Psychological Review, 97*(3), 315–331.

Panksepp, J. (1992). A critical role for "affective neuroscience" in resolving what is basic about basic emotions. *Psychological Review, 99*(3), 554–560.

Parrott, G. W., & Spackman, M. P. (2000). Emotion and memory. In M. Lewis & J. M. Haviland-Jones (Eds.), *Handbook of emotions* (2nd ed., pp. 476–490). New York: The Guilford Press.

Pecjak, V. (1970). Verbal synthesis of colors, emotions and days of the week. *Journal of Verbal Learning and Verbal Behavior, 9*, 623–626.

Peretti, P. O. (1974). Color model associations in young adults. *Perceptual and Motor Skills, 39*, 715–718.

Picard, R. W. (1997a). *Affective computing*. Cambridge, MA: The MIT Press.

Picard, R. W. (1997b). Does HAL cry digital tears? Emotions and computers. In D. G. Stork (Ed.), *Hal's Legacy: 2001's Computer as Dream and Reality* (pp. 279–303). Cambridge, MA: The MIT Press.

Picard, R. W., Vyzas, E., & Healey, J. (in press). Toward machine emotional intelligence: Analysis of affective physiological state. *IEEE Transactions on Pattern Analysis and Machine Intelligence*.

Plutchik, R., & Kellerman, H. (Eds.). (1989). *Emotion: Theory, research, and experience* (Vol. 4: The Measurement of Emotions). San Diego, CA: Academic Press, Inc.

Reeves, B., Biocca, F., Pan, Z., Oshagan, H., & Richards, J. (1989). *Unconscious processing and priming with pictures: Effects on emotional attributions about people on television*. Unpublished manuscript.

Reeves, B., & Nass, C. (1996). *The media equation: How people treat computers, television, and new media like real people and places*. New York: Cambridge University Press.

Reeves, B., Newhagen, J., Maibach, E., Basil, M. D., & Kurz, K. (1991). Negative and positive television messages: Effects of message type and message content on attention and memory. *American Behavioral Scientist, 34*, 679–694.

Reuter-Lorenz, P., & Davidson, R. J. (1981). Differential contributions of the two cerebral hemispheres to the perception of happy and sad faces. *Neuropsychologia, 19*(4), 609–613.

Rivera, K., Cooke, N. J., & Bauhs, J. A. (1996). The effects of emotional icons on remote communication. *CHI '96 interactive posters*, 99–100.

Roseman, I. J., Antoniou, A. A., & Jose, P. E. (1996). Appraisal determinants of emotions: Constructing a more accurate and comprehensive theory. *Cognition and Emotion, 10*(3), 241–277.

Rosenfeld, A. (1997). Eyes for computers: How HAL could "see". In D. G. Stork (Ed.), *Hal's legacy: 2001's Computer as dream and reality* (pp. 211–235). Cambridge, MA: The MIT Press.

Rosengren, K. E. (1974). Uses and gratifications: A paradigm outlined. In J. G. Blumler & E. Katz (Eds.), *The uses of mass communications: Current perspectives on gratifications research* (pp. 269–286). Beverly Hills, CA: Sage.

Rubin, A. M. (1986). Uses, gratifications, and media effects research. In J. Bryant & D. Zillman (Eds.), *Perspectives on media effects* (pp. 281–301). Hillsdale, NJ: Lawrence Erlbaum Associates.

Schachtel, E. J. (1943). On color and affect. *Psychiatry, 6*, 393–409.

Schachter, S., & Singer, J. E. (1962). Cognitive, social, and physiological determinants of emotional state. *Psychological Review, 69*(5), 379–399.

Scherer, K. (1981). Speech and emotional states. In J. K. Darby (Ed.), *Speech evaluation in psychiatry* (pp. 189–220). New York: Grune & Stratton.

Scherer, K. R. (1988). Criteria for emotion-antecedent appraisal: A review. In V. Hamilton, G. H. Bower, & N. H. Frijda (Eds.), *Cognitive perspectives on emotion and motivation* (pp. 89–126). Dordrecht, the Netherlands: Kluwer Academic.

Scherer, K. R. (1989). Vocal measurement of emotion. In R. Plutchik & H. Kellerman (Eds.), *Emotion: Theory, research, and experience* (Vol. 4, pp. 233–259). San Diego, CA: Academic Press, Inc.

Scheutz, M., Sloman, A., & Logan, B. (2000, November 11–12). *Emotional states and realistic agent behavior*. Paper presented at the GAME-ON 2000, Imperial College, London.

Schiano, D. J., Ehrlich, S. M., Rahardja, K., & Sheridan, K. (2000, April 1–6). *Face to interFace: Facial affect in (hu)man and machine*. Paper presented at the Human factors in computing systems: CHI'00 conference proceedings, New York.

Schwartz, N., & Bless, H. (1991). Happy and mindless, but sad and smart?: The impact of affective states on analytic reasoning. In J. P. Forgas (Ed.), *Emotion and social judgment* (pp. 55–71). Oxford, U.K.: Pergamon.

Shweder, R. A. (1994). "You're not sick, you're just in love": Emotions as an interpretive system. In P. Ekman & R. J. Davidson (Eds.), *The nature of emotions* (pp. 32–44). New York: Oxford University Press.

Skinner, E. A. (1995). *Perceived control, motivation, & coping*. Newbury Park, CA: Sage Publications.

Thayer, R. E. (1989). *The biopsychology of mood and arousal*. New York: Oxford University Press.

Thomas, F., & Johnson, O. (1981). *The illusion of life: Disney animation*. New York: Abbeville Press.

Thorson, E., & Friestad, M. (1985). The effects on emotion on episodic memory for television commercials. In P. Cafferata & A. Tybout (Eds.), *Advances in consumer psychology* (pp. 131–136). Lexington, MA: Lexington.

Tian, Y.-L., Kanade, T., & Cohn, J. F. (2001). Recognizing action units for facial expression analysis. *IEEE Transactions on Pattern Analysis and Machine Intelligence, 23*(2), 1–19.

Tooby, J., & Cosmides, L. (1990). The past explains the present: Emotional adaptations and the structure of ancestral environments. *Ethology and Sociobiology, 11*, 407–424.

Ucros, C. G. (1989). Modd state-dependent memory: A meta-analysis. *Cognition and Emotion, 3*(139–167).

Voelker, D. (1994). *The effects of image size and voice volume on he evlauaiton of represented faces*. Unpublished Dissertation, Stanford University, Stanford, CA.

Walters, J., Apter, M. J., & Svebak, S. (1982). Color preference, arousal, and theory of psychological reversals. *Motivation and Emotion, 6*(3), 193–215.

Watson, D., Clark, L. A., & Tellegen, A. (1988). Development and validation of brief measures of positive and negative affect: The PANAS scales. *Journal of Personality and Social Psychology, 54*, 128–141.

Wegner, D. M. (1994). Ironic processes of mental control. *Psychological Review, 101*, 34–52.

Wegner, D. M., Bargh, J. A., Gilbert, D. T., Fiske, S. T., et al. (1998). Control and automaticity in social life, *The handbook of social psychology* (4th ed., Vol. 1). Boston: McGraw-Hill.

Wexner, L. (1954). The degree to which colors (hues) are associated with mood-tones. *Journal of Applied Psychology, 38*, 432–435.

Wierzbicka, A. (1992). Talking about emotions: Semantics, culture, and cognition. *Cognition and Emotion, 6*(3/4), 285–319.

Yacoob, Y., & Davis, L. S. (1996). Recognizing human facial expressions from long image sequences using optical flow. *IEEE Transactions on Pattern Analysis and Machine Intelligence, 18*(6), 636–642.

Zillmann, D. (1991). Television viewing and physiological arousal. In J. Bryant & D. Zillman (Eds.), *Responding to the screen: Reception and reaction processes* (pp. 103–133). Hillsdale, NJ: Lawrence Erlbaum Associates.

·5·

COGNITIVE ARCHITECTURE

Michael D. Byrne
Rice University

Designing interactive computer systems to be efficient and easy to use is important so that people in our society may realize the potential benefits of computer-based tools. . . . Although modern cognitive psychology contains a wealth of knowledge of human behavior, it is not a simple matter to bring this knowledge to bear on the practical problems of design—to build an applied psychology that includes theory, data, and knowledge. (Card, Moran, & Newell, 1983, p. vii)

Integrating theory, data, and knowledge about cognitive psychology and human performance in a way that is useful for guiding design in HCI is still not a simple matter. However, there have been significant advances since Card, Moran, and Newell (1983) wrote the previous passage. One of the key advances is the development of cognitive architectures, the subject of this chapter. The chapter will first consider what it is to be a cognitive architecture and why cognitive architectures are relevant for HCI. In order to detail the present state of cognitive architectures in HCI, it is important to consider some of the past uses of cognitive architectures in HCI research. Then, three architectures actively in use in the research community (LICAI/CoLiDeS, EPIC, and ACT-R) and their application to HCI will be examined. The chapter will conclude with a discussion of the future of cognitive architectures in HCI.

WHAT ARE COGNITIVE ARCHITECTURES?

Most dictionaries will list several different definitions for the word *architecture*. For example, dictionary.com lists among them "a style and method of design and construction," e.g., Byzantine architecture; "orderly arrangement of parts; structure," e.g., the architecture of a novel; and one from computer science: "the overall design or structure of a computer system." What, then, would something have to be to qualify as a "cognitive architecture"? It is something much more in the latter senses of the word *architecture,* an attempt to describe the overall structure and arrangement of a very particular thing, the human cognitive system. A cognitive architecture is a broad theory of human cognition based on a wide selection of human experimental data, and implemented as a running computer simulation program. Young (Gray, Young, & Kirschenbaum, 1997; Ritter & Young, 2001) defined a cognitive architecture as an embodiment of "a scientific hypothesis about those aspects of human cognition that are relatively constant over time and relatively independent of task."

This idea has been a part of cognitive science since the early days of cognitive psychology and artificial intelligence, as manifested in the General Problem Solver or GPS (Newell & Simon, 1963), one of the first successful computational cognitive models. These theories have progressed a great deal since GPS and are gradually becoming broader. One of the best descriptions of the vision for this area is presented in Newell's (1990) book *Unified Theories of Cognition.* In it, Newell argued that the time has come for cognitive psychology to stop collecting disconnected empirical phenomena and begin seriously considering theoretical unification in the form of computer simulation models. Cognitive architectures are attempts to do just this.

Cognitive architectures are distinct from engineering approaches to artificial intelligence, which strive to construct intelligent computer systems by whatever technologies best serve that purpose. Cognitive architectures are designed to simulate human intelligence in a humanlike way (Newell, 1990). For example, the chess program that defeated Kasparov, Deep Blue, would not qualify as a cognitive architecture because it does not solve the problem (chess) in a humanlike way. Deep Blue uses massive search of the game space, while human experts generally look only a few moves ahead, but concentrate effectively on quality moves.

Cognitive architectures differ from traditional research in psychology in that work on cognitive architecture is integrative—that is, architectures include mechanisms to support attention, memory, problem solving, decision making, learning, and so forth. Most theorizing in psychology follows a divide-and-conquer strategy that tends to generate highly specific theories of a very limited range of phenomena; this has changed little since the 1970s (Newell, 1973). This limits the usefulness of such theories for an applied domain such as HCI, where users employ a wide range of cognitive capabilities in even simple tasks. Instead of asking, "How can we describe this isolated phenomenon?" people working with cognitive architectures can ask, "How does this phenomenon fit in with what we already know about other aspects of cognition?"

Another important feature of cognitive architectures is that they specify only the human "virtual machine," the fixed architecture. A cognitive architecture alone cannot do anything. Generally, the architecture has to be supplied with the knowledge needed to perform a particular task. The combination of an architecture and a particular set of knowledge is generally referred to as a "computational cognitive model," or just a "model." In general, it is possible to construct more than one model for any particular task. The specific knowledge incorporated into a particular model is determined by the modeler. Because the relevant knowledge must be supplied to the architecture, the knowledge engineering task facing modelers attempting to model performance on complex tasks can be formidable.

A third centrally important feature of cognitive architectures is that they are software artifacts constructed by human programmers. This has a number of relevant ramifications. First, a model of a task constructed in a cognitive architecture is runnable and produces a sequence of behaviors. These behavior sequences can be compared with the sequences produced by human users to help assess the quality of a particular model. They may also provide insight into alternate ways to perform a task—that is, they may show possible strategies that are not actually utilized by the people performing the task. This can be useful in guiding interface design as well. Another feature of many architectures is that they enable the creation of quantitative models. For instance, the model may say more than just "click on button A and then menu B," but may include the time between the two clicks as well. Models based on cognitive architectures can produce execution times, error rates, and even learning curves. This is a major strength of cognitive architectures as an approach to certain kinds of HCI problems and will be discussed in more detail in the next section.

On the other hand, cognitive architectures are large software systems, which are often considered difficult to construct and maintain. Individual models are also essentially programs, written in the "language" of the cognitive architecture. Thus, individual modelers need to have solid programming skills.

Finally, cognitive architectures are not in wide use among HCI practitioners. Right now, they exist primarily in academic research laboratories. One of the barriers for practitioners is that learning and using most cognitive architectures is itself generally a difficult task; however, this is gradually changing, and some of the issues being addressed in this regard will be discussed in section 4. Furthermore, even if cognitive architectures are not in wide use by practitioners, this does not mean that they are irrelevant to practitioners. The next section highlights why cognitive architectures are relevant to a wide HCI audience.

RELEVANCE TO HUMAN–COMPUTER INTERACTION

For some readers, the relevance of models that produce quantitative predictions about human performance will be obvious. For others, this may be less immediately clear. Cognitive architectures (a) are relevant to usability as an engineering discipline, (b) have several HCI-relevant applications in computing systems, and (c) serve an important role in HCI as a theoretical science.

At nearly all HCI-oriented conferences, and many online resources, there are areas where corporations recruit HCI professionals. A common job title in these forums is "usability engineer." Implicit in this title is the view that usability is, at least in part, an engineering enterprise. In addition, while people with this job title are certainly involved in product design, there is a sense in which most usability engineering would not be recognized as engineering by people trained in more traditional engineering disciplines, such as electrical or aerospace engineering. In traditional engineering disciplines, design is generally guided at least in part by quantitative theory. Engineers have at their disposal hard theories of the domain in which they work, and these theories allow them to derive quantitative predictions. Consider an aerospace engineer designing a wing. Like a usability engineer, the aerospace engineer will not start with nothing; a preexisting design often provides a starting point; however, when the aerospace engineer decides to make a change in that design, there is usually quantitative guidance about how the performance of the wing will change because of the change in design. This guidance, while quantitative, is not infallible, hence the need for evaluation tools such as wind tunnels. This is similar to the usability engineer's usability test. Unlike usability testing, however, the aerospace engineer has some quantitative idea about what the outcome of the test will be, and this is not guided simply by intuition and experience, but by a mathematical theory of aerodynamics. In fact, this theory is now so advanced that few wind tunnels are built anymore. Instead, they are being replaced by computer simulations based on "computational fluid dynamics," an outcome of the application of computational techniques to complex problems in aerodynamics. Simulations have not entirely replaced wind tunnels, but the demand for wind tunnel time has clearly been affected by this development.

For the most part, the usability engineer lacks the quantitative tools available to the aerospace engineer. Every design must be subjected to its own wind tunnel (usability) test, and the engineer has little guidance about what to expect other than from intuition and experience with similar tests. While intuition and experience can certainly be valuable guides, they often fall short of more "hard" quantitative methods. Perhaps the engineer can intuit that interface "X" will allow users to complete tasks faster than with interface "Y," but how much faster? Ten percent? Twenty percent? Even small savings in execution times can add up to large financial savings for organizations when one considers the scale of the activity. The paradigm example is the telephone operators studied by Gray, John, and Atwood (1993), where even a second saved on an average call would save the telephone company millions of dollars.

Computational models based on cognitive architectures have the potential to provide detailed quantitative answers, and for more than just execution times. Error rates, transfer of knowledge, learning rates, and other kinds of performance measures are all metrics than can often be provided by architecture-based models. Even if such models are not always precisely accurate in an absolute sense, they may still be useful in a comparative sense. For example, if a usability engineer is comparing interface A with interface B and the model at his or her disposal does not accurately predict the absolute times to complete some set of benchmark tasks, it may still accurately capture the ordinal difference between the two interfaces, which may be enough.

Additionally, there are certain circumstances when usability tests are either impractical, prohibitively costly, or both. For example, access to certain populations such as physicians or astronauts may be difficult or expensive, so bringing them in for repeated usability tests may not be feasible. While developing a model of a pilot or an air traffic controller performing an expert task with specialized systems may be difficult at first, rerunning that model to assess a change made to the user interface should be much more straightforward than performing a new usability test for each iteration of the system. This is possible only with a quantitatively realistic model of the human in the loop, one that can produce things such as execution times and error rates. Computational models can, in principle, act as surrogate users in usability testing, even for special populations.

Of course, some of these measures can be obtained through other methods such as GOMS (Goals, Operators, Methods, and Selection rules) analysis (Card et al., 1983; John & Kieras, 1996) or cognitive walkthrough (Polson, Lewis, Rieman, & Wharton, 1992); however, these techniques were originally grounded in the same ideas as some prominent cognitive architectures and are essentially abstractions of the relevant architectures for particular HCI purposes. In addition, architecture-based computational models provide things that GOMS models and cognitive walkthroughs do not. First, models are executable and generative. A GOMS analysis, on the other hand, is a description of the procedural knowledge the user must have and the sequence of actions that must be performed to accomplish a specific task instance, while the equivalent computational model actually generates the behaviors, often in real time or faster. Equally importantly, computational models have the capacity to be reactive in real time. So while it may be possible to construct a GOMS model that describes the knowledge necessary and the time it will take an operator to classify a new object on an air traffic controller's screen, a paper-and-pencil GOMS model cannot actually execute the procedure in response to the appearance of such an object. A running computational model, on the other hand, can.

Because of this property, architecture-based computational models have some other important uses beyond acting as virtual users in usability tests. One such use is in intelligent tutoring systems (ITSs). Consider the Lisp tutor (Anderson, Conrad, & Corbett, 1989). This tutoring system contained an architecture-based running computational model of the knowledge necessary to implement the relevant Lisp functions, as well as a module for assessing which pieces of this knowledge were mastered by the student. Because the model was executable, it could predict what action the student would take if the student had correct knowledge of how to solve the problem. When the student took a different action, this told the ITS that the student was missing one or more relevant pieces of knowledge. The student could then be given feedback about what knowledge was missing or incomplete, and problems that exercise this knowledge could be selected by the ITS. It is possible to generate more effective educational experiences by identifying students' knowledge, as well as the gaps in that knowledge. Problems that contain knowledge students have already mastered can be avoided, so as not to bore students with things they already know. This frees up students to concentrate on the material they have not yet mastered, resulting in improved learning (Anderson et al., 1989). While the Lisp tutor is an old research system, ITSs based on the same underlying cognitive architecture with the same essential methodology have been developed for more pressing educational needs such as algebra and geometry and are now sold commercially (see http://www.carnegielearning.com).

Another HCI-relevant application for high-fidelity cognitive models is populating simulated worlds or situations. For example, training an F-16 fighter pilot is expensive, even in a simulator, because that trainee needs to face realistic opposition. Realistic opposition consists of other trained pilots, so training one person requires taking several trained pilots away from their normal duties (e.g., flying airplanes on real missions). This is difficult and expensive. If, however, the other pilots could be realistically simulated, the trainee could face opposition that would have useful training value without having to remove already trained pilots from their duties. Many training situations such as this exist, where the only way to train someone is to involve multiple human experts who must all be taken away from their regular jobs. The need for expensive experts can potentially be eliminated (or at least reduced), however, by using architecturally based cognitive models in place of the human experts. The U.S. military has already started to experiment with just such a scenario (Jones, Laird, Nielsen, Coulter, Kenny, & Koss, 1999). Having realistic opponents is desirable in domains other than training as well, such as video games. Besides things like texture-mapped 3-D graphics, one of the features often used to sell games is network play. This enables players to engage opponents whose capabilities are more comparable to their own than typical computer-generated opponents; however, even with network play, it is not always possible for a game to find an appropriate opponent. If the computer-generated opponent were a more high-fidelity simulation of a human in terms of cognitive and perceptual-motor capabilities, then video game players would have no difficulty finding appropriate opponents without relying on network play. While this might not be the most scientifically interesting use of cognitive architectures, it seems inevitable that cognitive architectures will be used in this way.

Cognitive architectures are also theoretically important to HCI as an interdisciplinary field. Many people (including some cognitive psychologists) find terms from cognitive psychology such as "working memory" or "mental model" vague and ill defined. A harsh evaluation of explanations relying on such terms is found in Salthouse (1988, p. 3): "It is quite possible that interpretations relying on such nebulous constructs are only masquerading ignorance in what is essentially vacuous terminology." Computational cognitive architectures, on the other hand, require explicit specifications of the semantics of theoretical terms. Even if the architectures are imperfect descriptions of the human cognitive system, they are at a minimum well specified and therefore clearer in what they predict than strictly verbal theories.

A second theoretical advantage of computational theories such as cognitive architectures is that they provide a window into how the theory actually works. As theories grow in size and number of mechanisms, the interactions of those mechanisms becomes increasingly difficult to predict analytically. Computer simulations permit relatively rapid evaluations of complex mechanisms and their interactions. (For an excellent discussion of this topic, see Simon, 1996.) Another problem with theories based solely on verbal descriptions is their internal coherence can be very difficult to assess, while such assessment is much more straightforward with computational models. Verbal theories can easily hide subtle (and not so subtle) inconsistencies that make them poor scientific theories. Computational models, on the other hand, force explanations to have a high level of internal coherence; theories that are not internally consistent are typically impossible to implement on real machines.

Finally, HCI is an interdisciplinary field, and thus theories that are fundamentally interdisciplinary in nature are appropriate. Cognitive architectures are such theories, combining computational methods and knowledge from the artificial intelligence end of computer science with data and theories from cognitive psychology. While cognitive psychology and computer science are certainly not the only disciplines that participate in HCI, they are two highly visible forces in the field. Psychological theories that are manifested as executable programs should be less alien to people with a computer science background than more traditional psychological theories.

Thus, cognitive architectures are clearly relevant to HCI at a number of levels. This fact has not gone unnoticed by the HCI research community. In fact, cognitive architectures have a long history in HCI, dating back to the original work of Card, Moran, and Newell (1983).

BRIEF LOOK AT PAST SYSTEMS IN HCI

The total history of cognitive architectures and HCI would be far too long to document in a single chapter; however, it is possible to touch on some highlights. While not all of the systems described in this section qualify as complete cognitive architectures, they all share intellectual history with more current architectures and influenced their development and use in HCI. Finally, many of the concepts developed in these efforts are still central parts of the ongoing research on cognitive architecture. In addition, there is a natural starting point:

The Model Human Processor (MHP) and GOMS

The Psychology of Human–Computer Interaction (Card, Moran, & Newell, 1983) is clearly a seminal work in HCI, one of the defining academic works in the early days of the field. While that work did not produce a running cognitive architecture, it was clearly in the spirit of cognitive architectures and was quite influential in the development of current cognitive architectures. Two particular pieces of that work are relevant here, the Model Human Processor (MHP) and GOMS.

The MHP represents a synthesis of the literature on cognitive psychology and human performance up to that time, and sketches the framework around which a cognitive architecture could be implemented. The MHP is a system with multiple memories and multiple processors, and many of the properties of those processors and memories are described in some detail. (See Fig. 5.1.) Card, Moran, and Newell (1983) also specified the interconnections of the processors and a number of general operating principles. In this system, there are three processors:

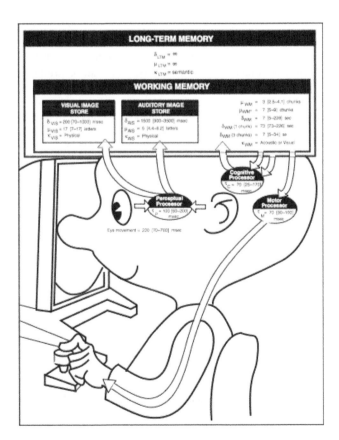

FIGURE 5.1. Model Human Processor. Based on Card, Moran, and Newell (1983).

(a) one cognitive, (b) one perceptual, and (c) one motor. In some cases, the system essentially behaves serially. For instance, in order for the system to press a key in response to the appearance of a light, the perceptual processor must detect the appearance of the light and transmit this information to the cognitive processor. The cognitive processor's job is to decide what the appropriate response should be, and then transmit that to the motor processor, which is responsible for actually executing the appropriate motor command. In this situation, the processors act serially, one after another; however, in more complex tasks such as transcription typing, all three processors often work in parallel.

Besides the specification of the timing for each processor and the connectivity of the processors, Card, Moran, and Newell named some general operating principles ranging from very general and qualitative to detailed and quantitative. For example, Principle P9, the Problem Space Principle, stated:

The rational activity in which people engage to solve a problem can be described in terms of (1) a set of states of knowledge, (2) operators for changing one state into another, (3) constraints on applying operators, and (4) control knowledge for deciding which operator to apply next. (Card et al. 1983, p. 27)

This is a particularly general and somewhat vague principle. In contrast, consider Principle P5, Fitts' Law:

The time T_{pos} to move the hand to a target of size S which lies a distance D away is given by:

$$T_{pos} = I_M \log_2(D/S + .5)$$

where $I_M = 100$ [70 ~ 120] ms/bit. (Card et al., 1983, p. 27).

This is a very specific principle that quantitatively describes hand movement behavior, which is highly relevant to, say, pointing with a mouse. Overall, the specification of the MHP is quite thorough, and it lays out a basis for a cognitive architecture able to do a wide variety of HCI-relevant tasks; however, Card et al. (1983) did not implement the MHP as a running cognitive architecture. This is likely for pedagogical reasons; it is not necessary to have a complete, running cognitive architecture for the general properties of that architecture to be useful for guiding HCI researchers and practitioners. At the time, computational modeling was the domain of a very specialized few in cognitive psychology.

Card et al. (1983) laid out another concept that has been highly influential throughout HCI and particularly in the community of computational modelers. This is GOMS, which stands for Goals, Operators, Methods, and Selection rules. GOMS is a framework for task analysis that describes routine cognitive skills in terms of the four listed components. Routine cognitive skills are those where the user knows what the task is and how to do the task without doing any problem solving. Text editing with a familiar editor is the prototypical case of this, but clearly, a great many tasks of interest in HCI could be classified as routine cognitive skills. Thus, the potential applicability of GOMS is quite broad. Indeed, GOMS has been applied to a variety of tasks; the website http://www.gomsmodel.org lists 143 GOMS-related papers in its bibliography.

What does a GOMS analysis provide? Essentially, a GOMS analysis of a task describes the hierarchical procedural knowledge a person must have to successfully complete that task. Based on that, and the sequence of operators that must be executed, it is possible to make quantitative predictions about the execution time for a particular task. Other analyses, such as predictions of error, functionality coverage, and learning time, are also sometimes possible. Since the original formulation presented in Card et al. (1983), a number of different forms of GOMS analysis have been developed, each with slightly different strengths and weaknesses (John & Kieras, 1996).

The core point as it relates to cognitive architectures is that GOMS analysis is originally based on a production rule analysis (Card, personal communication, 1999). Because this will come up several times, a brief introduction to production systems is warranted. Production rules are IF-THEN condition-action pairs, and a set of production rules (or simply "productions," or just "rules") and a computational engine that interprets those productions is called a "production system." In addition to productions, production systems contain some representation of the current state. This representation typically consists of a set of loosely structured data elements such as propositions or attribute-value pairs. This set is called the "working memory" or "declarative memory." Because "working memory" is also a psychological term with somewhat different meaning in that literature, "declarative memory" will be used in all further discussions.

The operation of a production system is cyclic. On each cycle, the system first goes through a pattern-matching process. The IF side of each production tests for the presence of a particular pattern in declarative memory. When the IF conditions of a production are met, the production is said to fire, and the actions specified on the THEN side are executed. The actions can be things such as pressing a button, or even some higher level abstraction of action (e.g., "turn left"). Actions also include modifying the contents of declarative memory, which usually means that a different production or productions will match on the next cycle. At this abstract and purely symbolic level, production systems are Turing complete and thus can compute anything that is computable (Newell, 1990), thus, they should be flexible enough to model the wide array of computations performed by the human cognitive system.

This is relevant to cognitive architectures because most cognitive architectures are (or contain) production systems. GOMS was actually abstracted from production rule analysis. Card et al. (1983) discovered that, for routine cognitive skills, the structure of the productions was quite similar across tasks and a more abstract representation was possible. This representation is the original GOMS formulation. Thus, translating a GOMS analysis into production rules, the language of most cognitive architectures, is generally straightforward. Similarly, for routine cognitive skills, it is often relatively simple to derive a GOMS analysis from the set of productions used to model the task. Models based on cognitive architectures can go well beyond routine cognitive skills, but this connection has certainly influenced the evolution of research on cognitive architecture and HCI. This connection has also fed back into research and development of GOMS techniques themselves, such as Natural GOMS Language (NGOMSL; Kieras, 1988). NGOMSL allows the prediction of learning time for the knowledge described in a GOMS model based on a theory of transfer of training referred to as cognitive complexity theory (CCT), which will be described in more detail in the next section.

Cognitive Complexity Theory (CCT)

When someone has learned to perform a task with a particular interface and must switch, doing the same task with a new interface, how much better off will they be than someone just learning to do the task with the new interface?—that is, how much is the knowledge gained from using the old interface "transferred" to using the new interface? This kind of question has intrigued psychologists for at least a century, and having some answers to this question has implications for training programs, user interface design, and many other areas. Cognitive complexity theory (Bovair, Kieras, & Polson, 1990; Kieras & Polson, 1985) is a psychological theory of transfer of training applied to HCI. Most relevant to the current discussion, this theory is based on production rules. The major points of CCT are as follows:

- Knowledge of the procedures that people need to execute to perform routine tasks can be represented with production rules. The relevant production rules can be generated based on a GOMS analysis of the task to be modeled.
- The complexity of a task will be reflected in the number and content of the production rules. When certain conventions are adopted about the style of those production rules, complexity is reflected almost entirely in the number of rules.
- The time it takes to execute a procedure can be predicted with a production system that interprets those rules along with a set of typical operator times, for example, the time it takes to type a three-letter command. The production interpreter used in this work was not intended to be a general cognitive architecture, but the production system framework is certainly consistent with current architectures.
- The time it takes to learn a task is a function of the number of new rules that the user must learn. "New" is clearly defined in this context. If the user already has a production, and a new task requires a rule that is similar (again, similarity is well defined based on the production rule syntax), then the rule for the new task need not be learned.
- Some predictions about errors and speedup with practice can also be gleaned from the contents of the production rules.

Obviously, this was an ambitious agenda, and there are many subtleties. For example, the notion of a "task" as the term was used in the description of CCT actually includes more than just the task at an abstract level. Consider a simple instance of a text-editing task, deleting the word "redux" from the middle of a sentence. The actual commands needed to accomplish this task could be very different in different text editors, thus, modeling the "delete word" task would require two different sets of productions, one for each editor—that is, the necessary knowledge, and thus the production rules for representing it, is actually a function both of the task from the user point of view (e.g., "delete word") and the interface provided to accomplish the task. Transfer from one text editor to another therefore depends a great deal on the particulars of each interface. CCT thus predicts asymmetrical transfer: learning editor A after editor B should not be the same as learning editor B after editor A.

CCT models, such as a GOMS analysis, omit modeling many details of user behavior. In general, anything that falls outside the domain of procedural knowledge (how-to-do-it knowledge) is not modeled. This means that the model does not model motor actions such as keypresses, and instead has a "DoKeystroke" primitive operator. Nor do CCT models model things such as natural language comprehension, clearly a requirement in text editing. CCT models also do not include any model of the perceptual processes required by users—the model was simply given information about the state of the display and did not have to, for example, look to see if the cursor was over a particular character. This is the same scope as a typical GOMS model, though a CCT model is more formalized and quantitative than the GOMS models described by Card et al. (1983).

In spite of these limitations (or perhaps in part because these limitations allowed the researchers to concentrate on the most central aspects of the phenomena), CCT fared very well. Numerous laboratory experiments provide empirical support for many of the claims of CCT (see especially Bovair et al., 1990). The CCT framework was developed and validated in greatest detail to pre-GUI text editing, but it has also been applied to menu-based systems (Polson, Muncher, & Engelbeck, 1986) and a control panel device (Kieras & Bovair, 1986). Singley and Anderson (1989) provided a strikingly similar analysis of transfer of training, as well as supporting empirical results, lending credence to the CCT analysis. CCT was certainly one of the most prominent early successes of computational modeling in HCI.

CAPS

CAPS (Collaborative Activation-based Production System; Just & Carpenter, 1992) is a cognitive architecture designed to model individual differences in working memory (WM) capacity and the effects of working memory load. This speciality is applicable to a number of HCI situations. Certainly, some kinds of user interfaces can create excessive working memory demands, for example, phone-based interfaces. In phone-based interaction (PBI), options do not remain on a screen or in any kind of available storage; rather, users are forced to remember the options presented. This seems like a prime candidate for modeling with a system designed to capture the effects of working memory demand, and this is exactly what Huguenard, Lerch, Junker, Patz, and Kass (1997) did. Their data showed that, contrary to guideline advice and most people's intuition, restricting phone menus to only a few (three) items each does not reduce error rates. The CAPS-based model provided a clear theoretical account of this phenomenon. The model showed that short menus are not necessarily better in PBI because of two side effects of designing menu hierarchies with few options at each level. First, for the same number of total items, this increases menu depth, which creates working memory demand. Second, with fewer items at each level, each individual item has to be more general and therefore more vague, especially at the top

levels of the hierarchy. This forces users to spend WM resources on disambiguating menu items when they are in a situation where WM demands outstrip supply.

Another application of CAPS that is HCI relevant is the account of postcompletion error provided by Byrne and Bovair (1997). What is a postcompletion error? Anecdotal evidence and intuition suggests that, when interacting with man-made artifacts, certain kinds of errors occur with greater frequency than others. In particular, one entire family of errors seems anecdotally common: errors that are made when some part of a task occurs after the main goal of the task has been accomplished (hence, "postcompletion"). Nearly everyone reports having made an error of this type at one time or another. Here are two prototypical examples:

1. Leaving the original on the glass of a photocopier. The main goal one generally has when using a photocopier is "get copies," and this goal is satisfied before one remove the original document. This error is less common now that many photocopiers include document feeders; the more current equivalent is leaving a document on the glass in a flatbed scanner.
2. Leaving one's bank card in an automated teller machine (ATM). Again, the main goal is something on the order of "get cash," and in many ATMs, card removal occurs after the cash is dispensed. This error was common enough in the first generation of ATMs that many ATMs are now designed in such a way that this error is now impossible to make.

Other postcompletion errors include leaving the gas cap off after filling up the car's gas tank, leaving change in vending machines, and more—most readers can probably think of several others. While numerous HCI researchers were aware of this class of error (e.g., Young, Barnard, Simon, & Whittington, 1989; Polson et al., 1992), no previous account explained why this type of error is persistent, yet not so frequent that it occurs every time. The CAPS model provides just such an account, and can serve as a useful example of the application of a cognitive architecture to an HCI problem.

Like most other production systems, CAPS contains two kinds of knowledge: (a) declarative memory and (b) productions. Declarative memory elements in CAPS also have associated with them an activation value, and elements below a threshold level of activation cannot be matched by productions' IF sides. Additionally, unlike most other production systems, the THEN side of a CAPS production may request that the activation of an element be incremented. For this to be truly useful in modeling working memory, there is a limit to the total amount of activation available across all elements. If the total activation exceeds this limit, then all elements lose some activation to bring the total back within the limit. This provides a mechanism for simulating human working memory limitations.

In Byrne and Bovair's (1997) postcompletion error model, there is a production that increments the activation of subgoals when the parent goal is active and unsatisfied. Therefore, to use the photocopier example, the "get copies" subgoal supplies activation to all the unfulfilled subgoals throughout the task; however, when the "get copies" goal is satisfied, the activation supply to the subgoals stops. Because the goal to remove the original is a subgoal of that goal, it loses its activation supply. Thus, what the model predicts is that the postcompletion subgoals are especially vulnerable to working memory load, and lower capacity individuals are more "at risk" than higher-capacity individuals. Byrne and Bovair conducted an experiment to test this prediction, and the data supported the model.

This is a good demonstration of the power of cognitive architectures. Byrne and Bovair neither designed nor implemented the CAPS architecture, but were able to use the theory to construct a model that made empirically testable predictions, and those predictions were borne out. Though it seems unlikely that CAPS will guide future HCI work (since its designers have gone in a different direction), it provides an excellent example case for a variety of reasons.

Soar

The development of Soar is generally credited to Allan Newell (especially, Newell, 1990), and Soar has been used to model a wide variety of human cognitive activity from syllogistic reasoning (Polk & Newell, 1995) to flying combat aircraft in simulated war games (Jones et al., 1999). Soar was Newell's candidate "unified theory of cognition" and was the first computational theory to be offered as such.

While Soar is a production system like CAPS, it is possible to think of Soar at a more abstract level. The guiding principle behind the design of Soar is Principle P9 from the Model Human Processor, the Problem Space Principle. Soar casts all cognitive activity as occurring in a problem space, which consists of a number of states. States are transformed through the applications of operators. Consider Soar playing a simple game like tic-tac-toe as player "X." The problem space is the set of all states of the tic-tac-toe board—not a very large space. The operators available at any given state of that space are placing an X at any of the available open spaces on the board. Obviously, this is a simplified example; the problem space and the available operators for flying an F-16 in a simulated war game are radically more complex.

Soar's operation is also cyclic, but the central cycle in Soar's operation is called a decision cycle. Essentially, on each decision cycle, Soar answers the question, "What do I do next?" Soar does this in two phases. First, all productions that match the current contents of declarative memory fire. This usually causes changes in declarative memory, so other productions may now match. Those productions are allowed to fire, and this continues until no new productions fire. At this time, the decision procedure is executed, in which Soar examines a special kind of declarative memory element, the preference. Preferences are statements about possible actions, for example, "operator o3 is better than o5 for the current operator," or "s10 rejected for supergoal state s7." Soar examines the available preferences and selects an action. Thus, each decision cycle may contain many production cycles. When modeling human performance, the convention is that each decision cycle lasts 50 ms, so productions in Soar are very low-level, encapsulating knowledge at a small grain size.

Other than the ubiquitous application of the problem space principle, Soar's most defining characteristics come from two

mechanisms developed specifically in Soar: (a) universal subgoaling and (b) a general-purpose learning mechanism. Because the latter depends on the former, universal subgoaling will be described first. One of the features of Soar's decision process is that it is not guaranteed to have an unambiguous set of preferences to work with. Alternately, there may be no preferences listing an acceptable action. Perhaps the system does not know any acceptable operators for the current state, or perhaps the system lacks the knowledge of how to apply the best operator. Whatever the reason, if the decision procedure is unable to select an action, an impasse is said to occur. Rather than halting or entering some kind of failure state, Soar sets up a new state in a new problem space with the goal of resolving the impasse. For example, if multiple operators were proposed, the goal of the new problem space is to choose between the proposed operators. In the course of resolving one impasse, Soar may encounter another impasse and create another new problem space, and so on. As long as the system is provided with some fairly generic knowledge about resolving degenerate cases (e.g., if all else fails, choose randomly between the two good operators), this universal subgoaling allows Soar to continue even in cases where there is little knowledge present in the system.

Learning in Soar is a by-product of universal subgoaling. Whenever an impasse is resolved, Soar creates a new production rule. This rule summarizes the processing that went on in the substate. The resolution of an impasse makes a change to the superstate (the state in which the impasse originally occurred); this change is called a result. This result becomes the action, or THEN, side of the new production. The condition, or IF, side of the production is generated through a dependency analysis by looking at any declarative memory item matched in the course of determining this result. When Soar learns, it learns only new production rules, and it only learns as the result of resolving impasses. It is important to realize that an impasse is not equated with failure or an inability to proceed in the problem solving, but may arise simply because, for example, there are multiple good actions to take and Soar has to choose one of them. Soar impasses regularly when problem solving, and thus learning is pervasive in Soar.

Not surprisingly, Soar has been applied to a number of learning-oriented HCI situations. One of the best examples is the recent work by Altmann (Altmann & John, 1999; Altmann, 2001). Altmann collected data from an experienced programmer while she worked at understanding and updating a large computer program by examining a trace. These data included verbal protocols (e.g., thinking aloud) as well as a log of the actions taken (keystrokes and scrolling). The programmer occasionally scrolled back to find a piece of information that had previously been displayed. Altmann constructed a Soar model of her activity. This model simply attempts to gather information about its environment; it is not a complete model of the complex knowledge of an experienced programmer. The model attends to various pieces of the display, attempts to comprehend what it sees, and then issues commands. "Comprehension" in this model is manifested as an attempt to retrieve information about the object being comprehended.

When an item is attended, this creates a production that notes that the object was seen at a particular time. Because learning is pervasive, Soar creates many new rules like this, but because of the dependency-based learning mechanism, these new productions are quite specific to the context in which the impasse originally occurred. Thus, the "index" into the model's rather extensive episodic memory consists of very specific cues, usually found on the display. Seeing a particular variable name is likely to trigger a memory for having previously seen that variable name. Importantly, this memory is generated automatically, without need for the model to deliberately set goals to remember particular items.

Altmann (2001) discussed some of the HCI ramifications for this kind of always-on episodic memory trace in terms of display clutter. While avoiding display clutter is hardly new advice, it is generally argued that it should be avoided for visual reasons (e.g., Tullis, 1983). What Altmann's model shows, however, is that clutter can also have serious implications for effective use of episodic memory. Clutter can create enormous demands for retrieval. Since more objects will generally be attended on a cluttered display, the episodic trace will be large, lowering the predictive validity for any single cue. It is unlikely that kind of analysis would have been generated if it had not been guided by a cognitive architecture that provided the omnipresent learning of Soar.

Soar has also been used to implement models of exploratory learning, somewhat in the spirit of LICAI (described later). There are two prominent models here, (a) one called IDXL (Rieman, Young, & Howes, 1996) and (b) a related model called Task-Action Learner (Howes & Young, 1996). These models both attempt to learn unfamiliar GUI interfaces. IDXL operates in the domain of using software to construct data graphs, while the Task-Action Leaner starts with even less knowledge and learns basic GUI operations such as how to open a file. For brevity, only IDXL will be described in detail.

IDXL goes through many of the same scanning processes as LICAI, but must rely on very different mechanisms for evaluation since Soar is fundamentally different from LICAI. IDXL models evaluation of various display elements as search through multiple problem spaces, one that is an internal search through Soar's internal knowledge and the other a search through the display. As items are evaluated in the search, Soar learns productions that summarize the products of each evaluation. At first, search is broad and shallow, with each item receiving a minimum of elaboration; however, that prior elaboration guides the next round of elaboration, gradually allowing IDXL to focus in on the "best" items. This model suggests a number of ways in which interface designers could thus help learners acquire the new knowledge needed to utilize a new interface. Like the LICAI work, the IDXL work highlights the need for good labels to guide exploration. A more radical suggestion is based on one of the more subtle behavior of users and IDXL. When exploring and evaluating alternatives, long pauses often occur on particular menu items. During these long pauses, IDXL is attempting to determine the outcome of selecting the menu item being considered. Thus, one suggestion for speeding up learning of a new menu-driven GUI is to detect such pauses, and show (in some easily undoable way) what the results of selecting that item would be. For instance, if choosing that item brings up a dialog box for specifying certain options, that dialog box could be shown in some grayed-out form, and would simply vanish if the user moved off of that menu item. This would make the

evaluation of the item much more certain and would be an excellent guide for novice users. This is not unlike ToolTips for toolbar icons, but on a much larger scale.

A model that does an excellent job of highlighting the power of cognitive architectures is NTD-Soar (Nelson, Lehman, & John, 1994). NTD stands for "NASA Test Director," who

> . . . is responsible for coordinating many facets of the testing and preparation of the Space Shuttle before it is launched. He must complete a checklist of launch procedures that, in its current form, consists of 3000 pages of looseleaf manuals . . . as well as graphical timetables describing the critical timing of particular launch events. To accomplish this, the NTD talks extensively with other members of launch team over a two-way radio. . . . In addition to maintaining a good understanding of the status of the entire launch, the NTD is responsible for coordinating troubleshooting attempts by managing the communication between members of the launch team who have the necessary expertise. (p. 658)

Constructing a model that is even able to perform this task at all is a significant accomplishment. Nelson, Lehman, and John (1994) were able to not only build such a model, but this model was able to produce a timeline of behavior that closely matched the timeline produced by the actual NTD being modeled—that is, the ultimate result was a quantitative model of human performance, and an accurate one at that.

It is unlikely that such an effort could have been accomplished without the use of an integrated cognitive architecture. The NTD model made extensive use of other Soar models. Nelson, Lehman, and John (1994) did not have to generate and implement theory of natural language understanding to model the communication between the NTD and others, or the NTD reading the pages in the checklist, because one had already been constructed in Soar (Lewis, 1993). They did not have to construct a model of visual attention to manage the scanning and visual search of those 3000 pages, because such a model already existed in Soar (Wiesmeyer, 1992). A great deal of knowledge engineering still took place in order to understand and model this complex task, but using an integrated architecture greatly eased the task of the modelers.

While this modeling effort was not aimed at a particular HCI problem, it is not difficult to see how it would be applicable to one. If one wanted to replace the 3000-page checklist with something like a personal digital assistant (PDA), how could the PDA be evaluated? There are very few NTDs in the world, and it is unlikely that they would be able to devote much time to participatory design or usability testing. Because an appropriate quantitative model of the NTD exists, however, it should be possible to give the model a simulated PDA and assess the impact of that change on the model's performance. Even if the model does not perfectly capture the effects of the change, it is likely that the model would identify problem areas and at least guide the developers in using any time they have with an actual NTD.

Soar has also been used as the basis for simulated agents in war games, as mentioned previously (Jones et al., 1999). This model (TacAir-Soar) participates in a virtual battle space in which humans also participate. TacAir-Soar models take on a variety of roles in this environment, from fighter pilots to helicopter crews to refueling planes. Their interactions are more complex than simple scripted agents, and they can interact with humans in the environment with English natural language. One of the major goals of the project is to make sure that TacAir-Soar produces humanlike behavior, because this is critical to their role, which is to serve as part of training scenarios for human soldiers. In large-scale simulations with many entities, it is much cheaper to use computer agents than to have humans fill every role in the simulation. While agents (other than the ubiquitous and generally disliked paper clip) have not widely penetrated the common desktop interface, this remains an active HCI research area, and future agents in other roles could also be based on cognitive architecture rather than more engineering-oriented AI models.

While the Soar architecture is still being used for applied AI work such as TacAir-Soar, there have been essentially no new HCI-oriented Soar efforts in the last several years. Nothing, in principle, has precluded such efforts, but for the time being, Soar is primarily an example of previous achievements using cognitive architectures in HCI, as opposed to a source of new insight.

CONTEMPORARY ARCHITECTURES

Current cognitive architectures are being actively developed, updated, and applied to HCI-oriented tasks. Two of the three most prominent systems (EPIC and ACT-R) are production systems, or rather are centrally built around production systems. While all contain production rules, the level of granularity of an individual production rule varies considerably from architecture to architecture. Each one has a different history and unique focus, but they share a certain amount of intellectual history; in particular, they have all been influenced one way or another by both the MHP and each other. At some level, they may have more similarities than differences; whether this is because they borrow from one another or because the science is converging is still an open question. The third system is somewhat different from these two production system models and will be considered first.

LICAI/CoLiDeS

LICAI (Kitajima & Polson, 1997) is a primary example of an HCI-oriented cognitive architecture not based on the production system framework. With the exception of the Soar work on exploratory learning, all of the work discussed up to this point more or less assumes that the modeled users are relatively skilled with the specific interface being used; these approaches do a poor job of modeling relatively raw novices. One of the main goals of LICAI is to address this concern. The paradigm question addressed by LICAI is, "How do users explore a new interface?"

Unlike the other architectures discussed, LICAI's central control mechanisms are not based on a production system. Instead, LICAI is built around an architecture originally designed to model human discourse comprehension: construction-integration (C-I; Kintsch, 1998). Like production systems, C-I's operation is cyclic. What happens on those cycles, however, is somewhat different from what happens in a production system. Each cycle is divided into two phases: (a) construction and (b) inte-

gration (hence, the name). In the construction phase, an initial input (e.g., the contents of the current display) is fed into a weakly constrained, rule-based process that generates a network of propositions. Items in the network are linked based on their argument overlap. For example, the goal of "graph data" might be represented with the proposition (PERFORM GRAPH DATA). Any proposition containing GRAPH or DATA would thus be linked to that goal.

Once the construction phase is completed, the system is left with a linked network of propositions. What follows is the integration phase, in which activation propagates through the network in a neural, network-like fashion. Essentially, this phase is a constraint-satisfaction phase, which is used to select one of the propositions in the network as the "preferred" one. For example, the system may need to select the next action to perform while using an interface. Action representations will be added to the network during the construction phase, and an action will be selected during the integration phase. The action will be performed, and the next cycle initiated. Various C-I models have used this basic process to select things other than actions. The original C-I system used these cycles to select between different interpretations of sentences.

LICAI has three main types of cycles: (a) one to select actions, (b) one to generate goals, and (c) one to select goals. This contrasts with how most HCI tasks have been modeled in production system architectures; in such systems, the goals are usually included in the knowledge given to the system. This is not true in LICAI; in fact, the knowledge given to LICAI by the modelers is minimal. For the particular application of LICAI, which was modeling users who knew how to use a Macintosh for other tasks (e.g., word processing) and were now being asked to plot some data using a Macintosh program called CricketGraph (one group of users actually worked with Microsoft Excel), it included some very basic knowledge about the Macintosh GUI and some knowledge about graphing. Rather than supply the model with the goal hierarchy, Kitajima and Polson (1997) gave the model the same minimal instructions that were given to the subjects. One of the major jobs of the LICAI model, then, was to generate the appropriate goals as they arose while attempting to carry out the instructions.

Again, this illustrates one of the benefits of using a cognitive architecture to model HCI tasks. Kitajima and Polson (1997) did not have to develop a theory of text comprehension for LICAI to be able to comprehend the instructions given to subjects; since LICAI is based on an architecture originally designed for text comprehension, they essentially got that functionality gratis. Additionally, they did not include just any text comprehension engine, but one that makes empirically validated predictions about how people represent the text they read. Thus, the claim that the model started out with roughly the same knowledge as the users is highly credible.

The actual behavior of the model is also revealing, as it exhibits many of the same exploratory behaviors seen in the users. First, the model pursues a general strategy that can be classified as label-following (Polson et al., 1992). The model, like the users, had a strong tendency to examine anything on the screen that had a label matching or nearly matching (e.g., a near synonym) a key word in the task instructions. When the particular subtask being pursued by the model contained well-labeled steps, the users were rapid, which was predicted by the model. While this prediction is not counter-intuitive, it is important to note that LICAI is not programmed with this strategy. This strategy naturally emerges through the normal operation of construction-integration through the linkages created by shared arguments. The perhaps less-intuitive result—modeled successfully by LICAI—is the effect of the number of screen objects. During exploration in this task, users were slower to make choices if there were more objects on the screen, but only if those items all had what were classified as "poor" labels. In the presence of "good" labels (literal match or near synonym), the number of objects on the screen did not affect decision times of either the users or LICAI.

The programmers who implemented the programs operated by the users put in several clever direct manipulation tricks. For example, to change the properties of a graph axis, one double-clicks on the axis and a dialog box specifying the properties of that axis appears. Microsoft Excel has some functionality that is most easily accessed by drag-and-drop. Franzke (1994) found that in a majority of first encounters with these types of direct manipulations, users required hints from the experimenter to be able to continue, even after two minutes of exploration. LICAI also fails at these interactions because no appropriate links are formed between any kind of task goal and these unlabeled, apparently static screen objects during the construction phase. Thus, these screen objects tend to receive little activation during the integration phase, and actions involving other objects are always selected.

Overall, LICAI does an excellent job of capturing many of the other empirical regularities in exploratory learning of a GUI interface. This is an important issue for many interfaces—particularly, any interface that is aimed at a walk-up-and-use audience. While currently common walk-up-and-use interfaces, such at ATMs, provide simple enough functionality that this is not always enormously difficult, this is not the case for more sophisticated systems, such as many information kiosks.

More recently, LICAI has been updated (and renamed to CoLiDeS, for Comprehension-based Linked model of Deliberate Search; Kitajima, Blackmon, & Polson, 2000) to handle interaction with web pages. This involves goals that are considerably less well elucidated and interfaces with a much wider range of semantic content. In order to help deal with these complexities, LICAI has been updated with a more robust attentional mechanism and a much more sophisticated notion of semantic similarity based on Latent Semantic Analysis (LSA; Landauer & Dumais, 1997). LSA is a technique for mathematically representing the meaning of words. While the details of the mathematics are complex, the underlying idea is straightforward. To perform LSA, one needs a large corpus of documents. An enormous matrix is created, showing which words appear in which contexts (documents). This matrix is then reduced by a mathematical technique called "singular value decomposition" (a cousin of factor analysis) to a smaller dimensional space of, say, 300 by 300 dimensions. Each word can then be represented by a vector in this space. The "meaning" of a passage is simply the sum of the vectors of the individual words in the passage. The similarity of two words (or two passages) can be quantified as the cosine between the vectors representing the words (or passages), with larger numbers signifying greater semantic similarity.

While LSA is certainly surrounded by a host of mathematical and theoretical issues (it should be noted that many similar techniques that differ in a variety of subtle ways have also been developed), the key result from an HCI point of view is that the technique works. Similarity measures generated by LSA generally agree very well with human judges, even for complex language understanding tasks such as grading essays. (In fact, an LSA-based system for grading essays and similar tasks has been turned into a successful commercial product; see http://www.pearsonkt.com.)

Based on the LSA-augmented CoLiDeS, Blackmon and colleagues (2002, 2003, 2005) have developed a web usability assessment and repair technique they call "Cognitive Walkthrough for the Web" (CWW). CWW is based on the idea that when users are exploring a large website, they go through a two-stage process of parsing and understanding the page, and ultimately generate descriptions of the available actions. Using those descriptions, they judge which action is most similar to their goal, and then select the link or widget that corresponds with that action. This is somewhat more complex than simply applying LSA to all of the labels on the page and comparing each one to the goal; the page is parsed into "regions," each of which is assessed, and smaller bits of text (say, link labels) are elaborated with a C-I-based process. It is those elaborations, rather than the raw link labels, that are compared via LSA to the description of the user's goal.

Based on this, CWW identifies where users with a particular goal are likely to have navigation problems and even generates problem severity estimates. These predictions have been validated in a series of experiments (Blackmon, Kitajima, & Polson, 2003; Blackmon, Kitajima, & Polson, 2005), and tools which partially automate the process are available on the web (see http://autocww.colorado.edu/). This is a significant advance in HCI, as the problem of predicting how people use semantic similarity to guide search on the web is not a trivial one and, prior to techniques like this, has generally required extensive iteration of user tests.

EPIC

With the possible exception of the NTD model, the models discussed up to this point have been primarily "pure" cognitive models—that is, the perception and action parts of the models have been handled in an indirect, abstract way. These models focus on the cognition involved, which is not surprising given the generally cognitive background of these systems; however, even the original formulation of the MHP included processors for perception and motor control. Additionally, in fact, user interfaces have also moved from having almost exclusively cognitive demands (e.g., one had to remember or problem-solve to generate command names) to relying much more heavily on perceptual-motor capabilities. This is one of the hallmarks of the GUI, the shift to visual processing and direct manipulation rather than reliance on complex composition of commands.

Providing accurate quantitative models for this kind of activity, however, requires a system with detailed models of human perceptual and motor capabilities. This is one of the major foci and contributions of EPIC (for executive process interactive control). EPIC is the brainchild of Kieras and Meyer (see especially 1996, 1997; Kieras, Wood, & Meyer, 1997). The overall structure of the processors and memories in EPIC is shown in Fig. 5.2. This certainly bears some surface similarity to the MHP, but EPIC is substantially more detailed. EPIC was explicitly designed to pair high-fidelity models of perception and motor mechanisms with a production system. The perceptual-motor processors represent a new synthesis of the human performance literature, while the production system is the same one used in the CCT work discussed earlier.

All EPIC processors run in parallel with one another. Therefore, while the Visual Processor is recognizing an object on the screen, the Cognitive Processor can decide what word should be spoken in response to some other input, while at the same time the Manual Motor processor is pressing a key. The information flow is typical of traditional psychological models, with information coming in through the eyes and ears, and outputs coming from the mouth and hands. More specifically, what is modeled in each of the processors is primarily time course. EPIC's Visual Processor does not take raw pixels as input and compute that those pixels represent a letter "A." Instead, it determines whether the object on the screen can be seen and at what level of detail, as well as how long it will take for a representation of that object to be delivered to EPIC's declarative memory once the letter becomes available to the Visual Proces-

sor. The appearance of the letter can actually cause a number of different elements to be deposited into EPIC's declarative memory at different times. For example, information about the letter's color will be delivered before information about the letter's identity.

Similarly, on the motor side, EPIC does not simulate the computation of the torques or forces needed to produce a particular hand movement. Instead, EPIC computes the time it will take for a particular motor output to be produced after the Cognitive Processor has requested it. This is complicated by the fact that movements happen in phases. Most importantly, each movement includes a preparation phase and an execution phase. The time to prepare a movement is dependent on the number of movement features that must be prepared for each movement and the features of the last movement prepared. Features that have been prepared for the previous movement can sometimes be reused to save time. EPIC can make repeated identical movements rapidly because no feature preparation time is necessary if the movements are identical. If they are not identical, the amount of savings is a function of how different the current movement is from the previous one. After being prepared, a movement is executed. The execution time for a movement corresponds roughly to the time it takes to physically execute the movement; the execution time for aimed movements of the hands or fingers are governed by Fitts' Law. EPIC's motor processors can only prepare one movement at a time, and can

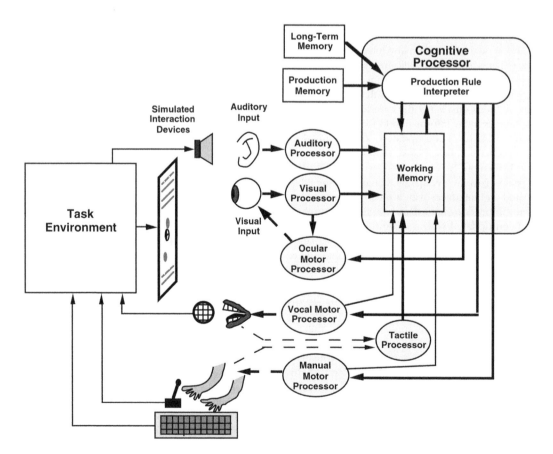

FIGURE 5.2.   Overall structure of the EPIC architecture. Based on Kieras and Meyer (1996).

only execute one movement at a time, but may be preparing one movement while executing another. Thus, in some tasks, it may be possible to effectively pipeline movements in order to generate very rapid sequences of movements.

EPIC's Cognitive Processor is a production system, the same one that was used for the earlier CCT work. One highly salient feature of this system is that multiple rules can fire on a production cycle. In fact, there is no upper bound on the number of productions that can fire on a cycle. Productions in this system are at a much higher grain size than productions in Soar, which gives EPIC a parallel quality at all levels—that is, all the processors work in parallel, and EPIC's cognitive processor is itself capable of parallel processing.

This allows EPIC particular leverage in multiple-task situations. When more than one task is being performed, the tasks can execute in parallel; however, many of the perceptual-motor processors are effectively serial. People only have one set of eyes that can only be aimed at one place at a time, so if multiple tasks are ongoing and they both require the eyes, there must be something that arbitrates. In EPIC, this additional knowledge about how to manage multiple tasks is termed "executive" knowledge, and the productions that implement this knowledge execute in parallel with the productions implementing the task knowledge.

Why is all this machinery and extra knowledge necessary? Because the world of HCI is changing. The GUI forced designers and analysts to consider more seriously the perceptual-motor constraints, and the propagation of computers with user interfaces away from the desktop and into mobile phones, kiosks, automobiles, and many, many other places creates a huge demand on people's ability to multitask. Multiple-task issues have largely gone unmodeled and have been outside the theoretical scope of most psychological accounts in HCI, at least before EPIC.

While LICAI and Soar have not been adequately equipped to deal with high-performance perception and action components of many tasks, EPIC is not equipped to handle some of the issues covered by other architectures. In particular, EPIC does not include any learning mechanisms, so it would be difficult to generate EPIC models for many of the domains Soar has approached successfully. This is not a fatal shortcoming, however, as there are a wide variety of domains in which learning is not a key component and where high-fidelity modeling of perception and action, along with multiple tasking, are central.

Naturally, these are the kinds of domains to which EPIC has been applied. One of the first major applications of EPIC was to a deceptively simple dual-task paradigm known as the "psychological refractory period" (or PRP; see Meyer & Kieras, 1997). In this task, laboratory subjects are typically confronted with two choice reaction time tasks, something on the order of, "Either a red light or a green light will appear. If it's red, hit the 'L' key. If it's green, hit the 'J' key." This sounds simple, but the empirical literature is rich and shows a variety of subtle effects, for which EPIC provides the first unified account. Critically, what the EPIC models of these experiments show is that people's low-level strategies for scheduling the tasks play a large role in determining performance in this simple paradigm.

EPIC has been used to model several tasks with a more directly HCI-oriented flavor. One of those tasks is menu selection (Kieras & Meyer, 1997; Hornof & Kieras, 1997, 1999), but for brevity, a detailed description of these models will be omitted. Another application of EPIC that definitely merits mention is the model of telephone assistance operators (TAOs), data originally presented in Gray et al. (1993). When a telephone customer dials "0," a TAO is the person who answers. The TAOs modeled here sat at a "dumb terminal" style workstation and assisted customers in completing telephone calls. In particular, TAOs determine how calls should be billed, and this is done by speaking to the customer. The detailed EPIC models (Kieras et al., 1997) covered a subset of the possible billing types.

This provided a good test of EPIC because the task was performed under time pressure, and seconds—actually, milliseconds—counted in task performance. Second, this task is multimodal. The TAO must speak, type, listen, and look at a display. Third, very fine-grained performance data were available to help guide model construction. By now, it should come as no surprise to the reader that it was possible to construct an EPIC model that did an excellent job of modeling the time course of the TAO's interaction with the customer; however, this modeling effort went beyond just that and provided some insight into the knowledge engineering problem facing modelers utilizing cognitive architectures as well.

Like other production system models, EPIC provides a certain amount of freedom to the modeler in model construction. While the architecture used provides certain kinds of constraints, and these constraints are critical in doing good science and affecting the final form of the model (Howes & Young, 1997), the modeler does have some leeway in writing the production rules in the model. This is true even when the production rule model is derived from another structured representation such as a GOMS model, which was the case in the TAO model. In EPIC, it is possible to write a set of "aggressive" productions that maximize the system's ability to process things in parallel, while it is also possible to write any number of less aggressive sets representing more conservative strategies. EPIC will produce a quantitative performance prediction regardless of the strategy, but which kind of strategy should the modeler choose? There is generally no a priori basis for such a decision, and it is not clear that people can accurately self-report on such low-level decisions.

Kieras et al. (1997) generated an elegant approach to this problem, later termed "bracketing" (Kieras & Meyer, 2000). The idea is this: construct two models, one of which is the maximally aggressive version. At this end of the strategy spectrum, the models contain very little in the way of cognition. The Cognitive Processor does virtually no deliberation and spends most of its cycles simply reading off perceptual inputs and immediately generating the appropriate motor output. This represents the "super-expert" whose performance is limited almost entirely by the rate of information flow through the perceptual-motor systems. At the other end of the spectrum, a model incorporating the slowest reasonable strategy is produced. The slowest reasonable strategy is one where the basic task requirements are met, but with no strategic effort made to optimize scheduling to produce rapid performance. The idea is that observed performance should fall somewhere in between these two extremes, hence, the data should be bracketed by the two versions of the model. Different users will tend to perform at different ends of this range for different tasks, so this is an excellent way to accommodate some of the individual differences that are always observed in real users.

What was discovered by employing this bracketing procedure to the TAO models was surprising. Despite the fact that the TAOs were under considerable time pressure and were extremely well practiced experts, their performance rarely approached the fastest possible model. In fact, their performance most closely matched a version of the model termed the "hierarchical motor-parallel model." In this version of the model, eye, hand, and vocal movements are executed in parallel with one another when possible; furthermore, the motor processor is used somewhat aggressively, preparing the next movement while the current movement was in progress. The primary place where EPIC could be faster but the data indicated the TAOs were not was in the description of the task knowledge. It is possible to represent the knowledge for this task as one single, flat GOMS method with no use of subgoals. On the other hand, the EPIC productions could represent the full subgoal structure or a more traditional GOMS model. Retaining the hierarchical representation—thus incurring time costs for goal management—provided the best fit to the TAOs performance. This provides solid evidence for the psychological reality of the hierarchical control structure inherent in GOMS analysis, since even well-practiced experts in fairly regular domains do not abandon it for some kind of faster knowledge structure.

The final EPIC model that will be considered is the model of the task first presented in Ballas, Heitmeyer, and Perez (1992). Again, this model first appeared in Kieras and Meyer (1997), but a richer version of the model is described in more detail later, in Kieras, Meyer, and Ballas (2001). The display used is a split screen. On the right half of the display, the user was confronted with a manual tracking task, which was performed using a joystick. The left half of the display was a tactical task in which the user had to classify objects as hostile or neutral based on their behavior. There were two versions of the interface to the tactical task, one a command-line style interface using a keypad and one a direct-manipulation-style interface using a touch screen. The performance measure of interest in this task is the time taken to respond to events (such as the appearance of a new object or a change in state of an object) on the tactical display.

This is again a task well suited to EPIC because the perceptual-motor demands are extensive. This is not, however, what makes this task so interesting. What is most interesting is the human performance data: in some cases, the keypad interface was faster than the touch screen interface, and in many cases, the two yielded almost identical performance, while in some other cases, the touch screen was faster. Thus, general claims about the superiority of GUIs do not apply to this case and a more precise and detailed account is necessary.

EPIC provides just the tools necessary to do this. Two models were constructed for each interface, again using the bracketing approach. The results were revealing. In fact, the fastest possible models showed no performance advantage for either interface. The apparent direct-manipulation advantage of the touch screen for initial target selection was almost perfectly offset by some type-ahead advantages for the keypad. The reason for the inconsistent results is that the users generally did not operate at the speed of the fastest possible model; they tended to work somewhere in between the brackets for both interfaces. However,

they tended to work more toward the upper (slowest-reasonable) bracket for the touch screen interface. This suggests an advantage for the keypad interface, but the caveat is that the slowest reasonable performance bound for the touch screen was faster than the slowest possible for the keypad. Thus, any strategy changes made by users in the course of doing the task, perhaps as a dynamic response to changes in workload, could affect which interface would be superior at any particular point in the task. Thus, results about which interface is faster are likely to be inconsistent—which is exactly what was found.

This kind of analysis would be impossible to conduct without a clear, quantitative human performance model. Constructing and applying such a model also suggested an alternative interface that would almost certainly be faster than either, which is one using a speech-driven interface. One of the major performance bottlenecks in the task was the hands, and so in this case, voice-based interaction should be faster. Again, this could only be clearly identified with the kind of precise quantitative modeling enabled by something like the EPIC architecture.

ACT-R 5.0

ACT-R 5.0 (Anderson et al., 2004) represents another approach to a fully unified cognitive architecture, combining a very broad model of cognition with rich perceptual-motor capabilities. ACT-R 5.0 is the most recent iteration of the ACT-R cognitive architecture (introduced in Anderson, 1993). ACT-R has a long history within cognitive psychology, as various versions of the theory have been developed over the years. In general, the ACT family of theories has been concerned with modeling the results of psychology experiments, and this is certainly true of the current incarnation, ACT-R. Anderson and Lebiere (1998) show some of this range, covering areas as diverse as list memory (chapter 7), choice (chapter 8), and scientific reasoning (chapter 11).

ACT-R is, like EPIC and Soar, a production system with activity centered on the production cycle, which is also set at 50 ms in duration; however, there are many differences between ACT-R and the other architectures. First, ACT-R can only fire one production rule per cycle. When multiple production rules match on a cycle, an arbitration procedure called conflict resolution comes into play. Second, ACT-R has a well-developed theory of declarative memory. Unlike EPIC and Soar, declarative memory elements in ACT-R are not simply symbols. Each declarative element in ACT-R also has an activation value associated with it, which determines whether and how rapidly it may be accessed. Third, ACT-R contains learning mechanisms, but is not a pervasive learning system in the same sense as Soar. These mechanisms are based on a "rational analysis" (Anderson, 1990) of the information needs of an adaptive cognitive system.

For example of rational analysis, consider conflict resolution. Each production in ACT-R has associated with it several numeric parameters, including numbers which represent the probability that if the production fires, the goal will be reached and the cost, in time, that will be incurred if the production fires. These values are combined according to a formula that trades off probability of success vs. cost and produces an "expected gain" for each production. The matching production with the highest expected gain is the one that gets to fire when conflict resolution is invoked. The expected gain values are noisy, so the system's behavior is somewhat stochastic, and the probability and cost values are learned over time so that ACT-R can adapt to changing environments.

Similarly, the activation of elements in declarative memory is based on a Bayesian analysis of the probability that a declarative memory element will be needed at a particular time. This is a function of the general utility of that element, reflected in what is termed its "base-level" activation, and that element's association with the current context. The more frequently and recently an element has been accessed, the higher its base-level activation will be, and thus the easier it is to retrieve. This value changes over time according to the frequency and recency of use, thus this value is learned. These kinds of mechanisms have helped enable ACT-R to successfully model a wide range of cognitive phenomena.

The current version of the theory, 5.0, incorporates several important changes over previous versions of the theory. First, the architecture is now a fully modular architecture, with fully separate and independent modules for different kinds of information processing (such as declarative memory and vision). Second, ACT-R 5.0 no longer contains a "goal stack" for automated goal management; goals are now simply items in declarative memory and subject to processes like learning and decay. Third, the new modular architecture allows processing in ACT-R to be mapped onto the human brain, with different modules identified with different brain regions. Finally, a production rule learning mechanism has been fully incorporated, so learning is now more pervasive in ACT-R. The 5.0 version of the theory subsumes all previous versions of the theory including ACT-R/PM, which is now considered obsolete.

In ACT-R 5.0, the basic production system is augmented with four EPIC-like peripheral modules, as well as a goal module and a declarative memory module, as depicted in Fig. 5.3. Like EPIC, all of these modules run in parallel with one another, giving ACT-R the ability to overlap processing. The peripheral modules come from a variety of sources. ACT-R's Vision Module is based on the ACT-R Visual Interface described in Anderson, Matessa, and Lebiere (1997). This is a feature-based attentional visual system, but does not explicitly model eye movements. Recently, the Vision Module has been extended to include an eye-movement model (Salvucci, 2001a) as well. The Motor Module is nearly identical to the Manual Motor Processor in EPIC and is based directly on the specification found in Kieras and Meyer (1996), and the Speech Module is similarly derived from EPIC. The Audition Module is a hybrid of the auditory system found in EPIC and the attentional system in ACT-R's Vision Module.

One other important property of ACT-R is that it is possible to have models interact with the same software as the human users being modeled. The software development conditions are fairly restrictive, but if these conditions are met, then both the user and the model are forced to use the same software. This reduces the number of degrees of freedom available to the modeler in that it becomes impossible to force any unpleasant modeling details into the model of the user interface, because there is no model of the user interface. More will be said about this issue in the next section.

ACT-R has been applied to numerous HCI domains. The first example comes from the dissertation work of Ehret (1999).

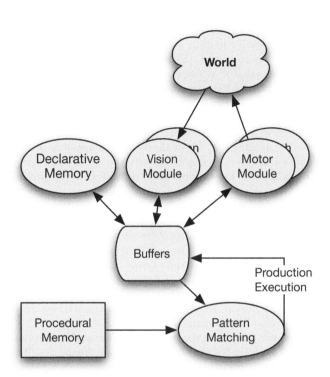

FIGURE 5.3. Overall structure of the ACT-R 5.0 architecture.

Among other things, Ehret developed an ACT-R/PM model of a fairly simple, but subtle, experiment. In that experiment, subjects were shown a target color and asked to click on a button that would yield that color. The buttons themselves had four types: (a) blank, (b) arbitrary icon, (c) text label, and (d) color. In the color condition, the task was simple: users simply found the color that matched the target and then clicked the button. In the text label condition, the task was only slightly more difficult: users could read the labels on the buttons and select the correct one because the description matched the color. In the arbitrary icon condition, more or less random pictures appeared on each icon (e.g., a mailbox). Users had to either memorize the picture-to-color mapping, which they had to discover by trial and error, or memorize the location of each color, since the buttons did not change their functions over time. In the hardest condition, the blank condition, users simply had to memorize the mapping between button location and color, which they had to discover through trial and error.

Clearly, the conditions will produce different average response times, and what Ehret (1999) found is that they also produced somewhat different learning curves over time. Ehret added an additional manipulation as well: after users performed the task for some time, all labeling was removed. Not surprisingly, the amount of disruption was different in the different conditions, reflecting the amount of incidental location learning that went on as subjects performed the task. The ACT-R model that Ehret constructed did an excellent job of explaining the results. This model represented the screen with the built-in visual mechanisms from ACT-R and learned the mappings between color and location via ACT-R's standard associative learning mechanisms. The initial difference between the various conditions was reproduced, as were the four learning curves. The model also suffered disruptions similar to those suffered by the human users when the labels were removed. This model is an excellent demonstration of the power of ACT-R, exercising both the perceptual-motor capabilities of the system as well as the graded learning in ACT-R's rational-analysis driven mechanisms.

Salvucci (2001b) described an ACT-R model that tests ACT-R's ability to handle multimodal, high-performance situations in a very compelling task: this model drives an automobile driving simulator. This is not a robotics project; the ACT-R model does not actually turn the steering wheel or manipulate the pedals; rather, it communicates with the automobile simulation software. The model's primary job is to maintain lane position as the car drives down the road. Salvucci (2001b) added an additional task that makes it particularly interesting: the model dials telephone numbers on a variety of mobile phone interfaces. Two factors were crossed: (a) whether the telephone was dialed manually via keypad vs. dialed by voice, and (b) whether the full telephone number needed to be dialed vs. a shortened "speed dial" system. The model was also validated by comparison with data from human users.

Both the model and the human users showed that dialing while not driving is faster than dialing while driving, and that steering performance can be disrupted by telephone dialing. Not surprisingly, the most disruptive interface was the "full-manual" interface, where the full phone numbers were dialed on a keypad. This is due largely to the fact that dialing with the keypad requires visual guidance, causing the model (and the users) to take their eyes off the road. There was very little disruption associated with the voice interfaces, regardless of whether full numbers or speed-dial was used.

This is a powerful illustration of the value of cognitive architectures for a number of reasons. First, the basic driving model could simply be reused for this task; it did not have to be reimplemented. Second, the model provides an excellent quantitative fit to the human data. Third, this is an excellent example of a situation where testing human users can be difficult. Testing human drivers with interfaces that degrade driving performance is dangerous, so simulators are generally used for this kind of evaluation; however, maintaining a driving simulator requires a great deal of space and is quite expensive. If someone wanted to test another variant of the telephone interface, it would be much faster and cheaper to give that interface to Salvucci's (2001b) model than it would be to recruit and test human drivers.

One of the long-standing criticisms of cognitive architectures in HCI is that they are theoretical tools designed to support research and are too difficult for the average practitioner to use in a more applied context. This criticism certainly holds some truth (though how much may be open to debate); however, some researchers are actively trying to address it. One of the most interesting developments in the cognitive architecture arena in the past few years has been the work of John and colleagues (2004) on something they term "CogTool" (see http://www.cogtool.org). CogTool allows interface designers to mock up an interface in HTML, demonstrate the tasks to be performed in a GUI environment, and have an ACT-R based cognitive model of the tasks automatically synthesized. Alternative interfaces can be mocked up, and predicted performance compared, giving the designer important information about how changes in the interface may impact user performance. This is not limited strictly to prediction of time of execution, either, as the full model output—a complete trace of the predicted user behavior—is also available for inspection.

This is an important advance, as it makes the power of complete and detailed computational cognitive modeling available to people with no experience with cognitive architectures. It also has important ramifications for the methodology of model building, since models generated by CogTool are not hand-tweaked in an attempt to optimize fit to data; in fact, the intent is to make predictions about nonexistent data. Some validation, however, has been performed, and early explorations with CogTool have found the time predictions to be quite accurate. As exciting a development as this is, it is important to note that there are clearly limitations. The language of available actions is based on the keystroke-level model (KLM) of Card, Moran, and Newell (1983) and is thus somewhat limited. For example, it does not provide a model for any kind of cognition more complex than very simple decisions about where to point. For example, if one wanted to model use of an ATM, CogTool would not model the process of the user trying to retrieve a security code from memory. (However, if the user was highly practiced such that this retrieval is very rapid and always successful, then CogTool could be appropriate.) Efforts are underway to make automatic translation of slightly more complex GOMS-style models into ACT-R models possible (St. Amant, Freed, & Ritter, 2005), but that still would not address the more complex cognition required by many interfaces. Nonetheless, CogTool represents a significant advance in making architecture-based modeling accessible to those who are not experts in the field.

The last application of ACT-R in HCI to be covered in detail is based on Pirolli and Card's (1999) theory of *information foraging*. This theory is based on the idea that people forage for information in much the same way that animals forage for food. The literature on the latter type of foraging is quite rich, including equations that describe when foragers will give up on a particular patch of information and move on to the next one. Foraging for food is often guided by proximal cues (e.g., scent) that indicate the presence of distal content (e.g., food); analogously, information foragers may use proximal cues (e.g., visible link labels) to guide their search for information that they cannot presently access. In the spirit of the analogy, Pirolli and Card term these proximal cues "information scent." That such a theory should be applicable to the web should be fairly obvious.

This theory has spawned a number of models based on modified versions of ACT-R. The original model from Pirolli and Card (1999) was called ACT-IF (for "information foraging"). Later versions were SNIF-ACT 1.0 (Card et al., 2001) and SNIF-ACT 2.0 (Pirolli & Fu, 2003), where SNIF stands for "scent-based navigation and information foraging." These models use a modified form of ACT-R's spreading activation to compute information scent, which drives the model's choices as it browses through web pages. SNIF-ACT 1.0 and later also uses an altered version of ACT-R's conflict resolution mechanism to determine when to abandon the current website (or information "patch") and move on to a new site. SNIF-ACT 2.0 adds a further improvement, which is the use of a new measure to compute the strengths of association (and therefore spreading activation) between words. This new measure, termed "pointwise mutual information" (PMI), is much easier to compute than LSA, but generally produces similar results. SNIF-ACT 2.0 was used to predict link selection frequency in a large ($N = 244$) Web behavior study with impressive results, accounting for 72% of the selection variance on one site and 90% of the variance on a second site.

Obviously, this work is clearly related to CWW. In fact, the foraging work has also led to a tool to help automate the process of web usability assessment, this one named "Bloodhound" (Chi et al., 2003). Bloodhound takes a site and a goal description and uses scent metrics combined with Monte Carlo methods to compute usage patterns and identify navigation problems. The two projects clearly have similar goals and similar outputs. What is less clear is the extent to which the differences in the underlying cognitive architecture and differences in other aspects of the techniques (such as LSA vs. PMI) actually lead to different predictions about user behavior. As of this writing, the two systems have never both been applied to the same website. However appealing such a "bake-off" might be, it is important not to lose sight of the advance even the possibility of such a comparison represents: it is only recently that the field has had psychologically informed and quantitative predictions about something as complex as Web behavior available for potential comparison. Thus, this is a concrete example of cognitive architectures paying off in HCI.

Other published and ongoing work applying ACT-R to HCI-oriented problems is available (e.g., Taatgen & Lee, 2003; Peebles & Cheng, 2003; Byrne, 2001; Schoelles & Gray, 2000), but space considerations prohibit a more exhaustive review. The vitality of the research effort suggests that ACT-R's combination of perceptual-motor modules and a strong theory of cognition will continue to pay dividends as an HCI research tool.

In fact, this is not limited to ACT-R; overall, cognitive architectures are an exciting and active area of HCI research. The systems described here all take slightly different approaches and focus on slightly different aspects of various HCI problems, but there is clearly a great deal of cross-pollination. Lessons learned and advancements made in one architecture often affect other

systems, for example, the development of EPIC's peripheral systems clearly impacted ACT-R.

Comparisons

An exhaustive and detailed comparison of the major cognitive architectures is beyond the scope of this chapter; however, an excellent comparison that includes a number of other architectures can be found in Pew and Mavor (1998). Certainly, the three production systems are related: (a) Soar, (b) EPIC, and (c) ACT-R. A major difference between them is their original focus; they were originally developed to model slightly different aspects of human cognition. As they develop, however, there appears to be more convergence than divergence. This is generally taken as a good sign that the science is cumulating. Still, there are differences, and certainly between the production systems and LICAI/CoLiDeS. Many of the relevant comparisons are summarized in Table 5.1.

Most of the entries on this table have been previously discussed, with the exception of the last two. Support for learning will be discussed in the next section. The presence and size of the user community has not been discussed, as it is not clear what role (if any) such a community plays in the veridicality of the predictions made by the system; however, it may be relevant to researchers for other reasons, particularly those trying to learn the system.

In addition, many of the values on this table are likely to change in the future. For example, the 6.0 version of ACT-R will be available by the time this chapter appears (though there are few major theoretical changes planned), and planning for a major revision for version 7.0 is already underway.

It is difficult to classify the value an architecture has on a particular attribute as an advantage or a disadvantage, because what constitutes an advantage for modeling one phenomenon may be a disadvantage for modeling others. For example, consider learning in Soar. Certainly, when attempting to model the improvement of users over time with a particular interface, Soar's learning mechanism is critical; however, there are many applications for which modeling learning is not critical, and Soar's pervasive learning feature occasionally causes undesired side effects that can make model construction more difficult.

TABLE 5.1. Architecture Feature Comparison

	LICAI/CoLiDeS	EPIC	ACT-R 5.0
Original focus	Text comprehension	Multiple-task performance	Memory and problem-solving
Basic cycle	Construction-integration	Production cycle (parallel)	Production cycle (serial)
Symbolic or activation-based?	Both	Symbolic	Both
Architectural goal management	Special cycle types, supports Vague goals	None	None
Detailed perceptual-motor systems	No	Yes	Yes
Learning mechanisms	No	No	Yes
Large, integrated models	No	No	Soon
Extensive natural language	Yes	No	No
Support for learning system school	None	None	Extensive tutorial materials, summer
User community*	None	None	Growing, primarily psychology

*Outside of the researchers who have developed the system.

THE FUTURE OF COGNITIVE ARCHITECTURES IN HCI

Beyond the incremental development and application of architectures such as EPIC and ACT-R, what will the future hold for cognitive architectures in HCI? What challenges are faced, and what is the ultimate promise? Indeed, a number of important limitations for cognitive architectures currently exist. There are questions they cannot yet address, and it is hard to see how they even would address some questions. Other limitations are more pragmatic than in principle, but these are relevant as well.

First, there is a wide array of HCI problems that are simply outside the scope of current cognitive architectures. Right now, these architectures focus on cognition and performance, but not on other aspects of HCI, such as user preference, boredom, aesthetics, fun, and so on. Another important challenge, though one that might be overcome, is that these architectures generally have not been applied to social situations, such as those encountered in groupware or online communities (Olson & Olson, this volume; Preece, this volume). In principle, it is possible to implement a model of social interaction in a cognitive architecture; however, the knowledge engineering problem here would certainly be a difficult one. How does one characterize and implement knowledge about social situations with enough precision to be implemented in a production system? It may ultimately be possible to do so, but it is unlikely that this will happen anytime soon.

One problem that will never entirely be resolved, no matter how diligent the modelers are, is the knowledge engineering problem. Every model constructed using a cognitive architecture still needs knowledge about how to use the interface and what the tasks are. By integrating across models, the knowledge engineering demands when entering a new domain may be reduced (the NTD is a good example), but they will never be eliminated. This requirement will persist even if an architecture was to contain a perfect theory of human learning—and there is still considerable work to be done to meet that goal.

Another barrier to the more widespread use of cognitive architectures in HCI is that the architectures themselves are large and complex pieces of software, and (ironically) little work has been done to make them usable or approachable for novices. For example, "EPIC is not packaged in a 'user-friendly' manner; full-fledged Lisp programming expertise is required to use the simulation package, and there is no introductory tutorial or user's manual" (Kieras & Meyer, 1997, p. 399). The situation is slightly better for ACT-R: tutorial materials, documentation, and examples for ACT-R are available, and most years there is a two-week "summer school" for those interested in learning ACT-R (see http://act.psy.cmu.edu).

Another limiting factor is implementation. In order for a cognitive architecture to accurately model interaction with an interface, it must be able to communicate with that interface. Because most user interfaces are "closed" pieces software with no built-in support for supplying a cognitive model with the information it needs for perception (e.g., what is on the screen where) or accepting input from a model, this creates a technical problem. Somehow, the interface and the model must be connected. An excellent summary of this problem can be found in

Ritter, Baxter, Jones, and Young (2000). A number of different approaches have been taken. In general, the EPIC solution to this problem has been to reimplement the interface to be modeled in Lisp (more recent versions of EPIC require C++), so the model and the interface can communicate via direction function calls. The ACT-R solution is not entirely dissimilar. In general, ACT-R has only been applied to relatively new experiments or interfaces that were initially implemented in Lisp, and thus ACT-R and the interface can communicate via function calls. In order to facilitate the construction of models and reduce the modeler's degrees of freedom in implementing a custom interface strictly for use by a model, ACT-R does provide some abilities for automatically managing this communication when the interface is built with the native GUI builder for Macintosh Common Lisp under MacOS and Allegro Common Lisp under Windows. If the interface is implemented this way, both human users and the models can interact with the same interface.

The most intriguing development along this line, however, is recent work by St. Amant and colleagues (St. Amant & Riedl, 2001; St. Amant, Horton, & Ritter, in press). They have implemented a system called SegMan, which directly parses the screen bitmap on Windows systems—that is, given a Windows display—any Windows display—SegMan can parse it and identify things such as text, scroll bars, GUI widgets, and the like. It also has facilities for simulating mouse and keyboard events. This is an intriguing development, because in principle, it should be possible to connect this to an architecture such as EPIC or ACT-R, which would enable the architecture to potentially work with any Windows application in its native form.

While a number of technical details would have to be worked out, this has the potential of fulfilling one of the visions held by many in the cognitive architecture community: a high-fidelity "virtual user" that could potentially use any application or even combination of applications. Besides providing a wide array of new domains to researchers, this could be of real interest to practitioners as well because this opens the door for at least some degree of automated usability testing. This idea is not a new one (e.g., Byrne, Wood, Sukaviriya, Foley, & Kieras, 1994; St. Amant, 2000), but technical and scientific issues have precluded its adoption on even a limited scale. This would not eliminate the need for human usability testing (see Ritter & Young, 2001, for a clear discussion of this point) for some of the reasons listed above, but it could significantly change usability engineering practice in the long run.

The architectures themselves will continue to be updated and applied to more tasks and interfaces. As mentioned, a new version of ACT-R (version 6.0) is currently under development, and this new version has definitely been influenced by issues raised by HCI concerns. New applications of EPIC result in new mechanisms (e.g., similarity-based decay in verbal working memory storage; Kieras, Meyer, Mueller, & Seymour, 1999) and new movement styles (e.g., click-and-point, Hornof & Kieras, 1999). Applications such as the World Wide Web are likely to drive these models into more semantically rich domain areas, and tasks that involve greater amounts of problem solving are also likely candidates for future modeling.

Despite their limitations, this is a particularly exciting time to be involved in research on cognitive architectures in HCI. There is a good synergy between the two areas, as cognitive architectures are certainly useful to HCI, so HCI is also useful for cognitive architectures. HCI is a complex and rich yet still tractable domain, which makes it an ideal candidate for testing cognitive architectures. HCI tasks are more realistic and require more integrated capabilities than typical cognitive psychology laboratory experiments, and thus cognitive architectures are the best theoretical tools available from psychology. Theories like EPIC-Soar (Chong & Laird, 1997) and ACT-R are the first complete psychological models that go from perception to cognition to action in detail. This is a significant advance and holds a great deal of promise.

The need for truly quantitative engineering models will only grow as user interfaces propagate into more and more places and more and more tasks. Cognitive architectures, which already have a firmly established niche in HCI, seem the most promising road toward such models. Thus, as the architectures expand their range of application and their fidelity to the humans they are attempting to model, this niche is likely to expand. HCI is an excellent domain for testing cognitive architectures as well, so this has been, and will continue to be, a fruitful two-way street.

References

Altmann, E. M. (2001). Near-term memory in programming: A simulation-based analysis. *International Journal of Human-Computer Studies, 54*(2), 189–210.

Altmann, E. M., & John, B. E. (1999). Episodic indexing: A model of memory for attention events. *Cognitive Science, 23*(2), 117–156.

Anderson, J. R. (1990). *The adaptive character of thought.* Hillsdale, NJ: Erlbaum.

Anderson, J. R. (1993). *Rules of the mind.* Hillsdale, NJ: Erlbaum.

Anderson, J. R., Bothell, D., Byrne, M. D., Douglass, S., Lebiere, C., & Quin, Y. (2004). An integrated theory of the mind. *Psychological Review, 111,* 1036–1060.

Anderson, J. R., Conrad, F. G., & Corbett, A. T. (1989). Skill acquisition and the LISP tutor. *Cognitive Science, 13*(4), 467–505.

Anderson, J. R., & Lebiere, C. (1998). *The atomic components of thought.* Mahwah, NJ: Erlbaum.

Anderson, J. R., Matessa, M., & Lebiere, C. (1997). ACT-R: A theory of higher level cognition and its relation to visual attention. *Human–Computer Interaction, 12*(4), 439–462.

Ballas, J. A., Heitmeyer, C. L., & Perez, M. A. (1992). Evaluating two aspects of direct manipulation in advanced cockpits. *Proceedings of ACM CHI 92 Conference on Human Factors in Computing Systems* (pp. 127–134). New York: ACM.

Blackmon, M. H., Kitajima, M., & Polson, P. G. (2003). Repairing usability problems identified by the Cognitive Walkthrough for the Web. In *Human factors in computing systems: Proceedings of CHI 2003* (pp. 497–504). New York: ACM.

Blackmon, M. H., Kitajima, M., & Polson, P. G. (2005). Tool for accurately predicting website navigation problems, non-problems, problem severity, and effectiveness of repairs. In *Human factors in computing systems: Proceedings of CHI 2005* (pp. 31–40). New York: ACM.

Blackmon, M. H., Polson, P. G., Kitajima, M., & Lewis, C. (2002). Cognitive walkthrough for the web. In *Human factors in computing systems: Proceedings of CHI 2002* (pp. 463–470). New York: ACM.

Bovair, S., Kieras, D. E., & Polson, P. G. (1990). The acquisition and performance of text-editing skill: A cognitive complexity analysis. *Human–Computer Interaction, 5*(1), 1–48.

Byrne, M. D. (2001). ACT-R/PM and menu selection: Applying a cognitive architecture to HCI. *International Journal of Human-Computer Studies, 55,* 41–84.

Byrne, M. D., & Bovair, S. (1997). A working memory model of a common procedural error. *Cognitive Science, 21*(1), 31–61.

Byrne, M. D., Wood, S. D., Sukaviriya, P. N., Foley, J. D., & Kieras, D. E. (1994). Automating interface evaluation. *ACM CHI '94 Conference on Human Factors in Computing Systems* (pp. 232–237). New York: ACM Press.

Card, S. K., Moran, T. P., & Newell, A. (1983). *The psychology of human-computer interaction.* Hillsdale, NJ: Erlbaum.

Card, S. K., Pirolli, P., Van Der Wege, M., Morrison, J. B., Reeder, R. W., Schraedley, P. K., et al. (2001). Information scent as a driver of Web behavior graphs: Results of a protocol analysis method for Web usability. In *Human factors in computing systems: Proceedings of CHI 2001* (pp. 498–505). New York: ACM.

Chi, E. H., Rosien, A., Supattanasiri, G., Williams, A., Royer, C., Chow, C., et al. (2003). The Bloodhound project: Automating discovery of web usability issues using the Infoscent simulator. In *Human factors in computing systems: Proceedings of CHI 2003* (pp. 505–512). New York: ACM.

Chong, R. S., & Laird, J. E. (1997). Identifying dual-task executive process knowledge using EPIC-Soar. In M. Shafto & P. Langley (Eds.), *Proceedings of the Nineteenth Annual Conference of the Cognitive Science Society* (pp. 107–112). Hillsdale, NJ: Erlbaum.

Ehret, B. D. (1999). Learning where to look: The acquisition of location knowledge in display-based interaction. Unpublished doctoral dissertation, George Mason University, Fairfax, VA.

Franzke, M. (1994). Exploration, acquisition, and retention of skill with display-based systems. Unpublished doctoral dissertation, University of Colorado, Boulder.

Gray, W. D., John, B. E., & Atwood, M. E. (1993). Project Ernestine: Validating a GOMS analysis for predicting and explaining real-world task performance. *Human–Computer Interaction, 8*(3), 237–309.

Gray, W. D., Young, R. M., & Kirschenbaum, S. S. (1997). Introduction to this special issue on cognitive architectures and human–computer interaction. *Human–Computer Interaction, 12,* 301–309.

Hornof, A., & Kieras, D. E. (1997). Cognitive modeling reveals menu search is both random and systematic. *Proceedings of ACM CHI 97 Conference on Human Factors in Computing Systems* (pp. 107–114). New York: ACM.

Hornof, A., & Kieras, D. (1999). Cognitive modeling demonstrates how people use anticipated location knowledge of menu items. *Proceedings of ACM CHI 99 Conference on Human Factors in Computing Systems* (pp. 410–417). New York: ACM.

Howes, A., & Young, R. M. (1996). Learning consistent, interactive, and meaningful task-action mappings: A computational model. *Cognitive Science, 20*(3), 301–356.

Howes, A., & Young, R. M. (1997). The role of cognitive architecture in modeling the user: Soar's learning mechanism. *Human–Computer Interaction, 12*(4), 311–343.

Huguenard, B. R., Lerch, F. J., Junker, B. W., Patz, R. J., & Kass, R. E. (1997). Working memory failure in phone-based interaction. *ACM Transactions on Computer-Human Interaction, 4,* 67–102.

John, B. E., & Kieras, D. E. (1996). The GOMS family of user interface analysis techniques: Comparison and contrast. *ACM Transactions on Computer-Human Interaction, 3,* 320–351.

John, B. E., Prevas, K., Savucci, D. D., & Koedinger, K. (2004). Predictive human performance modeling made easy. In *Human factors in computing systems: Proceedings of CHI 2004* (pp. 455–462). New York: ACM.

Jones, R. M., Laird, J. E., Nielsen, P. E., Coulter, K. J., Kenny, P., & Koss, F. V. (1999). Automated intelligent pilots for combat flight simulation. *AI Magazine, 20*(1), 27–41.

Just, M. A., & Carpenter, P. A. (1992). A capacity theory of comprehension: Individual differences in working memory. *Psychological Review, 99*(1), 122–149.

Kieras, D. E., & Bovair, S. (1986). The acquisition of procedures from text: A production-system analysis of transfer of training. *Journal of Memory & Language, 25*(5), 507–524.

Kieras, D. E., & Meyer, D. E. (1996). The EPIC architecture: Principles of operation. Retrieved July 4, 1996, from ftp://ftp.eecs.umich .edu/people/kieras/EPICarch.ps.

Kieras, D. E., & Meyer, D. E. (1997). An overview of the EPIC architecture for cognition and performance with application to human–computer interaction. *Human–Computer Interaction, 12*(4), 391–438.

Kieras, D. E., & Meyer, D. E. (2000). The role of cognitive task analysis in the application of predictive models of human performance. In J. M. Schraagen & S. F. Chipman (Eds.), *Cognitive task analysis* (pp. 237–260). Mahwah, NJ: Erlbaum.

Kieras, D. E., Meyer, D. E., Mueller, S., & Seymour, T. (1999). Insights into working memory from the perspective of the EPIC architecture for modeling skilled perceptual-motor and cognitive human performance. In A. Miyake & P. Shah (Eds.), *Models of working memory: Mechanisms of active maintenance and executive control* (pp. 183–223). New York: Cambridge University Press.

Kieras, D. E., Meyer, D. E., & Ballas, J. A. (2001). Towards demystification of direct manipulation: cognitive modeling charts the gulf of execution, *Proceedings of ACM CHI 01 Conference on Human Factors in Computing Systems* (pp. 128–135). New York: ACM.

Kieras, D., & Polson, P. G. (1985). An approach to the formal analysis of user complexity. *International Journal of Man-Machine Studies, 22*(4), 365–394.

Kieras, D. E., Wood, S. D., & Meyer, D. E. (1997). Predictive engineering models based on the EPIC architecture for multimodal high-performance human–computer interaction task. *Transactions on Computer-Human Interaction, 4*(3), 230–275.

Kintsch, W. (1998). *Comprehension: A paradigm for cognition.* New York: Cambridge University Press.

Kitajima, M., & Polson, P. G. (1997). A comprehension-based model of exploration. *Human–Computer Interaction, 12*(4), 345–389.

Kitajima, M., Blackmon, M. H., & Polson, P. G. (2000). A comprehension-based model of web navigation and its application to web usability analysis. In S. McDonald & Y. Waern & G. Cockton (Eds.), *People and Computers XIV-Usability or Else!* (Proceedings of HCI 2000) (pp. 357–373). New York: Springer.

Landauer, T. K., & Dumais, S. T. (1997). A solution to Plato's problem: The latent semantic analysis theory of acquisition, induction, and representation of knowledge. *Psychological Review, 104*(2), 211–240.

Lewis, R. L. (1993). An architecturally-based theory of human sentence comprehension. Unpublished doctoral dissertation, University of Michigan, Ann Arbor.

Meyer, D. E., & Kieras, D. E. (1997). A computational theory of executive cognitive processes and multiple-task performance: I. Basic mechanisms. *Psychological Review, 104*(1), 3–65.

Nelson, G., Lehman, J. F., & John, B. E. (1994). Integrating cognitive capabilities in a real-time task. In A. Ram & K. Eiselt (Eds.), *Proceedings of the Sixteenth Annual Conference of the Cognitive Science Society* (pp. 353–358). Hillsdale, NJ: Erlbaum.

Newell, A. (1973). You can't play 20 questions with nature and win: Projective comments on the papers of this symposium. In W. G. Chase (Ed.), *Visual information processing* (pp. 283–308). New York: Academic Press.

Newell, A. (1990). *Unified theories of cognition.* Cambridge, MA: Harvard University Press.

Newell, A., & Simon, H. A. (1963). GPS, a program that simulates human thought. In E. A. Feigenbaum & J. Feldman (Eds.), *Computers and thought* (pp. 279–293). Cambridge, MA: MIT Press.

Peebles, D., & Cheng, P. C. H. (2003). Modeling the effect of task and graphical representation on response latency in a graph reading task. *Human Factors, 45*, 28–45.

Pew, R. W., & Mavor, A. S. (Eds.). (1998). *Modeling human and organizational behavior: Application to military simulations.* Washington, DC: National Academy Press.

Pirolli, P., & Card, S. (1999). Information foraging. *Psychological Review, 106*(4), 643–675.

Pirolli, P., & Fu, W. (2003). SNIF-ACT: A model of information foraging on the world wide web. In P. Brusilovsky, A. Corbett, & F. de Rosis (Eds.), *Proceedings of the ninth international conference on user modeling* (pp. 45–54). Johnstown, PA: Springer-Verlag.

Polk, T. A., & Newell, A. (1995). Deduction as verbal reasoning. *Psychological Review, 102*(3), 533–566.

Polson, P. G., Lewis, C., Rieman, J., & Wharton, C. (1992). Cognitive walkthroughs: A method for theory-based evaluation of user interfaces. *International Journal of Man-Machine Studies, 36*(5), 741–773.

Polson, P. G., Muncher, E., & Engelbeck, G. (1986). A test of a common elements theory of transfer. *Proceedings of ACM CHI'86 Conference on Human Factors in Computing Systems* (pp. 78–83). New York: ACM.

Rieman, J., Young, R. M., & Howes, A. (1996). A dual-space model of iteratively deepening exploratory learning. *International Journal of Human-Computer Studies, 44*(6), 743–775.

Ritter, F. E., Baxter, G. D., Jones, G., & Young, R. M. (2000). Cognitive models as users. *ACM Transactions on Computer-Human Interaction, 7*, 141–173.

Ritter, F. E., & Young, R. M. (2001). Embodied models as simulated users: Introduction to this special issue on using cognitive models to improve interface design. *International Journal of Human-Computer Studies, 55*, 1–14.

Salthouse, T. A. (1988). Initiating the formalization of theories of cognitive aging. *Psychology & Aging, 3*(1), 3–16.

Salvucci, D. D. (2001a). An integrated model of eye movements and visual encoding. *Cognitive Systems Research, 1*(4), 201–220.

Salvucci, D. D. (2001b). Predicting the effects of in-car interface use on driver performance: An integrated model approach. *International Journal of Human-Computer Studies, 55*, 85–107.

Schoelles, M. J., & Gray, W. D. (2000). Argus Prime: Modeling emergent microstrategies in a complex simulated task environment. In N. Taatgen & J. Aasman (Eds.), *Proceedings of the Third International Conference on Cognitive Modeling* (pp. 260–270). Veenendal, NL: Universal Press.

Simon, H. A. (1996). *The sciences of the artificial* (3rd ed.). Cambridge, MA: MIT Press.

Singley, M. K., & Anderson, J. R. (1989). *The transfer of cognitive skill.* Cambridge, MA: Harvard University Press.

St. Amant, R. (2000, Summer). Interface agents as surrogate users. *Intelligence* magazine, *11*(2), 29–38.

St. Amant, R., Horton, T. E., & Ritter, F. E. (in press). Model-based evaluation of expert cell phone menu interaction. *ACM Transactions on Human–Computer Interaction.*

St. Amant, R., & Riedl, M. O. (2001). A perception/action substrate for cognitive modeling in HCI. *International Journal of Human-Computer Studies, 55*, 15–39.

St. Amant, R., Freed, A. R., & Ritter, F. E. (2005). Specifying ACT-R models of user interaction with a GOMS language. *Cognitive Systems Research, 6,* 71–88.

Taatgen, N. A., & Lee, F. J. (2003). Production compilation: A simple mechanism to model skill acquisition. *Human Factors, 45,* 61–76.

Tullis, T. S. (1983). The formatting of alphanumeric displays: A review and analysis. *Human Factors, 25*(6), 657–682.

Wiesmeyer, M. (1992). An operator-based model of covert visual attention. Unpublished doctoral dissertation, University of Michigan, Ann Arbor.

Young, R. M., Barnard, P., Simon, T., & Whittington, J. (1989). How would your favorite user model cope with these scenarios? *ACM SIGCHI Bulletin, 20,* 51–55.

TASK LOADING AND STRESS IN HUMAN–COMPUTER INTERACTION: THEORETICAL FRAMEWORKS AND MITIGATION STRATEGIES

J. L. Szalma and P. A. Hancock
University of Central Florida

TASKS AS STRESSORS: THE CENTRALITY OF STRESS IN HUMAN–COMPUTER INTERACTION

For those whose professional lives revolve around human–computer interaction (HCI), they may ask themselves why they should even glance at a chapter on stress. While it is evident that many computer systems have to support people operating in stressful circumstances, there are important design issues concerning how to present information in these very demanding circumstances. However, one could still ask, are these of central interest to those in the mainstream of HCI? Indeed, if these were the only issues, we would agree and would recommend the reader to pass quickly onto something of much more evident relevance. However, we hope to persuade you that the relevance of stress research to HCI is not limited to such concerns. Indeed, we hope to convince the reader that stress, in the form of task loading, is central to all HCI. To achieve this, we first present a perspective that puts stress front and center in the HCI realm. Traditionally, stress has been seen as exposure to some adverse environmental circumstances, such as excessive heat, cold, noise, vibration, and so forth, and its effects manifest themselves primarily in relation to the physiological system most perturbed by the stress at hand. However, Hancock and Warm (1989) observed that stress effects are virtually all mediated through the brain, but for the cortex such effects are almost always of secondary concern since the brain is primarily involved with the goals of ongoing behavior or, more simply, the current task. Therefore, we want to change the orientation of concern so that stress is not just a result of peripheral interference but rather that the primary source of stress comes from the ongoing task itself. If we now see the task itself as the primary driving influence then stress concerns are central to all HCI issues.

It is one of the most evident paradoxes of modern work that computer-based systems, which are designed to reduce task complexity and cognitive workload, actually often impose even greater demands and stresses on the very individuals they are supposed to be helping. How individuals cope with such stress has both immediate and protracted effects on their performance and well-being. Although operational environments and their associated tasks vary considerably (e.g., air traffic control, baggage screening, hospital patient monitoring, power plant operations, command and control, and banking and finance), there are certain mechanisms that are common to the stress appraisal of all task demands. Thus, there are design and HCI principles for stress that generalize across multiple domains (Hancock & Szalma, 2003a). In this chapter we explore such principles to understand stress effects in the HCI domain.

The structure of our chapter flows from these fundamental observations. First, we provide the reader with a brief overview of stress theory and its historical development to set our observations in context. Second, we articulate areas for future research needed to more completely understand how stress and workload impact human–computer interaction and how to exploit the positive effects while mitigating their negative effects.

TRADITIONAL APPROACHES TO STRESS RESEARCH

Traditionally, stress has been conceived of as either (a) an external, aversive stimulus (constituted of either physical, cognitive, or social stimulation patterns) imposed upon an individual or (b) response to such perturbations. Each of these views presents certain intrinsic operational difficulties. Considering stress as external stimulation is useful for categorizing effects of the physical environments (e.g., heat, noise, vibration), but such an approach cannot explain why the same stimulus pattern has vastly different effects on different individuals. Physiological interpretations (e.g., Selye, 1976) have tried to promulgate arousal explanations of stress. However, the more recent recognition that different sources of stress are associated with different patterns of cognitive effects made clear that adaptation or arousal theories of stress do not completely address the issue either (Hockey, R., 1984; Hockey, R. & Hamilton, 1983; Hockey, G. R. J., Gaillard, & Coles, 1986).

Thus to understand stress effects, we now have to embrace an even wider, multidimensional perspective (e.g., Matthews, 2001). Here we emphasize a view of stress as primarily an outcome of the appraisal of environmental demands as either taxing or exceeding an individual's resources to cope with that demand. These person-environment transactions (Lazarus & Folkman, 1984) occur at multiple levels within the organism (Matthews, 2001; van Reekum & Scherer, 1997). Further, these processes represent efforts by organism to adapt to demands imposed via regulation of both the internal state and the external environment. In the following section, we describe the theoretical frameworks that guide our observations on HCI. These perspectives emerge from the work of Hancock and Warm (1989), G. R. J. Hockey (1997), and Lazarus (1999; see also Lazarus & Folkman, 1984).

THEORETICAL FRAMEWORKS

Appraisal Theory

Among the spectrum of cognitive theories of stress and emotion, perhaps the best known is the relational theory proposed by Richard Lazarus and his colleagues (see Lazarus, 1991, 1999; Lazarus & Folkman, 1984). This theory is cognitive in that stress and emotion each depend upon an individual's cognitive appraisals of internal and external events, and these appraisals depend in part on the person's knowledge and experience (cf., Bless, 2001). The theory is motivational in that emotions in general, including stress responses, are reactions to one's perceived state of progress toward or away from one's goals (see Carver & Scheier, 1998). The relational aspect emphasizes the importance of the transaction between individuals and their environment. Together these three components shape the emotional and stress state of an individual. The outcomes of these processes are patterns of appraisal that Lazarus (1991) referred to as "core relational themes." For instance, the core relational theme for anxiety is uncertainty and existential threat, while that for happiness is evident progress toward goal achievement. Thus, when individuals appraise events relative to their desired outcomes (goals), these can produce negative, goal-incongruent emotions and stress if such events are appraised as hindering progress. Conversely, promotion of well-being and pleasure occur when events are appraised as facilitating progress toward a goal (e.g., goal-congruent emotions). Promotion of pleasure and happiness (see Hancock, Pepe, & Murphy, 2005; Ryan & Deci, 2001) therefore requires the design of environments and tasks themselves that afford goal-congruent emotions. The understanding of interface characteristics in HCI that facilitate positive appraisals and reduce negative appraisals is thus a crucial issue and an obvious avenue in which HCI and stress research can fruitfully interact.

A major limitation of all appraisal theories, however, is the neglect of understanding how task parameters influence resulting coping response. While the appraisal mechanism itself may be similar across individuals and contexts (e.g., see Scherer, 1999), the specific content (e.g., which events are appraised as a threat to well-being) does vary across individuals and contexts. One would expect that the criteria for appraisal (e.g., personal relevance, self-efficacy for coping) would be similar across individuals for specific task parameters as for any other stimulus or event. Individual differences occur in the specific content of the appraisal (e.g., one person's threat is another's challenge) and in the resultant response. An understanding of stress effects in HCI therefore requires understanding the task and person factors, and treating the transaction between the human and the system as the primary unit of analysis (see Lazarus & Folkman, 1984). This entails knowing how different individuals appraise specific task parameters and how changes in knowledge structures might ameliorate negative stress effects and promote positive affect in human-technology interaction. A visual representation of this emergent unit of analysis that comes from the interaction of person and environment, including the task, is shown in Fig. 6.1 (Hancock, 1997).

Adaptation Under Stress

A theoretical framework developed specifically for stress as it relates to performance is the maximal adaptability model presented by Hancock and Warm (1989). They distinguished the three facets of stress distinguished above, and they labeled these the trinity of stress, shown in Fig. 6.2. The first, "Input," refers to the environmental events to which the individual is exposed, which include information (e.g., displays) as well as traditional input categories such as temperature, noise, vibration, and so forth (e.g., Conway, Szalma, & Hancock, 2007; Hancock, Ross, Szalma, in press; Pilcher, Nadler, & Busch, 2002). The second, "Adaptation," encompasses the appraisal mechanisms referred to previously. The third and final component, "Output," is the level that indicates how the organism behaves in respect to goal achievement. A fundamental tenet of the Hancock and Warm (1989) model is that in the large majority of situations (and even in situations of quite high demand) individuals do adapt effectively to the input disturbance. That is, they can tolerate high levels of either overload or underload without

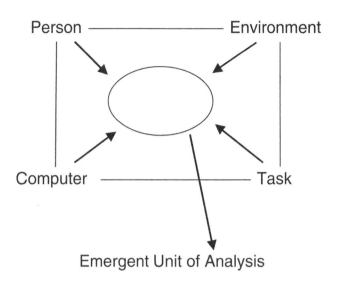

FIGURE 6.1. An illustration of the emergence of a supra-ordinate unit of analysis that derives from the interaction of the individual (person), the tool they use (computers), the task they have to perform, and the context (environment) in which action occurs. From Hancock (1997)

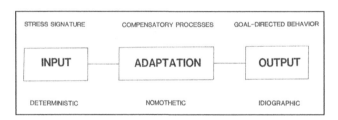

FIGURE 6.2. The trinity of stress which identifies three possible 'loci' of stress. It can be viewed as an *input* from the physical environment, which can be described deterministically. Since such a profile is by definition unique, it is referred to as a stress signature. The second locus is *adaptation*, which describes the populational or nomothetic reaction to the input itself. It is most evidently measurable in the processes of compensation. The third and final locus is the *output*, which is expressed as the impact on the on-going stream of behavior. Since the goals of different individuals almost always vary, this output is largely idiographic or person-specific. It is this facet of stress that has been very much neglected in prior and contemporary research. From Hancock and Warm (1989)

enormous change to their performance capacity. Adaptive processes occur at multiple levels, some being the physiological, behavioral (e.g., performance), and subjective/affective levels. These adaptations are represented in the model as a series of nested, extended, inverted-U functions (see Fig. 6.3) that reflect the fact that under most conditions the adaptive state of the organism is stable. However, under extremes of environmental underload or overload, failures in adaptation do occur. Thus, as the individual is perturbed by the input, the first threshold they traverse is subjective comfort. This is followed by behavioral effects, and finally failure of the physiological system (e.g., loss of consciousness). Examples of such extreme failures

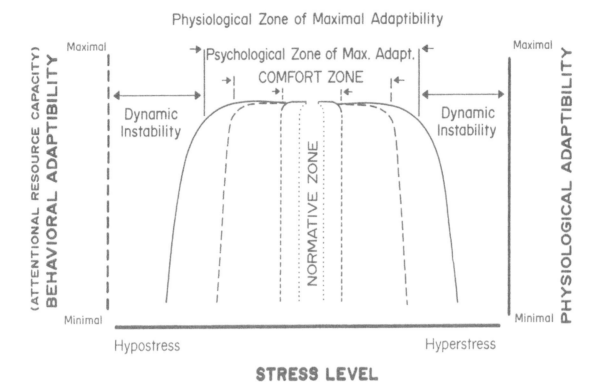

FIGURE 6.3. The extended-U relationship between stress level and response capacity. As is evident, the form of degradation is common across the different forms of response. At the center of the continuum is the normative zone which reflects optimal subjective functioning. Outside of this is the comfort zone which reflects the behavioral recognition of a state of satisfaction. Beyond this lies the reaction of psychological or cognitive performance capacity. Finally, the outer envelope is composed of physiological functioning. There are proposed strong linkages between the deviation from stability at one level being matched to the onset of radical failure at the more vulnerable level which is nested within it. The model is symmetrical in that underload (hypostress) has mirror effects to overload (hyperstress). The latter is considered the commonly perceived interpretation of stress. From Hancock and Warm (1989)

are relatively rare in most work settings, although when they do occur they are often catastrophic for the individual and the system they are operating (e.g., Harris, Hancock, & Harris, 2005).

This model is unique in that it provides explicit recognition that the proximal form of stress in almost all circumstances is the task itself. Task characteristics are incorporated in the model by two distinct base axes representing spatial and temporal components of any specified task. Information structure (the spatial dimension) represents how task elements are organized, including challenges to such psychological capacities such as working memory, attention, decision making, response capacity, and the like. The temporal dimension is represented by information rate. Together these dimensions can be used to form a vector (see Fig. 6.4) that serves to identify the current state of adaptation of the individual. Thus, if the combination of task characteristics and an individual's stress level can be specified, a vector representation can be used to predict behavioral and physiological adaptation. The challenge lies in quantifying the information processing components of cognitive work (see Hancock, Szalma, & Oron-Gilad, 2005).

Although the model shown in Fig. 6.4 describes the level of adaptive function, it does not articulate the mechanisms by which such adaptation occurs. Hancock and Warm (1989) argued that one way in which individuals adapt to stress is to narrow their attention by excluding task irrelevant cues (Easterbrook, 1959). Such effects are known to occur in spatial perception (e.g., Bursill, 1958; Cornsweet, 1969), and narrowing can occur at levels of both the central and peripheral neural systems (Dirkin & Hancock, 1984; 1985; Hancock & Dirkin, 1983). More recently Hancock and Weaver (2005) argued that distortions of temporal perception under stress are also related to this narrowing effect. However, recent evidence suggests that these two perceptual dimensions (space and time) may not share common perceptual mechanisms (see Ross, Szalma, Thropp, & Hancock, 2003; Thropp, Szalma, & Hancock, 2004).

The Cognitive-Energetic Framework

The Hancock and Warm (1989) model accounts for the levels of adaptation and adaptation changes under the driving forces

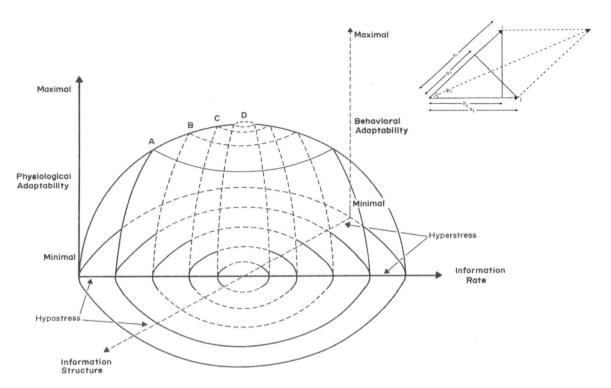

FIGURE 6.4. The description given in Figure 6.3 is now expanded into a three-dimensional representation by parsing the base "hypostress-hyperstress" axis into its two component elements. These divisions are composed of information rate (the temporal axis) and information structure (the spatial axis). Note that any one source of input stress can be described as a scalar on the base axis, and these scalars can be summed to provide a multi-input stress vector, which then provides a prediction of both performance and physiological adaptability, which are the primary descriptors on the vertical axis. From Hancock and Warm (1989)

of stress. However, it does not articulate how effort is allocated under stress or the mechanisms by which individuals appraise the task parameters that are the proximal source of stress. The effort allocation issue is address by a cognitive-energetical framework described by G. R. J. Hockey (1997). The compensatory control model is based upon three assumptions: behavior is goal directed; self-regulatory processes control goal states; and regulatory activity has energetic costs (e.g., consumes resources). In this model a compensatory control mechanism allocates resources dynamically according to the goals of the individual and the environmental constraints. The mechanisms operate at two levels (see Fig. 6.5). The lower level is more or less automatic and represents established skills. Regulation at this level requires few energetic resources or active regulation and effort (cf., Schneider & Shiffrin, 1977). The upper level is a supervisory controller, which can shift resources (effort) strategically to maintain adaptation and reflects effortful and controlled processing. The operation of the automatic lower loop is regulated by an effort monitor, which detects changes in the regulatory demands placed on the lower loop. When demand increases beyond the capacity of the lower loop control is shifted to the higher, controlled processing loop. Two strategic responses of the supervisory system are increased effort and changing the goals. Goals can be modified in their kind (change the goal

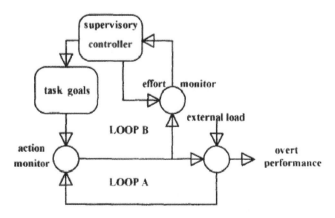

FIGURE 6.5. The two-level effort regulation model by Hockey. This model provides a mechanism by which an individual allocates limited cognitive resources to different aspects of performance. From Hockey (1997)

itself) or in strength (e.g., lowering the criterion for performance). Essentially, this is adjusting the discrepancy between goal state and current state by increasing effort or changing the goal (and see Carver & Scheier, 1998).

RELATIONSHIP BETWEEN STRESS AND COGNITIVE WORKLOAD

Cognitive Workload as a Form of Stress

The Hancock and Warm (1989) model explicitly identified the task itself as the proximal source of stress. In operational environments, this is often manifested as increases or decreases in cognitive workload (Moray, 1979). As in the case of stress, workload is easy to define anecdotally but difficult to define operationally. Workload can manifest in terms of the amount of information to be processed (an aspect of information structure), and the time available for processing (information rate). Thus, the base axes of the Hancock and Warm model captured dimensions of workload as well as stress (see Hancock & Caird, 1993). Indeed, physiological measures of workload (O'Donnell & Eggemeier, 1986) are often the same as those measures used to assess physiological stress. Similarly, subjective measures of workload and stress reflect appraisals of the task environment and of its perceived effect on the individual (Hart & Staveland, 1988). Although the two concepts developed in separate research traditions, the artificial boundary between them should be dissolved, as each term refers to similar processes. The implication for HCI is that computer-based tasks that impose either too much or too little demand will likely be appraised as stressful. Thus, the design process for development of computer interfaces should include assessment of perceived workload as well as affective state.

Performance and Workload: Associations/Dissociations

It is often the case that performance is maintained under increased workload and stress, which is reflected in the extended-U model described by Hancock and Warm (1989) and in the mechanisms of Hockey's (1997) energetic model of compensatory control. Maintaining performance under stress has costs, both physiologically and cognitively. Further, one would expect that in easier tasks performance is not as costly and that there should therefore be a direct association between task difficulty and perceived workload. Such performance-workload associations do occur, most prevalently in vigilance (Warm, Dember, & Hancock, 1996; see also Szalma et al., 2004). However, other forms of workload-performance relations can occur. For instance, perceived workload may change as a function of changes in task demand, but performance remains constant. Hancock (1996) referred to these situations as "insensitivities," which can be diagnostic with respect to the relation between the individual and the task (see also Parasuraman & Hancock, 2001). Thus, consistent with both Hancock and Warm and G. R. J. Hockey's (1997) frameworks, one response to increased task demand is to exert more effort, thereby maintaining performance but increasing perceived workload. Alternatively, one could have a situation in which task demands increase, performance decreases, but perceived workload does not change. This suggests that the appraisals of the task are not sensitive to actual changes in that task.

Interesting corollaries are the performance-workload dissociations that sometimes occur (Hancock, 1996; Yeh & Wickens, 1988). In such cases, decreased performance is accompanied by decreased workload. One possible reason for such a result might be disengagement of the individual from the task (e.g., the person gives up; see Hancock, 1996). In the case where increased performance is observed to be accompanied by increased perceived workload, the pattern suggests effective improvement of performance at the cost of increased effort allocation. An area of much needed research is establishing which task parameters control the patterns of performance-workload associations and dissociations, and how these change dynamically as a function of time on task. It may well be that reformulating the task by innovations in the interface itself may well address these crucial concerns (see Hancock, 1997). Indeed, the structure and organization of computer interfaces will be a major factor in both performance under stress and in the relation of performance to perceived workload.

MITIGATION OF STRESS

If changing the fundamental nature of the demand is one solution, we now look at other approaches to mitigation of the negative effects of stress and workload. These strategies include skill development (e.g., Hancock, 1986) and specific display design (Hancock & Szalma, 2003a; Wickens, 1996), as well as technologies employing adaptive automation and decision aids (Hancock & Chignell, 1987). Developing skills so that they are relatively automatic rather than controlled processing (Schneider & Shiffrin, 1977) and developing expertise can mitigate some of the negative effects of stress. In regard to display design, simple, easily perceivable graphics can permit quick, direct extraction of information when cognitive resources are reduced by stress and workload (Hancock & Szalma, 2003a). Adaptive automation can be employed by adjusting the level of automation and the management of automation according to stress state (e.g., Scerbo, Freeman, & Mikulka, 2003). In addition, adapting the form of automation (e.g., level, management type) to the operator based on individual differences can serve to improve its utility for aiding performance and reducing stress and workload (see Thropp, Oron-Gilad, Szalma, & Hancock, 2007).

Changing the Person

Training/skill development. Clearly, the greater the skill of the individual the more resilient their performance will be under stress (Hancock, 1986). This well-established phenomenon is incorporated into the energetic theories of stress and performance (Hancock & Warm, 1989; Hockey, 1997) and is an approach most often taken to mitigate workload and stress effects. However, training on relevant tasks is only one method of training for stress. There are also techniques for training individuals to cope more effectively with stress, essentially building stress coping skills. An example of such an approach is stress exposure training (SET; Johnston & Cannon-Bowers, 1996), a three phase procedure in which individuals are provided with information regarding the stress associated with task performance, are provided with training on the task, and then practice their task skills under simulated stress conditions. This technique has been shown to be effective in reducing anxiety and enhancing performance (Saunders, Driskell, Johnston, & Salas, 1996) and there is preliminary evidence that coping skills learned with a particular type of stressor and task can transfer to novel stressors and tasks (Driskell, Johnston, & Salas, 2001). For such an intervention to succeed, however, it is crucial that the training be designed based on an analysis of the task environment (Johnston & Cannon-Bowers, 1996). If the task parameters that are most responsible for the workload and stress are identified, these can be targeted for attention in training.

An additional issue for training for more effective stress coping is to modify the individual's appraisal of events. By inducing automaticity in some skills, not only are more resources freed for stress coping, but the task environment itself will be appraised as less threatening. Even if the event is appraised as a threat to an individual's psychological or physical well-being, the highly skilled individual will appraise his or her coping ability as sufficient to handle the increased demand. However, there has been limited research on how individuals who develop expertise also develop the capacity to effectively cope with the stress that accompanies performance in a given domain, and the extent to which stress coping skills in one domain transfer to other domains. Deliberate practice generally facilitates skill development (Ericsson, 2007). If one considers coping with stress to be a skill, then in principle deliberate practice should permit the development of expertise in coping. This will likely involve parsing the task into components, based on cognitive task analysis, and designing training procedures that target the stressful aspects of the task. However, such efforts will require understanding of how different forms of stress affect different forms of information processing. Since these variables are difficult to quantify, establishing these linkages must be theory driven. Elucidation of these issues will provide the groundwork for future development of stress mitigation tools during training and skill development.

Personnel selection. Selection techniques have been a popular choice for matching individuals to specific jobs, but the focus has typically been on intellectual skills (e.g., Yerkes, 1918). Selecting individuals for their stress-coping capability has been applied to the selection criteria for police officers, who therefore tend to be as stable as or more emotionally stable than the rest of the population (for a review, see Brown & Campbell, 1994). However, research is needed that links particular traits to stress coping skills for specific task environments. The effectiveness of general life stress coping, such as that observed in individuals who are extraverted (McCrae & Costa, 1986; Penley & Tomaka, 2002) or optimistic (Aspinwall, Richter, & Hoffman, 2002; Scheier & Carver, 1985), may not predict effective coping in specific task domains. Understanding which individuals will likely cope effectively with a particular task therefore requires

a thorough understanding of the perceptual, cognitive, and psychomotor demands of the task, and then linking these parameters to trait profiles. By far, the most research on the relation of affective traits to task performance has been in Extraversion and trait anxiety/Neuroticism (for a review, see Matthews, Deary, & Whiteman, 2003). However, the characteristics of greatest interest may vary somewhat across domains, although some general traits (e.g., emotional stability, conscientiousness) would be expected to moderate performance across a variety of task environments.

Changing the Task

Display design. Although training and selection can mitigate stress effects, the tasks themselves should be redesigned, for two reasons. First, there will be many instances where selection is not possible and expenditure of significant resources on training is undesirable. Second, there are instances in which one wishes to design an interface that requires little or no training and that can be used by any member of a large population of individuals (e.g., consumers). Particularly in light of the observation that the task represents the proximal source of stress, future work in stress mitigation for HCI should focus on redesign of the task and the interface itself. We have previously argued that existing display design techniques that are simple and easily perceived would be the best choice for an interface that will be used in stressful environments (Hancock & Szalma, 2003a). Specifically, configural or object displays can represent complex, multivariable systems as simple geometric shapes or emergent features if those features are well-mapped to system dynamics (see Bennett & Flach, 1992). Under stress, the complex problem solving and analytical skills are the most vulnerable and decline first. A display that allows fast extraction of information with minimal cost in working memory load can mitigate stress effects (Hancock & Szalma, 2003a; Wickens, 1996). A combination of training to automaticity and displays of information that can be perceived directly with a minimum of information processing requirements is currently one of the best approaches for stress mitigation in cognitively complex environments.

Adaptive automation. Another approach for stress mitigation is the allocation of function to automated systems (Hancock & Chignell, 1987). The advent of modern automated systems allows for automation to adapt to the state of the individual (Scerbo et al., 2003). Thus, at points in time when an operator is overtaxed, the system can assume control of some task functions, thereby freeing resources to effectively cope with increased task demand. Two potential problems for automated systems are that over reliance can occur and operator skills can atrophy. However, a dynamic (adaptive) automated system that permitted or required the operator to perform functions at different points in time could reduce the probability of skill atrophy while still relieving the workload and stress of task performance.

However, the introduction of automation can itself induce stress. Operators who work with automated systems, particularly static, inflexible automation, are relegated to the role of monitors who must respond only when untoward events occur. Sustained attention requirements are in fact quite stressful (Szalma, 1999; Warm, 1993), and paradoxically induce higher levels of perceived workload (Warm et al., 1996). Adaptive automation can mitigate this problem by dynamically assigning tasks to the machine or the human depending on the environmental conditions and the state of the operator (Hancock & Chignell, 1987). Indeed, efforts to use operator neurological state to adjust automation are currently underway (e.g., Scerbo, 2007).

Hedonomics: Promoting Enjoyable Human–Computer Interaction

Stress research has traditionally followed the edict of ergonomics and human factors in general, to do no harm and to prevent pain and injury. As with the rest of behavioral science, stress researchers sought to treat the symptoms of stress and mitigate its negative effects on performance. However, with the advent of positive psychology (Seligman & Csikszentmihalyi, 2000) there has been a movement to incorporate the promotion of pleasure and well-being rather than restricting efforts to pain prevention. Hancock coined the term *hedonomics* and defined it as that branch of science that facilitates the pleasant or enjoyable aspects of human-technology interaction (Hancock, Pepe, & Murphy, 2005). In short, the goal for hedonomics is to design for happiness. Hedonomics is a fairly new research area, but during the last 10 years, there has been a rapid growth in research concerning affect and pleasure. Affective evaluations provide a new and different perspective in Human Factors Engineering. It is not how to evaluate users—it is how the user evaluates (Hancock, Pepe, et al., 2005). The research on hedonic values and seductive interfaces is in fact a welcome contrast to safety and productivity, which have dominated human factors and ergonomics. Note, however, that pleasurable interaction with technology is not necessarily conducive to happiness. Indulging pleasures can sometimes interfere with happiness and well-being (see Fromm, 1976; Kasser, 2002; Ryan & Deci, 2001).

Our argument is not that we should discard current methods in human factors and ergonomics. Clearly functionality and usability are necessary conditions for pleasurable interaction with technology. If an interface does not function in a way congruent with the user's goals, so that the user appraises the technology as an agent that is interfering with goal achievement, that interaction is likely to be stressful and system performance more vulnerable to decline. However, function and usability are not sufficient conditions for pleasurable interactions with technology. The interface should be designed such that it affords appraisals of the technology as a convivial tool (Illich, 1973) or aid. One can also utilize the human tendency to anthropomorphize technology to facilitate appraisals of the technology as helpful and supportive rather than as an enemy (Luczak, Roetting, & Schmidt, 2003).

Hedonomic design will be of obvious importance for development of consumer products, but in principle, it can also transform the very nature of work, rendering it fun. Although there will be some tasks which will never be enjoyable, there are many individuals who have jobs that could be made more enjoyable by designing the tasks such that they promote teletic work (Csikszentmihalyi, 1990) and facilitate intrinsic motivation (Deci & Ryan, 2000).

Teletic work and intrinsic motivation. A useful theoretical framework for hedonomics is Self-Determination Theory (SDT; Deci & Ryan, 1985, 2000; Ryan & Deci, 2000, 2001). From this perspective there are three organismic needs that are essential for facilitating intrinsic motivation for task activity and the positive affect that can accompany such states. These needs are for competence (self-efficacy; see also Bandura, 1997), autonomy (personal agency, not independence per se), and relatedness. An important difference between this theory and other theories of motivation is the recognition that there are qualitatively different forms of motivation (Gagne & Deci, 2005). Thus, in SDT five categories of motivated behavior are identified that vary in the degree to which the motivation is self-determined. Four of the categories reflect extrinsic motivation and one category is intrinsic motivation. In the latter case, an individual is inherently motivated to engage in activity for its own sake or for the novelty and challenge. The four extrinsic motivation categories vary in the degree to which regulation of behavior is internalized by the individual and therefore more autonomous and self-determined (Ryan & Deci, 2000). The process of internalization involves transforming an external regulation or value into one that matches one's own values. The development of such autonomous motivation is crucial to skill development, since the person must maintain effort throughout a long and arduous process. Individuals who are autonomously motivated to learn are those who develop a variety of effective self-regulation strategies, have high self-efficacy, and who set a number of goals for themselves (Zimmerman, 2000). Further, effective self-regulation develops in four stages: observation, emulation, self-control, and self-regulation. Successful skill development involves focus on process goals in early stages of learning and outcome goals in the fourth stage (Zimmerman, 2000).

Intrinsic motivation and skill development. Research has established that intrinsic motivation is facilitated by conditions promoting autonomy, competence, and relatedness (see Deci & Ryan, 2000). Three factors that support autonomy are (a) meaningful rationale for doing the task, (b) acknowledgement that the task might not be interesting, and (c) an emphasis on choice rather than control. It is important to note that externally regulated motivation predicts poorer performance on heuristic tasks (Gagne & Deci, 2005), suggesting that as experts develop better knowledge representations it will be crucial to promote internal regulation of motivation. Although intrinsic motivation has been linked to how task activities and environmental contexts meet psychological needs, it is not clear why skilled performers are able to meet these needs, or why individuals choose a particular computer interface. It is likely that interest in activities codevelops with abilities and traits (see Ackerman & Heggestad, 1997), but this issue needs thorough investigation in the context of complex computer environments that require highly skilled workers.

In addition to the issue of efficacy and self-regulation, there is a need to examine the process by which individuals internalize extrinsic motivation as they gain experience with a particular interface or system. In particular, Gagne and Deci (2005) noted that little research has examined the effect of reward structures and work environments on the internalization process. It is likely that those environments that are structured to meet basic needs will more likely facilitate internalization processes and inoculate learners against the trials and tribulations that face them as they interact with new technologies.

Teletic work and motivational affordances. Teletic, or autoteletic, work refers to "work" that is experienced as enjoyable and is associated with flow or optimal experience characterized by a sense of well being and harmony with one's surroundings (Csikszentmihalyi, 1990). There is variation in both tasks and individuals with respect to the degree to which the human-technology interaction is teletic. There are four categories in which individuals tend to fall with respect to their relation to work. First, there is a small proportion of the population that are always happy in life, regardless of their activity, which Csikszentmihalyi referred to as individuals with an autotelic personality. There are also individuals who are predisposed to happiness about a specific task. They appraise such tasks as enjoyable, and seek out these activities. The third group consists of individuals who enjoy specific activities but cannot do them professionally, such as amateur athletes. The vast majority of people, however, do work for purely functional reasons (e.g., security). For these individuals work is boring and grinding because the task itself is aversive. A goal for hedonomics, then, is to design work that can be enjoyed to the greatest extent possible. This means structuring the environment as an entire system, ranging from the specific cognitive and psychomotor demands to the organization in which the person works. Even in jobs that are not inherently enjoyable, some degree of positive affect can be experienced by workers if they their environment is structured to facilitate a sense of autonomy (personal agency), competence, and relatedness (Deci & Ryan, 2000; see also Gagne & Deci, 2005). From an ecological perspective (e.g., Flach, Hancock, Caird, & Vicente, 1995), this means identifying the *motivational affordances* in the task and work environment, and designing for these affordances. Thus, just as one might analyze the affordance structure of an interface using ecological interface design methods (e.g., Vicente & Rasmussen, 1992), one can design an environment so that the elements of the physical and social environment afford stress reduction and enhanced intrinsic motivation. Note that although an affordance is a property of the environment, it does not exist independently of the individual in the environment. Affordances therefore share conceptual elements of person-environment transactions that drive emotion and stress. They differ in that the classical definition of an affordance is a physical property of the environment (Gibson, 1966, 1979), while a transaction emphasizes the individual's subjective appraisal of the environment. In both cases, however, one cannot define the concept by isolating either the individual or the context. Thus, although affordances and invariants are considered physical properties of the environment, these concepts are still relevant for motivational processes (and see Reed, 1996).

Motivational affordances may be conceived as elements of the work environment that facilitate and nurture intrinsic motivation. The key for design is to identify motivational invariants or environmental factors that consistently determine an individual's level of intrinsic motivation across contexts. There are some aspects of work that have been identified as important for facilitating intrinsic motivation and would thus be considered motivational invariants. For instance, providing feedback that is perceived as controlling rather than informative tends to undermine a sense of autonomy and competence and thereby reduces intrinsic motivation (Deci, Ryan, & Koestner, 1999). Careful analyses of the motivational affordance structure will permit design of tasks that are more likely to be enjoyable by rendering the tools convivial (Illich, 1973) and thereby facilitating human-machine synergy (and see Hancock, 1997).

PROBLEMS FOR FUTURE RESEARCH

In this section, we will identify the areas for future research. These include a better understanding of resources and quantifying task dimensions defined in the Hancock and Warm (1989) model, which will likely reduce to the thorny problem of quantifying human information processing (see Hancock, Szalma, & Oron-Gilad, 2005). Further, we will discuss the need for research on performance-workload associations and dissociations, and the evident need for programmatic investigation of individual differences in performance, workload, and stress.

The Hancock and Warm (1989) model of stress explicitly identified task dimensions that influence stress state and behavioral adaptability. However, the metrics for these dimensions, and how specific task characteristics map to them, have yet to be fully understood. Thus, future research should aim to examine how different task components relate to performance and subjective and physiological state. Development of a quantitative model of task characteristics will permit the derivation of vectors for the prediction of adaptability under stress. Cognitive Neuroscience and Neuroergonomics in particular offer one promising approach to such development. An additional step in this direction, however, will be facilitated by improved quantitative models of how humans process information (Hancock, Szalma, & Oron-Gilad, 2005).

Understanding Mental Resources

One of the challenges for quantifying human information processing is that there is little understanding or consensus regarding the capacities that process the information. A central concept in energetic models of human performance is mental resources. Resource theory replaced arousal and served as an intervening variable to explain the relations between task demand and performance. However, a continual problem for the resource concept is to operationally define what it is. Most treatments of resources use that term metaphorically (Navon & Gopher, 1979; Wickens, 1980, 1984), and failures to specify what resources are have led some to challenge the utility of the concept (Navon, 1984). As resource theory is a central concept in the theories of stress discussed herein, and represents one of the most important issues to be resolved in future research on stress and performance, we now turn to the definitional concerns associated with the resource construct and the imperative for future research to refine the concept.

Resource metaphors. Two general categories of resource metaphors may be identified: structural metaphors and energetic metaphors. One of the earliest conceptualizations of resource capacity used a computer-based metaphor (Moray, 1967). Thus, cognitive capacity was viewed as analogous to the RAM and processing chip of a computer, consisting of information processing units that can be deployed for task performance. However, the structural metaphor has been applied more to theories of working memory than to attention and resource theory.[1] Most early resource theories, including Kahneman's (1973) original view and modifications by Norman and Bobrow (1975), Navon and Gopher (1979), and Wickens (1980, 1984), applied energetic metaphors to resources, and conceptualized them as commodities or as pools of energy to be spent on task performance. In general, energetic approaches tend to employ either economic or thermodynamic/hydraulic metaphors. The economic model is reflected in the description of resources in terms of supply and demand: Performance on one or more tasks suffers when the resource demands of the tasks exceed available supply. Presumably, the total amount of this supply fluctuates with the state of the individual, with the assets diminishing with increases in the intensity or duration of stress. Although Kahneman's original conception allowed for dynamic variation available resource capacity, most early models assumed a fixed amount of resources (see Navon & Gopher, 1979). In thermodynamic analogies, resources are a fuel that is consumed, or a tank of liquid to be divided among several tasks, and under stressful conditions the amount of resources available is depleted and performance suffers. In discussing his version of resource theory, Wickens (1984) warned that the hydraulic metaphor should not be taken too literally, but most subsequent descriptions of resources have employed visual representations of resources as just this form (e.g., a tank of liquid). Similarly, many discussions of resource availability and expenditure adopt the economic language of supply and demand, and Navon and Gopher explicitly adopted principles of microeconomics in developing their approach. An additional problem for resource theory is that in most cases (e.g., Navon & Gopher, 1979; Wickens, 1980, 1984), the structural and energetic metaphors were treated as interchangeable, a further testament to the ambiguity of the construct.

A problem with using nonbiological metaphors to represent biological systems is that such models often fail to capture the complexity and the unique dynamic characteristics (e.g. adaptive responses) of living systems. For instance, a hydraulic model of resources links the activity of a tank of liquid, governed by thermodynamic principles, to the action of arousal mechanisms or energy reserves that are allocated for task performance. However, a thermodynamic description of the physiological processes underlying resources is at a level of explanation that may not adequately describe the psychological processes that govern performance. Thermodynamic principles can be applied to the chemical processes that occur within and between neurons, but they may be less useful in describing the behavior of large networks of neurons.[2] Similarly, economic metaphors of supply and demand may not adequately capture the relation between cognitive architecture and energy allocated for their function. Economic models of resources define them as commodities to be spent on one or more activities, and they assume an isomorphism between human cognitive activity and economic activity, an assumption which may not be tenable. Indeed, Navon and Gopher (1979) admitted that their static economic metaphor for multiple resources may need to be replaced by a dynamic one that includes temporal factors (e.g. serial versus parallel processing; activity of one processing unit being contingent upon the output of another). Such concerns over the metaphors used to describe resources are hardly new (Navon, 1984; Wickens, 1984), but their use has become sufficiently ingrained in thinking about resources and human performance that reevaluation of the metaphors is warranted. A regulatory model based on physiology may serve as a better metaphor (and, in the future may serve to describe resources themselves to the extent that they can be established as a hypothetical construct) to describe the role of resources in human cognition and performance. However, even a physiologically-based theory of resources must be tempered by the problems inherent in reducing psychological processes to physiological activity.

Function of resources. Another problem for resource theory is the absence of a precise description of how resources control different forms of information processing. Do resources determine the energy allocated to an information processor (Kahneman, 1973), do they provide the space within which the processing structure works (Moray, 1967), or does the processor draw on the resources as needed (and available)? In the latter case, the cognitive architecture would drive energy consumption and allocation, but the locus of control for the division of resources remains unspecified in any case. Presumably, an executive function that either coordinates information processors drawing on different pools of resources or decides how resources will be allocated must itself consume resources, in terms of both energy required for decision making and mental space or structure required. Hence, resource theory does not solve the homunculus problem for theories of attention, nor does it adequately describe resource allocation strategies behind performance of information processing tasks.

Empirical tests of the model. Navon and Gopher (1979) commented on the problem of empirically distinguishing declines in performance due to insufficient supply from those resulting from increases in demand. They asked, "When the performance of a task deteriorates, is it because the task now gets fewer resources or because it now requires more?" (p. 243). Navon and Gopher characterized the problem as distinguishing between changes in resources and changes in the subject-task parameters that constrain resource utilization, and they offered two approaches to avoid this difficulty. One approach is to define the fixed constraints of the task and observe how the information processing system manages the processes within those constraints. The degree of freedom of the system, in this view,

---

[1]This is a curious historical development, since these relatively separate areas of research converge on the same psychological processes.

[2]The argument here is not that neural structures are not constrained by the laws of thermodynamics—clearly they are—but that thermodynamic principles implied by the metaphor are not sufficient for the development of a complete description of resources and their relation to cognitive activity.

is the pool of resources available, in which the term *resource* is interpreted broadly to include quality of information, number of extracted features, or visual resolution. The subject-task parameters define what is imposed on the system (the demands) and the resources refer to what the system does in response to the demands (allocation of processing units). From this perspective resources can be manipulated by the information processing system within the constraints set by the subject-task parameters. A second approach is to distinguish the kind of control the system exerts on resources, between control on the use of processing devices (what we have called "structure") and the control of the properties of the inputs that go into these devices. The devices are processing resources. The other kind of control is exerted on input resources, which represents the flexibility the person has for determining which inputs are operated on, as determined by subject-task parameters. Processing resources are limited by the capacities of the information processors, while the input resources are limited by subject-task parameters (and allocation strategies that determine which information the operator attends to). Presumably, the individual would have some control over the allocation strategy, in terms of the processing resources devoted to a task, although these can also be driven by task demands (e.g. a spatial task requires spatial processing units). Navon and Gopher did not advocate either approach, but presented them as alternatives for further investigation. The implication for examining the resource model of stress is that one must manipulate both the subject-task parameters (e.g. by varying the psychophysical properties of the stimulus, manipulating the state of the observer, or varying the kind of information processing demanded by the task) as well as the allocation strategies the operator uses (the input resources—e.g. payoff matrices, task instructions). This would provide information regarding how specific stressors impair specific information processing units and how they change the user's resource allocation strategies in the presence of stress that is continuously imposed on operators of complex computer-based systems.

In a later article, Navon (1984) moved to a position less favorable toward resources than the earlier approach, asserting that predictions derived by resource theory could be made, and results explained, without appealing to the resource concept (see also Rugg, 1986). One could instead interpret effects in terms of the outputs of information processors. Most manipulations, such as difficulty (which in his view influences the efficiency of a unit of resources) or complexity (which affects the load, or the number of operations required) influence the demand for processing, with supply having no impact upon their interaction. However, this approach assumes a clear distinction between outputs of a processing system and the concept of a resource, and Navon's notion of specific processors seems blurred with the notion of a resource, as both are utilized for task performance. Nevertheless, his critique regarding the vagueness of the resource concept is relevant, and Navon argued that if resources are viewed as an intervening variable rather than a hypothetical construct, the concept has utility for describing the process.

Structural mechanisms. If different kinds of information processing draw on different kinds of resources, in terms of the

information processors engaged in a task, stressors may have characteristic effects on each resource. In addition, as Navon and Gopher (1979) noted, an aspect of resource utilization is the efficiency of each resource unit. It may be that stress degrades the efficiency of information processing units, independent of energy level or allocation strategy (cf., Eysenck, M. W. & Calvo, 1992). Investigation of such effects could be accomplished by transitioning between tasks requiring different kinds of information processing and determining if the effects of stress on one structure impacts the efficiency of a second structure.

The quality of resources can vary not only in terms of the kind of information processing unit engaged, but also in terms of the kind of task required. Following Rasmussen's (1983) classification system for behavior as a heuristic for design, some tasks require knowledge-based processing, in which the operator must consciously rely on his or her mental model of the system in order to achieve successful performance. Other tasks fall under the category of rule-based behavior, in which a set of rules or procedures define task performance. The third category is skill-based behavior, in which the task is performed with a high degree of automaticity. Presumably, each kind of task requires different amounts of resources, but they may also represent qualitatively different forms of resource utilization. In other words, these tasks may differ in the efficiency of a unit of resources as well as different effort allocation strategies. As task performance moves from knowledge to rule to skill based processing (e.g. with training), the cognitive architecture may change such that fewer information processing units are required, and those that are engaged become more efficient. Moreover, the way in which each of these systems degrade with time under stress may be systematic, with the more fragile knowledge-based processing degrading first, followed by rule based processing, with skill based processing degrading last (at this point, one may begin to see breakdown of not only psychological processes but physiological ones as well; see Hancock & Warm, 1989). This degradation may follow a hysteresis function, such that a precipitous decline in performance occurs as the operator's resource capacity is reduced below a minimum threshold for performance. Moreover, these processes may recover in an inverse form, with skill-based processing recovering first, followed by rule and knowledge-based processing.

Note that it may be difficult to distinguish pure knowledge-based processing from rule- or skill-based activity. An alternative formulation is the distinction between controlled and automatic processing (Schneider & Shiffrin, 1977). Although originally conceived as categories, it is likely that individuals engaged in real-world tasks utilize both automatic and controlled processing for different aspects of performance and that for a given task there are levels of automaticity possible. Treating skills as a continuum rather than as discrete categories may be a more theoretically useful framework for quantifying resources and information processing, and thereby elucidating the effects of stress on performance.

Energetic mechanisms. To investigate the energetic aspects of resources, one must manipulate environmentally-based perturbations, in the form of external stressors (noise, heat) and task demands, to systematically affect inflow versus outflow of energy. Presumably, inflow is controlled by arousal levels, physiological energy reserves, and effort. One could examine performance under manipulations of energetic resources under dual task performance (e.g., What happens to performance on two tasks under sleep deprivation or caffein consumption?). For example, the steady state can be perturbed by increasing (e.g. caffeine) or decreasing (e.g. sleep deprivation) energy while systematically varying the demands for two tasks.

Structure and energy. Another empirical challenge is to distinguish resources as structure from resources as energy. Given the definitional problems associated with the resource concept, it is not clear whether performance declines because of reduction in energy level or degradation in structures (e.g., failures or declines in the efficiency of the processing units), or a combination of both. If structure and energy are distinct elements of resources, it is hypothetically possible to manipulate one while holding the other constant, although the validity of that assumption is questionable. Is it possible to manipulate specific forms of information processing under constant energy level? Is it possible to manipulate energy level independent of which cognitive processes are utilized? If the decline in available resources is, at least in part, due to the degradation of particular information processing units, then transferring to a task requiring the same processor should lead to worse performance than transferring to one that is different (cf., Wickens, 1980, 1984). For instance, if a person engages in a task requiring verbal working memory while under stress, then transitions to a task requiring spatial discrimination, performance on the latter should depend only on energetic factors, not on structural ones. Note, however, that in this case the effects of different mental capacities would be confounded with the effects of novelty and motivation on performance.

Application of neuroergonomics. The burgeoning field of Neuroergonomics seeks to identify the neural bases of psychological processes involved in real-world human-technology interaction (Parasuraman, 2003). As we have stated elsewhere (Hancock & Szalma, 2007), recent advances in Neuroergonimcs promises to identify cognitive processes and their link to neurological processes. This may permit a more robust and quantitative definition of resources, although we caution that a reductionistic approach is not likely to be fruitful (and see Hancock & Szalma, 2003b). In addition, the stress concept itself rests in part on more precise definitions of resources (Hancock & Szalma, 2007). Thus, resolution of the resource issue in regard to cognitive processing and task performance would also clarify the workload and stress concepts. We view Neuroergonomics as one promising avenue for future research to refine the workload and stress and resource concepts.

Development of the Adaptation under Stress Model

Quantify the task dimensions. A major challenge for the Hancock and Warm (1989) model is the quantification of the base axes representing task dimensions. Specification of these dimensions is necessary if the vector representation postulated by Hancock and Warm is to be developed and if the resource construct is to be more precisely defined and quantified. However, task taxonomies that are general across domains present a theoretical challenge, because they require an understanding and quantification of how individuals process information along the spatial and temporal task dimensions, and how these change under stressful conditions. Quantification of information processing, and subsequent quantification of the base axes in the Hancock and Warm model, permit the formalization of the vector representation of adaptive state under stress (see Fig. 6.4).

Attentional narrowing. Recall that Hancock and Weaver (2005) argued that the distortions of spatial and temporal perception have a common attentional mechanism. Two implications of this assertion are (a) that events (internal or external) that distort one dimension will distort the other, and (b) that these distortions are unlikely to be orthogonal. With very few exceptions, little research has addressed the possibility of an interaction between distortions of spatial and temporal perceptions in stressful situations on operator performance. Preliminary evidence suggests that these two dimensions may in fact not share a common mechanism (Ross et al., 2003; Thropp et al., 2004), although further research is needed to confirm these findings. An additional important issue for empirical research is whether we are dealing with time-in-memory or time-in-passing (and to some extent space-in-memory vs. space-in-passing). Thus, the way in which perceptions of space and time interact to influence operator state will depend upon how temporal perceptions (and spatial perception, for that matter) are measured.

A possible explanation for perceptual distortions under conditions of heavy workload and stress concerns the failure to switch tasks when appropriate. Switching failures may be responsible for the observation in secondary task methodology that some participants have difficulty dividing their time between tasks as instructed (e.g., 70% to the primary task and 30% to the secondary task). This difficulty may result from the participant's inability to accurately judge how long he or she has attended to each task during a given time period. The degree to which distortions in perception of space-time are related to impairments in task switching under stressful conditions, and the degree to which these distortions are related to attention allocation strategies in a secondary task paradigm, are questions for empirical resolution.

Stressor characteristics. Even if space and time do possess a common mechanism, it may be that specific stressors do not affect spatial and temporal perceptions in the same way. For instance, heat and noise may distort perception of both space and time, but not to the same degree or in the same fashion. It is important to note that spatial and temporal distortions may be appraised as stressful, as they might interfere with the information processing requirements of a task. Consequently, some kinds of information processing might be more vulnerable to one or the other kind of perceptual distortion. Clearly, performance on tasks requiring spatial abilities, such as mental rotation, could suffer as a result of spatial distortion, but they might be unaffected (or, in some cases, facilitated) by temporal distortion. Other tasks, such as those that rely heavily on working memory, mathematical ability, or tasks requiring target detection, could each show different patterns of change in response to space-time distortion.

Potential benefits of space-time distortion. Under certain conditions, the narrowing of spatial attention can benefit performance through the elimination of irrelevant cues. The precise conditions under which this occurs, however, remain unclear. In addition, it is important to identify the circumstances under which time distortion might actually prove beneficial. Here, operators perceive that they have additional time to complete the task at hand (Hancock & Weaver, 2005). This would have great benefit in task performance situations where attentional narrowing is less likely to have deleterious effects. At this point, this is an empirical question that might be amenable to controlled testing.

Changes in adaptation: the roles of time and intensity. The degree to which a task or the physical and social environment imposes stress is moderated by the characteristics of the stimuli as well as the context in which events occur. However, two factors that seem to ubiquitously influence how much stress impairs adaptation are the (appraised) intensity of the stressor and the duration of exposure. We have recently reported meta-analytic evidence that these two factors jointly impact task performance across different orders of task (e.g., vigilance, problem solving, tracking; see Hancock, Ross, & Szalma, in press). Duration is further implicated in information processing itself, and may be a central organizing principle for information processing in the brain (Hancock, Szalma, & Oron-Gilad, 2005). Empirical research is needed, however, to programmatically explore the interactive effects of these two variables across multiple forms of information processing.

Understanding Performance-Workload Associations/Dissociations

Task factors. Although Hancock (1996) and Yeh and Wickens (1988) articulated the patterns of performance-workload relations and how these are diagnostic with respect to processing requirements, there has been little systematic effort to further investigate these associations/dissociations. The primary question is what factors drive dissociations and insensitivities when they occur. For instance, for vigilance mostly associations are observed, while for other tasks, such as those with high working memory demand, dissociations are more common (Yeh & Wickens, 1988). Enhanced understanding of these relations would inform the Hancock and Warm (1989) model by permitting specification of the conditions under which individuals pass over the thresholds of failure at each level of person-environment transaction/adaptation.

Multidimensionality of workload. To date, consideration of performance-workload dissociations has been primarily concerned with global measures of perceived workload. However, there is clear evidence that perceived workload is in fact multidimensional. For instance, vigilance tasks are characterized by high levels of mental demand and frustration (Warm, Dember, & Hancock, 1996). It is likely that the pattern of performance-

workload links will be different for different orders of performance (different tasks) but also for different dimensions of workload. One approach to addressing this question would be to systematically manipulate combinations of these two variables. For instance, if we consider performance in terms of detection sensitivity, memory accuracy, speed of response, and consider the dimensions of workload defined by the NASA Task Load Index (Hart & Staveland, 1988), one could examine how variations in memory load or discrimination difficulty link to each subscale.

INDIVIDUAL DIFFERENCES IN PERFORMANCE, WORKLOAD, AND STRESS

Elsewhere, we reviewed the relations between individual differences in state and trait to efforts to quantify human information processing (Szalma & Hancock, 2005). Here, we address how individual differences (state and trait) are related to stress and coping.

Trait Differences

Individual differences research has been a relatively neglected area in human factors and experimental psychology. Much of the early work on individual differences was done by individuals not concerned with human-technology interaction, to the extent that a bifurcation between two kinds of psychology occurred (Cronbach, 1957). There is evidence, however, that affective traits influence information processing and performance. Thus, extraversion is associated with superior performance in working memory tasks and divided attention, but also with poorer sustained attention (cf., Koelega, 1992). Trait anxiety is associated with poorer performance, although results vary across task types and contexts (Matthews et al., 2003). A possible next step for such research will be to systematically vary task elements, as discussed previously in the context of the Hancock and Warm (1989) model, and test hypotheses regarding how trait anxiety relates to specific task components (for an example applied to Extraversion, see Matthews, 1992). The theoretical challenge for such an undertaking is that it requires a good taxonomic scheme for tasks as well as a well-articulated theory of traits and performance. However, trait theories have neglected specific task performance, focusing instead on global measures (e.g., see Barrick, Mount, & Judge, 2001), and there is a lack of a comprehensive theory to account for trait-performance relations (Matthews et al., 2003). Most current theories are more like frameworks that do not provide specific mechanisms for how personality impacts cognition and performance (e.g., see McCrae & Costa, 1999). Although H. J. Eysenck (1967) proposed a theory of personality based on arousal and activation, which has found some support (Eysenck & Eysenck, 1985), there has also been evidence that arousal and task difficulty fail to interact as predicted (Matthews, 1992). H. J. Eysenck's (1967) theory was also weakened by the general problems associated with arousal theory accounts for stress effects (Hockey, R., 1984). An alternative formulation is that of Gray (1991), who argued for two systems, one responding to reward signals and one with punishment. The behavioral activation system (BAS) is associated with positive affect, while the behavioral inhibition system with negative affect. In a review and comparisons of the H. J. Eysenck and Gray theories, Matthews and Gilliland (1999) concluded that both theories have only been partially supported, but that Gray's BAS/BIS distinction provides a superior match to positive and negative affect relative to H. J. Eysenck's arousal dimensions. Further, the BAS/BIS accords with theories of approach/avoidance motivation (e.g., Elliot & Covington, 2001). There are also theories that focus on a particular trait, such as Extraversion (Humphreys & Revelle, 1984) or trait anxiety (Eysenck, M. W., & Calvo, 1992). While useful, such specific theories do not encompass other traits or interactions among traits. Such interactive effects can influence cognitive performance and perceived stress and workload (Szalma, Oron-Gilad, Stafford, & Hancock, 2005). These interactions should be further studied with an eye to linking them to information processing theories.

Affective State Differences

It is intuitive that stress would induce more negative affective states, and that traits would influence performance via an ef-

fect on states. For instance, one would expect that trait anxiety would influence performance because high trait anxious individuals experience state anxiety more frequently than those low on that trait. While such mediation effects are observed, there is also evidence that, for certain processes, such as hyper vigilance to threat, trait anxiety is a better predictor of performance than state anxiety (Eysenck, M. W., 1992). In terms of appraisal theory, traits may influence the form and content of appraisal, as well as the coping skills the individual can deploy to deal with the stress. In regard to the adaptation, it is likely that individual differences in both trait and state will influence adaptation, both behavioral and physiological, by affecting the width of the plateau of effective adaptation at a given level, and by changing the slope of decline in adaptation when the adaptation threshold has been reached. That is, higher skill levels protect from declines in adaptive function by increasing the threshold for failure at a given level (e.g., comfort, performance, physiological response). The modification of the Hancock and Warm (1989) model, illustrating these individual differences effects, is shown in Fig. 6.6. Multiple frameworks of state dimensions exist, but most focus on either two (e.g., Thayer, 1989; Watson & Tellegen, 1985), or three (Matthews et al., 1999, 2002). In the context of task performance, Matthews and his colleagues identified three

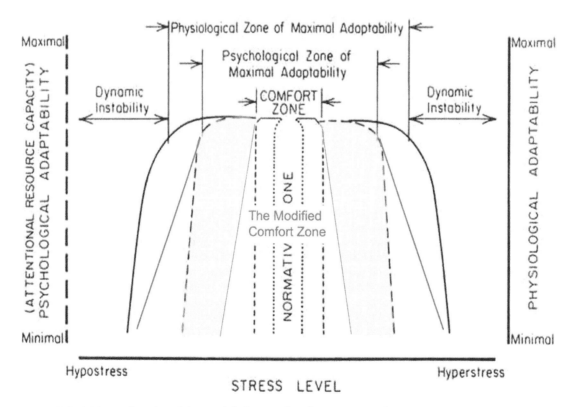

FIGURE 6.6. The adaptability model of Hancock and Warm (1989) shown in Figure 6.3 has been modified to illustrate how individual differences may influence stress and adaptation. It is likely that cognitive and affective traits influence both the width of the comfort and performance zones (i.e., the 'thresholds' for declines in adaptation) as well as the rate of decline in adaptability when a threshold has been crossed. For instance, individuals high in trait anxiety would likely have a narrower plateau of stability and would therefore manifest lower thresholds for discomfort and performance degradation than individuals low on that trait. Further, the rate of decline in adaptation may increase as a function of trait anxiety.

broad state dimensions reflecting the cognitive, affective, and motivational aspects of an individual's current psychological state. These dimensions are "worry," which reflects the cognitive dimension of stress, and "task engagement" and "distress," which reflect the affective, cognitive, and motivational components of state. Specifically, a high level of distress is indicative of overload in processing capacity, while task engagement reflects a theme of commitment to effort (Matthews et al., 2002). Matthews and his colleagues (2002) demonstrated that changes in task demand influence the pattern of stress state. It is therefore important to incorporate assessment of operator state into the interface design process so that the interaction with the technology fosters task engagement and minimizes distress and worry.

Attentional narrowing and adaptive response. As with other aspects of perception, there are individual differences in the perception of space and time (Hancock & Weaver, 2002; Wachtel, 1967). Further, because the subjective experience of stress is often multidimensional, it may be that two individuals are subjectively stressed by the same situation but that their stress state profiles differ. Individuals are also likely to differ in the strategies they employ to cope with the distortions of space-time they experience while in a stressful environment, and these coping differences, if they exist, might depend on the quality (e.g. noise, heat, low signal salience) and source (e.g. environment, the task) of the stress and the personality traits of the individual.

Hedonomics and individual differences. In addition to application of individual differences research to development of training or selection procedures, individual of relevant individual different variables can promote hedonomic approaches to design and facilitate individuation in interface design. Thus, if the traits that influence the subjective experience of an interaction with technology are identified, that interface can then be configured to meet the preferences and the trait/state profile of the individual user and promote positive affective states. However, for such efforts to succeed, the relations among traits and cognitive, perceptual, and motor performance will need to be established via theory-guided empirical research.

IMPLICATIONS OF STRESS FOR RESEARCHERS AND PRACTITIONERS

For both research and design application, the extant research on stress and performance indicates that assessment of workload and affective state are important for a more complete understanding of HCI. Such assessments can aid in identifying which components of an interface or task are appraised as stressful and thereby design to mitigate their negative effects. For instance, research is needed to establish which task parameters control the patterns of performance-workload associations and dissociations, and how these change dynamically as a function of time on task. The Hancock and Warm (1989) model of stress established general task dimensions (space-time) that influence stress state and behavioral adaptability, but the metrics

for these dimensions remain elusive. This problem results from the central issue regarding how to quantify human information processing (Hancock, Szalma, & Oron-Gilad, 2005) and define mental resources more precisely (Hancock & Szalma, 2007). Efforts to resolve these definitional problems would improve stress theory and its application to interface design. Future research should therefore examine the relations between task dimensions and user characteristics, and how these change over time and under high-stress conditions.

In addition to changing the task, there are other techniques that can be applied to the design of HCIs for use in stressful environments. These include skill development (e.g., Hancock, 1986) and use of configural displays (Hancock & Szalma, 2003a; Wickens, 1996), as well as technologies employing adaptive automation and decision aids (Hancock & Chignell, 1987). In regard to skill development in particular, an area in need of research is how individuals who develop expertise also learn how to cope with stress while performing the task. Understanding how individuals accomplish this will require advances in understanding how different forms of stress influence different forms of information processing.

It is also important for both researchers and practitioners to consider the characteristics of the user and to consider how these characteristics interact with the task or interface to influence performance. Understanding how individual differences influence human–computer interaction can facilitate development of tailored training regimens as well as interfaces that more effectively adapt to the user. Systems that can respond to changes in operator affective state will achieve the desired human-machine synergy in HCI (c.f., Hancock, 1997). Realizing these goals, however, will require adequate theory development and subsequent empirical research to determine the nature of the relations among the person and environmental variables. It will be particularly important to design interfaces that permit autonomous motivation (Deci & Ryan, 2000), and to understand how operators of computer-based systems can internalize extrinsic motivation as they gain experience with the task. (Gagne & Deci, 2005). We suggest here that researchers and designers identify the motivational affordances in the task environment and utilize these to enhance the experience of HCI and improve overall system performance under stress. Motivational affordances will be elements of the work environment that facilitate and nurture intrinsic motivation. Particularly important for design will be to identify motivational invariants, which are those environmental factors that consistently determine an individual's level of intrinsic (or extrinsic) motivation across contexts. Careful analyses of the motivational affordance structure will permit design of tasks that are more likely to be enjoyable by rendering the tools convivial (Illich, 1973) and thereby facilitating human-machine synergy (and see Hancock, 1997).

SUMMARY AND CONCLUSIONS

In this chapter, we reviewed three theories of stress and performance and their relevance for human-technology interaction. We also showed that despite separate research traditions, workload and stress might be viewed as different perspectives on the same problem. We outlined some general principles for stress mitigation, and issues requiring further research. Of particular importance will be establishing sound measures of information processing and mental resources, as well as articulating the relevant task dimensions and how they relate to self-regulatory mechanisms. Given that stress can only be understood in relation to the transaction between an individual and the environment, it will be crucial to establish how traits and states of the individual influence their appraisals of their environments. Finally, it will be important in practical application to treat stress at multiple levels, ranging from the physiological to the organizational sources of adverse performance effects. Traditional attempts to treat stress problems unidimensionally will continue to fail until the person, task, and physical, social, and organizational environments are treated as a system. Researchers and practitioners in HCI should therefore expand their efforts beyond the design of the displays and controls of interfaces and include assessment of the person factors that influence performance as well as the design of the physical and social environment in which the human–computer interaction occurs.

The argument here is not that neural structures are not constrained by the laws of thermodynamics—clearly they are—but that thermodynamic principles implied by the metaphor are not sufficient for the development of a complete description of resources and their relation to cognitive activity.

References

Ackerman, P. L., & Heggestad, E. D. (1997). Intelligence, personality, and interests: Evidence for overlapping traits. *Psychological Bulletin, 121*, 219–245.

Aspinwall, L. G., Richter, L., & Hoffman, R. R. (2002). Understanding how optimism works: An examination of optimists' adaptive moderation of belief and behavior. In E. C. Chang (Ed.), *Optimism and pessimism: Implications for theory, research, and practice* (pp. 217–238). Washington, DC: American Psychological Association.

Bandura, A. (1997). *Self-efficacy: The exercise of control.* NY: W.H. Freeman and Company.

Barrick, M. R., Mount, M. K., & Judge, T. A. (2001). Personality and performance at the beginning of the new millennium: What do we know and where do we go next? *Personality and Performance, 9*, 9–29.

Bennett, K. B., & Flach, J. M. (1992). Graphical displays: Implications for divided attention, focused attention, and problem solving. *Human Factors, 34*, 513–552.

Bless, H. (2001). Mood and the use of general knowledge structures. In L. L. Martin & G. L. Clore (Eds.), *Theories of mood and cognition: A user's guidebook* (pp. 9–26). Mahwah, NJ: Erlbaum.

Brown, J. M., & Campbell, E. A. (1994). *Stress and policing: Sources and strategies.* Chichester, UK: Wiley.

Bursill, A. E. (1958). The restriction of peripheral vision during exposure to hot and humid conditions. *Quarterly Journal of Experimental Psychology, 10*, 113–129.

Carver, C. S., & Scheier, M. F. (1998). *On the self-regulation of behavior.* NY: Cambridge University Press.

Cornsweet, D. M. (1969). Use of cues in the visual periphery under conditions of arousal. *Journal of Experimental Psychology, 80*, 14–18.

Cronbach, L. J. (1957). The two disciplines of scientific psychology. *American Psychologist, 12*, 671–684.

Csikszentmihalyi, M. (1990). *Flow: The psychology of optimal experience.* NY: Harper & Row.

Deci, E. L., & Ryan, R. M. (1985). *Intrinsic motivation and self-determination in human behavior.* NY: Plenum Press.

Deci, E. L., & Ryan, R. M. (2000). The "what" and "why" of goal pursuits: Human needs and the self-determination of behavior. *Psychological Inquiry, 11*, 227–268.

Deci, E. L., Ryan, R. M., & Koestner, R. (1999). A meta-analytic review of experiments examining the effects of extrinsic rewards on intrinsic motivation. *Psychological Bulletin, 125*, 627–668.

Dirkin, G. R., & Hancock, P. A. (1984). Attentional narrowing to the visual periphery under temporal and acoustic stress. *Aviation, Space and Environmental Medicine, 55*, 457.

Dirkin, G. R., & Hancock, P. A. (1985). An attentional view of narrowing: The effect of noise and signal bias on discrimination in the peripheral visual field. In I. D. Brown, R. Goldsmith, K. Coombes, & M. A. Sinclair (Eds.), *Ergonomics International 85: Proceedings of the Ninth Congress of the International Ergonomics Association,* (pp. 751–753). Bournemouth, England, September.

Driskell, J. E., Johnston, J. H., & Salas, E. (2001). Does stress training generalize to novel settings? *Human Factors, 43*, 99–110.

Easterbrook, J. A. (1959). The effect of emotion on cue utilization and the organization of behavior. *Psychological Review, 66*, 183–201.

Elliot, A. J., & Covington, M. V. (2001). Approach and avoidance motivation. *Educational Psychology Review, 13*, 73–92.

Ericsson, K. A. (2006). The influence of experience and deliberate practice on the development of superior expert performance. In K. A. Ericsson, N. Charness, R. R. Hoffman, & P. J. Feltovich (Eds.), *The Cambridge handbook of expertise and expert performance* (pp. 683–703). Cambridge: Cambridge University Press.

Eysenck, H. J. (1967). *The biological basis of personality.* Springfield, IL: Charles C. Thomas.

Eysenck, H. J., & Eysenck, M. W. (1985). *Personality and individual differences: A natural science approach.* New York: Plenum Press.

Eysenck, M. W. (1992). *Anxiety: The cognitive perspective.* Hillsdale, NJ: Erlbaum.

Eysenck, M. W., & Calvo, M. G. (1992). Anxiety and performance: The processing efficiency theory. *Cognition and Emotion, 6*, 409–434.

Flach, J., Hancock, P., Caird, J., & Vicente, K. (Eds.). (1995). *Global perspectives on the ecology of human-machine systems vol 1.* Hillsdale, NJ: Erlbaum.

Fromm, E. (1976). *To have or to be?* NY: Harper & Row.

Gagne, M., & Deci, E. L. (2005). Self-determination theory and work motivation. *Journal of Organizational Behavior, 26*, 331–362.

Gibson, J. J. (1966). *The senses considered as perceptual systems.* Boston: Houghton Mifflin.

Gibson, J. J. (1979). *The ecological approach to visual perception.* Boston: Houghton Mifflin.

Gray, J. A. (1991). Neural systems, emotion, and personality. In J. Madden, IV (Ed.), *Neurobiology of learning, emotion, and affect.* (pp. 273–306). NY: Raven Press.

Hancock, P. A. (1986). The effect of skill on performance under an environmental stressor. *Aviation, Space, and Environmental Medicine, 57*, 59–64.

Hancock, P. A. (1996). Effects of control order, augmented feedback, input device and practice on tracking performance and perceived workload. *Ergonomics, 39*, 1146–1162.

Hancock, P. A. (1997). *Essays on the future of human-machine systems.* Eden Prairie, MN: BANTA Information Services Group.

Hancock, P. A., & Caird, J. K. (1993). Experimental evaluation of a model of mental workload. *Human Factors, 35*, 413–429.

Hancock, P. A., & Chignell, M. H. (1987). Adaptive control in human-machine systems. In P. A. Hancock (Ed.), *Human factors psychology* (pp. 305–345). Amsterdam: Elsevier.

Hancock, P. A., & Dirkin, G. R. (1983). Stressor induced attentional narrowing: Implications for design and operation of person-machine systems. *Proceedings of the Human Factors Association of Canada, 16*, 19–21.

Hancock, P. A., Pepe, A. A., & Murphy, L. L. (2005). Hedonomics: The power of positive and pleasurable ergonomics. *Ergonomics in Design, 13*, 8–14.

Hancock, P. A., Ross, J. M., & Szalma, J. L. (in press). A meta-analysis of performance response under thermal stressors. *Human Factors.*

Hancock, P. A., & Szalma, J. L. (2003a). Operator stress and display design. *Ergonomics in Design, 11*, 13–18.

Hancock, P. A., & Szalma, J. L. (2003b) The future of neuroergonomics. *Theoretical Issues in Ergonomics Science, 4*, 238–249.

Hancock, P. A., & Szalma, J. L. (2007). Stress and neuroergonomics. In R. Parasuraman & M. Rizzo (Eds.), *Neuroergonomics: The brain at work* (pp. 195–206). Oxford: Oxford University Press.

Hancock, P. A., Szalma, J. L., & Oron-Gilad, T. (2005). Time, emotion, an the limits to human information processing. In D. K. McBride & D. Schmorrow (Eds.), *Quantifying Human Information Processing* (pp. 157–175). Lanham, MD: Lexington Books.

Hancock, P. A., & Warm, J. S. (1989). A dynamic model of stress and sustained attention. *Human Factors, 31*, 519–537.

Hancock, P. A., & Weaver, J. L. (2005). On time distortion under stress. *Theoretical Issues in Ergonomics Science, 6*, 193–211.

Harris, W. C., Hancock, P. A., & Harris, S. C. (2005) Information processing changes following extended stress. *Military Psychology, 17*, 115–128.

Hart, S. G., & Staveland, L. E. (1988). Development of NASA-TLX (Task Load Index): Results of empirical and theoretical research. In P. A. Hancock & N. Meshkati (Eds.), *Human mental workload* (pp. 139–183). Amsterdam: Elsevier.

Hockey, G. R. J. (1997). Compensatory control in the regulation of human performance under stress and high workload: A cognitive-energetical framework. *Biological Psychology, 45*, 73–93.

Hockey, G. R. J., Gaillard, A. W. K., & Coles, M. G. H. (Eds.). (1986). *Energetic aspects of human information processing.* The Netherlands: Nijhoff.

Hockey, R. (1984). Varieties of attentional state: The effects of environment. In R. Parasuraman & D. R. Davies (Eds.), *Varieties of attention* (pp. 449–483). NY: Academic Press.

Hockey, R., & Hamilton, P. (1983). The cognitive patterning of stress states. In G. R. J. Hockey (Ed.), *Stress and fatigue in human performance* (pp. 331–362). Chichester: Wiley.

Humphreys, M. S., & Revelle, W. (1984). Personality, motivation, and performance: A theory of the relationship between individual differences and information processing. *Psychological Review, 91*, 153–184.

Illich, I. (1973). *Tools for conviviality.* NY: Harper & Row.

Johnston, J. H., & Cannon-Bowers, J. A. (1996). Training for stress exposure. In J. E. Driskell & E. Salas (Eds.), *Stress and human performance* (pp. 223–256) Mahwah, NJ: Erlbaum.

Kahneman, D. (1973). *Attention and effort.* Englewood Cliffs, NJ: Prentice Hall.

Kasser, T. (2002). The *high price of materialism.* Cambridge, MA: MIT Press.

Koelega, H. S. (1992). Extraversion and vigilance performance: Thirty years of inconsistencies. *Psychological Bulletin, 112*, 239–258.

Lazarus, R. S. (1991). *Emotion and adaptation.* Oxford: Oxford University Press.

Lazarus, R. S. (1999). *Stress and emotion: A new synthesis.* NY: Springer.

Lazarus, R. S., & Folkman, S. (1984). *Stress, appraisal, and coping.* NY: Springer Verlag.

Luczak, H., Roettling, M., & Schmidt, L. (2003). Let's talk: Anthropomorphization as means to cope with stress of interacting with technical devices. *Ergonomics, 46*, 1361–1374.

McCrae, R. R., & Costa, P. T. (1986). Personality, coping, and coping effectiveness in an adult sample. *Journal of Personality, 54*, 385–405.

McCrae, R. R., & Costa, P. T. (1999). A five-factor theory of personality. In L. A. Pervin & O. P. John (Eds.), *Handbook of personality: Theory and research, 2nd ed.* (pp. 139–153). NY: Guilford Press.

Matthews, G. (1992). Extraversion. In A. P. Smith & D. M. Jones (Eds.), *Handbook of human performance, Vol. 3: State and Trait* (pp. 95–126). London: Academic Press.

Matthews, G. (2001). Levels of transaction: A cognitive science framework for operator stress. In P. A. Hancock & P. A. Desmond (Eds.), *Stress, workload, and fatigue* (pp. 5–33). Mahwah, NJ: Erlbaum.

Matthews, G., Deary, I. J., & Whiteman, M. C. (2003). *Personality traits, 2nd ed.* Cambridge: Cambridge University Press.

Matthews, G., & Gilliland, K. (1999). The personality theories of H. J. Eysenck and J. A. Gray: A comparative review. *Personality and Individual Differences, 26*, 583–626.

Matthews, G., Joyner, L., Gilliland, K., Campbell, S., Falconer, S., & Huggins, J. (1999). Validation of a comprehensive stress state questionnaire: Towards a state 'big three'? In I. Mervielde, I. J. Deary, F. De Fruyt, & F. Ostendorf (Eds.), *Personality Psychology in Europe: Vol. 7* (pp. 335–350). Tilburg, the Netherlands: Tilburg University Press.

Matthews, G., Campbell, S. E., Falconer, S., Joyner, L. A., Huggins, J., Gilliland, et al. (2002). Fundamental dimensions of subjective state in performance settings: Task engagement, distress, and worry. *Emotion, 2*, 315–340.

Moray, N. (1967). Where is capacity limited? A survey and a model. *Acta Psychologica, 27*, 84–92.

Moray, N. (Ed.). (1979). *Mental workload: Its theory and measurement.* NY: Plenum Press.

Navon, D. (1984). Resources—A theoretical soupstone? *Psychological Review, 91*, 216–234.

Navon, D., & Gopher, D. (1979). On the economy of the human information processing system. *Psychological Review, 86*, 214–255.

Norman, D., & Bobrow, D. (1975). On data-limited and resource-limited processing. *Journal of Cognitive Psychology, 7*, 44–60.

O'Donnell, R. D., & Eggemeier, F. T. (1986). Workload assessment methodology. In K. R. Boff, L. Kaufman, & J. P. Thomas (Eds.), *Handbook of human perception and performance, Vol. II: Cognitive processes and performance.* (pp. 42-1–42-49). NY: Wiley.

Parasuraman, R. (2003). Neuroergonomics: Research and practice. *Theoretical Issues in Ergonomics Science, 4*, 5–20.

Parasuraman, R., & Hancock, P. A. (2001). Adaptive control of mental workload. In P. A. Hancock & P. A. Desmond (Eds.), *Stress, workload, and fatigue.* (pp. 305–320). Mahwah, NJ: Erlbaum.

Penley, J. A., & Tomaka, J. (2002). Associations among the big five, emotional responses, and coping with acute stress. *Personality and Individual Differences, 32*, 1215–1228.

Pilcher, J. J., Nadler, E., & Busch, C. (2002). Effects of hot and cold temperature exposure on performance: A meta-analytic review. *Ergonomics, 45*, 682–698.

Rasmussen, J. (1983). Skills, rules, and knowledge; Signals, signs, and symbols, and other distinctions in human performance models. *IEEE Transactions on Systems, Man, and Cybernetics, SMC-13*, 257–266.

Reed, E. S. (1996). *Encountering the world: Toward an ecological psychology.* Oxford: Oxford University Press.

Ross, J. M., Szalma, J. L., Thropp, J. E., & Hancock, P. A. (2003). Performance, workload, and stress correlates of temporal and spatial task demands. *Proceedings of the Human Factors and Ergonomics Society, 47*, 1712–1716.

Rugg, M. D. (1986). Constraints on cognitive performance: Some problems with and alternatives to resource theory. In G. R. J. Hockey,

A. W. K. Gaillard, & M. G. H. Coles (Eds.), *Energetics and human information processing* (pp. 353–371). Dordrecht: Martinus Nijhoff Publishers.

Ryan, R. R., & Deci, E. L. (2000). Self-determination theory and the facilitation of intrinsic motivation, social development, and well-being. *American Psychologist, 55,* 66–78.

Ryan, R. R., & Deci, E. L. (2001). On happiness and human potentials: A review of research on hedonic and eudaimonic well-being. *Annual Review of Psychology, 52,* 141–166.

Saunders, T., Driskell, J. E., Johnston, J., & Salas, E. (1996). The effect of stress inoculation training on anxiety and performance. *Journal of Occupational Health Psychology, 1,* 170–186.

Scerbo, M. W. (2007). Adaptive automation. In: R. Parasuraman and M. Rizzo (Eds.). *Neuroergonomcs: The brain at work* (pp. 239–252). Oxford: Oxford University Press.

Scerbo, M. W., Freeman, F. G., & Mikulka, P. J. (2003). A brain-based system for adaptive automation. *Theoretical Issues in Ergonomics Science, 4,* 200–219.

Scheier, M. F., & Carver, C. S. (1985). Optimism, coping, and health: Assessment and implications of generalized outcome expectancies. *Health Psychology, 4,* 219–247.

Scherer, K. R. (1999). Appraisal theory. In T. Dalgleish & M. Power (Eds.), *Handbook of cognition and emotion* (pp. 638–663). NY: Wiley.

Schneider, W., & Shiffrin, R. M. (1977). Controlled and automatic human information processing I: Detection, search, and attention. *Psychological Review, 84,* 1–66.

Seligman, M. E. P., & Csikszentmihalyi, M. (2000). Positive psychology: An introduction. *American Psychologist, 55,* 5–14.

Selye, H. (1976). *The stress of life, revised edition.* NY: McGraw-Hill.

Szalma, J. L. (1999). *Sensory and temporal determinants of workload and stress in sustained attention.* Unpublished doctoral dissertation, University of Cincinnati: Cincinnati, OH.

Szalma, J. L., Warm, J. S., Matthews, G., Dember, W. N., Weiler, E. M., Meier, A., et al. (2004). Effects of sensory modality and task duration on performance, workload, and stress in sustained attention. *Human Factors, 46,* 219–233.

Szalma, J. L., & Hancock, P. A. (2005). Individual differences in information processing. In D. K. McBride & D. Schmorrow (Eds.), *Quantifying human information processing* (pp. 177–193). Lanham, MD: Lexington Books.

Szalma, Oron-Gilad, Stafford, & Hancock (2007). Police firearms training: Individual differences in performance, workload, and stress submitted.

Thayer, R. E. (1989). *The biopsychology of mood and arousal.* NY: Oxford University Press.

Thropp, J. E., Oron-Gilad, Szalma, & Hancock (2007). *Individual preferences using automation.* Manuscript in preparation.

Thropp, J. E., Szalma, J. L., & Hancock, P. A. (2004). Performance operating characteristics for spatial and temporal discriminations: Common or separate capacities? *Proceedings of the Human Factors and Ergonomics Society, 48,* 1880–1884.

van Reekum, C., & Scherer, K. R. (1997). Levels of processing in emotion-antecedent appraisal. In G. Matthews (Ed.), *Cognitive science perspectives on personality and emotion* (pp. 259–300). Amsterdam: Elsevier.

Vicente, K. J., & Rasmussen, J. (1992). Ecological interface design: Theoretical foundations. *IEEE Transactions on Systems, Man, and Cybernetics, 22,* 589–606.

Wachtel, P. L. (1967). Conceptions of broad and narrow attention. *Psychological Bulletin, 68,* 417–429.

Warm, J. S. (1993). Vigilance and target detection. In B. M. Huey & C. D. Wickens (Eds.), *Workload transition: Implications for individual and team performance* (pp. 139–170). Washington, DC: National Academy Press.

Warm, J. S., Dember, W. N., & Hancock, P. A. (1996). Vigilance and workload in automated systems. In R. Parasuraman & M. Mouloua (Eds.), *Automation and human performance: Theory and applications* (pp. 183–200). Hillsdale, NJ: Erlbaum.

Watson, D., & Tellegen, A. (1985). Toward a consensual structure of mood. *Psychological Bulletin, 98,* 219–235.

Wickens, C. D. (1980). The structure of attentional resources. In R. Nickerson (Ed.), *Attention and performance VIII* (pp. 239–257). Hillsdale, NJ: Erlbaum.

Wickens, C. D. (1984). Processing resources in attention. In R. Parasuraman & D. R. Davies (Eds.), *Varieties of attention* (pp. 63–102). New York, NY: Academic Press.

Wickens, C. D. (1996). Designing for stress. In J. E. Driskell & E. Salas (Eds.), *Stress and human performance* (pp. 279–295). Mahwah, NJ: Erlbaum.

Yeh, Y., & Wickens, C. D. (1988). Dissociation of performance and subjective measures of workload. *Human Factors, 30,* 111–120.

Yerkes, R. (1918). Psychology in relation to the war. *Psychological Review, 25,* 85–115.

Zimmerman, B. J. (2000). Attaining self-regulation: A social cognitive perspective. In M. Boekaerts, P. R. Pintrch, & M. Zeidner (Eds.), *Handbook of self-regulation* (pp. 13–39). San Diego: Academic Press.

MOTIVATING, INFLUENCING, AND PERSUADING USERS: AN INTRODUCTION TO CAPTOLOGY

B.J. Fogg,* Gregory Cuellar, David Danielson
Stanford University

*B.J. Fogg directs the Stanford Persuasive Technology Lab and is on the consulting faculty for Stanford's Computer Science Department.

Since the advent of modern computing in 1946, the uses of computing technology have expanded far beyond their initial role of performing complex calculations (Denning & Metcalfe, 1997). Computers are not just for scientists any more; they are an integral part of workplaces and homes. The diffusion of computers has led to new uses for interactive technology; including the use of computers to change people's attitudes and behavior—in a word: persuasion. Computing pioneers of the 1940s probably never imagined computers being used to persuade.

Today, creating successful human–computer interactions (HCIs) requires skills in motivating and persuading people. However, interaction designers don't often view themselves as agents of influence. They should. The work they perform often includes crafting experiences that change people—the way people feel, what they believe, and the way in which they behave. Consider these common challenges: How can designers motivate people to register their software? How can they get people to persist in learning an online application? How can they create experiences that build product loyalty? Often, the success of today's interactive product hinges on changing people's attitudes or behaviors.

Sometimes the influence elements in HCI are small, almost imperceptible, such as creating a feeling of confidence or trust in what the computing product says or does. Other times, the influence element is large, even life altering, such as motivating someone to quit smoking. Small or large, elements of influence are increasingly present on Websites, in productivity tools, in video games, in wearable devices, and in other types of interactive computing products. Due to the growing use of computing products and to the unparalleled ability of software to scale, interaction designers may well become leading change agents of the future. Are we ready?

The study and design of computers as persuasive technologies, referred to as *captology*, is a relatively new endeavor when compared to other areas of HCI (Fogg, 1997, 1998, 2003). Fortunately, understanding in this area is growing. HCI professionals have established a foundation that outlines the domains of applications, useful frameworks, methods for research, design guidelines, best-in-class examples, as well as ethical issues (Berdichevsky & Neuenschwander, 1999; Fogg, 1999; Khaslavsky & Shedroff, 1999; King & Tester, 1999; Tseng & Fogg, 1999). This chapter will not address all these areas in-depth, but it will share some key perspectives, frameworks, and design guidelines relating to captology.

DEFINING PERSUASION AND GIVING HIGH-TECH EXAMPLES

What is "persuasion"? As one might predict, scholars do not agree on the precise definition. For the sake of this chapter, *persuasion is a noncoercive attempt to change attitudes or behaviors*. There are some important things to note about this definition. First, persuasion is *noncoercive*. Coercion—the use of force—is not persuasion; neither is manipulation or deceit. These methods are shortcuts to changing how people believe or behave, and for interaction designers these methods are rarely justifiable.

Next, persuasion requires an *attempt* to change another person. The word *attempt* implies intentionality. If a person changes someone else's attitude or behavior without intent to do so, it is an accident or a side effect; it is not persuasion. This point about intentionality may seem subtle, but it is not trivial. Intentionality distinguishes between a *side effect* and a *planned effect* of a technology. At its essence, captology focuses on the planned persuasive effects of computer technologies.

Finally, persuasion deals with *attitude changes* or *behavior changes* or both. While some scholars contend persuasion pertains only to attitude change, other scholars would concur with our view: including behavior change as a target outcome of persuasion. Indeed, these two outcomes—attitude change and behavior change—are fundamental in the study of computers as persuasive technologies.

Note how attitude and behavior changes are central in two examples of persuasive technology products. First, consider the CD-ROM product *5 A Day Adventures* (www.dole5aday.com). Created by Dole Foods, this computer application was designed to persuade kids to eat more fruits and vegetables. Using *5 A Day Adventures*, children enter a virtual world with characters like "Bobby Banana" and "Pamela Pineapple," who teach kids about nutrition and coach them to make healthy food choices. The program also offers areas where children can practice making meals using fresh produce, and the virtual characters offer feedback and praise. This product clearly aims to change the attitudes children have about eating fruits and vegetables. However, even more important, the product sets out to change their eating behaviors.

Next, consider a more mundane example: Amazon.com. The goal of this Website is to persuade people to buy products again and again from Amazon.com. Everything on the Website contributes to this result: user registration, tailored information, limited-time offers, third-party product reviews, one-click shopping, confirmation messages, and more. Dozens of persuasion strategies are integrated into the overall experience. Although the Amazon online experience may appear to be focused on providing mere information and seamless service, it is really about persuasion—buy things now and come back for more.

THE FOURTH WAVE: PERSUASIVE INTERACTIVE TECHNOLOGY

Computing systems did not always contain elements of influence. It has only been in recent years that interactive computing became mature enough to spawn applications with explicit elements of influence. The dramatic growth of technologies designed to persuade and motivate represents the fourth wave of focus in end-user computing. The fourth wave leverages advances from the three previous waves (Fig. 7.1).

The first wave of computing began over 50 years ago and continues today. The energy and attention of computer professionals mainly focused on getting computing devices to work properly, and then to make them more and more capable. In short, the first wave is function.

The second wave of computing began in the 1970s with the emergence of digital gaming, first represented by companies like Atari and with products like Pong. This wave is entertainment, and it continues to swell because of continued attention and energy devoted to computer-based fun.

The third wave of computing came in the 1980s when human factors specialists, designers, and psychologists sought to create computers for ordinary people. This third wave is ease of use. Although new developments, like the computer mouse and the graphical-user interface came before 1980, a consumer product—the Apple Macintosh—generated widespread attention and energy to making computers easier to use. Like the previous two waves, the third wave keeps rolling today. It provides the foundation for most work in HCI arenas.

In addition, this brings us to the fourth wave: computers designed to persuade. Early signs of this wave appeared in the 1970s and 1980s with a handful of computing systems designed to motivate health behaviors and work productivity. However, it wasn't until the late-1990s—specifically during the rise of the World Wide Web—that more than a handful of people began to devote attention and energy to making interactive systems capable of motivating and influencing users. This fourth wave—persuasion—is new and could be as significant as the three waves that have come before it.

DOMAINS WHERE PERSUASION AND MOTIVATION MATTER

Captology is relevant to systems designed for many facets of human life. The most obvious domain is in promoting commerce—buying and branding, especially via the Web. While promoting commerce is perhaps the most obvious and lucrative application, at least 11 other domains are potential areas for persuasive technology products. The various domains, along with a sample target behavior change, are summarized in Table 7.1.

The domains in the table reflect how much persuasion is part of ordinary human experience, from personal relationships to environmental conservation. Interactive technologies have been—and will continue to be—created to influence people in these 12 domains, as well as in others that are less apparent. The way various computing products incorporate persuasion and motivation principles will evolve as computing technology

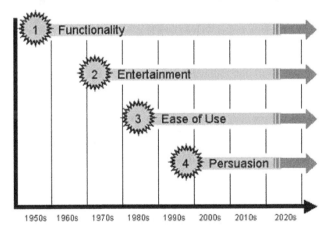

Major Waves in Computing

FIGURE 7.1. Persuasion is the fourth major wave of computing.

TABLE 7.1. 12 Domains for Persuasive Technology

Domains for Persuasive Technologies	Example
Commerce—Buying and Branding	To buy a certain product
Education, Learning, and Training	To engage in activities that promote learning
Safety	To drive more safely
Environmental Conservation	To reuse shopping bags
Occupational Productivity	To set and achieve goals at work
Preventative Health Care	To quit smoking
Fitness	To exercise with optimal intensity/frequency
Disease Management	To manage diabetes better
Personal Finance	To create and adhere to a personal budget
Community Involvement/Activism	To volunteer time at a community center
Personal Relationships	To keep in touch with their aging parents
Personal Management and Improvement	To avoid procrastination

advances and as people adopt a wider array of interactive systems for a wider range of human activities. The influence elements in these systems can be readily apparent, or they can be woven into the fabric of an interactive experience, a distinction explored in the next section.

PERSUASION AND INTERACTIVE TECHNOLOGY: TWO LEVELS OF ANALYSIS

One key insight in captology is to see that persuasion in computing products takes place on two levels: macro and micro. On the macro level, one finds products designed for an overall persuasive outcome. For example, the Dole *5 A Day* CD-ROM and the Amazon.com Website are designed specifically for persuasion. For these and other products, persuasion and motivation are the sole reason these products exist. We use the word "macrosuasion" to describe this type of big-picture target outcome.

On the other hand, one finds computing products with what I call "microsuasion." These products could be word-processing programs or spreadsheets; they do not necessarily have a persuasive outcome as the overall goal of the product. However, they will incorporate smaller elements of influence to achieve other goals. Microsuasion can be incorporated into dialogue boxes, visual elements, interactions sequences, and more. In productivity software, microsuasion can lead to increased productivity or stronger brand loyalty. The following examples will help clarify the distinction between macrosuasion and microsuasion (Fogg, 2003).

Examples of Macrosuasion

One notable example of macrosuasion is a product named Baby Think It Over. A U.S. company (www.btio.com) designed this computerized doll to simulate the time and energy required to care for a baby, with the overall purpose of persuading teens to avoid becoming parents prematurely. Used as part of many school programs in the United States, the Baby Think It Over infant simulator looks, weighs, and cries something like a real baby. The computer embedded inside the doll triggers a crying sound at random intervals; in order to stop the crying sound, the teen caregiver must pay immediate attention to the doll. If the caregiver fails to respond appropriately, the computed embedded inside the doll records the neglect. After a few days of caring for the simulated infant, teenagers generally report less interest in becoming a parent in the near future (see www.btio .com), which—along with reduced teen pregnancy rates—is the intended outcome of the device.

Next, consider Scorecard.org as another example of macrosuasion. Created by the Environmental Defense Foundation, this Website helps people find information about pollution threats in their neighborhoods. When users enter their zip code, the site lists names of the polluting institutions in their area, gives data on chemicals being released, and outlines the possible health consequences. But that's not all. Scorecard.org then encourages users to take action against the polluting or-

ganizations and makes it easy to contact policymakers to express concerns. This Website aims to increase community activism in order to pressure officials and offending institutions into cleaning up the environment. The entire point of this Website is to get people to take action against polluting institutions in their neighborhoods. This is macrosuasion.

Examples of Microsuasion

Most computing products were not created with persuasion as the main focus. Larger software categories include applications for productivity, entertainment, and creativity. Yet these same products often use influence elements, microsuasion, as part of the overall experience. Examples of interactive products using microsuasion are plentiful—and sometimes subtle. A word-processing program may encourage users to spell check text, or a Website devoted to high-school reunions may reward alumni for posting a current photograph online. This is persuasion on a microlevel.

For a deeper look at microsuasion, consider the personal finance application Quicken, created by Intuit (www.intuit.com). Quicken is a productivity software product. Its overall goal is to simplify the process of managing personal finances. Quicken uses microsuasion to accomplish this overall goal. For example, the application reminds users to take financial responsibility, such as paying bills on time. In addition, the software tracks personal spending habits and shows results in graphs, allowing projections into future financial scenarios. In addition, the software praises users for doing necessary but menial tasks, like balancing their online check registry. These microsuasion elements—reminders, visualizations, and praise—are influence elements embedded in the Quicken experience in order to change what users think and how they act. Ideally, when these microsuasion elements succeed, users benefit from Quicken's approach to managing personal finances.

Like Quicken, educational software often uses microsuasion. The overall point of most educational applications and interactive experiences is to teach facts and skills, not to persuade. However, in order to get users to stick with the program or to believe what is presented, many products will incorporate motivational elements as well as building credibility perceptions of the content. The product may seek to persuade the learner that the content is important, that the learner is able to successfully master it, and that following the guidelines of the program will lead to the greatest success. Note how these smaller elements of the program—the microsuasions—contribute to the overall goal: learning. Furthermore, interactive educational products will often incorporate elements of games, which leads to a large area related to microsuasion: computer-based gaming.

Video games are typically rich in microsuasion elements. The overall goal of most games is to provide entertainment, not to persuade. However, during the entertainment experience players can be bombarded with microsuasion elements, sometimes continuously. Video games can leverage the seven basic intrinsic motivators: challenge, curiosity, fantasy, control, competition, cooperation, and recognition (Maline & Lepper, 1987). Video games can also incorporate other categories of microsuasion, such as social-influence dynamics.

Captology is relevant to computing products designed with macrosuasion in mind—like Baby Think It Over—and to those that simply use microsuasion in order to make the product more successful—like Quicken. In both cases, designers must understand how to create interactive experiences that change the way people think and behave, whether it is for a single overall outcome or for near-term outcomes that are the building blocks of a larger experience.

NO UNIVERSAL THEORY OF PERSUASION

Creating interactive technology experiences that motivate and influence users would be easy if persuasion were fully understood. It's not. Our understanding is limited, despite the fact that the study of persuasion extends back at least 2,000 years. The fields of psychology, marketing, advertising, public-information campaigns, and others have developed theories and perspectives on how to influence and motivate people, but all approaches have limitations. The reality is this: we have no universal theory or framework for persuasion. In other words, no single set of principles fully explains what motivates people, what causes them to adopt certain attitudes, and what leads them to perform certain behaviors (Fogg, 2003; Ford, 1992). In some ways, this is not a surprise. Human psychology is complex, and persuasion is a large domain, often with fuzzy boundaries. Without a universal theory of persuasion, we must draw from a set of theories and models that describe influence, motivation, or behavior change in specific situations and for specific types of people. This limitation creates an additional challenge for designers of persuasive technology products.

Because computing technology creates new possibilities for influencing people, work in captology can lead to new frameworks, which, although not perfect, enhance the knowledge and practice in HCI. One such framework is the "Functional Triad" (Fogg, 2003).

THE FUNCTIONAL TRIAD: A FRAMEWORK FOR PERSUASIVE TECHNOLOGY

Computers play many roles, some of which go unseen and unnoticed. From a user's perspective, computers function in three basic ways: as (a) tools, as (b) media, and as (c) social actors. In the last two decades, researchers and designers have discussed variants of these functions, usually as metaphors for computer use (i.e., Kay, 1984; Verplank, Fulton, Black, & Moggridge, 1993). However, these three categories are more than metaphors; they are basic ways that people view or respond to computing technologies. These categories also represent three basic types of experiences that motivate and influence people.

Described in more detail elsewhere (Fogg, 1999, 2000, 2003), the Functional Triad is a framework that makes explicit these three computer functions—tools, media, and social actors. First, as this framework suggests, computer applications or systems function as tools, providing users with new abilities or powers. Using computers as tools, people can do things they could not do before, or they can do things more easily.

The Functional Triad also suggests that computers function as media, a role that has grown dramatically during the 1990s as computers became increasingly powerful in displaying graphics and in exchanging information over a network such as the Internet. As a medium, a computer can convey either symbolic content (i.e., text, data graphs, icons) or sensory content (i.e., real-time video, virtual worlds, simulation).

Finally, computers also function as social actors. Empirical research demonstrates that people form social relationships with technologies (Reeves & Nass, 1996). The precise causal factors for these social responses have yet to be outlined in detail, but I propose that users respond socially when computers do at least one of the following: (1) adopt animate characteristics (i.e., physical features, emotions, voice communication), (2) play animate roles (i.e., coach, pet, assistant, opponent), or (3) follow social rules or dynamics (i.e., greetings, apologies, taking turns) (Fogg, 2003).

The Functional Triad is not a theory; it is a framework for analysis and design. In all but the most extreme cases, a single interactive technology is a mix of these three functions, combining them to create an overall user experience.

In captology the Functional Triad is useful because it helps show how computer technologies can employ different techniques for changing attitudes and behaviors. For example, computers as tools persuade differently than computers as social actors. The strategies and theories that apply to each function differ. The paragraphs that follow use the Functional Triad to highlight aspects of persuasive technology, including general design strategies and approaches for creating computing products that persuade and motivate.

Computers as Persuasive Tools

In general, computers as persuasive tools affect attitude and behavior changes by increasing a person's abilities or making something easier to do (Tombari, Fitzpatrick, & Childress, 1985). Although one could propose numerous possibilities for persuasion in this manner, below are four general ways in which computers persuade as tools: by (a) increasing self-efficacy, (b) providing tailored information, (c) triggering decision making, and (d) simplifying or guiding people through a process.

Computers That Increase Self-Efficacy

Computers can increase self-efficacy (Lieberman, 1992), an important contributor to attitude and behavior change processes. Self-efficacy describes individuals' beliefs in their ability to take successful action in specific domains (Bandura, 1997; Bandura, Georgas, & Manthouli, 1996). When people perceive high self-efficacy in a given domain, they are more likely to take action. In addition, because self-efficacy is a perceived quality, even if individuals merely believe that their actions are more effective and productive (perhaps because they are using a specific computing technology), they are more likely to perform a particular behavior (Bandura, 1997; Bandura, Georgas, & Manthousli, 1996). As a result, functioning as tools, computing technologies can make individuals feel more efficient, productive, in control, and generally more effective (DeCharms, 1968; Kernal, 1999; Pancer, George, & Gebotys, 1992). For example, a heart-rate monitor may help people feel more effective in meeting their exercise goals when it provides ongoing information on heart rate and calories burned. Without the heart-rate monitor, people could still take their pulse and calculate calories, but the computer device—whether it be worn or part of the exercise machinery—makes these tasks easier. The ease of tracking heart rate and calories burned likely increases self-efficacy in fitness behavior, making it more likely the individual will continue to exercise (Brehm, 1997; Strecher, DeVellis, Becker, & Rosenstock, 1986; Thompson, 1992).

Computers That Provide Tailored Information

Next, computers act as tools when they tailor information, offering people content that is pertinent to their needs and contexts. Compared to general information, tailored information increases the potential for attitude and behavior change (Beniger, 1987; Dijkstra, Librand, & Timminga, 1998; Jimison, Street, & Gold, 1997; Nowak, Shamp, Hollander, Cameron, Schumann, & Thorson, 1999; Strecher, 1999; Strecher, Kreuter, Den Boer, Kobrin, Hospers, & Skinner, 1994).

One notable example of a tailoring technology is the Website discussed earlier, Chemical Scorecard (www.scorecard.org), which generates information according to an individual's geographical location in order to achieve a persuasive outcome. After people enter their zip code into this Website, the Web technology reports on chemical hazards in their neighborhood, identifies companies that create those hazards, and describes the potential health risks. Although no published studies document the persuasive effects of this particular technology, outside research and analysis suggests that making information relevant to individuals increases their attention and arousal, which can ultimately lead to increased attitude and behavior change (Beniger, 1987; MacInnis & Jaworski, 1989; MacInnis, Moorman, & Jaworski, 1991; Strecher, 1999).

Computers That Trigger Decision-Making

Technology can also influence people by triggering or cueing a decision-making process. For example, today's web browsers launch a new window to alert people before they send information over insecure network connections. The message window serves as a signal to consumers to rethink their planned actions. A similar example exists in a very different context. Cities concerned with automobile speeding in neighborhoods can use a stand-alone radar trailer that senses the velocity of an oncoming automobile and displays that speed on a large screen. This technology is designed to trigger a decision-making process regarding driving speed.

Computers That Simplify or Guide People Through a Process

By facilitating or simplifying a process for users, technology can minimize barriers that may impede a target behavior. For ex-

ample, in the context of web commerce, technology can simplify a multistep process down to a few mouse clicks. Typically, in order to purchase something online, a consumer needs to select an item, place it in a virtual shopping cart, proceed to checkout, enter personal and billing information, and verify an order confirmation. Amazon.com and other e-commerce companies have simplified this process by storing customer information so that consumers need not reenter information every transaction. By lowering the time commitment and reducing the steps to accomplish a goal, these companies have reduced the barriers for purchasing products from their sites. The principle used by Web and other computer technology (Todd & Benbasat, 1994) is similar to the dynamic Ross and Nisbett (1991) discussed on facilitating behaviors through modifying the situation.

In addition to reducing barriers for a target behavior, computers can also lead people through processes to help them change attitudes and behaviors (Muehlenhard, Baldwin, Bourg, & Piper, 1988; Tombari, Fitzpatrick, & Childress, 1985). For example, a computer nutritionist can guide individuals through a month of healthy eating by providing recipes for each day and grocery lists for each week. In general, by following a computer-led process, users (a) are exposed to information they may not have seen otherwise, and (b) are engaged in activities they may not have done otherwise (Fogg, 2000, 2003).

Computers as Persuasive Media

The next area of the Functional Triad deals with computers as persuasive media. Although "media" can mean many things, here the focus is on the power of computer simulations. In this role computer technology provides people with experiences, either first-hand or vicarious. By providing simulated experiences, computers can change people's attitudes and behaviors. Outside the world of computing, experiences have a powerful impact on people's attitudes, behaviors, and thoughts (Reed, 1996). Experiences offered via interactive technology have similar effects (Bullinger, Roessler, Mueller-Spahn, Riva, & Wiederhold, 1998; Fogg, 2000).

Three types of computer simulations are relevant to persuasive technologies:

- simulated cause-and-effect scenarios
- simulated environments
- simulated objects

The paragraphs that follow discuss each simulation type in turn. (Note that other taxonomies for simulations exist. For example, see Gredler (1986), de Jong (1991), and Alessi (1991)).

Computers That Simulate Cause and Effect

One type of computer simulation allows users to vary the inputs and observe the effects (Hennessy & O'Shea, 1993)—what one could call "cause-and-effect simulators." The key to effective cause-and-effect simulators is their ability to demonstrate the consequence of actions immediately and credibly (Alessi, 1991;

Balci, 1998; Balci, Henrikson, & Roberts, 1986; Crosbie & Hay, 1978; de Jong, 1991; Hennessy & O'Shea, 1993; Zietsman & Hewson, 1986). These computer simulations give people first-hand insight into how inputs (such as putting money in a savings account) affect an output (such as accrued retirement savings). By allowing people to explore causes and effects of situations, these computer simulations can shape attitudes and behaviors.

Computers That Simulate Environments

A second type of computer simulation is the environment simulator. These simulators are designed to provide users with new surroundings, usually through images and sound. In these simulated environments, users have experiences that can lead to attitude and behavior change (Bullinger et al., 1998), including experiences that are designed as games or explorations (Lieberman, 1992; Schlosser & Kanifer, 1999; Schneider, 1985; Woodward, Carnine, & Davis, 1986).

The efficacy of this approach is demonstrated by research on the Tectrix Virtual Reality Bike (an exercise bike that includes a computer and monitor that shows a simulated world). Porcari and colleagues (1998) found that people using an exercise device with computer simulation of a passing landscape exercised harder than those who used an exercise device without simulation. Both groups, however, felt that they had exerted themselves a similar amount. This outcome caused by simulating an outdoor experience mirrors findings from other research: people exercise harder when outside than inside a gym (Ceci & Hassmen, 1991).

Environmental simulators can also change attitudes. Using a virtual reality environment in which the people saw and felt a simulated spider, Carlin and colleagues (1997) were able to decrease the fear of spiders in his participants. In this research, participants wore a head-mounted display that immersed them into a virtual room, and they were able to control both the number of spiders and their proximity. In this case study, Carlin found that the virtual reality treatment reduced the fear of spiders in the real world. Other similar therapies have been used for fear of flying (Klein, 1999; Wiederhold, Davis, Wiederhold, & Riva, 1998), agoraphobia (Ghosh & Marks, 1987), claustrophobia (Bullinger et al., 1998), and fear of heights (Bullinger), among others (Kirby, 1996).

Computers That Simulate Objects

The third type of computer simulations are "object simulators." These computerized devices simulate an object (as opposed to an environment). The Baby Think It Over infant simulator described earlier in this chapter is one such device. Another example is a specially equipped car created by Chrysler Corporation, designed to help teens experience the effect of alcohol on their driving. Used as part of high-school programs, teen drivers first navigate the special car under normal conditions. Then the operator activates an onboard computer system, which simulates how an inebriated person would drive—breaking sluggishly, steering inaccurately, and so on. This computer-enhanced care provides teens with an experience designed to

change their attitudes and behaviors about drinking and driving. Although the sponsors of this car do not measure the impact of this intervention, the anecdotal evidence is compelling (i.e., see Machrone, 1998).

Table 7.2 lists the three types of simulations just discussed above and outlines what advantage each type of simulation offers as far as persuasion and motivation are concerned.

Computers as Persuasive Social Actors

The final corner of the Functional Triad focuses on computers as "persuasive social actors," a view of computers that has only recently become widely recognized. Past empirical research has shown that individuals form social relationships with technology, even when the stimulus is rather impoverished (Fogg, 1997; Marshall & Maguire, 1971; Moon & Nass, 1996; Muller, 1974; Nass, Fogg, & Youngme, 1996; Nass, Moon, Fogg, Reeves, & Dryer, 1995; Nass & Steuer, 1993; Nass, Youngme, Morkes, Eun-Young, & Fogg, 1997; Parise, Kiesler, Sproull, & Waters, 1999; Quintanar Crowell, & Pryor 1982; Reeves & Nass, 1996). For example, individuals share reciprocal relationships with computers (Fogg & Nass, 1997a; Parise, Keisler, Sproull, & Waters, 1999), can be flattered by computers (Fogg & Nass, 1997b), and are polite to computers (Nass, Moon, & Carney, 1999).

In general we propose that computers as social actors can persuade people to change their attitudes and behaviors by (a) providing social support, (b) modeling attitudes or behaviors, and (c) leveraging social rules and dynamics (Fogg, 2003).

Computers That Provide Social Support

Computers can provide a form of social support in order to persuade, a dynamic that has long been observed in human-

TABLE 7.2. Captology Includes Three Types of Persuasive Simulations

Simulation Type	Key Advantages
Cause-and-effect simulators	• Allow users to explore and experiment • Show cause-and-effect relationships clearly and quickly • Persuade without being overly didactic
Environment simulators	• Can create situations that reward and motivate people for a target behavior • Allow rehearsal: practicing a target behavior • Can control exposure to new or frightening situations • Facilitate role playing: adopting another person's perspective
Object simulators	• Fit into the context of a person's normal life • Are less dependent on imagination or suspension of disbelief • Make clear the impact on normal life

human interactions (Jones, 1990). While the potential for effective social support from computer technology has yet to be fully explored, a small set of empirical studies provide evidence for this phenomenon (Fogg, 1997; Fogg & Nass, 1997b; Nass et al., 1996; Reeves & Nass, 1996). For example, computing technology can influence individuals by providing praise or criticism, thus manipulating levels of social support (Fogg & Nass, 1997b; Muehlenhard et al., 1988).

Outside the research context, various technology products use the power of praise to influence users. For example, the Dole *5 A Day* CD-ROM, discussed earlier, uses a cast of over 30 onscreen characters to provide social support to users who perform various activities. Characters such as "Bobby Banana" and "Pamela Pineapple" praise individuals for checking labels on virtual frozen foods, for following guidelines from the food pyramid, and for creating a nutritious virtual salad.

Computers That Model Attitudes and Behaviors

In addition to providing social support, computer systems can persuade by modeling target attitudes and behaviors. In the natural world, people learn directly through first-hand experience and indirectly through observation (Bandura, 1997). When a behavior is modeled by an attractive individual or is shown to result in positive consequences, people are more likely to enact that behavior (Bandura). Lieberman's research (1997) on a computer game designed to model health-maintenance behaviors shows the positive effects that an onscreen cartoon model had on those who played the game. In a similar way, the product "Alcohol 101" (www.centurycouncil.org/underage/education/a101.cfm) uses navigable onscreen video clips of human actors dealing with problematic situations that arise during college drinking parties. The initial studies on the Alcohol 101 intervention show positive outcomes (Reis, 1998). Computer-based characters, whether artistically rendered or video images, are increasingly likely to serve as models for attitudes and behaviors.

Computers That Leverage Social Rules and Dynamics

Computers have also been shown to be effective persuasive social actors when they leverage social rules and dynamics (Fogg, 1997; Friedman & Grudin, 1998; Marshall & Maguire, 1971; Parise et al., 1999). These rules include turn taking, po-

liteness norms, and sources of praise (Reeves & Nass, 1996). The rule of reciprocity— that we must return favors to others—is among the most powerful social rules (Gouldner, 1960) and is one that has also been shown to have force when people interact with computers. Fogg and Nass (1997a) showed that people performed more work and better work for a computer that assisted them on a previous task. In essence, users reciprocated help to a computer. On the retaliation side, the inverse of reciprocity, the research showed that people performed lower quality work for a computer that had served them poorly in a previous task. In a related vein, Moon (1998) found that individuals followed rules of impression management when interacting with a computer. Specifically, when individuals believed that the computer interviewing them was in the same room, they provided more honest answers, compared to interacting with a computer believed to be a few miles away. In addition, subjects were more persuaded by the proximate computer.

The previous paragraphs outline some of the early demonstrations of computers as social actors that motivate and influence people in predetermined ways, often paralleling research from long-standing human-human research.

Functional Triad Summary

Table 7.3 summarizes the Functional Triad and the persuasive affordances that each element offers.

In summary, the Functional Triad can be a useful framework in captology, the study of computers as persuasive technologies. It makes explicit how a technology can change attitudes and behaviors—either by increasing a person's capability, by providing users with an experience, or by leveraging the power of social relationships. Each of these paths suggests related persuasion strategies, dynamics, and theories. One element that is common to all three functions is the role of credibility. Credible tools, credible media, and credible social actors will all lead to increased power to persuade. This is the focus of the next section.

COMPUTERS AND CREDIBILITY

One key issue in captology is computer credibility, a topic that suggests questions such as, "Do people find computers to be credible sources?," "What aspects of computers boost credibility?,"

TABLE 7.3. Captology Outlines Three Ways That Computers Influence People

Function	Essence	Persuasive Affordances
Computer as **tool** or **instrument**	Increases capabilities	• Reduces barriers (time, effort, cost) • Increases self-efficacy • Provides information for better decision making • Changes mental models
Computer as **medium**	Provides experiences	• Provides first-hand learning, insight, visualization, resolve • Promotes understanding of cause-and-effect relationships • Motivates through experience, sensation
Computer as **social actor**	Creates relationship	• Establishes social norms • Invokes social rules and dynamics • Provides social support or sanction

and "How do computers gain and lose credibility?" Understanding the elements of computer credibility promotes a deeper understanding of how computers can change attitudes and behaviors, as credibility is a key element in many persuasion processes (Gahm, 1986; Lerch & Prietula, 1989; Lerch, Prietula, & Kulik, 1997).

Credibility has been a topic of social science research since the 1930s (for reviews, see Petty & Cacioppo, 1981; Self, 1996). Virtually all credibility researchers have described credibility as a perceived quality made up of multiple dimensions (i.e., Buller & Burgoon, 1996; Gatignon & Robertson, 1991; Petty & Cacioppo, 1981; Self, 1996; Stiff, 1994). This description has two key components germane to computer credibility. First, credibility is a perceived quality; it does not reside in an object, a person, or a piece of information. Therefore, in discussing the credibility of a computer product, one is always discussing the *perception* of credibility for the computer product.

Next, researchers generally agree that credibility perceptions result from evaluating multiple dimensions simultaneously. Although the literature varies on exactly how many dimensions contribute to the credibility construct, the majority of researchers identify trustworthiness and expertise as the two key components of credibility (Self, 1996). Trustworthiness, a key element in the credibility calculus, is described by the terms *well intentioned, truthful, unbiased,* and so on. The trustworthiness dimension of credibility captures the perceived goodness or morality of the source. Expertise, the other dimension of credibility, is described by terms such as *knowledgeable, experienced, competent,* and so on. The expertise dimension of credibility captures the perceived knowledge and skill of the source.

Extending research on credibility to the domain of computers, we have proposed that *highly credible computer products will be perceived to have high levels of both trustworthiness and expertise* (Fogg & Tseng, 1999). In evaluating credibility, a computer user will assess the computer product's trustworthiness and expertise to arrive at an overall credibility assessment.

When Does Credibility Matter?

Credibility is a key component in bringing about attitude change. Just as credible people can influence other people, credible computing products also have the power to persuade. Computer credibility is not an issue when there is no awareness of the computer itself or when the dimensions of computer credibility—trustworthiness and expertise—are not at stake. In these cases computer credibility does not matter to the user. However, in many cases credibility is key. The following seven categories outline when credibility matters in HCI (Tseng & Fogg, 1999).

1. When computers act as a knowledge repository.
Credibility matters when computers provide data or knowledge to users. The information can be static information, such as simple web pages or an encyclopedia on CD-ROM. But computer information can also be dynamic. Computers can tailor information in real time for users, such as providing information that matches interests, personality, or goals. In such cases, users may question the credibility of the information provided.

2. When computers instruct or tutor users.
Computer credibility also matters when computers give advice or provide instructions to users. Sometimes it's obvious why computers give advice. For example, auto-navigation systems give advice about which route to take, and online help systems advise users on how to solve a problem. These are clear instances of computers giving advice. However, at times the advice from a computing system is subtle. For example, interface layout and menu options can be a form of advice. Consider a default button on a dialogue box. The fact that one option is automatically selected as the default option suggests that certain paths are more likely or profitable for most users. One can imagine that if the default options are poorly chosen, the computer program could lose some credibility.

3. When computers report measurements.
Computer credibility is also at stake when computing devices act as measuring instruments. These can include engineering measurements (i.e., an oscilloscope), medical measurements (i.e., a glucose monitor), geographical measurements (i.e., devices with GPS technology), and others. In this area we observed an interesting phenomenon in the 1990s when digital test and measurement equipment was created to replace traditional analog devices. Many engineers, usually those with senior status, did not trust the information from the digital devices. As a result, some engineers rejected the convenience and power of the new technology because their old analog equipment gave information they found more credible.

4. When computers report on work performed.
Computers also need credibility when they report to users on work the computer has performed. For example, computers report the success of a software installation or the eradication of viruses. In these cases and others, the credibility of the computer is at issue if the work the computer reports does not match what actually happened. For example, suppose a user runs a spell check and the computer reports no misspelled words. If the user later finds a misspelled word, then the credibility of the program will suffer.

5. When computers report about their own state.
Computers also report their own state, and these reports have credibility implications. For example, computers may report how much disk space they have left, how long their batteries will last, how long a process will take, and so on. A computer reporting about its own state raises issues about its competence in conveying accurate information about itself, which is likely to affect user perceptions of credibility.

6. When computers run simulations.
Credibility is also important when computers run simulations. This includes simulations of aircraft navigation, chemical processes, social dynamics, nuclear disasters, and so on. Simulations can show cause-and-effect relationships, such as the progress of a disease in a population or the effects of global warming. Similarly, simulations can replicate the dynamics of an experience, such as piloting an aircraft or caring for a baby. Based on rules that humans provide, computer simulations can be flawed or biased. Even if the bias is not intentional, when users perceive that the

computer simulation lacks veridicality, the computer application will lose credibility.

7. *When computers render virtual environments.* Related to simulations is the computer's ability to create virtual environments for users. Credibility is important in making these environments believable, useful, and engaging. However, virtual environments don't always need to match the physical world; they simply need to model what they propose to model. For example, like good fiction or art, a virtual world for a fanciful arcade game can be highly credible if the world is internally consistent.

Web Credibility Research and Guidelines for Design

When it comes to credibility, the Web is unusual. The Web can be the most credible source of information, and the Web can be among the least credible sources. Limitations inherent to traditional media—most notably modality, interactivity, and space limitations—are often avoidable on the Web. As a result, online information has the potential to be more complete and enhanced by interactive options for users to more thoughtfully process what they read.

However, this potential is accompanied by several features of the Web that can erode its credibility as a medium (Danielson, 2005). First, the Web lacks the traditional gate keeping and quality-control mechanisms that are commonplace to more traditional publishing, such as editing and fact checking. Second, because digital information can be manipulated, disseminated, and published with relative ease, online information seekers must learn to account for incorrect information being widely and quickly duplicated (Metzger, Flanagin, & Zwarun, 2003), as in the case of ubiquitous "Internet hoaxes." Third, where in most media environments prior to the Web and in face-to-face interactions the speaker or writer of proposed ideas and facts was typically clear to the listener or reader, source *ambiguity* is often the rule rather than the exception in web information seeking. Finally, as with any new media technology, the Web requires users to develop new skills when evaluating various claims (Greer, 2003), as in the case of checking Uniform Resource Locators (URLs) as an indicator of site credibility.

Many Websites today offer users low-quality—or outright misleading— information. As a result, credibility has become a major concern for those seeking or posting information on the Web (Burbules, 2001; Caruso, 1999; Johnson & Kaye, 1998; Kilgore, 1998; McDonald, Schumann, & Thorson, 1999; Nielsen, 1997; Sullivan, 1999). Web users are becoming more skeptical of what they find online and may be wary of Web-based experiences in general.

There's a direct connection between web credibility and persuasion via the Web. When a site gains credibility, it also gains the power to change attitudes, and, at times, behaviors. When a Website lacks credibility, it will not be effective in persuading or motivating users. In few arenas is this connection more direct than in e-commerce, where various online claims and promises about products and services provide the primary or sole basis for buying decisions.

As part of the Persuasive Technology Lab, we have been investigating factors influencing Website credibility and user

strategies for making such assessments. A general framework for research in this relatively young field is captured in Fogg's Prominence-Interpretation Theory (Fogg, 2002, 2003). Credibility assessment is an iterative process driven by (a) the likelihood that particular Website elements (such as its privacy policy, advertisements, attractiveness, etc.) will be noticed by an information seeker (*prominence*), and (b) the value that element will be assigned by the user in making a credibility judgment (i.e., increases or decreases perceived credibility) (*interpretation*). Several factors can influence the likelihood of an element being noticed, including the user's level of involvement, the significance of the information sought, and the user's level of Web experience, domain expertise, and other individual differences. Similarly, interpretation is influenced by such individual and contextual factors. Noticeable Website elements are evaluated until either the user is satisfied with an overall credibility assessment, or a constraint (often associated with lack of time or motivation) is reached.

Perhaps more than with any other medium, Web-interaction designers face increasing challenges to design Web experiences that first and foremost hold the attention and motivation of information seekers; the second hill to climb is in persuading Web users to adopt specific behaviors, such as the following:

- register personal information
- purchase things online
- fill out a survey
- click on the ads
- set up a virtual community
- download software
- bookmark the site and return often

If web designers can influence people to perform these actions, they have been successful. These are key behavioral outcomes. But what do users notice when evaluating web content, and how are those noticed elements interpreted? What makes a Website credible? We offer the following broad guidelines, arising out of our lab's experimental work:

Guideline #1: Design websites to convey the "real world" aspect of the organization. Perhaps the most effective way to enhance the credibility of a Website is to include elements that highlight the brick-and-mortar nature of the organization it represents. Despite rampant source ambiguity on the Web, web users show a strong reliance on indicators of identity (Rieh, 2002), including credentials, photos, and contact information (Fogg, Marshall, Laraki, Osipovich, Varma, Fang, et al., 2001). The overall implication seems clear: To create a site with maximum credibility, designers should highlight features that communicate the legitimacy and accessibility of the organization.

Guideline #2: Invest resources in visual design. Web users depend to a surprisingly large degree on the visual design of Websites when making credibility judgments. In one study, we found "design look" to be the single most mentioned category by a sample of more than 2,800 users when evaluating the credibility of sites across a wide variety of domains (Fogg,

Soohoo, Danielson, Marable, Stanford, & Tauber, 2003). Similar to the assessment of human communicators, the attractiveness and professional design of a Website is often used as a first indicator of credibility.

Guideline #3: Make Websites easy to use. In the HCI community we have long emphasized ease of use, so a guideline advocating ease of use is not new. However, our work adds another important reason for making Websites usable: it will enhance the site's credibility. In one study (Fogg et al., 2001), people awarded a Website credibility points for being usable (i.e., "The site is arranged in a way that makes sense to you"), and they deducted credibility points for ease-of-use problems (i.e., "the site is difficult to navigate"). While this information should not change how we, as HCI professionals, design user experiences for the Web, it does add a compelling new reason for investing time and money in usable design—it makes a site more credible. Going beyond the data, one could reasonably conclude that a simple, usable Website would be perceived as more credible than a site that has extravagant features but is lacking in usability.

Guideline #4: Include markers of expertise. Expertise is a key component in credibility, and our work supports the idea that Websites that convey expertise can gain credibility in users' eyes. Important "expertise elements" include listing an author's credentials and including citations and references. It's likely that many other elements also exist. Many Websites today miss opportunities to convey legitimately expertise to their users.

Guideline #5: Include markers of trustworthiness. Trustworthiness is another key component in credibility. As with expertise, Website elements that convey trustworthiness will lead to increased perceptions of credibility. Such elements include linking to outside materials and sources, stating a policy on content, and so on. Making information verifiable on a Website increases credibility despite the fact that users are unlikely to follow through on verification (Metzger et al., 2003). Thus, the mere presence of some design elements will influence user perceptions. We propose that Website designers who concentrate on conveying the honest, unbiased nature of their Website will end up with a more credible—and therefore more effective—Website.

Guideline #6: Tailor the user experience. Although not as vital as the previous suggestions, tailoring does make a difference. Our work shows that tailoring the user experience on a Website leads to increased perceptions of web credibility. For example, people think a site is more credible when it acknowledges that the individual has visited it before. To be sure, tailoring and personalization can take place in many ways. Tailoring extends even to the type of ads shown on the page: ads that match what the user is seeking seem to increase the perception of Website credibility.

Guideline #7. Avoid overly commercial elements on a Website. Although most Websites, especially large Websites, exist for commercial purposes, our work suggests that users penalize sites that have an aggressively commercial flavor. For example, web pages that mix ads with content to the point of confusing readers will be perceived as not credible. Fogg et al. (2001) found that mixing ads and content received the most negative response of all. However, it is important to note that ads don't always reduce credibility. In this study and elsewhere (Kim, 1999), quantitative research shows that banner ads done well can enhance the perceived credibility of a site. It seems reasonable that, as with other elements of people's lives, we accept commercialization to an extent but become wary when it is overdone.

Guideline #8. Avoid the pitfalls of amateurism. Most web designers seek a professional outcome in their work. Organizations that care about credibility should be ever vigilant—and perhaps obsessive—to avoid small glitches in their Websites. These "small" glitches seem to have a large impact on web credibility perceptions. Even one typographical error or a single broken link is damaging. While designers may face pressures to create dazzling technical features on Websites, failing to correct small errors undermines that work.

Despite the growing body of research, much remains to be discovered about web credibility. The study of web credibility needs to be an ongoing concern because three things continue to evolve: (a) Web technology, (b) the type of people using the Web, and (c) people's experiences with the Web. Fortunately, what researchers learn about designing for web credibility can translate into credible experiences in other high-tech devices that share information, from mobile phones to gas pumps.

POSITIVE AND NEGATIVE APPLICATIONS OF PERSUASIVE TECHNOLOGY

As the power of persuasive techniques becomes more understood, we are beginning to see more examples of persuasive technologies being created. Many have positive goals in mind, but there are also many technologies designed to negatively influence attitudes and behaviors.

An example of a positive technology is a mobile application called MyFoodPhone. Mobile persuasive devices have several unique properties that may improve their abilities to persuade. First, they are personal devices: people carry their mobile phones everywhere, customize them, and store personal information in them. Second, intrinsic to them being mobile, these devices have the potential to intervene at the right moment, a concept called *kairos*.

MyFoodPhone is an application for the camera phone that helps people watch what they eat—whether they want to change their weight or just eat right. Whenever a user is concerned with an item they are about to eat, they simply take a picture of it with their camera, then use MyFoodPhone to send it to a system that shares the images with a expert dietician. The user receives expert evaluation and feedback. In this case, the simplicity and appropriate timing of the application make it a powerful persuasive tool.

On the Web, GoDaddy (www.godaddy.com), a popular Web-hosting company, attempts to persuade users to purchase more expensive hosting solutions by "disguising" links to their less-expensive plans with plain text links, while links to more pricey upgrades are in large, brightly colored buttons.

A more negative example can be found in the rise in "Pro Anorexia" Websites, encouraging self-starvation and sharing tips for losing weight. Though they reached their height in earlier part of the decade, many of these sites are still being operated. By creating social networks around it, people suffering from anorexia are supported and encouraged to continue their unhealthy habits. Many of these Websites use the web credibility techniques discussed earlier: the sites are well designed and contain expert advice.

As the power of persuasive technologies becomes more understood, the consideration of the ethical ramifications of these technologies becomes essential.

THE ETHICS OF COMPUTING SYSTEMS DESIGNED TO PERSUADE

In addition to research and design issues, captology addresses the ethical issues that arise from design or distributing persuasive interactive technologies. Persuasion is a value-laden activity. By extension, creating or distributing an interactive technology that attempts to persuade is also value laden. Ethical problems arise when the values, goals, and interests of the creators don't match with those of the people who use the technology. HCI professionals can ask a few key questions to get insight into possible ethical problem areas:

- Does the persuasive technology advocate what's good and fair?
- Is the technology inclusive, allowing access to all, regardless of social standing?
- Does it promote self-determination?
- Does it represent what's thought to be true and accurate?

Answering no to any of these questions suggests the persuasive technology at hand could be ethically questionable and perhaps downright objectionable (for a longer discussion on ethics, see Friedman & Kahn, later in this volume).

While it's clear that deception and coercion are unethical in computing products, some behavior change strategies such as conditioning, surveillance, and punishment are less cut and dry. For example, Operant conditioning—a system of rewards—can powerfully shape behaviors. By providing rewards, a computer product could get people to perform new behaviors without their clear consent or without them noticing the forces of influence at work.

Surveillance is another common and effective way to change behavior. People who know they are being watched behave differently. Today, computer technologies allow surveillance in ways that were never before possible, giving institutions remarkable new powers. Although advocates of computer-based employee surveillance (i.e., DeTienne, 1993) say that monitoring can "inspire employees to achieve excellence," they and opponents agree that such approaches can hurt morale or create a more stressful workplace. When every keystroke and every restroom break is monitored and recorded, employees may feel they are part of an electronic sweatshop.

Another area of concern is when technologies use punishment—or threats of punishment—to shape behaviors. Although punishment is an effective way to change outward behaviors in the short term, punishment has limited outcomes beyond changing observable behavior, and many behavior change experts frown on using it. The problems with punishment increase when a computer product punishes people. The punishment may be excessive or inappropriate to the situation. Also, the long-term effects of punishment are likely to be negative. In these cases, who bears responsibility for the outcome?

Discussed elsewhere in more detail (Berdichevsky, 1999; Fogg, 1998, 2003), those who create or distribute persuasive technologies have a responsibility to examine the moral issues involved.

PERSUASIVE TECHNOLOGY: POTENTIAL AND RESPONSIBILITY

Computer systems are now becoming a common part of everyday life. Whatever the form of the system, from a desktop computer to a smart car interior to a mobile phone, these interactive experiences can be designed to influence our attitudes and affect our behaviors. They can motivate and persuade by merging the power of computing with the psychology of persuasion.

We humans are still the supreme agents of influence—and this won't change any time soon. Computers are not yet as effective as skilled human persuaders are, but at times computing technology can go beyond what humans can do. Computers never forget, they don't need to sleep, and they can be programmed to never stop trying. For better or worse, computers provide us with a new avenue for changing how people think and act.

To a large extent, we as a community of HCI professionals will help create the next generation of technology products, including those products designed to change people's attitudes and behaviors. If we take the right steps—raising awareness of persuasive technology in the general public and encouraging technology creators to follow guidelines for ethical interactive technologies—we may well see persuasive technology reach its potential, enhancing the quality of life for individuals, communities, and society.

References

Alessi, S. M. (1991). Fidelity in the design of instructional simulations. *Journal of Computer-Based Instruction, 15*, 40–47.

Balci, O. (1998, December 13–16). *Verification, validation, and accreditation*. Paper presented at the Winter Simulation Conference, Washington, United States.

Balci, O., Henrikson, J. O., & Roberts, S. D. (1986). Credibility assessment of simulation results. In J. R. Wilson (Ed.), Winter Simulation Conference Proceedings.

Bandura, A. (1997). *Self-efficacy: The exercise of control*. New York: W. H. Freeman.

Bandura, A., Georgas, J., & Manthouli, M. (1996). Reflections on human agency, *Contemporary psychology in Europe: Theory, research, and applications.* Seattle, WA: Hogrefe & Huber.

Beniger, J. R. (1987). Personalization of mass media and the growth of pseudo-community. *Communication Research, 14*(3), 352–371.

Berdichevsky, D., & Neunschwander, E. (1999). Towards an ethics of persuasive technology. *Communications of the ACM, 42*(5), 51–8.

Brehm, B. (1997, December). Self-confidence and exercise success. *Fitness Management,* 22–23.

Buller, D. B., & Burgoon, J. K. (1996). Interpersonal Deception Theory. *Communication Theory, 6*(3), 203–242.

Bullinger, A. H., Roessler, A., Mueller-Spahn, F., Riva, G., & Wiederhold, B. K. (1998). From toy to tool: The development of immersive virtual reality environments for psychotherapy of specific phobias, *Studies in Health Technology and Informatics, 58.* Amsterdam: IOS Press.

Burbules, N. C. (2001). Paradoxes of the Web: The ethical dimensions of credibility. *Library Trends, 49,* 441–453.

Carlin, A. S., Hoffman, H. G., & Weghorst, S. (1997). Virtual reality and tactile augmentation in the treatment of spider phobia: A case report. *Behaviour Research & Therapy, 35*(2), 153–158.

Caruso, D. (1999, November 22). Digital commerce: Self indulgence in the Internet industry. *The New York Times.*

Ceci, R., & Hassmen, P. (1991). Self-monitored exercise at three different PE intensities in treadmill vs. field running. *Medicine and Science in Sports and Exercise, 23,* 732–738.

Crosbie, R. E., & Hay, J. L. (1978). The credibility of computerised models. In R. E. Crosbie (Ed.), *Toward real-time simulation: Languages, models and systems.* La Jolla, CA: Society of Computer Simulation.

Danielson, D. R. (2005). Web credibility. To appear in C. Ghaoui (Ed.), *Encyclopedia of human–computer interaction.* Idea Group.

DeCharms, R. (1968). *Personal causation: The internal affective determinants of behavior.* New York: Academic.

de Jong, T. (1991). Learning and instruction with computer simulations. *Education & Computing, 6,* 217–229.

Denning, P., & Metcalfe, R. (1997). *Beyond calculation: The next fifty years of computing.* New York: Springer-Verlag.

DeTienne, Kristen Bell. (1993, September/October). Big Brother or friendly coach? Computer monitoring in the 21st century. *The Futurist.*

Dijkstra, J. J., Liebrand, W. B. G., & Timminga, E. (1998). Persuasiveness of expert systems. *Behaviour and Information Technology, 17*(3), 155–63.

Fogg, B.J. (1997). *Charismatic computers: Creating more likable and persuasive interactive technologies by leveraging principles from social psychology* (Doctoral Thesis. Stanford University).

Fogg, B.J. (1998). Persuasive computers: Perspectives and research directions. *Proceedings of the Conference on Human Factors in Computing Systems.* CHI98, Los Angeles, USA.

Fogg, B.J. (1999). Persuasive technologies. *Communications of the ACM, 42*(5), 26–29.

Fogg, B.J. (2002). *Prominence-interpretation theory: Explaining how people assess credibility.* Research report from the Stanford Persuasive Technology Lab, Stanford University. Retrieved (date), from at: http://credibility.stanford.edu/

Fogg, B.J. (2003a). *Persuasive technology: Using computers to change what we think and do.* San Francisco: Morgan-Kaufmann.

Fogg, B.J. (2003b). Prominence-interpretation theory: Explaining how people assess credibility online. *Proceedings of CHI'03, Extended Abstracts on Human Factors in Computing Systems* (pp. 722–723).

Fogg, B.J., Marshall, J., Laraki, O., Osipovich, A., Varma, C., Fang, N., et al. (2000). Elements that affect web credibility: Early results from a self-report study. *Proceedings of ACM CHI 2000 Conference on Human Factors in Computing Systems.* New York: ACM Press.

Fogg, B.J., Marshall, J., Laraki, O., Osipovich, A., Varma, C., Fang, N., et al. (2001). What makes a Website credible? A report on a large quantitative study. *Proceedings of ACM CHI 2001 Conference on Human Factors in Computing Systems.* New York: ACM Press.

Fogg, B.J., & Nass, C. (1997a). How users reciprocate to computers: An experiment that demonstrates behavior change. *Proceedings of the Conference on Human Factors in Computing Systems,* CHI 97. New York: ACM.

Fogg, B.J., & Nass, C. (1997b). Silicon sycophants: The effects of computers that flatter. *International Journal of Human-Computer Studies, 46*(5), 551–61.

Fogg, B.J., Soohoo, C., Danielson, D. R., Marable, L., Stanford, J., & Tauber, E. R. (2003). How do users evaluate the credibility of Websites? A study with over 2,500 participants. *Proceedings of DUX2003, Designing for User Experiences Conference.*

Fogg, B.J., & Tseng, H. (1999). The elements of computer credibility. *Proceedings of the Conference on Human Factors and Computing Systems,* CHI99. Pittsburgh, PA USA.

Ford, M. E. (1992). *Motivating humans: Goals, emotions, personal agency beliefs.* Newbury Park: Sage.

Friedman, B., & Grudin, J. (1998). Trust and accountability: Preserving human values in interactional experience. *Proceedings of the Conference on Human Factors in Computing Systems,* CHI98, Los Angeles, USA.

Gahm, G. A. (1986). *The effects of computers, source salience and credibility on persuasion.* Doctoral thesis. State University of New York.

Gatignon, H., & Robertson, T. S. (1991). *Innovative Decision Processes.* Englewood Cliffs, NJ: Prentice-Hall.

Ghosh, A., & Marks, I. M. (1987). Self-treatment of agoraphobia by exposure. *Behavior Therapy, 18*(1), 3–16.

Gouldner, A. W. (1960). The Norm of reciprocity: A preliminary statement. *American Sociological Review, 25,* 161–178.

Gredler, M. B. (1986). A taxonomy of computer simulations. *Educational Technology, 26,* 7– 12.

Greer, J. D. (2003). Evaluating the credibility of online information: A test of source and advertising influence. *Mass Communication & Society, 6*(1), 11–28.

Hennessy, S., & O'Shea, T. (1993). Learner perceptions of realism and magic in computer simulations. *British Journal of Educational Technology, 24*(2), 125–38.

Jimison, H. B., Street, R. L., Jr., & Gold, W. R. (1997). Patient-specific interfaces to health and decision-making information, *LEA's communication series.* Mahwah, NJ: Lawrence Erlbaum.

Johnson, T. J., & Kaye, B. K. (1998). Cruising is believing?: Comparing Internet and traditional sources on Media Credibility Measures. *Journalism and Mass Communication Quarterly, 75*(2), 325–40.

Jones, E. E. (1990). *Interpersonal perception.* New York: W. H. Freeman.

Kay, A. (1984). Computer software. *Scientific American, 251,* 53–59.

Kernal, H. K. (1999). *Effects of design characteristics on evaluation of a home control system: A comparison of two research methodologies.* Paper presented to SIGGRAPH Annual Conference 2000, New Orleans, LA, USA.

Khaslavsky, J., & Shedroff, N. (1999). Understanding the seductive experience. *Communications of the ACM, 42*(5), 45–9.

Kilgore, R. (1998). Publishers must set rules to preserve credibility. *Advertising Age, 69*(48), 31.

Kim, N. (1999). *World Wide Web credibility: What effects do advertisements and typos have on the perceived credibility of web page information?* Unpublished honors thesis, Stanford University, California.

King, P., & Tester, J. (1999). The landscape of persuasive technologies. *Communications of the ACM, 42*(5), 31–8.

Kirby, K. C. (1996). Computer-Assisted Treatment of Phobias. *Psychiatric Services, 4*(2), 139–140, 142.

Klein, R. A. (1999). Treating fear of flying with virtual reality exposure therapy. *Innovations in Clinical Practice: A Source Handbook, 17,* 449–465.

Lerch, F. J., & Prietula, M. J. (1989). How do we trust machine advice? Designing and using human-computer interfaces and knowledge based systems. *Proceedings of the Third Annual Conference on Human Computer Interaction.*

Lerch, F. J., Prietula, M. J., & Kulik, C. T. (1997). The Turing Effect: The nature of trust in expert system advice. *Expertise in context: Human and machine.* Cambridge, MA: The MIT Press.

Lieberman, D. (1992). The computer's potential role in health education. *Health Communication, 4,* 211–225.

Lieberman, D. (1997). Interactive video games for health promotion. In W. G. R. Street, & T. Mannin (Eds.), *Health promotion and interactive technology (pp.).* Mahwah, NJ: Lawrence Earlbaum.

Machrone, B. (1998, July 1). Driving drunk. *PC Magazine.*

MacInnis, D. J., & Jaworski, B. J. (1989). Information processing from advertisements: Toward an integrative framework. *Journal of Marketing, 53*(4), 1–23.

MacInnis, D. J., Moorman, C., & Jaworski, B. J. (1991). Enhancing and measuring consumers' motivation, opportunity, and ability to process brand information from ads. *Journal of Marketing, 55*(4), 32–53.

Malone, T., & Lepper, M. (1987). Making learning fun: A taxonomy of intrinsic motivation for learning. In R. E. Snow & M. J. Farr (Eds.), *Aptitude, learning, and instruction.* Hillsdale, N.J.: Lawrence Earlbaum.

Marshall, C., & Maguire, T. O. (1971). The computer as social pressure to produce conformity in a simple perceptual task. *AV Communication Review, 19*(1), 19–28.

McDonald, M., Schumann, D. W., & Thorson, E. (1999). Cyberhate: Extending persuasive techniques of low credibility sources to the World Wide Web. In D. W. Schumann & E. Thorson (Eds.), *Advertising and the World Wide Web.* Mahwah: Lawrence Erlbaum.

Metzger, M. J., Flanagin, A. J., & Zwarun, L. (2003). College student Web use, perceptions of information credibility, and verification behavior. *Computers & Education, 41,* 271–290.

Moon, Y. (1998). The effects of distance in local versus remote human–computer interaction. *Proceedings of the Conference on Human Factors in Computing Systems,* CHI 98, (pp. 103–108). New York: ACM.

Moon, Y., & Nass, C. (1996). How "Real" are computer personalities? Psychological responses to personality types in human–computer interaction. *Communication Research, 23*(6), 651–674.

Muehlenhard, C. L., Baldwin, L. E., Bourg, W. J., & Piper, A. M. (1988). Helping women "break the ice" A computer program to help shy women start and maintain conversations with men. *Journal of Computer-Based Instruction, 15*(1), 7–13.

Muller, R. L. (1974). *Conforming to the computer: Social influence in computer-human interaction.* Doctoral thesis, Syracuse University, New York.

Nass, C., Fogg, B.J., & Youngme, M. (1996). Can computers be teammates? *International Journal of Human-Computer Studies, 45*(6), 669–78.

Nass, C., Moon, Y., & Carney, P. (1999). Are people polite to computers? Responses to computer-based interviewing systems. *Journal of Applied Social Psychology, 29*(5), 1093–1110.

Nass, C., Moon, Y., Fogg, B.J., Reeves, B., & Dryer, D. C. (1995). Can computer personalities be human personalities? *International Journal of Human-Computer Studies, 43*(2), 223–39.

Nass, C., & Steuer, J. (1993). Voices, boxes, and sources of messages: Computers and social actors. *Human Communication Research, 19*(4), 504–527.

Nass, C. I., Youngme, M., Morkes, J., Eun-Young, K., & Fogg, B.J. (1997). Computers are social actors: A review of current research. In B. Friedman (Ed.), *Human values and the design of computer technology.* Stanford, CA: CSLI and Cambridge Press.

Nielsen, J. (1997). *How Users Read on the Web.* Retrieved (date), from www.useit.com/alertbox/9710a.html.

Nowak, G. J., Shamp, S., Hollander, B., Cameron, G. T., Schumann, D. W., & Thorson, E. (1999). Interactive media: A means for more meaningful advertising? *Advertising and consumer psychology.* Mahwah: Lawrence Erlbaum.

Pancer, S. M., George, M., & Gebotys, R. J. (1992). Understanding and predicting attitudes toward computers. *Computers in Human Behavior, 8,* 211–222.

Parise, S., Kiesler, S., Sproull, L., & Waters, K. (1999). Cooperating with life-like interface agents. *Computers in Human Behavior, 15*(2), 123–142.

Petty, R. E., & Cacioppo, J. T. (1981). *Attitudes and persuasion: Classic and contemporary approaches.* Dubuque, IA: W.C. Brown.

Porcari, J. P., Zedaker, M. S., & Maldari, M. S. (1998). Virtual Motivation. *Fitness Management,* 48–51.

Quintanar, L., Crowell, C., & Pryor, J. (1982). Human–computer interaction: A preliminary social psychological analysis. *Behavior Research Methods and Instrumentation, 14*(2), 210–220.

Reed, E. (1996). *The Necessity of Experience.* New Haven, CT: Yale University Press.

Reeves, B., & Nass, C. I. (1996). *The media equation: How people treat computers, television, and new media like real people and places.* New York: Cambridge University Press.

Reis, J. (1998). *Research Results: National Data Anaylsis.* Retrieved (Date), from the Century Council Website: www.centurycouncil .org/alcohol101/dem_nat.cfm

Rieh, S. Y. (2002). Judgment of information quality and cognitive authority in the Web. *Journal of the American Society for Information Science and Technology, 53,* 145–161.

Ross, L., & Nisbett, R. E. (1991). *The person and the situation: Perspectives of social psychology.* New York: McGraw-Hill.

Schlosser, A. E., & Kanifer, A. (1999). Current Advertising on the Internet: The Benefits and Usage of Mixed-Media Advertising Strategies. In D. W. Schumann & E. Thorson (Eds.), *Advertising and the World Wide Web* (pp. 41–62). Mahwah, NJ: Lawrence Erlbaum.

Schneider, S. J. (1985). Computer technology and persuasion: the case of smoking cessation, *Proceedings of COMPINT 85: Computer Aided Technologies.* Washington, DC: IEEE.

Self, C. S. (1996). Credibility. In M. Salwen & D. Stacks (Eds.), *An integrated approach to communication theory and research.* Mahwah, NJ: Erlbaum.

Stiff, J. B. (1994). *Persuasive communication.* New York: The Guilford Press.

Strecher, V. J. (1999). Computer-tailored smoking cessation materials: A review and discussion. Special Issue: Computer-tailored education. *Patient Education & Counseling, 36*(2), 107–117.

Strecher, V. J., DeVellis, B. M., Becker, M. H., & Rosenstock, I. M. (1986). The role of self-efficacy in achieving health behavior change. *Health Education Quarterly, 13*(1), 73–92.

Strecher, V. J., Kreuter, M., Den Boer, D.-J., Kobrin, S., Hospers, H. J., & Skinner, C. S. (1994). The effects of computer-tailored smoking cessation messages in family practice settings. *Journal of Family Practice, 39*(3), 262–270.

Sullivan, C. (1999). Newspapers must retain credibility. *Editor & Publisher, 4.*

Thompson, C. A. (1992). *Exercise adherence and performance: Effects on self-efficacy and outcome expectations.* Doctoral thesis.

Todd, P. A., & Benbasat, I. (1994). The influence of decision aids on choice strategies under conditions of high cognitive load. *IEEE Transactions on Systems, Man, & Cybernetics, 24*(4), 537–547.

Tombari, M. L., Fitzpatrick, S. J., & Childress, W. (1985). Using computers as contingency managers in self-monitoring interventions: A case study. *Computers in Human Behavior, 1*(1), 75–82.

Tseng, S., & Fogg, B.J. (1999). Credibility and computing technology. *Communications of the ACM, 42*(5), 39–44.

Verplank, B., Fulton, J., Black, A., & Moggridge, B. (1993). *Observation and invention: Uses of scenarios in interaction design.* Paper presented at INTERCHI'93.

Wiederhold, B. K., Davis, R., Wiederhold, M. D., & Riva, G. (1998). The effects of immersiveness on physiology, *Studies in health technology and informatics, 58.* Amsterdam: IOS.

Woodward, J. P., Carnine, D., & Davis, L. G. (1986). Health ways: A computer simulation for problem solving in personal health management [Special issue: Technological advances in community health]. *Family & Community Health, 9*(2), 60–63.

Zietsman, A. I., & Hewson, P. W. (1986). Effect of Instruction Using Microcomputer Simulations and Conceptual Change Strategies on Science Learning. *Journal of Research in Science Teaching, 23*(1), 27–39.

·8·

HUMAN-ERROR IDENTIFICATION IN HUMAN–COMPUTER INTERACTION

Neville A. Stanton
Brunel University
Runnymede Campus

HUMAN ERROR

We are all familiar with the annoyance of errors we make with everyday devices, such as turning on the heat under on an empty kettle, or making mistakes in the programming sequence of our videocassette recorders. People have a tendency to blame themselves for "human error." However, the use and abuse of the term has led some to question the very notion of "human error" (Wagenaar & Groeneweg, 1988). "Human error" is often invoked in the absence of technological explanations. Chapanis (1999) wrote that back in the 1940s he noted that "pilot error" was really "designer error." This was a challenge to contemporary thinking, and showed that design is all-important in human-error reduction. Chapanis became interested in why pilots often retracted the landing gear instead of the landing flaps after landing the aircraft. He identified the problem as designer error rather than pilot error, as the designer had put two identical toggle switches side by side—one for the landing gear, the other for the flaps. Chapanis proposed that the controls should be separated and coded. The separation and coding of controls is now standard human-factors practice. Half a century after Chapanis's original observations, the idea that one can design error-tolerant devices is beginning to gain credence (Baber & Stanton, 1994). One can argue that human error is not a simple matter of one individual making one mistake, so much as the product of a design which has permitted the existence and continuation of specific activities which could lead to errors (Reason, 1990).

Human error is an emotive topic and psychologists have been investigating its origins and causes since the dawn of the discipline (Reason, 1990). Traditional approaches have attributed errors to individuals. Indeed, so-called "Freudian slips" were considered the unwitting revelation of intention, errors revealing what a person was really thinking but did not wish to disclose. More recently, cognitive psychologists have considered the issues of error classification and explanation (Senders & Moray, 1991). The taxonomic approaches of Norman (1988) and Reason (1990) have fostered the development and formal definition of several categories of human error (e.g., capture errors, description errors, data driven errors, associated activation errors, and loss of activation errors) while the work of Reason (1990) and Wickens (1992) attempted to understand the psychological mechanisms which combine to cause errors (e.g., failure of memory, poor perception, errors of decision making, and problems of motor execution). Reason (1990) in particular has argued that we need to consider the activities of the individual if we are to be able to identify what may go wrong. Rather than viewing errors as unpredictable events, this approach regards them to be wholly predictable occurrences based on an analysis of an individual's activities. Reason's definition proposes that errors are "those occasions in which a planned sequence of mental or physical activities fail to achieve its intended outcome, [and] when these failures cannot be attributed to the intervention of some chance agency." (p. 9)

If errors are no longer to be considered as random occurrences, then it follows that we should be able to identify them and predict their likelihood. The impetus to achieve this has been fueled in the wake of several recent and significant inci-

dents, most notably in the nuclear industry where there now exists several human-error identification (HEI) techniques. The aims of this chapter are to,

1. Consider human-error classifications;
2. Look at systems approaches to human error;
3. Consider how human error can be predicted;
4. Examine the validation evidence; and
5. Look at human error in the context of design.

HUMAN-ERROR CLASSIFICATION

The development of formal human-error classification schemes has assisted in the anticipation and analysis of error. The anticipation of error has come about through the development of formal techniques for predicting error, which is dealt with in the Predicting Human Error section. The analysis of error is assisted by taxonomic systems and the interpretation of underlying psychological mechanisms. Three contemporary systems were presented in the work of Norman (1981), Reason (1990), and Wickens (1992).

Norman's (1981) research focused on the categorization of action slips, in which he presented the analysis of 1,000 incidents. Underpinning his analysis was a psychological theory of schema activation. He argued that action sequences are triggered by knowledge structures (organized as memory units and called "schemas"). Within the mind is a hierarchy of schemas that are invoked (or triggered) if particular conditions are satisfied or certain events occur. The theory seems particularly pertinent as a description of skilled behavior. The classification scheme is presented in Table 8.1.

In Neisser's (1976) seminal work "Cognition and Reality," he contended that human thought is closely coupled with a person's interaction with the world. He argued that knowledge of how the world works (e.g., mental models) leads to the anticipation of certain kinds of information, which in turn directs behavior to seek out certain kinds of information and to provide a ready means of interpretation. During this process, as the environment is sampled, the information garnered serves to update and modify the internal, cognitive schema of the world, which will again direct further search. An illustration of the perceptual cycle is shown in Fig. 8.1.

The perceptual cycle can be used to explain human information processing in control rooms. For example—assuming that an individual has the correct knowledge of a videocassette recorder he is programming—his mental model will enable him to anticipate events (such as the menu items he expects to see), search for confirmatory evidence (look at the panel on the video machine), direct a course of action (select a channel, day of the week, start time, end time, etc.), and continually check that the outcome is as expected (menu item and data field respond as anticipated). If the individual uncovers some data he does not expect (such as a menu item not previously encountered, or the data field not accepting his input), he is required to access a source of a wider knowledge of the world to consider possible explanations that will direct future search activities. The completeness of the model is in the description of process

TABLE 8.1.   Taxonomy of Slips with Examples

Taxonomy of Slips	Examples of Error Types
Slips that result from errors in the formation of intention	Mode errors: erroneous classification of the situation
	Description errors: ambiguous or incomplete specification of intention
Slips that result from faulty activation of schemas	Capture errors: similar sequences of action, where stronger sequence takes control
	Data-driven activation errors: external events that cause the activation of schemas
	Association-activation errors: currently active schemas that activate other schemas with which they are associated
	Loss-of-activation errors: schemas that lose activation after they have been activated
Slips that result from faulty triggering of active schemas	Blend errors: combination of components from competing schemas
	Premature activation errors: schemas that are activated too early
	Failure to activate errors: failure of the trigger condition or event to activate the schema

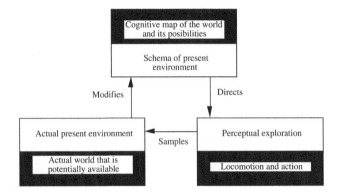

FIGURE 8.1.   The perceptual cycle.

(the cyclical nature of sampling the world) and product (the updating of the world model at any point in time).

This interactive schema model works well for explaining how we act in the world. As Norman's (1981) research has shown, it may also explain why errors occur as they do. If, as schema theory predicts, action is directed by schema, then faulty schemas or faulty activation of schemas will lead to erroneous performance. As Table 8.1 shows, this can occur in at least three ways. First, we can select the wrong schema due to misinterpretation of the situation. Second, we can activate the wrong schema because of similarities in the trigger conditions. Third, we can activate schemas too early or too late. Examples of these types of errors are presented in Table 8.1.

Of particular interest is the problem of mode errors. Norman (1981) singled out this type of error as requiring special attention in the design of computing systems. He pointed out that the misclassification of the computing system mode could lead to input errors which may result in serious consequences. In word processors this may mean the loss of documents; in video recorders this may mean the loss of recordings; on flight decks this may mean damage to aircraft.

Casey (1993) described a case in which an apparently simple mode error by a radiotherapy technician working in a cancer care center led to the death of a patient. The Therac-25 she was operating was a state-of-the-art, million-dollar machine that could be used as both a high-power x-ray machine (25 million electron volts delivered by typing an "x" on the keyboard) and low-power electron beam machine (200 rads delivered by typing an "e" on the keyboard). After preparing the patient for radiation therapy, the radiotherapist went to her isolated control room. She accidentally pressed the "x" key instead of the "e" key, but quickly realized this and selected the edit menu so that she could change the setting from x-ray to electron-beam. Then she returned to the main screen to wait for the "beam ready" prompt. All of this occurred within eight seconds. Unknown to her, this rapid sequence of inputs had never been tested on the machine before and it had actually entered a hybrid mode, delivering blasts of 25,000 rads, which was more than 125 times the prescribed dose. When the "beam ready" prompt was displayed, the radiotherapist pressed the "b" key to fire the beam. The high-energy beam was delivered to the patient and the computer screen displayed the prompt "Malfunction 54." Unaware that the machine had already fired, the operator reset it and pressed "b" again. This happened for a third time until the patient ran out of the room reporting painful electric shocks. On investigation, the problem with the machine modes was found, but not before other overdoses had been given. This case study demonstrates the need to consider the way in which design of a system can induce errors in users.

A thorough understanding of human error is required by the design team. Error classification schemes can certainly help, but they need to be supported by formal error-prediction techniques within a user-centered design approach.

Reason (1990) developed a higher-level error classification system, incorporating slips, lapses, and mistakes. Slips and lapses are defined by attentional failures and memory failures, respectively. Both slips and lapses are examples of where the action was unintended, whereas mistakes are associated with intended action. This taxonomy is presented in Table 8.2.

Wickens (1992), taking the information processing framework, considered the implications of psychological mechanisms at work in error formation. He argued that with mistakes, the situation assessment and/or planning are poor while the retrieval action execution is good; with slips, the action execution is poor whereas the situation assessment and planning are good; and finally, with lapses, the situation assessment

TABLE 8.2. Basic Error Types with Examples

Basic Error Type	Example of Error Type
Slip	Action intrusion
	Omission of action
	Reversal of action
	Misordering of action
	Mistiming of action
Lapse	Omitting of planned actions
	Losing place in action sequence
	Forgetting intended actions
Mistake	Misapplication of good procedure
	Application of a bad procedure
	Misperception
	Poor decision making
	Failure to consider alternatives
	Overconfidence

TABLE 8.3. Error Types and Associated Psychological Mechanisms

Error Type	Associated Psychological Mechanism
Slip	Action execution
Lapse and mode errors	Memory
Mistake	Planning and intention of action
Mistake	Interpretation and situation assessment

and action execution are good but memory is poor. A summary of these distinctions is shown in Table 8.3.

Wickens (1992) was also concerned with mode errors, with particular reference to technological domains. He suggested that a pilot raising the landing gear while the aircraft is still on the runway is an example of a mode error. Wickens proposed that mode errors are the result of poorly conceived system design that allows the mode confusion to occur and the operation in an inappropriate mode. Chapanis (1999) argued that the landing gear switch could be rendered inoperable if the landing gear could be configured to detect weight on the wheels, as the aircraft would be on the ground.

Taxonomies of errors can be used to anticipate what might go wrong in any task. Potentially, every task or activity could be subject to a slip, lapse, or mistake. The two approaches represented within the taxonomies are a schema-based approach and an error-list-based approach. Examples of these two approaches will be presented next in the form of formal human-error identification techniques.

PREDICTING HUMAN ERROR

An abundance of methods for identifying human error exists; some of these methods may be appropriate for the analysis of consumer products. In general, most of the existing techniques have two key problems. The first of these problems relates to the lack of representation of the external environment or objects. Typically, human-error analysis techniques do not represent the activity of the device nor the material that the human interacts with in more than a passing manner. Hollnagel (1993) emphasized that Human Reliability Analysis (HRA) often fails to take adequate account of the context in which performance occurs. Second, there tends to be an overdependence on the analyst's judgment. Different analysts with different experiences may make different predictions regarding the same problem (intra-analyst reliability). Similarly, the same analyst may make different judgments on different occasions (inter-analyst reliability). This subjectivity of analysis may weaken the confidence that can be placed in any predictions made. The analyst is required to be

an expert in the technique as well as in the operation of the device being analyzed if the analysis has any potential for validity.

Two techniques are considered here because of the inherent differences in the way the methods work. Systematic Human Error Reduction and Prediction Approach (SHERPA) is a divergent error-prediction method; it works by associating up to 10 error modes with each action. In the hands of a novice, it is typical for there to be an over-inclusive strategy for selecting error modes. The novice user would rather play-safe-than-sorry and would tend to predict many more errors than would actually occur. This might be problematic; "crying wolf" too many times might ruin the credibility of the approach. Task Analysis For Error Identification (TAFEI), by contrast, is a convergent error-prediction technique; it works by identifying the possible transitions between the different states of a device and uses the normative description of behavior (provided by the Hierarchical Task Analysis (HTA)) to identify potentially erroneous actions. Even in the hands of a novice the technique seems to prevent the individual from generating too many false alarms, certainly no more than they do using heuristics. In fact, by constraining the user of TAFEI to the problem space surrounding the transitions between device states, it should exclude extraneous error prediction. Indeed, this was one of the original aims for the technique when it was originally developed (Baber & Stanton, 1994).

Systematic Human-Error Reduction and Prediction Approach (SHERPA)

SHERPA represents the error-list approach. At its core is an error taxonomy that is not unlike the classification schemes presented in the previous section. The idea is that each task can be classified into one of five basic types. SHERPA (Embrey, 1986; Stanton, 2005) uses Hierarchical Task Analysis (HTA; Annett, Duncan, Stammers, & Gray, 1971) together with an error taxonomy to identify credible errors associated with a sequence of human activity. In essence, the SHERPA technique works by indicating which error modes are credible for each task step in turn, based on an analysis of work activity. This indication is based on the judgment of the analyst, and requires input from a subject matter expert to be realistic. A summary of the procedure is shown in Fig. 8.2.

The process begins with the analysis of work activities, using Hierarchical Task Analysis. HTA (Annett et al., 1971; Annett, 2004, 2005) is based on the notion that task performance can be expressed in terms of a hierarchy of goals (what the person is seeking to achieve), operations (the activities executed to achieve the goals), and plans (the sequence in which the operations are executed). An example of HTA for the programming of a videocassette recorder is shown in Fig. 8.3.

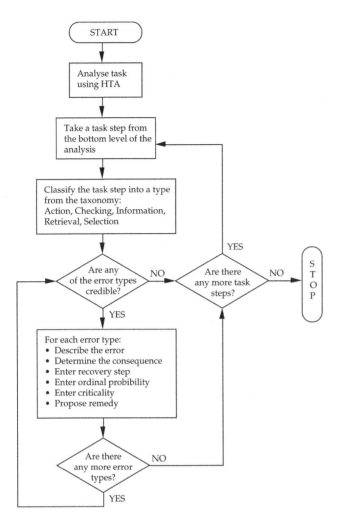

FIGURE 8.2. The SHERPA procedure.

For the application of SHERPA, each task step from the bottom level of the analysis is taken in turn. First, each task step is classified into one of the following types from the taxonomy:

- Action (e.g., pressing a button, pulling a switch, opening a door)
- Retrieval (e.g., getting information from a screen or manual)
- Checking (e.g., conducting a procedural check)
- Information communication (e.g., talking to another party)
- Selection (e.g., choosing one alternative over another)

This classification of the task step then leads the analyst to consider credible error modes associated with that activity, as shown in Table 8.4.

For each credible error (e.g., those judged by a subject matter expert to be possible) a description of the form that the error would take is given, as illustrated in Table 8.5. The consequence of the error on the system needs to be determined next, as this has implications for the criticality of the error. The last four steps consider the possibility for error recovery, the ordinal probability of the error (high, medium of low), its criticality

(either critical or not critical), and potential remedies. Again, these are shown in Table 8.5.

As Table 8.5 shows, there are six basic error types associated with the activities of programming a VCR. These are,

1. Failing to check that the VCR clock is correct;
2. Failing to insert a cassette;
3. Failing to select the program number;
4. Failing to wait;
5. Failing to enter programming information correctly;
6. Failing to press the confirmatory buttons.

The purpose of SHERPA is not only to identify potential errors with the current design, but also to guide future design considerations. The structured nature of the analysis can help to focus the design remedies for solving problems, as shown in the remedial strategies column. As this analysis shows, quite a lot of improvements could be made. It is important to note, however, that the improvements are constrained by the analysis. This does not address radically different design solutions that may remove the need to program at all.

Task Analysis For Error Identification (TAFEI)

TAFEI represents the schema-based approach. It explicitly analyzes the interaction between people and machines (Baber & Stanton, 1994; Stanton & Baber, 1996; Stanton & Baber, 2005a). TAFEI analysis is concerned with task-based scenarios. This analysis is done by mapping human activity onto machine states. An overview of the procedure is shown in Fig. 8.4. TAFEI analysis consists of three principal components: Hierarchical Task Analysis (HTA); State-Space Diagrams (SSD), which are loosely based on finite state machines (Angel & Bekey, 1968); and Transition Matrices (TM). HTA provides a description of human activity, SSDs provide a description of machine activity, and TM provides a mechanism for determining potential erroneous activity through the interaction of the human and the device. In a similar manner to Newell and Simon (1972), legal and illegal operators (called "transitions" in the TAFEI methodology) are identified.

In brief, the TAFEI methodology is as follows. First, the system needs to be defined. Next, the human activities and machine states are described in separate analyses. The basic building blocks are HTA (describing human activity; see Fig. 3) and state space diagrams (describing machine activity). These two types of analysis are then combined to produce the TAFEI description of human-machine interaction, as shown in Fig. 8.5.

From the TAFEI diagram, a transition matrix is compiled and each transition is scrutinized. In Table 8.6, each transition has been classified as "impossible" (e.g., the transition cannot be performed), "illegal" (the transition can be performed but it does not lead to the desired outcome), or "legal" (the transition can be performed and is consistent with the description of error-free activity provided by the HTA), until all transitions have been analyzed. Finally, "illegal" transitions are addressed in turn as potential errors, to consider changes that may be introduced.

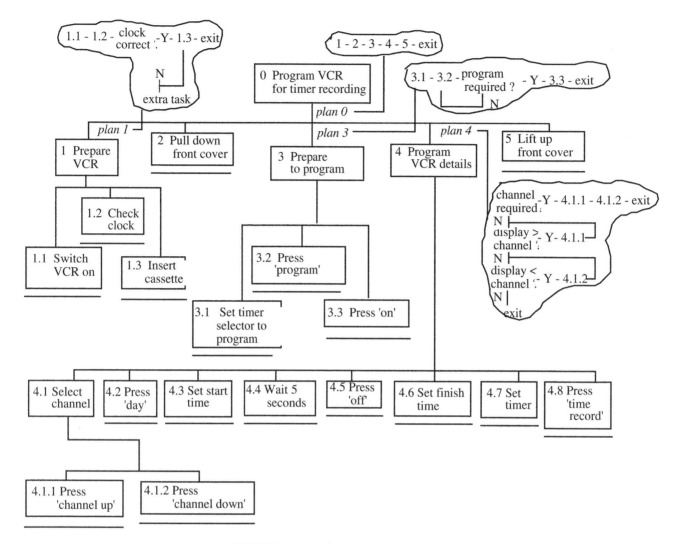

FIGURE 8.3.   HTA for programming a VCR.

Thirteen of the transitions are defined as "illegal," and these can be reduced to a subset of six basic error types:

1. Switch VCR off inadvertently
2. Insert cassette into machine when switched off
3. Program without cassette inserted
4. Fail to select program number
5. Fail to wait for "on" light
6. Fail to enter programming information

In addition, one legal transition has been highlighted because it requires a recursive activity to be performed. These activities seem to be particularly prone to errors of omission. These predictions then serve as a basis for the designer to address the redesign of the VCR. A number of illegal transitions could be dealt with relatively easily by considering the use of modes in the operation of the device, such as switching off the VCR without stopping the tape and pressing play without inserting the tape. As with the SHERPA example, the point of the analysis is to help guide design efforts to make the product error-tolerant (Baber & Stanton, 2004).

VALIDATION OF HUMAN-ERROR IDENTIFICATION

There have been a few attempts to validate HEI techniques (Williams, 1989; Whalley & Kirwan, 1989; Kirwan, 1992a, 1992b; Kennedy, 1995; Baber & Stanton, 1996). For instance, Whalley and Kirwan (1989) evaluated six HEI techniques (Heuristics, PHECA, SRK, SHERPA, THERP, and HAZOP) for their ability to account for the errors known to have contributed to four genuine incidents within the nuclear industry. More recently, Kirwan (1992b) has developed a comprehensive list of eight criteria to evaluate the acceptability of these techniques at a more qualitative level. In an unpublished study, Kennedy (1995) has included Kirwan's criteria when examining the ability of the techniques to predict ten actual incidents retrospectively. While

TABLE 8.4. Error Modes and their Description

Error Mode	Error Description
Action	
A1	Operation too long/short
A2	Operation mistimed
A3	Operation in wrong direction
A4	Operation too much/little
A5	Misalign
A6	Right operation on wrong object
A7	Wrong operation on right object
A8	Operation omitted
A9	Operation incomplete
A10	Wrong operation on wrong object
Information Retrieval	
R1	Information not obtained
R2	Wrong information obtained
R3	Information retrieval incomplete
Checking	
C1	Check omitted
C2	Check incomplete
C3	Right check on wrong object
C4	Wrong check on right object
C5	Check mistimed
C6	Wrong check on wrong object
Information Communication	
I1	Information not communicated
I2	Wrong information communicated
I3	Information communication incomplete
Selection	
S1	Selection omitted
S2	Wrong selection made

these studies failed to identify a clear favorite from among these HEI techniques, all three studies indicated impressive general performance using the SHERPA method. SHERPA achieved the highest overall rankings and Kirwan (1992b) recommended a combination of expert judgment together with the SHERPA technique as the most valid approach.

The strength of these studies lies in the high level of ecological or face validity that they achieve. The methodologies make use of the opinions of expert assessors for the prediction of errors contributing to real world events. However, these studies do raise several methodological concerns. Specifically, the number of assessors using each technique is small (typically one to three), and the equivalence of assessment across techniques is brought into question because different people are assessing each HEI technique. A second methodological concern centers on the use of subjective rating scales. It is doubtful that the assessors will share the same standards when rating the acceptability or usefulness of an HEI technique. This factor, combined with the small number of assessors for each technique, means that these data should be accepted with some degree of caution.

In light of these criticisms, Baber and Stanton (1996) aimed to provide a more rigorous test for the predictive validity of SHERPA and TAFEI. Predictive validity was tested by comparing the errors identified by an expert analyst with those observed during 300 transactions with a ticket machine on the London Underground. Baber and Stanton (1996) suggested that SHERPA and TAFEI provided an acceptable level of sensitivity based on the data from two expert analysts ($d' = 0.8$). The strength of this latter study over Kirwan's was that it reports the use of the method in detail as well as the error predictions made using SHERPA and TAFEI. Stanton and Baber (2002) reported a study on the performance of SHERPA and TAFEI using a larger pool of analysts, including novice analysts, to examine the important issue of ease of acquiring the method. They reported reliability values of between 0.4 and 0.6 and sensitivity values of between 0.7 and 0.8. This compares favorably with Hollnagel, Kaarstad, and Lee's (1998) analysis of the Cognitive Reliability and Error Analysis Method (CREAM), for which they claim a 68.6% match between predicted outcomes and actual outcomes.

The research into HEI techniques suggested that the methods enable analysts to structure their judgment (Stanton & Baber, 2005b). However, the results run counter to the literature in some areas (such as usability evaluation), which suggest the superiority of heuristic approaches (Nielsen, 1993). The views of Lansdale and Ormerod (1994) may help us to reconcile these findings. They suggested that, to be applied successfully, a heuristic approach needs the support of an explicit methodology to "ensure that the evaluation is structured and thorough" (p. 257). Essentially, SHERPA and TAFEI provide a semi-structured approach that forms a framework for the judgment of the analyst without constraining it. It seems to succeed precisely because of its semi-structured nature, which alleviates the burden otherwise placed on the analyst's memory while allowing them room to use their own heuristic judgment.

APPLYING TAFEI TO INTERFACE DESIGN

A study by Baber and Stanton (1999) has shown how TAFEI can be used as part of the interface design process for a computer workstation. Their case study is based on a design project for a medical imaging software company. The software was used by cytogeneticists in research and hospital environments. The existing software was a menu-driven, text-based interface. The task of metaphase finding has six main subtasks: (a) the set-up task (where the computer is fired up, the microscope is calibrated, the scan is defined, and the cells are prepared); (b) the capture task (where the drug is applied to the cells on the slide using a pipette and the images are captured); (c) the processing task (where the background is subtracted from the cell image and the image processing is performed); (d) the analysis task (where the data are graphed and tabulated); (e) the storage task (where the images are assessed and selected images and data are saved); and (f) the shut-down task (where the computer and microscope are shut down). Using TAFEI to model the existing system, Baber and Stanton (1999) found that the sequence of activities required to conduct the metaphase-finding tasks were not supported logically by the computer interface. The two main problems were the confusing range of choices offered to users in each system state and the number of recursions in the task sequence required to perform even the simplest

TABLE 8.5. The SHERPA description.

Task Step	Error Mode	Error Description	Consequence	Recovery	P	C	Remedial Strategy
1.1	A8	Fail to switch VCR on	Cannot proceed	Immediate	L		Press any button to switch VCR on
1.2	C1	Omit to check clock	VCR Clock time may be incorrect	None	L	!	Automatic clock setting and adjust via radio transmitter
	C2	Incomplete check					
1.3	A3	Insert cassette wrong way around	Damage to VCR	Immediate	L	!	Strengthen mechanism On-screen prompt
	A8	Fail to insert cassette	Cannot record	Task 3	L		
2	A8	Fail to pull down front cover	Cannot proceed	Immediate	L		Remove cover to programming
3.1	S1	Fail move timer selector	Cannot proceed	Immediate	L		Separate timer selector from programming function
3.2	A8	Fail to press PROGRAM	Cannot proceed	Immediate	L		Remove this task step from sequence
3.3	A8	Fail to press ON button	Cannot proceed	Immediate	L		Label button START TIME
4.1.1	A8	Fail to press UP button	Wrong channel selected	None	M	!	Enter channel number directly from keypad
4.1.2	A8	Fail to press DOWN button	Wrong channel selected	None	M	!	Enter channel number directly from keypad
4.2	A8	Fail to press DAY button	Wrong day selected	None	M	!	Present day via a calendar
4.3	I1	No time entered	No program recorded	None	L	!	Dial time in via analogue clock
	I2	Wrong time entered	Wrong program recorded	None	L	!	Dial time in via analogue clock
4.4	A1	Fail to wait	Start time not set	Task 4.5	L		Remove need to wait
4.5	A8	Fail to press OFF button	Cannot set finish time				Label button FINISH TIME
4.6	I1	No time entered	No program recorded	None	L	!	Dial time in via analogue clock
	I2	Wrong time entered	Wrong program recorded	None	L	!	Dial time in via analogue clock
4.7	A8	Fail to set timer	No program recorded	None	L	!	Seperate timer selector from programming function
4.8	A8	Fail to press TIME RECORD button	No program recorded	None	L	!	Remove this task step from sequence
5	A8	Fail to lift up front cover	Cover left down	Immediate	L		Remove cover to programming

of tasks. Interviews with the cytogeneticists revealed some fundamental problems with the existing software, such as: (a) the existing system was not error tolerant, (b) there was a lack of consistency between commands, (c) the error and feedback messages were not meaningful to users, and (d) the same data had to be entered several times in the procedure. The lack of trust that the users had in the system meant that they also kept a paper-based log of the information they entered into the computer, meaning even greater duplication of effort.

From their analysis of the tasks sequences, Baber and Stanton (1999) developed a TAFEI diagram of the ideal system states that would lead the user through the metaphase-finding task in a logical manner. One of the software engineers described the analysis as "a sort of video playback of someone actually doing things with the equipment." The TAFEI analysis was a revelation to the company, who had some vague idea that all was not well with the interface, but had not conceived of the problem as defining the task sequence and user actions. As a modeling approach, TAFEI does this rather well. The final step in the analysis was to define and refine the interface screen. This was based on the task sequence and TAFEI description, to produce a prototype layout for comment with the users and software engineers. Specific task scenarios were then tried out to see if the new interface would support user activity. Following some minor modification, the company produced a functional prototype for user performance trials.

Baber and Stanton (1999) reported successful completion of the interface design project using TAFEI. They argued that the method supports analytical prototyping activity. TAFEI enables designers to focus attention on the task sequence, user activity, and interface design. It also highlights potential problem points in the interaction, where errors are likely to occur. While it is accepted that these considerations might also form part of other methods (e.g., storyboards), by contrast TAFEI does this in a structured, systematic, rigorous, and auditable approach for the consideration of human error potential in system design.

APPLYING SHERPA TO SAFETY CRITICAL SYSTEMS

Studies reported on SHERPA show how it can be applied to the evaluation of energy distribution (Glendon & Stanton, 1999) and oil extraction (Stanton & Wilson, 2000) in the safety-critical industries. Both of these represent multiperson systems. The study by Glendon and Stanton (1999) compared the impact of the implementation of a new safety management system on error potential in an energy-distribution company. The first assessment of human-error potential in electrical switching operations was undertaken in 1994. Following this, the company undertook major organization-restructuring activities. This

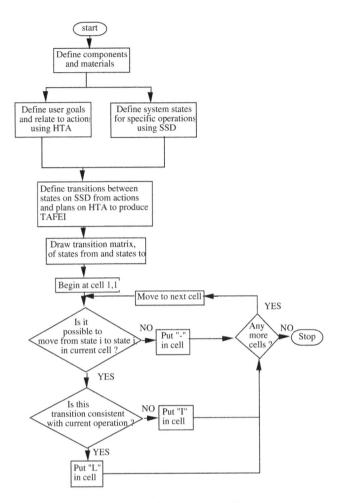

FIGURE 8.4. The TAFEI procedure.

included a major safety-management initiative, which was to separate operational control of the energy distribution system from safety management. In the original assessment of the system in 1994 it became obvious that safety was often secondary to operational performance. In order to remove this conflict, a separate safety-management center was established. This enabled the operational center to pass control of part of the energy-distribution system over to the safety-management center when safety issues arose. Thus operational control and safety management were not in conflict. When the safety issues were resolved, control of the energy-distribution system was passed back to the operational center. The effect on reducing error potential in the system is illustrated in Fig. 8.6.

Statistical analyses, using the Binomial test, were conducted to determine the difference between the error rates in 1994 with those in 1997. All comparisons between critical (low and medium likelihood) errors proved statistically significant ($p < 0.001$). It should be noted that there were no "high" likelihood errors in the analyses of this system, all of which had probably been designed out of the system as it evolved. Fig. 8.6 shows a reduction in the error potential by over 50% for critical errors from 1994 to 1997. This suggests that the system is likely to be safer. The SHERPA technique assisted in the

redesign process after the 1994 analysis by proposing design-reduction strategies. The results of these proposals are seen in the 1997 analysis.

Stanton and Wilson (2000) described the assessment of an oil drilling task undertaken offshore using the SHERPA technique. Surprisingly, the oil industry lacked a method for assessing the robustness of its working practices. SHERPA scrutinized the minutiae of human action, and provided some basis for assessing the magnitude of increased risk from human activity, which could then be used to inform which intervention strategies were likely to make the working environment safer. Drilling is a highly integrated team activity comprised of a site controller, drilling supervisor, geologist, toolpusher, directional driller, assistant driller, mudlogger, mud engineer, derrickman, crane operator, roughnecks, and roustabouts. It is also a fairly dangerous occupation, as the incidents at Piper Alpha and Ocean Odyssey show. It is argued that the risks arising from human operations can be controlled in the same way as engineering risk. Stanton and Wilson (2000) conducted a detailed task analysis of drilling operations, covering monitoring the status of the well (monitor), detecting abnormalities (detect), and dealing with kicks (deal). The results of the SHERPA analysis are shown in Fig. 8.7.

As Fig. 8.7 shows, the error potential in the detection of abnormalities task and the dealing with kicks task are considerably greater than the monitor the well task. This should be some cause for concern, and certainly explains some of the problems that oil companies have observed. The proposed remedies from the SHERPA analysis included strategies such as computer-prompted checks of drilling variables, placement of drilling parameters on separate displays, redesign of trend displays, computer-prompted alarm levels, automatic transmission of information, electronic links between the drilling team members, computer-generated tables of mud loss, computer-based procedures, online weather systems, and automated shutdown procedures. All of these proposals are really about the design of the human-computer interface. Stanton and Wilson (2000) argued that adopting these approaches should help organizations design safer systems and help prevent catastrophic disasters in the future.

CONCLUSIONS

Despite most HEI techniques, and ergonomics methods resulting from them being used in an evaluative and summative manner, it is entirely feasible for them to be used in a formative design approach. Stanton and Young (1999) argued that the methods may have the greatest impact at the prototyping stage, particularly considering one of the key design stages: *analytic prototyping*. Although in the past it may have been costly, or even impossible, to alter design at the structural prototyping stage, with the advent of computer-aided design it is made much simpler. It may even be possible to compare alternative designs at this stage with such technology. In terms of the analytical prototyping of human interfaces, Baber and Stanton (1999) argued that there are three main forms: functional analysis (e.g., consideration of the range of functions the device

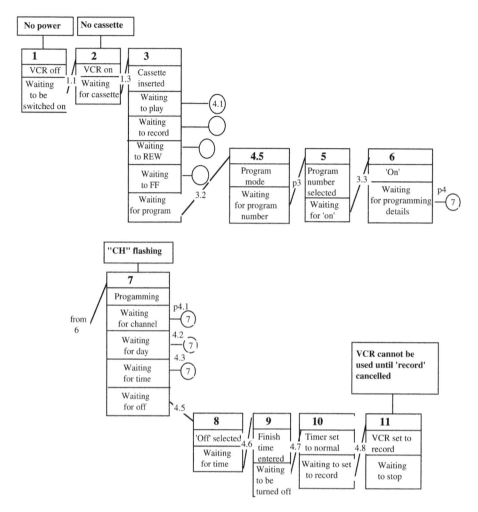

FIGURE 8.5. The TAFEI description.

TABLE 8.6. The Transition Matrix

From state:	To state:										
		1	2	3	4.5	5	6	7	8	9	10	11
1	-	L	I	-	-	-	-	-	-	-	-
2	L	-	L	I	-	-	-	-	-	-	-
3	L	-	-	L	-	-	-	-	-	-	-
4.5	I	-	-	-	L	I	-	-	-	-	-
5	I	-	-	-	-	L	I	I	-	-	-
6	I	-	-	-	-	-	L	I	-	-	-
7	I	-	-	-	-	-		L	-	-	-
8	I	-	-	-	-	-	-	-	L	-	-
9	I	-	-	-	-	-	-	-	-	L	-
10	I	-	-	-	-	-	-	-	-	-	L
11	-	-	-	-	-	-	-	-	-	-	-

supports), scenario analysis (e.g., consideration of the device with regard to a particular sequence of activities), and structural analysis (e.g., non-destructive testing of the interface from a user-centered perspective). The case studies of how TAFEI and SHERPA were applied on interface design projects demonstrate how system improvements may be made. Stanton and Young (1999) have extended this analysis to other ergonomics methods. There are three take-home messages from this chapter, which are that

1. Most technology-induced errors are entirely predictable;
2. Structured methods, such as SHERPA and TAFEI, produce reliable and valid error data;
3. Ergonomics methods should be used as part of the formative design process, to improve design and reduce errors.

Successful design will require expertise in the subject of analysis as well as expertise in the methodology being used. Exploring design weaknesses through SHERPA and TAFEI will help in developing more error-tolerant devices and products.

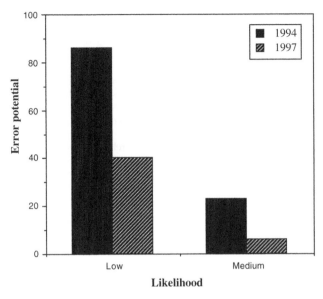

FIGURE 8.6. SHERPA analysis of an energy-distribution system.

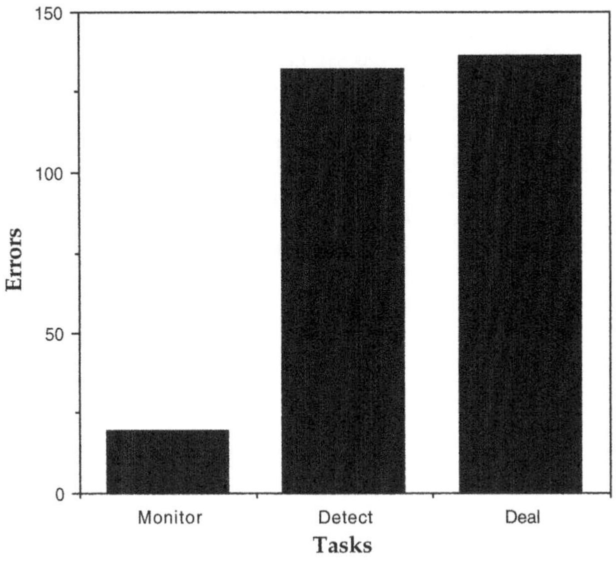

FIGURE 8.7. SHERPA analysis of drilling for oil tasks.

References

Annett, J. (2004). Hierarchical task analysis. In D. Diaper, & N. A. Stanton (Eds.), *The handbook of task analysis for human-computer iInteraction* (pp. 67–82). Mahwah, New Jersey: Lawrence Erlbaum.

Annett, J. (2005). Hierarchical task analysis. In N. A. Stanton, A. Hedge, E. Salas, H. Hendrick, & K. Brookhaus (Eds.), *Handbook of human factors and ergonomics methods* (pp. 33-1–33-7). London: Taylor & Francis,.

Annett, J., Duncan, K. D., Stammers, R. B., & Gray, M. J. (1971). *Task analysis. Training information, No. 6*. London: HMSO.

Angel, E. S., & Bekey, G. A. (1968). Adaptive finite state models of manual control systems. *IEEE Transactions on Man-Machine Systems*, March, 15–29.

Baber, C., & Stanton, N. A. (1994). Task analysis for error identification: a methodology for designing error tolerant consumer products. *Ergonomics, 37*(11), 1923–1941.

Baber, C., & Stanton, N. A. (1996). Human error identification techniques applied to public technology: Predictions compared with observed use. *Applied Ergonomics, 27*(2), 119–131.

Baber, C., & Stanton, N. A. (1999). Analytical prototyping. In J. M. Noyes & M. Cook (Eds.), *Interface technology: The leading edge* (pp. 175–194). Baldock, UK: Research Studies Press.

Baber, C., & Stanton, N. A. (2004). Task analysis for error identification. In D. Diaper, & N. A. Stanton (Eds.), *The handbook of task analysis for human–computer interaction* (pp. 367–379). Mahwah, New Jersey: Lawrence Erlbaum Associates.

Casey, S. (1993). *Set Phasers on Stun and other true tales of design, technology and human error*. Santa Barbara, California: Aegean.

Chapanis, A. (1999). *The Chapanis Chronicles: 50 years of human factors research, education, and design*. Santa Barbara, California: Aegean.

Embrey, D. E. (1986). *SHERPA: A Systematic Human Error Reduction and Prediction Approach*. Paper presented at the International Meeting on Advances in Nuclear Power Systems, Knoxville, Tennessee.

Glendon, A. I., & Stanton, N. A. (1999). Safety culture in organisations. *Safety Science, 34,* 193–214.

Hollnagel, E. (1993). *Human Reliability Analysis: context and control*. London: Academic Press.

Hollnagel, E., Kaarstad, M., & Lee, H. C. (1998). Error mode prediction. *Ergonomics, 42,* 1457–1471.

Kennedy, R. J. (1995). Can human reliability assessment (HRA) predict real accidents? A case study analysis of HRA. In A. I. Glendon, & N. A. Stanton (Eds.), *Proceedings of the risk assessment and risk reduction conference*. Birmingham: Aston University..

Kirwan, B. (1992a). Human error identification in human reliability assessment. Part 1: overview of approaches. *Applied Ergonomics, 23,* 299–318.

Kirwan, B. (1992b). Human error identification in human reliability assessment. Part 2: detailed comparison of techniques. *Applied Ergonomics, 23,* 371–381.

Lansdale, M. W., & Ormerod, T. C. (1994). *Understanding interfaces*. London: Academic Press.

Neisser, U. (1976). *Cognition and reality: Principles and implications of cognitive psychology*. San Francisco: Freeman.

Newell, A., & Simon, H. A. (1972). *Human problem solving*. Englewood Cliffs, NJ: Prentice Hall.

Nielsen, J. (1993). *Usability Engineering*. Boston: Academic Press.

Norman, D. A. (1981). Categorisation of action slips. *Psychological Review, 88*(1) 1–15.

Norman, D. A. (1988). *The psychology of everyday things*. New York: Basic Books.

Reason, J. (1990). *Human error*. Cambridge, UK: Cambridge University Press.

Senders, J. W., & Moray, N. P. (1991). *Human error*. Hillsdale, NJ: LEA.

Stanton, N. A. (2005). Systematic human error reduction and prediction approach. In N. A. Stanton, A. Hedge, E. Salas, H. Hendrick, & K. Brookhaus (Eds.), *Handbook of human factors and ergonomics methods* (pp. 371–378). London: Taylor & Francis.

Stanton, N. A., & Baber, C. (1996). A systems approach to human error identification. *Safety science, 22,* 215–228.

Stanton, N. A., & Baber, C. (2002). Error by design: Methods for predicting device usability. *Design Studies, 23*, 363–384.

Stanton, N. A., & Baber, C. (2005a). Task analysis for error identification. In N. A. Stanton, A. Hedge, E. Salas, H. Hendrick, & K. Brookhaus (Eds.), *Handbook of human factors and ergonomics methods* (pp. 381–389). London: Taylor & Francis.

Stanton, N. A., & Baber, C. (2005b). Validating task analysis for error identification: Reliability and validity of a human error prediction technique. *Ergonomics, 48*, 1097–1113.

Stanton, N. A., & Wilson, J. (2000). Human factors: Step change improvements in effectiveness and safety. *Drilling Contractor,* Jan/Feb, 46–51.

Stanton, N. A., & Young, M. S. (1999). *A guide to methodology in ergonomics.* London: Taylor and Francis.

Wagenaar, W. A., & Groeneweg, J. (1988). Accidents at sea: Multiple causes, impossible consequences. *International Journal of Man-Machine Studies, 27*, 587–598.

Whalley, S. J., & Kirwan, B. (1989, June). *An evaluation of five human error identification techniques.* Paper presented at the 5th International Loss Prevention Symposium, Oslo, Norway.

Wickens, C. D. (1992). *Engineering psychology and human performance.* New York: Harper Collins.

Williams, J. C. (1989). Validation of human reliability assessment techniques. *Reliability Engineering, 11*, 149–162.

Part

·II·

COMPUTERS IN HCI

·9·

INPUT TECHNOLOGIES AND TECHNIQUES

Ken Hinckley
Microsoft Research

WHAT'S AN INPUT DEVICE ANYWAY?

Input devices sense physical properties of people, places, or things. Yet any treatment of input devices without regard to the corresponding visual feedback is like trying to use a pen without paper. Small-screen devices with integrated sensors underscore the indivisibility of input and output. This chapter treats input technologies at the level of *interaction techniques*, which provide a way for users to accomplish tasks by combining input with appropriate feedback. An interaction designer must consider (a) the physical sensor, (b) the feedback presented to the user, (c) the ergonomic and industrial design of the device, and (d) the interplay between all of the interaction techniques supported by a system.

This chapter enumerates properties of input devices and provides examples of how these properties apply to common pointing devices, as well as mobile devices with touch or pen input. We will discuss how to use input signals in applications, as well as models and theories that help to evaluate interaction techniques and reason about design options. We will also discuss discrete symbolic entry, including mobile and keyboard-based text entry. The chapter concludes with some thoughts about future trends.

UNDERSTANDING INPUT TECHNOLOGIES

A designer who understands input technologies and the task requirements of users has a better chance of designing interaction techniques that match a user's natural workflow. Making an optimal choice for tasks in isolation leads to a poor design, so the designer must weigh competing design requirements, as well as transitions between tasks.

Input Device Properties

The variety of pointing devices is bewildering, but a few important properties characterize most input sensors. These properties help a designer understand a device and anticipate potential problems. We will first consider these device properties in general, and then show how they apply to some common input devices.

Property sensed. Most devices sense linear position, motion, or force; rotary devices sense angle, change in angle, and torque (Buxton, 1995c; Card, Mackinlay, & Robertson, 1991). For example, tablets sense position of a pen, mice sense motion (change in position), and isometric joysticks sense force. The property sensed determines the mapping from input to output, or *transfer function*, that is most appropriate for the device. Position sensing devices are *absolute input devices*, whereas motion-sensing devices are *relative input devices*. A relative device, such as the mouse, requires visual feedback in the form of a cursor to indicate a screen location. With absolute devices, the *nulling problem* (Buxton, 1983) arises if the position of a physical intermediary, such as a slider on a mixing console, is not in agreement with a value set in software. This prob-

lem cannot occur with relative devices, but users may waste time *clutching*: the user must occasionally lift a mouse to reposition it.

Number of dimensions. Devices sense one or more input dimensions. For example, a mouse senses two linear dimensions of motion, a knob senses one angular dimension, and a six degree-of-freedom magnetic tracker measures three position dimensions and three orientation dimensions. A pair of knobs or a mouse with a scroll wheel sense separate input dimensions and thus form a "1D + 1D" device, or a "2D + 1D" *multichannel device*, respectively (Zhai, Smith, & Selker, 1997). *Multidegree-of-freedom devices (3D input devices)* sense three or more simultaneous dimensions of spatial position or orientation (Bowman, Kruijff, LaViola, & Poupyrev, 2004; Hinckley, Pausch, Goble, & Kassell, 1994; Hinckley, Sinclair, Hanson, Szeliski, & Conway, 1999; Zhai, 1998).

Indirect versus direct. A mouse is an indirect input device because the user must move the mouse to indicate a point on the screen, whereas a direct input device has a unified input and display surface. Direct devices such as touchscreens, or display tablets operated with a pen, are not necessarily easier to use than indirect devices. Direct devices lack buttons for state transitions. Occlusion is also a major design challenge. The finger or pen occludes the area at which a user is pointing, so the user may not realize that he or she has activated a control; occlusion by the hand and arm also may cause the user to overlook pop-up menus, dialogs, or status indicators.

Device acquisition time. The average time to move one's hand to a device is known as *acquisition time. Homing time* is the time to return from a device to a "home" position (e.g., return from mouse to keyboard). For common desktop workflows that involve switching between text entry and pointing, the effectiveness of a device for pointing tends to dominate acquisition and homing time costs (Douglas & Mithal, 1994). Thus, integration of a pointing device with the keyboard may not improve overall performance, but evaluations still must assess any influence of acquisition times (Dillon, Eday, & Tombaugh, 1990; Hinckley et al., 2006).

Gain. Also known as control-to-display (C:D) gain or C:D ratio, *gain* is defined as the distance moved by an input device divided by the distance moved on the display. Gain confounds what should be two measurements—(a) device size and (b) display size—with one arbitrary metric (Accot & Zhai, 2001; MacKenzie, 1995), and is therefore suspect as a factor for experimental study. In experiments, gain typically has little or no effect on the time to perform pointing movements, but variable gain functions may provide some benefit by reducing the required *footprint* (physical movement space) of a device.

Other metrics. System designers must weigh other performance metrics, including pointing speed and accuracy, error rates, learning time, footprint, user preference, comfort, and cost (Card, Mackinlay, & Robertson, 1990). Other important engineering parameters include sampling rate, resolution, accuracy, and linearity (MacKenzie, 1995).

A Brief Tour of Pointing Devices

Most operating systems treat all input devices as *virtual devices*, which tempts one to believe that pointing devices are interchangeable; however, the details of what the input device senses, such as how it is held, the presence or absence of buttons, and many other properties, can significantly impact the interaction techniques—and hence, the end-user tasks—that a device can effectively support. The following tour discusses important properties of several common pointing devices.

Mice. Douglas Englebart and colleagues (English, Englebart, & Berman, 1967) invented the mouse in 1967 at the Stanford Research Institute. Forty years later, the mouse persists because its properties provide a good match between human performance and the demands of graphical interfaces (Balakrishnan, Baudel, Kurtenbach, & Fitzmaurice, 1997). For typical pointing tasks on a desktop computer, one can point with the mouse about as well as with the hand itself (Card, English, & Burr, 1978). Because the mouse stays put when the user releases it (unlike a stylus, for example), it is quick for users to reacquire and allows designers to integrate multiple buttons or other controls on its surface. Users exert force on mouse buttons in a direction orthogonal to the mouse's plane of motion, thus minimizing inadvertent motion. Finally, with mice, all of the muscle groups of the hand, wrist, arm, and shoulder contribute to pointing, allowing high performance for both rapid, coarse movements as well as slow, precise movements (Guiard, 1987; Zhai, Milgram, & Buxton, 1996). These advantages suggest the mouse is hard to beat; it will remain the pointing device of choice for desktop graphical interfaces.

Trackballs. A trackball senses the relative motion of a partially exposed ball in two degrees of freedom. Trackballs have a small working space (*footprint*), and afford use on an angled surface. Trackballs may require frequent clutching movements because users must lift and reposition their hand after rolling the ball a short distance. The buttons are located to the side of the ball, which can make them awkward to hold while rolling the ball (MacKenzie, Sellen, & Buxton, 1991). A trackball engages different muscle groups than a mouse, offering an alternative for users who experience discomfort when using a mouse.

Isometric joysticks. An isometric joystick (e.g., the IBM Trackpoint) is a force-sensing joystick that returns to center when released. Most isometric joysticks are stiff, offering little feedback of the joystick's displacement. The rate of cursor movement is proportional to the force exerted on the stick; as a result, users must practice in order to achieve good cursor control. Isometric joysticks may offer the only pointing option when space is at a premium (Douglas & Mithal, 1994; Rutledge & Selker, 1990; Zhai et al., 1997).

Isotonic joysticks. Isotonic joysticks sense angle of deflection. Some hybrid designs blur the distinctions of isometric and isotonic joysticks, but the main questions are, "Does the joystick sense force or angular deflection?" "Does the stick return to center when released?" and "Does the stick move from the starting position?" For a discussion of the complex design space of joysticks, see Lipscomb and Pique (1993).

Indirect tablets. Indirect tablets report the absolute position of a pointer on a sensing surface. *Touch tablets* sense a bare finger, whereas *graphics tablets* or *digitizing tablets* typically sense a stylus or other physical intermediary. Tablets can operate in *absolute mode*, with a fixed C : D gain between the tablet surface and the display, or in *relative mode*, in which the tablet responds only to motion of the stylus. If the user touches the stylus to the tablet in relative mode, the cursor resumes motion from its previous position; in absolute mode, it would jump to the new position. Absolute mode is generally preferable for tasks such as drawing, handwriting, tracing, or digitizing, but relative mode may be preferable for typical desktop interaction tasks such as selecting graphical icons or navigating through menus. Tablets thus allow coverage of many tasks (Buxton, Hill, & Rowley, 1985), whereas mice *only* operate in relative mode.

Touchpads. Touchpads are small, touch-sensitive tablets often found on laptop computers. Touchpads use relative mode for cursor control because they are too small to map to an entire screen, but most touchpads also have an absolute mode to allow features such as sliding along the edge of the pad to scroll. Touchpads support clicking by recognizing tapping or double-tapping gestures, but accidental contact (or loss of contact) can erroneously trigger such gestures (MacKenzie & Oniszczak, 1998). Like trackballs, the small size of touchpads necessitates frequent clutching, and touchpads can be awkward to use while holding down a button, unless the user employs his or her other hand.

Touchscreens and pen-operated devices. Touchscreens are transparent, touch-sensitive tablets mounted on a display. Some touchscreens can only sense a bare finger; others can sense either a plastic stylus or a bare finger. Transparent electromagnetic digitizers, such as those found on the Tablet PC, cannot sense touch, and require the use of a special pen. *Parallax error* is a mismatch between the sensed input position and the apparent input position due to viewing angle; look to minimize the displacement between the sensing and display surfaces to avoid this problem. Depending on the mounting angle, a touch or pen-operated display may result in arm or neck fatigue (Sears, Plaisant, & Shneiderman, 1992). A touchscreen that is integrated with a mobile device can be prone to accidental contact when the user picks up the device. Yet mobile devices that only sense a pen require the user to unsheathe the stylus to perform any interaction; the user cannot quickly poke at the screen with a finger. The limited states and events sensed by pen or touch-operated devices raise additional design challenges, as discussed below.

Input Device States

There is a fundamental mismatch between the demands of graphical interfaces and the states and events that can be sensed by devices such as touchscreens and pen-operated devices, which makes it difficult to support the full set of graphical interface primitives, including click, drag, double-click, and right-click. There is no easy solution when using nonstandard pointing devices that does not involve design compromises. When considering such devices, to make device limitations and differences concrete, one of the first things a designer should do is diagram all of these states and transitions.

Input devices taken in general support three possible states (Fig. 9.1): (a) out-of-range, (b) tracking, and (c) dragging; practitioners refer to these as State 0, State 1, and State 2, respectively, of the three-state model (Buxton, 1990). This model is useful for reasoning about the relationship between the events sensed by an input device and the demands of interaction techniques.

The three-state model describes the mouse as a two-state device, supporting State 1, the cursor tracking state, as well as State 2, the dragging state. State 1 provides cursor feedback of the screen position that the device will act upon, while State 2 allows the user to drag an object by holding down the primary mouse button while moving the mouse. The mouse senses movement in both the tracking and dragging states, as represented by the *dx, dy* in each state (Fig. 9.2, left), indicating relative motion tracking capability.

Many touch-activated devices such as touchscreens, touchpads, and PDA screens are also two-state devices, but *do not sense the same two states as the mouse* (Fig. 9.2, right). For example, a PDA can sense a finger when it is in contact with the screen; this is the equivalent of the mouse dragging state (State 2). The PDA can also sense when the finger is removed from the screen, but once the finger breaks contact, this enters State 0 (out-of-range), where no motion can be detected (emphasized by the *nil* in state 0 of Fig. 9.2, right). Thus, although the mouse and PDA screen both sense two states, the lack of a second motion-sensing state on the PDA means that it will be difficult to support the same interaction techniques as a mouse. For example, should sliding one's finger on the screen move a cursor, or drag an object? The designer must choose one; the PDA screen cannot support both behaviors at the same time.

The Tablet PC is an example of a pen-operated device that senses all three states of the three-state model (Fig. 9.3). The Tablet PC senses the location of the stylus when it is proximate to the screen. The pen triggers an event when it makes contact with the screen, as well as when it enters or leaves proximity.

Unfortunately, even with all three states, it is still difficult for a pen to support all the interaction techniques offered by a mouse. To help illustrate why this is the case, we can extend the three-state model to more fully characterize the interaction techniques at the core of graphical user interfaces (Fig. 9.4). The

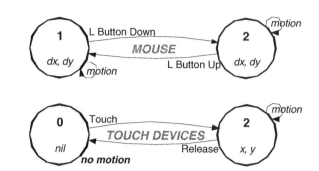

FIGURE 9.2. States sensed by a mouse (left) versus states sensed by touch-operated devices such as touchpads (right). Adapted from (Hinckley, Czerwinski, & Sinclair, 1998a)

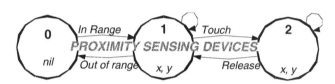

FIGURE 9.3. States sensed by a Tablet PC pen. Adapted from (Hinckley et al., 1998a).

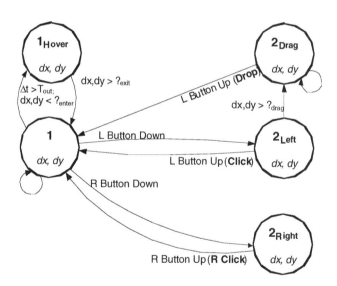

FIGURE 9.4. The five states and transition rules at the core of graphical user interfaces.

resulting five states of graphical user interfaces are (a) Tracking (1), (b) Hover (1_H), (c) Left Click (2_L), (d) Dragging (2_D), and (e) Right Click (2_R).

This diagram suggests will be difficult to support all of these interactions on a pen or touch-based device; there is no elegant solution in the literature. In this five-state model, a click is the series of transitions *1-2_L-1* with no motion in state 2_L, and a double click is *1-2_L-1-2_L-1* with no motion between the two

State	Description
0	*Out Of Range*: The device is not in its physical tracking range.
1	*Tracking*: Device motion moves only the cursor.
2	*Dragging*: Device motion moves objects on the screen.

FIGURE 9.1. Summary of states in Buxton's three-state model. Adapted from (Buxton, 1990).

clicks. Even if the device can sense State 1, hovering (1_H) for help text requires holding the pointer motionless above the display. These gestures are all difficult to perform with a pen or finger because touching or lifting disturb the cursor position (Buxton et al., 1985). Furthermore, because the pen may be out of range, and because the user must move the pen through the tracking zone to enter State 2, pen operated input devices lack a well-defined current cursor position.

Pen and touch devices also lack a second button for right-click. A finger obviously has no buttons, but even a barrel button on a pen is slow to access and is prone to inadvertent activation (Li, Hinckley, Guan, & Landay, 2005). Some mobile devices use dwelling with the pen or finger as a way to simulate right-click, but the timeout introduces an unavoidable delay; for rapid activation, it should be short as possible, but to avoid inadvertent activation (e.g., if the user is thinking about the next action to take while resting the pen on the screen), the timeout must be as long as possible. A 500-millisecond timeout offers a reasonable compromise (Hinckley, Baudisch, Ramos, & Guimbretiere, 2005a). Even techniques designed for pen-operated devices (Apitz & Guimbretiere, 2004; Kurtenbach & Buxton, 1991; Moran, Chiu, & van Melle, 1997) require rapid and unambiguous activation of one of several possible actions as fundamental building blocks; otherwise, inefficient or highly modal interactions become necessary and may reduce the appeal of such devices (Hinckley et al., 2006).

Similar issues plague other interaction modalities, such as (a) motion-sensing mobile devices (Hinckley, Pierce, Horvitz, & Sinclair, 2005b), (b) camera-based tracking of the hands (Wilson & Oliver, 2003), and (c) 3D input devices (Hinckley et al., 1994). All of these techniques require a method for users to move the device or their hands without accidentally performing an action. Thus state transitions form fundamental indications of intent that are essential for rapid and dependable interaction.

WHAT'S AN INPUT DEVICE FOR? THE COMPOSITION OF USER TASKS

One way of reasoning about input devices and interaction techniques is to view a device or technique in light of the tasks that it can express. But what sort of tasks are there?

Elemental Tasks

While computers can support many activities, at the input level, some subtasks appear repeatedly in graphical user interfaces, such as pointing at a target on the screen or typing a character. Foley and colleagues (Foley, Wallace, & Chan, 1984) identified elemental tasks including (a) *text* (entering symbolic data), (b) *select* (indicating objects from a set of alternatives), (c) *position* (pointing to a screen coordinate), and (d) *quantify* (specifying an exact numeric value). If these are elemental tasks, however, then where do devices such as global positioning system (GPS) readers, cameras, or fingerprint scanners fit in? These offer new "elemental" data types (e.g., location, images, and identity). Advances in technology will continue to yield data types that enable new tasks and scenarios of use.

Compound Tasks and Chunking

Another problem with the elemental task approach is that the level of analysis for "elemental" tasks is not well defined. For example, a mouse indicates an *(x, y)* position on the screen, but an Etch A Sketch® separates positioning into two subtasks by providing a single knob for *x* and a single knob for *y* (Buxton, 1986b). If *position* is an elemental task, why must we subdivide this task for some devices but not others? One way to resolve this puzzle is to view all tasks as hierarchies of subtasks (Fig. 9.5). Whether or not a task is "elemental" depends on the input device being used: the Etch A Sketch supports separate *QuantifyX* and *QuantifyY* tasks, whereas the mouse supports a compound *2D Position* task (Buxton, 1986a).

From the user's perspective, a series of elemental tasks may seem like a single task. For example, scrolling a web page to click on a link could be conceived as an elemental 1D positioning task followed by a 2D selection task, or, it can be viewed as a compound *navigation/selection* task (Buxton & Myers, 1986). An interaction technique can encourage the user to work at the higher level of the compound task, for example, by scrolling with one hand while pointing to the link with the other hand. This is known as *chunking*.

These examples show that the choice of device influences the level at which the user is required to think about the individual actions that must be performed to achieve a goal. The design of input devices and interaction techniques can help to structure the interface such that there is a more direct match between the user's tasks and the low-level syntax of the individual actions that must be performed to achieve those tasks. The choice of device and technique thus directly influences the steps required of the user and hence the apparent complexity of an interface design (Buxton, 1986a).

EVALUATION AND ANALYSIS OF INPUT DEVICES

Beyond standard usability engineering techniques (see Part VI: The Development Process), a number of techniques tailored to the study of input devices also exist. Representative tasks (Buxton, 1995c), such as target acquisition, pursuit tracking, freehand drawing, and dragging versus tracking performance (MacKenzie et al., 1991), can be used to formally or informally evaluate devices. Here, we focus on formal analysis using Fitts' Law, the Steering Law, and the Keystroke-Level Model.

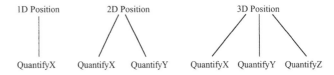

FIGURE 9.5.   Task hierarchies for 1D, 2D, and 3D position tasks.

Fitts' Law and Hick's Law

Fitts' Law (Fitts, 1954) is an experimental paradigm that has been widely applied to the comparison and optimization of pointing devices. Fitts' Law is used to measure how effectively a pointing device can acquire targets on the screen. Fitts' Law was first applied to the study of input devices by Card, English, and Burr (1978); it is now a standard for device comparisons (Douglas, Kirkpatrick, & MacKenzie, 1999). Fitts' Law applies to remarkably diverse task conditions, including (a) rate-controlled devices (MacKenzie, 1992a), (b) area cursors (Kabbash & Butxon, 1995), (c) scrolling (Hinckley et al., 2002), and (d) zooming (Guiard, Buourgeois, Mottet, & Beaudouin-Lafon, 2001). For further guidance on conducting Fitts' Law studies, see Douglas and colleagues (1999), MacKenzie (1992b), and Raskin (2000).

The standard Fitts' task paradigm measures the movement time MT between two targets separated by amplitude A, with a width W of error tolerance (Fig. 9.6).

Fitts' Law states that a logarithmic function of the ratio of A to W predicts the *average* movement time MT. The Fitts' Law formulation typically used for input device studies is

$$MT = a + b \log_2(A/W + 1) \qquad (9.1)$$

Here, the constants a and b are coefficients fit to the average of all observed MT for each combination of A and W in the experiment. One calculates a and b via linear regression using a statistical package or spreadsheet. The constants a and b depend heavily on the exact task setting and input device, so be wary of substituting "typical" values for these constants, or of comparing constants derived from different studies.

Psychomotor interpretations for Fitts' Law have been proposed (Douglas & Mithal, 1997); however, since the law characterizes the central tendency of a large number of pointing movements, the law may simply reflect information-theoretic entropy (MacKenzie, 1989). For example, Hick's Law, a model of decision time for a set of choices (e.g., in a menu), has a general form almost identical to Fitts' Law:

$$H = \log_2(n + 1) \qquad (9.2)$$

Here, n is the number of equally probable alternatives, and H is the entropy of the decision. If we view Fitts' task (0) as a "decision"

along the amplitude A between n discrete targets of width W, this raises the possibility that Fitts' Law and Hick's Law are fundamentally the same law where $n = A/W$.

The Steering Law and Minimum Jerk Law

Steering a cursor through a narrow tunnel, as required to navigate a pull-down menu, is not a Fitts' task because steering requires a continuous accuracy constraint: the cursor must stay within the tunnel at all times. For a straight-line tunnel (Fig. 9.7) of width W and length A, for example, the Steering Law predicts that movement time is a linear function of A and W:

$$MT = a + b A/W \qquad (9.3)$$

The Steering Law can also model arbitrary curved paths, as well as instantaneous velocity (Accot & Zhai, 1997). A limitation of the Steering Law is that it only models successful completion of the task; errors are not considered.

The Minimum Jerk Law (Viviani & Flash, 1995) characterizes the dynamics of motions that may follow a curved path but do not have a continuous accuracy constraint. The law states that unconstrained human movement trajectories tend to minimize the derivative of acceleration (jerk); one of its implications is that there is a two-thirds power law linking tangential velocity and path curvature; however, no one has yet formulated a universal law that handles varying accuracy constraints and curvature (Lank & Saund, 2005).

The Keystroke-Level Model (KLM) and GOMS Analysis

The KLM is an engineering and analysis tool that can be used to estimate the time needed for experts to complete a routine task (Card, Moran, & Newell, 1980). To apply the KLM, count the elemental inputs required to complete a task, including (a) keystrokes, (b) homing times to acquire input devices, (c) pauses for mental preparation, and (d) pointing at targets. For each elemental input, substitute a constant estimate of the average time required using the values from Card et al. (1980), or by collecting empirical data (Hinckley et al., 2006), and sum them to yield an overall time estimate. The model assumes error-free execution, so it cannot estimate time for the problem-solving behaviors of novices, but it does employ several heuristics to model mental pauses (Raskin, 2000).

GOMS (Goals, Objects, Methods, and Selection rules) models extend the keystroke-level model (John & Kieras, 1996). Some GOMS models can account for user knowledge and

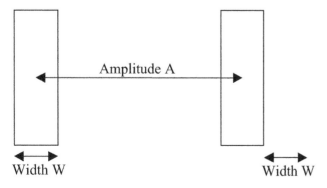

FIGURE 9.6.   Fitts' task paradigm (Fitts, 1954).

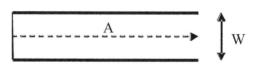

FIGURE 9.7.   The Steering Law for a straight tunnel. The user must follow the dotted line without moving beyond the borders. Adapted from Accot & Zhai, 1999.

interleaving of tasks, but are more difficult to apply than the KLM. Both GOMS and KLM models are engineering tools that produce *estimates* for expert completion time of routine tasks. These models do not replace the need for usability testing and evaluation, but they do offer a means to assess a design without implementing software, training end users, and evaluating their performance (Olson & Olson, 1990). Physical articulation times derived from KLM or GOMS analyses can also be used to help interpret results of empirical studies (Hinckley et al., 2006; see also chapter 60, "Model-based Evaluations"; and chapter 5, "Cognitive Architecture".

TRANSFER FUNCTIONS: HOW TO TRANSFORM AN INPUT SIGNAL

A transfer function is a mathematical transformation that scales the data from an input device. Typically, the goal is to provide more stable and more intuitive control, but one can easily design a poor transfer function that hinders performance. A transfer function that matches the properties of an input device is known as an *appropriate mapping*. For force sensing input devices, the transfer function should be a *force-to-velocity* function; for example, the force one exerts on the IBM Trackpoint isometric joystick controls the speed at which the cursor moves. Other appropriate mappings include *position-to-position* or *velocity-to-velocity* functions, used with tablets and mice, respectively.

A common example of an inappropriate mapping is calculating a velocity based on the position of the mouse cursor, such as to scroll a document. The resulting input is difficult to control, and this inappropriate rate mapping is only necessary because the operating system clips the cursor to the screen edge. A better solution would be to ignore the cursor position and instead use the relative position information reported by the mouse to directly control the change of position within the document.

Self-centering devices. Rate mappings suit force-sensing devices or other devices that return to center when released (Zhai, 1993; Zhai et al., 1997). This property allows the user to stop quickly by releasing the device. The formula for a nonlinear rate mapping is

$$dx = K x^{\alpha} \qquad (9.4)$$

Where x is the input signal, dx is the resulting rate, K is a gain factor, and α is the nonlinear parameter. The best values for K and α depend on the details of the device and application, and appropriate values must be identified by experimentation or optimal search (Zhai & Milgram, 1993). Many commercial devices use more complex mappings (Rutledge & Selker, 1990).

Motion-sensing devices. Desktop systems use an exponential transformation of the mouse velocity, known as an *acceleration* function, to modify the cursor response (Microsoft Corp., 2002). Acceleration functions do not directly improve pointing performance, but do limit the footprint required by a device (Jellinek & Card, 1990), which may lead to greater comfort or less frequent clutching (Hinckley et al., 2002).

Absolute devices. It is possible to temporarily violate the 1 : 1 control-to-display mapping of absolute devices such as touchscreens by damping small motions to provide fine adjustments; large motions revert to an absolute 1 : 1 mapping (Sears & Shneiderman, 1991). A drawback of this technique is that cursor feedback in the tracking state becomes the default behavior of the device, rather than dragging (Buxton, 1990), but researchers are exploring ways to overcome this (Benko, Wilson, & Baudisch, 2006).

FEEDBACK: WHAT HAPPENS IN RESPONSE TO AN INPUT?

From the technology perspective, one can consider feedback as active or passive. Active feedback is under computer control; passive feedback is not, and may result from internal sensations within the user's own body, such as muscle tension from holding down a button, or physical properties of the device, such as the feel of clicking its buttons.

The industrial design suggests how to use a device before a user touches it (Norman, 1990). Mechanical sounds and vibrations produced by a device provide positive feedback for the user's actions (Lewis, Potosnak, & Magyar, 1997). The shape of the device and the presence of landmarks can help users acquire a device without having to look at it (Hinckley, Pausch, Proffitt, & Kassell, 1998b).

Proprioceptive and Kinesthetic Feedback

Internal sensations of body posture, motion, and muscle tension (Burdea, 1996; Gibson, 1962) may allow users to feel how they are moving an input device without either looking at the device or receiving visual feedback on a display. This is important when the user's attention is divided between multiple tasks and devices (Balakrishnan & Hinckley, 1999; Fitzmaurice & Buxton, 1997; Mine, Brooks, & Sequin, 1997). Muscular tension can help to phrase together multiple related inputs (Buxton, 1986a) and may make mode transitions more salient to the user (Hinckley et al., 2006; Raskin, 2000; Sellen, Kurtenbach, & Buxton, 1992).

Kinesthetic Correspondence

Graphical feedback on the screen should correspond to the direction that the user moves the input device (Britton, Lipscomb, & Pique, 1978). If the user moves a device to the left, then the object on the screen should likewise move left; however, users can easily adapt to certain kinds of noncorrespondences: when the user moves a mouse forward and backward, the cursor actually moves up and down on the screen; if the user drags a scrollbar downward, the text on the screen scrolls upwards. Researchers have also found that the dimensions of an input device should match the perceptual structure of a task (Jacob, Sibert, McFarlane, & Mullen, 1994).

Snapping Behaviors and Active Haptic Feedback

Software constraints, such as snapping (Baudisch, Cutrell, Hinckley, & Eversole, 2005), often suffice to support a user's tasks. Active force or tactile feedback (Burdea, 1996) can provide attractive forces for a target, or additional feedback for the boundaries of a target, but such feedback typically yields little or no performance advantage even for isolated target selection (Akamatsu & Mackenzie, 1996; MacKenzie, 1995). Such techniques must evaluate selection among multiple targets, because haptic feedback or snapping behavior for one target interferes with the selection of others (Grossman & Balakrishnan, 2005; Oakley, Brewster, & Gray, 2001). *Visual dominance* refers to the tendency for vision to dominate other modalities (Wickens, 1992); haptic feedback typically must closely match visual feedback, which limits its utility as an independent modality (Campbell, Zhai, May, & Maglio, 1999). One promising use of tactile feedback is to improve state transitions (Poupyrev & Maruyama, 2003; Snibbe & MacLean, 2001). For further discussion of active feedback modalities, see chapter 12, "Haptic Interfaces," and chapter 13, "Nonspeech Auditory Output."

KEYBOARDS AND TEXT ENTRY

Typewriters have been in use for over 100 years; the QWERTY key layout dates to 1868 (Yamada, 1980). Despite the antiquity of the design, QWERTY keyboards are extremely well suited to human performance and are unlikely to be supplanted by new key layouts, speech recognition technologies, or other techniques any time soon. Many factors can influence typing performance, including the size, shape, activation force, key travel distance, and the tactile and auditory feedback provided by striking the keys (Lewis et al., 1997), but these well-established design details are not our focus here.

Procedural Memory

Procedural memory allows performance of complex sequences of practiced movements, seemingly without any cognitive effort (Anderson, 1980). Procedural memory enables touch typing on a keyboard with minimal attention while entering commonly used symbols. As a result, users can focus attention on mental composition and verification of the text appearing on the screen. Dedicated or chorded key presses for frequently used commands (hotkeys) likewise allow rapid command invocation (McLoone, Hinckley, & Cutrell, 2003). The automation of skills in procedural memory is described by the *power law of practice*:

$$T = aP^b \qquad (9.5)$$

Here, T is the time to perform a task, P is the amount of practice, and the multiplier a and exponent b are fit to the observed data (Anderson, 1980). For a good example of applying the power law of practice to text entry research, see MacKenzie, Kober, Smith, Jones, and Skepner (2001).

Alternative keyboard layouts such as Dvorak offer about a 5% performance gain (Lewis et al., 1997), but the power law of practice suggests this small gain comes at a substantial cost for retraining time; however, ergonomic QWERTY keyboards do preserve much of a user's skill for typing. These split-angle keyboards are not faster, but some can help maintain neutral posture of the wrist and thus avoid ulnar deviation (Honan, Serina, Tal, & Rempel, 1995; Marklin, Simoneau, & Monroe, 1997; Smutz, Serina, Bloom, & Rempel, 1994), which has been associated with increased pressure in the carpal tunnel (Putz-Anderson, 1988; Rempel, Bach, Gordon, & Tal, 1998).

Mobile Text Entry, Character Recognition, and Handwriting Recognition

The difficulty of entering text on handheld devices and cell phones has led to many new text-entry techniques (MacKenzie, 2002), but most offer only 10–20 words-per-minute (wpm) typing rates, compared to approximately 60 wpm for a touch typist.

Many designs for cell phones and other handheld devices, such as the RIM Blackberry, offer two-thumb keyboards with QWERTY key layouts. The principal virtue of QWERTY is that common pairs of letters tend to occur on opposite hands. This alternation is a very efficient movement pattern for both standard and two-thumb keyboards, since one hand completes a key press while the other hand moves to the next key (MacKenzie & Soukoreff, 2002). A recent study found that two-thumb keyboards offer text entry rates approaching 60 wpm (Clarkson, Clawson, Lyons, & Starner, 2005). This suggests that one-handed text entry rates are fundamentally limited due to the serial nature of character entry, despite novel improvements (MacKenzie et al., 2001; Wigdor & Balakrishnan, 2003). Word prediction may help, but also requires overhead for users to monitor and decide whether to use the predictions.

Soft keyboards and character recognition techniques are popular for pen-operated devices, but likewise are limited to serial entry. Soft keyboards depict keys in a graphical user interface to allow typing with a touchscreen or stylus. Design issues for soft keyboards differ from mechanical keyboards. Soft keyboards demand visual attention because the user *must look at the keyboard* to aim the pointing device. Only one key at a time can be touched, so much of the time is spent moving back and forth between keys (Zhai, Hunter, & Smith, 2000). A soft keyboard can allow the user to draw gestures across multiple keys; in combination with a language model, this can allow entire words to be entered with a single pen gesture (Kristensson & Zhai, 2004).

Handwriting (even on paper, with no "recognition" involved) proceeds at about 15 wpm. Ink has significant value as a natural data type without recognition: it offers an expressive mix of writing, sketches, and diagrams. Although recognition technology continues to improve, recognizing natural handwriting remains difficult and error prone for computers, and demands error correction input from the user. To make performance more predictable for the user, some devices instead rely on single-stroke gestures, known as *unistrokes* (Goldberg & Richardson, 1993), including *graffiti* for the PalmPilot. Unistroke alphabets attempt to strike a design balance such that each letter is easy for a computer to distinguish, yet also straightforward for users to learn (MacKenzie & Zhang, 1997).

MODALITIES OF INTERACTION

In the search for designs that enhance interfaces and enable new usage scenarios, researchers have explored several strategies that transcend any specific type of device.

Speech and Multimodal Input

Speech has substantial value without recognition. Computers can augment human-human communication across both time and space by allowing users to record, edit, replay, or transmit digitized speech and sounds (Arons, 1993; Buxton, 1995b; Stifelman, 1996). Systems can also use microphone input to detect ambient speech and employ this as a cue to help prioritize notifications (Horvitz, Jacobs, & Hovel, 1999; Sawhney & Schmandt, 2000; Schmandt, Marmasse, Marti, Sawhney, & Wheeler, 2000).

Computer understanding of human speech does not enable users to talk to a computer as one would converse with another person (but see also chapter 19, Conversational Interfaces and Technologies). Speech recognition can succeed for a limited vocabulary, such as speaking the name of a person from one's contact list to place a cell phone call; however, error rates increase as the vocabulary and complexity of the grammar grows, if the microphone input is poor, or if users employ "out-of-vocabulary" words. It is difficult to use speech to refer to spatial locations, so it cannot eliminate the need for pointing (Cohen & Sullivan, 1989; Oviatt, DeAngeli, & Kuhn, 1997; see also chapter 21, "Multimodal Interfaces"). Currently, keyboard-mouse text entry for the English language is about twice as fast as automatic speech recognition (Karat, Halverson, Horn, & Karat, 1999); furthermore, speaking can interfere with one's ability to compose text and remember words (Karl, Pettey, & Shneiderman, 1993). Finally, speech is inherently nonprivate in public situations. Thus, speech has an important role to play, but claims that speech will soon supplant manual input devices should be considered with skepticism.

Bimanual Input

People use both hands to accomplish most real-world tasks (Guiard, 1987), but computer interfaces make little use of the nonpreferred hand for tasks other than typing. Bimanual input enables compound input tasks such as navigation/selection tasks, where the user can scroll with the nonpreferred hand while using the mouse in the preferred hand (Buxton & Myers, 1986). This assignment of roles to the hands corresponds to Guiard's kinematic chain theory (Guiard, 1987): the nonpreferred hand sets a frame of reference (scrolling to a location in the document) for the action of the preferred hand (selecting an item within the page using the mouse). Other applications for bimanual input include command selection (Bier, Stone, Pier, Buxton, & DeRose, 1993; Kabbash, Buxton, & Sellen, 1994), drawing tools (Kurtenbach, Fitzmaurice, Baudel, & Buxton, 1997), and virtual camera control and manipulation (Balakrishnan & Kurtenbach, 1999; Hinckley et al., 1998b). Integrating additional buttons and controls with keyboards to encourage bimanual interaction can also improve the efficiency of some common tasks (MacKenzie & Guiard, 2001; McLoone et al., 2003).

Pen and Gesture Input

The Palm Pilot and Tablet PC have led to a renaissance in pen and gesture research. Pens lend themselves to command gestures analogous to proofreader's marks, such as crossing out a word to delete it. Note that in this example, the gesture integrates the selection of a *delete* command with the selection of the word to be deleted. Another example is moving a paragraph by circling it and drawing a line to its new location. This integrates the verb, object, and indirect object by specifying the command, the extent of text to move, and the new location for the text (Hinckley et al., 2005a; Kurtenbach & Buxton, 1991). *Marking menus* use straight-line gestures along the primary compass directions for rapid command selection (Kurtenbach, Sellen, & Buxton, 1993; Zhao & Balakrishnan, 2004).

Pen interfaces must decide whether to treat pen strokes as *ink* content or as *gesture* commands. Some applications avoid this recognition problem by treating all strokes as commands (Kurtenbach & Buxton, 1991), but for a free-form drawing or note-taking application, users need to interleave ink content and command input. The status-quo solution presents commands in a toolbar or menu at the edge of the screen; however, this necessitates round trips between the work area and the command area (Fitzmaurice, Khan, Pieke, Buxton, & Kurtenbach, 2003a), which becomes inconvenient in direct proportion to the display size. Pressing a button with the nonpreferred hand is a fast and robust means to switch between ink and gesture modes (Li et al., 2005).

Techniques to automatically distinguish ink and gestures have been proposed, but only for highly restricted gesture sets (Saund & Lank, 2003). Punctuation (tapping) has also been explored as a way to both identify and delimit command phrases (LaViola, 2004). A fundamental problem with both of these approaches is that the system cannot classify a set of strokes as a gesture or as ink until after the user has finished drawing the entire command phrase. This makes it difficult to provide interactive feedback or to prompt the user with the available commands before the user commits to an operation.

While moving the pen to toolbars at the edge of the screen seems slow on a tablet computer, in practice, this "round trip strategy" (Fitzmaurice et al., 2003a) is difficult to improve upon. On a tablet the size of a standard 8.5 × 11 inch sheet of paper, a round trip requires approximately 1.5 seconds; however, the user can mentally prepare for the next step of the interaction while moving the pen. A locally drawn gesture (such as a straight-line marking menu command) may take less time to articulate, but thinking about what command to select requires additional time unless the task is a routine one. Pressing a button for gesture mode also requires some overhead, as does lifting the pen at the end of the gesture. Also note that performing a sequence of gestures (e.g., tagging words in a document as keywords by circling them) requires time to travel between screen locations. The round-trip strategy absorbs this travel time into the round trip itself, but with gestures, this is an extra cost that reduces the benefit of keeping the interaction localized.

Thus, on a tablet-sized device, it is difficult to realize a substantial time savings just by reducing round trips. For our hypothetical task of tagging keywords in a document, Fig. 9.8 illustrates this predicament for average task times drawn from recent studies (Hinckley et al., 2006; Li et al., 2005). The chart shows two successive command selections, and assumes some mental preparation is required before issuing each command. Thus, the potential benefit of pen gestures depends on the sequence of operations as well as the elimination of *multiple* round trips, as may be possible with techniques that integrate selection of verb, object, and indirect object (Hinckley et al., 2005a; Kurtenbach & Buxton, 1991). Localized interaction may also offer indirect benefits by reducing physical effort and by keeping the user's visual attention focused on their work (Grossman, Hinckley, Baudisch, Agrawala, & Balakrishnan, 2006; Kabbash et al., 1994).

Whole Hand Input

Humans naturally gesture and point using their hands during verbal communication, which has motivated research into free-hand gestures, often in combination with speech recognition (Bolt, 1980; Hauptmann, 1989; Wilson & Oliver, 2003). Cadoz categorized hand gestures as semiotic, ergotic, or epistemic. Semiotic gestures, such as "thumbs up," communicate information (Rime & Schiaratura, 1991). Ergotic gestures manipulate physical objects. Epistemic gestures are exploratory motions that gather information (Kirsh, 1995; Kirsh & Maglio, 1994). The interaction literature focuses on empty-handed semiotic gestures (Freeman & Weissman, 1995; Jojic, Brumitt, Meyers, & Harris, 2000; Maes, Darrell, Blumberg, & Pentland, 1997). A major challenge is to correctly identify when a gesture, as opposed to an incidental hand movement, starts and stops (Baudel & Beaudouin-Lafon, 1993; Wilson & Oliver, 2003). The lack of deterministic state transitions (Buxton, 1990; Vogel & Balakrishnan, 2005) can lead to errors of user intent or errors of computer interpretation (Bellotti, Back, Edwards, Grinter, Lopes, & Henderson, 2002). Other problems include fatigue from extending one's arms for long periods, and the imprecision of pointing at a distance. By contrast, tangible interaction techniques (Ishii & Ullmer, 1997) and augmented devices (Harrison, Fishkin, Gujar, Mochon, & Want, 1998) sense ergotic gestures via a physical intermediary (Hinckley et al., 1998b; Zhai et al., 1996). The emergence of cameras, cell phones, and tablets augmented with accelerometers and other sensors suggest this is a promising design space.

Background Sensing Techniques

Sensors can enable a mobile device to sense when the user picks up, puts down, looks at, holds, or walks around with the device. These actions give a device information about the context of its use, and represent a hidden vocabulary of naturally

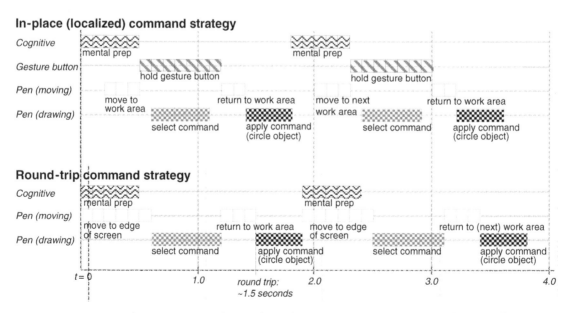

FIGURE 9.8. Chart comparing task times for performing two successive commands on a tablet, with either in-place command selection (top) or round-trip command selection (bottom). The individual boxes in each subtask represent 100 millisecond intervals.

occurring gestures that people spontaneously exhibit in day-to-day activity. For example, commercially available digital cameras now employ a tilt sensor to detect the orientation of the camera, and use this to automatically save photos in the correct orientation, as well as to interactively switch the display between portrait/landscape formats (Hinckley et al., 2005b; Hinckley, Pierce, Sinclair, & Horvitz, 2000). Here the sensor allows the device to adapt its behavior to the user's needs, rather than requiring the user to take extra steps to control the photo orientation and display format (Buxton, 1995a).

Sensors can also be embedded in the environment. When one walks into a modern department store, there is no explicit command to open the doors: the doors sense motion and automatically open. Researchers are investigating new ways to leverage such contextual sensing to enrich and simplify interaction with devices and digital environments (Abowd & Mynatt, 2000; Schilit, Adams, & Want, 1994).

Multitouch Tables and Screens

Technical advances have led to much recent research in touch-sensitive tables and projection screens. These technologies blur the distinction between whole-hand gestural input and traditional single-point touchscreens and touch tablets. Recent prototype systems demonstrate capture of the shape formed by the hand(s) in contact with a surface (Wilson, 2005), multiple points of contact (Han, 2005), or even images of objects placed on or near a surface (Matsushita & Rekimoto, 1997; Wilson, 2004). The DiamondTouch (Dietz & Leigh, 2001) is unique in its ability to determine which user produces each point of contact, which has led to a number of innovative applications and techniques (Shen, Everitt, & Ryall, 2003; Wu & Balakrishnan, 2003). For an overview of design issues for tabletop systems, see Scott, Grant, and Mandryk (2003).

A Society of Devices

Wireless networking is the technology that will most disrupt traditional approaches to human–computer interaction in the coming years, because it breaks down barriers between devices and enables a new society of devices that fill specific roles, yet can still coordinate their activities (Fitzmaurice, Khan, Buxton, Kurtenbach, & Balakrishnan, 2003b; Want & Borriello, 2000). The interaction designer thus must consider the full range of scale for display sizes and form-factors that may embody an interaction task, as well as the interactions between different types of devices. How can interaction migrate from watches, cell phones, handheld devices, and tablets, all the way up to desktop monitors, digital whiteboards, and interactive wall-sized displays?

As digital devices become physically smaller, the displays and input mechanisms they offer are shrinking. Considerable effort has been devoted to supporting web browsing in limited screen space (Buyukkokten, Garcia-Molina, & Paepcke, 2001; Jones, Marsden, Mohd-Nasir, Boone, & Buchanan, 1999; Trevor, Hilbert, Schilit, & Koh, 2001). Techniques to make small displays virtually larger include (a) peephole displays (Fitzmaurice, 1993; Yee, 2003), (b) transparent overlays (Harrison, Ishii, Vicente, & Bux-

ton, 1995; Kamba, Elson, Harpold, Stamper, & Sukaviriya, 1996), and (c) using on-screen visuals to suggest the locations of off-screen objects (Baudisch & Rosenholtz, 2003). Touchscreens and touch-sensitive controls minimize the vertical profile of devices (Wherry, 2003), but may suffer from inadvertent activation. Physical manipulations such as tilting that use the device itself as an interface seem particularly well suited to small devices (Harrison et al., 1998; Hinckley et al., 2000; Rekimoto, 1996). Tiny, bright, and inexpensive laser or LED projectors are just around the corner; progress on computer vision techniques suggests that interactive projection may allow small devices to project large displays and sensing surfaces (Raskar et al., 2004; Wilson, 2005).

At the other end of the spectrum, large-format displays are now affordable and common. Large displays lend themselves to collaboration and sharing of information with groups (Funkhouser & Li, 2000; Swaminathan & Sato, 1997), as well as giving a substantial physical presence to virtual activities (Buxton, Fitzmaurice, Balakrishnan, & Kurtenbach, 2000; Trimble, Wales, & Gossweiler, 2003). Researchers have explored pen and touch-screen interaction techniques for large displays (Guimbretiere, Stone, & Winograd, 2001; Moran et al., 1997). Unfortunately, many technologies sense only a single point of contact. For interaction at a distance with large displays, it remains unclear what interaction techniques work best (Olsen & Nielsen, 2001; Vogel & Balakrishnan, 2004; Vogel & Balakrishnan, 2005); even when a user is close to a large display, interacting with portions of the display that are out view or beyond arm's length raises challenges (Bezerianos & Balakrishnan, 2005; Khan, Matejka, Fitzmaurice, & Kurtenbach, 2005).

Displays of various sizes support different activities and social conventions; one of the principle challenges of ubiquitous computing (Weiser, 1991) is finding techniques that make it easy for users to work within a digital ecology that supports a range of tasks spanning multiple computers, displays, and interactive surfaces. Even on a single computer, users do not treat multiple monitors as one large display, but instead employ the boundary between displays to partition their tasks (Grudin, 2001). Several projects have probed how to use small displays as adjuncts to large ones (Myers, Lie, & Yang, 2000; Myers, Stiel, & Garguilo, 1998; Rekimoto, 1998), allowing simultaneous interaction with private information on a personal device, and a shared or public context on a larger display. Users need techniques that allow them to access and share information across the boundaries of individual devices, as well as to dynamically bind together multiple devices and displays to accomplish their tasks (Hinckley, Ramos, Guimbretiere, Baudisch, & Smith, 2004; Rekimoto, Ayatsuka, & Kohno, 2003). Such interaction techniques inherently involve multiple persons, and thus must consider how people use physical proximity and relative body orientation, a field known as *proxemics* (Deasy & Lasswell, 1985; Hinckley et al., 2004; Sommer, 1965).

CURRENT AND FUTURE TRENDS FOR INPUT

The designer of an interactive system should take a broad view of input, and consider not only traditional pointing techniques and graphical user interface widgets, but also issues such as search strategies to access information in the first place, sensor inputs that enable entirely new data types, synthesis techniques to extract meaningful structure from data, and integration of traditional technologies such as paper that offer fundamental strengths.

Good search tools may reduce the many inputs needed to manually search and navigate file systems. Knowledge work requires integration of external information from web pages or databases (Yee, Swearingen, Li, & Hearst, 2003) as well as reuse of personal information from documents, electronic mail messages, and other content authored or viewed by a user (Dumais, Cutrell, Cadiz, Jancke, Sarin, & Robbins, 2003; Lansdale & Edmonds, 1992). Unified full-text indexing allows users to quickly query their personal information across multiple information silos, and can present information in the context of memory landmarks such as the date a message was sent, who authored the text, or the application used to create a document (Cutrell, Robbins, Dumais, & Sarin, 2006).

New sensor inputs such as location and tagging technologies are coming to fruition. Radio-frequency identification (RFID) tags (Want, Fishkin, Gujar, & Harrison, 1999) allow computers to identify tagged physical objects, thus enabling manipulation of ordinary objects as an input for computers. A mobile tag reader can identify tagged physical locations (Smith, Davenport, & Hwa, 2003). Wireless communication technologies are poised to deliver ubiquitous location-based services (Schilit et al., 1994; Want, Hopper, Falcao, & Gibbons, 1992). Cell phones and low-power radios for wireless networking can sense their location or proximity to other devices via analysis of signal strengths (Bahl & Padmanabhan, 2000; Krumm & Hinckley, 2004). As another example, attempting to type a secure password on a mobile phone keypad quickly convinces one that biometric sensors or some other convenient means for establishing identity is essential for these devices to succeed. Such sensors could also make services such as personalization of interfaces much simpler.

The need to extract models and synthesize structure from large quantities of low-level inputs suggests that data mining and machine learning techniques will become important adjuncts to interaction (Fails & Olsen, 2003; Fitzmaurice, Balakrisnan, & Kurtenbach, 1999; Horvitz, Breese, Heckerman, Hovel, & Rommelse, 1998). Whenever a system considers automatic actions on behalf of the user, however, an important design principle in the face of uncertainty is to "do less, but do it well" (Horvitz, 1999).

With so many new technologies, it is easy to forget that paper remains an important input and display medium (Sellen & Harper, 2002). Paper is inexpensive, easy to annotate, rapid to access, comfortable to read, light to carry, and can be accessed after tens or even hundreds of years. Because technology will not replace paper any time soon, researchers are studying how to bring paperlike interfaces to digital devices (Schilit, Golovchinsky, & Price, 1998; Wolf, Rhyne, & Ellozy, 1989), as well as how to augment paper with digital capabilities (Guimbretiere, 2003; Liao, Guimbretiere, & Hinckley, 2005; Stifelman, 1996).

We must make substantial progress in all of these areas to advance human interaction with technology. The forms and capabilities of these and other technologies will continue to advance, but human senses and cognitive skills will not. We will continue to interact with computers using our hands and physical intermediaries, not necessarily because our technology requires us to do so, but because touching, holding, and moving physical objects is the foundation of the long evolution of tool use in the human species (Wilson, 1998).

References

Abowd, G., & Mynatt, E. (2000). Charting past, present, and future research in ubiquitous computing. *ACM Transactions on Computer-Human Interaction, 7*(1), 29–58.

Accot, J., & Zhai, S. (1997). Beyond Fitts' Law: Models for trajectory-based HCI tasks. *Proceedings of the CHI '97 ACM Conference on Human Factors in Computing Systems* (pp. 295–302).

Accot, J., & Zhai, S. (1999). Performance evaluation of input devices in trajectory-based tasks: An application of the Steering Law. *Proceedings of the CHI '99* (pp. 466–472).

Accot, J., & Zhai, S. (2001). Scale effects in Steering Law tasks. *Proceedings of CHI 2001 ACM Conference on Human Factors in Computing Systems,* (pp. 1–8).

Akamatsu, M., & Mackenzie, I. S. (1996). Movement characteristics using a mouse with tactile and force feedback. *International Journal of Human-Computer Studies, 45,* 483–493.

Anderson, J. R. (1980). Cognitive skills. In *Cognitive psychology and its implications* (pp. 222–254). San Francisco: W. H. Freeman.

Apitz, G., & Guimbretiere, F. (2004). CrossY: A crossing based drawing application. *UIST 2004,* 3–12.

Arons, B. (1993). SpeechSkimmer: Interactively skimming recorded speech. In *UIST'93 Symposium on User Interface Software & Technology* (pp. 187–195).

Bahl, P., & Padmanabhan, V. (2000). RADAR: An in-building RF-based user location and tracking system. *Proceedings of the IEEE 19th Annual Joint Conference of the IEEE Computer and Communications Societies (INFOCOM 2000),* 775–784.

Balakrishnan, R., Baudel, T., Kurtenbach, G., & Fitzmaurice, G. (1997). The Rockin'Mouse: Integral 3D manipulation on a plane. *Proceedings of the CHI '97 Conference on Human Factors in Computing Systems,* 311–318.

Balakrishnan, R., & Hinckley, K. (1999). The role of kinesthetic reference frames in two-handed input performance. *Proceedings of the ACM UIST'99 Symposium on User Interface Software and Technology,* 171–178.

Balakrishnan, R., & Kurtenbach, G. (1999). Exploring bimanual camera control and object manipulation in 3D graphics interfaces. *Proceedings of the CHI'99 ACM Conference on Human Factors in Computing Systems,* 56–63.

Baudel, T., & Beaudouin-Lafon, M. (1993). Charade: Remote control of objects using hand gestures. *Communications of the ACM, 36*(7), 28–35.

Baudisch, R., Cutrell, E., Hinckley, K., & Eversole, A. (2005). Snap-and-go: Helping users align objects without the modality of traditional snapping. *Proceedings of the CHI 2005,* 301–310.

Baudisch, P., & Rosenholtz, R. (2003). Halo: a technique for visualizing off-screen objects. *Proceedings of the CHI '03,* 481–488.

Bellotti, V., Back, M., Edwards, W. K., Grinter, R., Lopes, C., & Henderson, A. (2002). Making sense of sensing systems: Five questions for designers and researchers. *Proceedings of the ACM CHI 2002 Conference on Human Factors in Computing Systems*, 415–422.

Benko, H., Wilson, A., & Baudisch, P. (2006). Precise selection techniques for multi-touch screens. *Proceedings of the CHI 2006*.

Bezerianos, A., & Balakrishnan, R. (2005). The vacuum: Facilitating the manipulation of distant objects. *Proceedings of the CHI '05*, 361–370.

Bier, E., Stone, M., Pier, K., Buxton, W., & DeRose, T. (1993). Toolglass and magic lenses: The see-through interface. *Proceedings of SIGGRAPH 93*, 73–80.

Bolt, R. (1980, August). Put-That-There: Voice and gesture at the graphics interface. *Computer Graphics*, 262–270.

Bowman, D., Kruijff, E., LaViola, J., & Poupyrev, I. (2004). 3D user interfaces: Theory and practice, Addison-Wesley, Boston.

Britton, E., Lipscomb, J., & Pique, M. (1978). Making nested rotations convenient for the user. *Computer Graphics, 12*(3), 222–227.

Burdea, G. (1996). *Force and Touch Feedback for Virtual Reality*. New York: John Wiley & Sons.

Buxton, W. (1983). Lexical and pragmatic considerations of input structure. *Computer Graphics, 17*(1), 31–37.

Buxton, W. (1986a). Chunking and phrasing and the design of human-computer dialogues. *Proceedings of the IFIP 10th World Computer Congress*, 475–480.

Buxton, W. (1986b). There's more to interaction than meets the eye. In D. Norman and S. Draper (Eds.), *User Centered System Design: New Perspectives on Human–Computer Interaction* (pp. 319–337). Hillsdale, NJ.

Buxton, W. (1990). A three-state model of graphical input. *Proceedings of the INTERACT'90*, 449–456.

Buxton, W. (1995a). Integrating the periphery and context: A new taxonomy of telematics. *Proceedings of Graphics Interface '95*, 239–246.

Buxton, W. (1995b). Speech, language and audition. In R. Baecker, J. Grudin, W. Buxton, & S. Greenberg (Eds.), *Readings in human–computer interaction: Toward the year 2000* (pp. 525–537). Morgan Kaufmann Publishers, San Francisco.

Buxton, W. (1995c). Touch, gesture, and marking. In R. Baecker, J. Grudin, W. Buxton, & S. Greenberg (Eds.), *Readings in human–computer interaction: Toward the year 2000* (pp. 469–482). Morgan Kaufmann Publishers, San Francisco.

Buxton, W., Fitzmaurice, G., Balakrishnan, R., & Kurtenbach, G. (2000, July–August). Large displays in automotive design. *IEEE Computer Graphics and Applications*, 68–75.

Buxton, E., Hill, R., & Rowley, P. (1985). Issues and techniques in touch-sensitive tablet input. *Computer Graphics, 19*(3), 215–224.

Buxton, W., & Myers, B. (1986). A study in two-handed input. *Proceedings of CHI '86: ACM Conference on Human Factors in Computing Systems*, 321–326.

Buyukkokten, O., Garcia-Molina, H., & Paepcke, A. (2001). Accordion summarization for end-game browsing on PDAs and cellular phones. *Proceedings of the ACM CHI 2001 Conference on Human Factors in Computing Systems*.

Campbell, C., Zhai, S., May, K., & Maglio, P. (1999). What you feel must be what you see: Adding tactile feedback to the trackpoint. *Proceedings of INTERACT '99: 7th IFIP conference on Human Computer Interaction*, 383–390.

Card, S., English, W., & Burr, B. (1978). Evaluation of mouse, rate-controlled isometric joystick, step keys, and text keys for text selection on a CRT. *Ergonomics, 21*, 601–613.

Card, S., Mackinlay, J., & Robertson, G. (1990). The design space of input devices. *Proceedings of ACM CHI '90 Conference on Human Factors in Computing Systems*, 117–124.

Card, S., Mackinlay, J., & Robertson, G. (1991). A morphological analysis of the design space of input devices. *ACM Transactions on Information Systems, 9*(2), 99–122.

Card, S., Moran, T., & Newell, A. (1980). The keystroke-level model for user performance time with interactive systems. *Communications of the ACM, 23*(7), 396–410.

Clarkson, E., Clawson, J., Lyons, K., & Starner, T. (2005). An empirical study of typing rates on mini-QWERTY keyboards. *CHI Extended Abstracts*, 1288–1291.

Cohen, P. R., & Sullivan, J. W. (1989). Synergistic use of direct manipulation and natural language. *Proceedings of the ACM CHI '89 Conference on Human Factors in Computing Systems*, 227–233.

Cutrell, E., Robbins, D., Dumais, S., & Sarin, R. (2006). Fast, flexible filtering with phlat: Personal search and organization made easy. *Proceedings of the CHI 2006*.

Deasy, C. M., & Lasswell, T. E. (1985). *Designing places for people: A handbook on human behavior for architects, designers, and facility managers*. New York: Whitney Library of Design.

Dietz, P., & Leigh, D. (2001). DiamondTouch: A multi-user touch technology. *CHI Letters, 3*(2), 219–226.

Dillon, R. F., Eday, J. D., & Tombaugh, J. W. (1990). Measuring the true cost of command selection: Techniques and results. *Proceedings of the ACM CHI '90 Conference on Human Factors in Computing Systems*, 19–25.

Douglas, S., Kirkpatrick, A., & MacKenzie, I. S. (1999). Testing pointing device performance and user assessment with the ISO 9241, Part 9 Standard. *Proceedings of the ACM CHI '99 Conference on Human Factors in Computing Systems*, 215–222.

Douglas, S., & Mithal, A. (1994). The effect of reducing homing time on the speed of a finger-controlled isometric pointing device. *Proceedings of the ACM CHI '94 Conference on Human Factors in Computing Systems*, 411–416.

Douglas, S. A., & Mithal, A. K. (1997). *Ergonomics of Computer Pointing Devices*, Springer-Verlag.

Dumais, S., Cutrell, E., Cadiz, J., Jancke, G., Sarin, R., & Robbins, D. (2003). Stuff I've seen: A system for personal information retrieval and re-use. *SIGIR 2003*, 72–79.

English, W. K., Englebart, D. C., & Berman, M. L. (1967). Display-selection techniques for text manipulation. *Transactions on Human Factors in Electronics, 8*(1), 5–15.

Fails, J. A., & Olsen, D. R. (2003). Interactive machine learning. *Proceedings from the ACM Intelligent User Interfaces (IUI '03)*, 39–45.

Fitts, P. (1954). The information capacity of the human motor system in controlling the amplitude of movement. *Journal of Experimental Psychology, 47*, 381–391.

Fitzmaurice, G., & Buxton, W. (1997). An empirical evaluation of graspable user interfaces: Towards specialized, space-multiplexed input. *Proceedings of CHI '97: ACM Conference on Human Factors in Computing Systems*, 43–50.

Fitzmaurice, G., Khan, A., Pieke, R., Buxton, B., & Kurtenbach, G. (2003a). Tracking menus. *UIST 2003*, 71–79.

Fitzmaurice, G. W. (1993). Situated information spaces and spatially aware palmtop computers. *Communications of the ACM, 36*(7), 39–49.

Fitzmaurice, G. W., Balakrisnan, R., & Kurtenbach, G. (1999). Sampling, synthesis, and input devices. *Communications of the ACM, 42*(8) 54–63.

Fitzmaurice, G. W., Khan, A., Buxton, W., Kurtenbach, G., & Balakrishnan, R. (2003, November). Sentient Data. *Queue, 1*(8), 52–62.

Foley, J. D., Wallace, V., & Chan, P. (1984, November). The human factors of computer graphics interaction techniques. *IEEE Computer Graphics and Applications*, 13–48.

Freeman, W. T., & Weissman, C. (1995). Television control by hand gestures. *International Workshop on Automatic Face and Gesture Recognition*, 179–183.

Funkhouser, T., & Li, K. (2000). Large Format Displays. *IEEE Computer Graphics and Applications*. [Special Issue], (July–August), 20–75.

Gibson, J. (1962). Observations on active touch. *Psychological Review, 69*(6), 477–491.

Goldberg, D., & Richardson, C. (1993). Touch-typing with a stylus. *Proceedings of the INTERCHI '93 Conference on Human Factors in Computing Systems*, 80–87.

Grossman, T., & Balakrishnan, R. (2005). The bubble cursor: enhancing target acquisition by dynamic resizing of the cursor's activation area. *CHI '05*, 281–290.

Grossman, T., Hinckley, K., Baudisch, P., Agrawala, M., & Balakrishnan, R. (2006). Hover widgets: Using the tracking state to extend the capabilities of pen-operated devices. *CHI 2006*.

Grudin, J. (2001). Partitioning digital worlds: Focal and peripheral awareness in multiple monitor use. *CHI 2001*, 458–465.

Guiard, Y. (1987). Asymmetric division of labor in human skilled bimanual action: The kinematic chain as a model. *The Journal of Motor Behavior, 19*(4), 486–517.

Guiard, Y., Buourgeois, F., Mottet, D., & Beaudouin-Lafon, M. (2001). Beyond the 10-bit barrier: Fitts' law in multi-scale electronic worlds. *IHM-HCI 2001*.

Guimbretiere, F. (2003). Paper augmented digital documents. *UIST 2003*.

Guimbretiere, F., Stone, M. C., & Winograd, T. (2001). Fluid interaction with high-resolution wall-size displays. *Proceedings of the UIST 2001 Symposium on User Interface Software and Technology*, 21–30.

Han, J. Y. (2005). Low-cost multi-touch sensing through frustrated total internal reflection. *UIST 2005*, 115–118.

Harrison, B., Fishkin, K., Gujar, A., Mochon, C., & Want, R. (1998). Squeeze me, hold me, tilt me! An exploration of manipulative user interfaces. *Proceedings of the ACM CHI '98 Conference on human factors in computing systems*, 17–24.

Harrison, B., Ishii, H., Vicente, K., & Buxton, W. (1995). Transparent layered user interfaces: An evaluation of a display design to enhance focused and divided attention. *Proceedings of the CHI '95 ACM Conference on Human Factors in Computing Systems*, 317–324.

Hauptmann, A. (1989). Speech and gestures for graphic image manipulation. *Proceedings of the CHI '89 ACM Conference on Human Factors in Computing Systems*, 241–245.

Hinckley, K., Baudisch, P., Ramos, G., & Guimbretiere, F. (2005a). Design and analysis of delimiters for selection-action pen gesture phrases in Scriboli. *Proceedings from the CHI 2005*, 451–460.

Hinckley, K., Cutrell, E., Bathiche, S., & Muss, T. (2002). Quantitative analysis of scrolling techniques. *CHI 2002*.

Hinckley, K., Czerwinski, M., & Sinclair, M. (1998a). Interaction and modeling techniques for desktop two-handed input. *Proceedings of the ACM UIST '98 Symposium on User Interface Software and Technology*, 49–58.

Hinckley, K., Guimbretiere, F., Baudisch, P., Sarin, R., Agrawala, M., & Cutrell, E. (2006). The springboard: Multiple modes in one springloaded control. *CHI 2006*.

Hinckley, K., Pausch, R., Goble, J. C., & Kassell, N. F. (1994). A survey of design issues in spatial input. *Proceedings of the ACM UIST '94 Symposium on User Interface Software and Technology*, 213–222.

Hinckley, K., Pausch, R., Proffitt, D., & Kassell, N. (1998b). Two-Handed Virtual Manipulation. *ACM Transactions on Computer-Human Interaction, 5*(3), 260–302.

Hinckley, K., Pierce, J., Horvitz, E., & Sinclair, M. (2005b). Foreground and background interaction with sensor-enhanced mobile devices [Special Issue]. *ACM TOCHI, 12*(1), 31–52.

Hinckley, K., Pierce, J., Sinclair, M., & Horvitz, E. (2000). Sensing techniques for mobile interaction. *ACM UIST 2000 Symposium on User Interface Software & Technology*, 91–100.

Hinckley, K., Ramos, G., Guimbretiere, F., Baudisch, P., & Smith, M. (2004). Stitching: Pen gestures that span multiple displays. *ACM 7th International Working Conference on Advanced Visual Interfaces (AVI 2004)*, 23–31.

Hinckley, K., Sinclair, M., Hanson, E., Szeliski, R., & Conway, M. (1999). The VideoMouse: A camera-based multi-degree-of-freedom input device. *ACM UIST '99 Symposium on User Interface Software & Technology*, 103–112.

Honan, M., Serina, E., Tal, R., & Rempel, D. (1995). Wrist postures while typing on a standard and split keyboard. *Proceedings of the HFES Human Factors and Ergonomics Society 39th Annual Meeting*, 366–368.

Horvitz, E. (1999). Principles of mixed-initiative user interfaces. *Proceedings of the ACM CHI '99 Conference on Human Factors in Computing Systems*, 159–166.

Horvitz, E., Breese, J., Heckerman, D., Hovel, D., & Rommelse, K. (1998, July). The Lumiere Project: Bayesian user modeling for inferring the goals and needs of software users. *Proceedings of the Fourteenth Conference on Uncertainty in Artificial Intelligence* (pp. 256–265). San Francisco: Morgan Kaufmann.

Horvitz, E., Jacobs, A., & Hovel, D. (1999). Attention-sensitive alerting. *Proceedings of the UAI '99 Conference on Uncertainty and Artificial Intelligence*, 305–313.

Ishii, H., & Ullmer, B. (1997). Tangible bits: Towards seamless interfaces between people, bits, and atoms. *Proceedings of the CHI '97 ACM Conference on Human Factors in Computing Systems*, 234–241.

Jacob, R., Sibert, L., McFarlane, D., & Mullen, Jr., M. (1994). Integrality and separability of input devices. *ACM Transactions on Computer-Human Interaction, 1*(1), 3–26.

Jellinek, H., & Card, S. (1990). Powermice and user performance. *Proceedings of the ACM CHI '90 Conference on Human Factors in Computing Systems*, 213–220.

John, B. E., & Kieras, D. (1996). Using GOMS for user interface design and evaluation: Which technique? *ACM Transactions on Computer-Human Interaction, 3*(4), 287–319.

Jojic, N., Brumitt, B., Meyers, B., & Harris, S. (2000). Detecting and estimating pointing gestures in dense disparity maps. *Proceedings of the IEEE International Conference on Automatic Face and Gesture Recognition*.

Jones, M., Marsden, G., Mohd-Nasir, N., Boone, K., & Buchanan, G. (1999). Improving web interaction on small displays. *Computer Networks, 3*(11–16), 1129–1137.

Kabbash, P., & Butxon, W. (1995). The "Prince" Technique: Fitts' Law and selection using area cursors. *Proceedings of ACM CHI '95 Conference on Human Factors in Computing Systems*, 273–279.

Kabbash, P., Buxton, W., & Sellen, A. (1994). Two-handed input in a compound task. *Proceedings of the CHI '94 ACM Conference on Human Factors in Computing Systems*, 417–423.

Kamba, T., Elson, S. A., Harpold, T., Stamper, T., & Sukaviriya, P. (1996). Using small screen space more efficiently. *Proceedings of the Conference on Human Factors in Computing Systems*, 383.

Karat, C., Halverson, C., Horn, D., & Karat, J. (1999). Patterns of entry and correction in large vocabulary continuous speech recognition systems. *Proceedings of the ACM CHI '99 Conference on Human Factors in Computing Systems*, 568–575.

Karl, L., Pettey, M., & Shneiderman, B. (1993). Speech-activated versus mouse-activated commands for word processing applications: An empirical evaluation. *International Journal of Man-Machine Studies, 39*(4), 667–687.

Khan, A., Matejka, J., Fitzmaurice, G., & Kurtenbach, G. (2005). Spotlight: Directing users' attention on large displays. *Proceedings from the CHI '05*, 791–798.

Kirsh, D. (1995). Complementary strategies: Why we use our hands when we think. *Proceedings of the 7th Annual Conference of the Cognitive Science Society*, 212–217.

Kirsh, D., & Maglio, P. (1994). On distinguishing epistemic from pragmatic action. *Cognitive Science, 18*(4), 513–549.

Kristensson, P., & Zhai, S. (2004). SHARK2: A large vocabulary shorthand writing system for pen-based computers. *Proceedings of the UIST 2004 Symposium on User Interface Software and Technology*, 43–52.

Krumm, J., & Hinckley, K. (2004). The NearMe Wireless Proximity Server. *Proceedings from the Ubicomp 2004*.

Kurtenbach, G., & Buxton, W. (1991). Issues in combining marking and direct manipulation techniques. *Proceedings from the UIST '91*, 137–144.

Kurtenbach, G., Fitzmaurice, G., Baudel, T., & Buxton, B. (1997). The design of a GUI paradigm based on tablets, two-hands, and transparency. *Proceedings of the CHI '97 ACM Conference on Human Factors in Computing Systems*, 35–42.

Kurtenbach, G., Sellen, A., & Buxton, W. (1993). An empirical evaluation of some articulatory and cognitive aspects of 'marking menus'. *J. Human Computer Interaction, 8*(1), 1–23.

Lank, E., & Saund, E. (2005). Sloppy selection: Providing an accurate interpretation of imprecise selection gestures. *Computers and Graphics, 29*(4), 490–500.

Lansdale, M., & Edmonds, E. (1992). Using memory for events in the design of personal filing systems. *International Journal of Man-Machine Studies, 36*(1), 97–126.

LaViola, R. (2004). MathPad2: A system for the creation and exploration of mathematical sketches. *ACM Transactions on Graphics, 23*(3), 432–440.

Lewis, J., Potosnak, K., & Magyar, R. (1997). Keys and keyboards. In M. Helander, T. Landauer, & P. Prabhu (Eds.), *Handbook of human-computer interaction* (pp. 1285–1316). Amsterdam: North-Holland.

Li, Y., Hinckley, K., Guan, Z., & Landay, J. A. (2005). Experimental analysis of mode switching techniques in pen-based user interfaces. *Proceedings from the CHI 2005*, 461–470.

Liao, C., Guimbretiere, F., & Hinckley, K. (2005). PapierCraft: A command system for interactive paper. *UIST 2005*, 241–244.

Lipscomb, J., & Pique, M. (1993). Analog input device physical characteristics. *SIGCHI Bulletin, 25*(3), 40–45.

MacKenzie, I. S. (1989). A note on the information-theoretic basis for Fitts' Law. *Journal of Motor Behavior, 21*, 323–330.

MacKenzie, I. S. (1992a). Fitts' Law as a research and design tool in human–computer interaction. *Human–Computer Interaction, 7*, 91–139.

MacKenzie, I. S. (1992b). Movement time prediction in human-computer interfaces. *Proceedings from the Graphics Interface '92*, 140–150.

MacKenzie, I. S. (1995). Input devices and interaction techniques for advanced computing. In W. Barfield, & T. Furness (Eds.), *Virtual environments and advanced interface design* (pp. 437–470). Oxford, UK: Oxford University Press:.

MacKenzie, I. S. (2002). Introduction to this special issue on text entry for mobile computing. *Human–Computer Interaction, 17*, 141–145.

MacKenzie, I. S., & Guiard, Y. (2001). The two-handed desktop interface: Are we there yet? *Proceedings from the ACM CHI 2001 Conference on Human Factors in Computing Systems, Extended Abstracts*, 351–352.

MacKenzie, I. S., Kober, H., Smith, D., Jones, T., & Skepner, E. (2001). LetterWise: Prefix-based disambiguation for mobile text input. *Proceedings from the ACM UIST 2001 Symposium on User Interface Software & Technology*, 111–120.

MacKenzie, I. S., & Oniszczak, A. (1998). A comparison of three selection techniques for touchpads. *Proceedings from the ACM CHI '98 Conference on Human Factors in Computing Systems*, 336–343.

MacKenzie, I. S., Sellen, A., & Buxton, W. (1991). A Comparison of input devices in elemental pointing and dragging tasks. *Proceedings from ACM CHI '91 Conference on Human Factors in Computing Systems*, 161–166.

MacKenzie, I. S., & Soukoreff, R. W. (2002). A model of two-thumb text entry. *Proceedings of the Graphics Interface*, 117–124.

MacKenzie, I. S., & Zhang, S. (1997). The immediate usability of graffiti. *Proceedings of the Graphics Interface '97*, 129–137.

Maes, P., Darrell, T., Blumberg, B., & Pentland, A. (1997). The ALIVE system: Wireless, full-body interaction with autonomous agents. *Multimedia Syst. 5*, 2 (Mar. 1997), 105–112.

Marklin, R., Simoneau, G., & Monroe, J. (1997). The effect of split and vertically-inclined computer keyboards on wrist and forearm posture. *Proceedings of the HFES Human Factors and Ergonomics Society 41st Annual Meeting*, 642–646.

Matsushita, N., & Rekimoto, J. (1997). Holo Wall: Designing a finger, hand, body, and object sensitive wall. *Proceedings of the ACM UIST '97 Symposium on User Interface Software and Technology*, 209–210.

McLoone, H., Hinckley, K., & Cutrell, E. (2003). Bimanual interaction on the Microsoft Office keyboard. *Proceedings from the INTERACT 2003*.

Microsoft Corp. (2002). Windows XP Pointer Ballistics (January 17, 2006). http://www.microsoft.com/whdc/device/input/pointerbal.mspx

Mine, M., Brooks, F., & Sequin, C. (1997). Moving objects in space: Expoiting proprioception in virtual-environment interaction. *Computer Graphics, 31* (Proc. SIGGRAPH'97): 19–26.

Moran, T., Chiu, P., & van Melle, W. (1997). Pen-based interaction techniques for organizing material on an electronic whiteboard. *Proceedings of the UIST '97*, 45–54.

Myers, B., Lie, K., & Yang, B. (2000). Two-handed input using a PDA and a mouse. *Proceedings from the CHI 2000*, 41–48.

Myers, B., Stiel, H., & Gargiulo, R. (1998). Collaboration using multiple PDAs connected to a PC. *Proceedings from the ACM CSCW '98 Conference on Computer Supported Cooperative Work*, 285–294.

Norman, D. (1990). *The design of everyday things*. New York: Doubleday.

Oakley, I., Brewster, S., & Gray, P. (2001). Solving multi-target haptic problems in menu interaction. *Proceedings of the ACM CHI 2001 Conference on Human Factors in Computing Systems, Extended Abstracts*, 357–358.

Olsen, D. R., & Nielsen, T. (2001). Laser pointer interaction. *Proceedings of the ACM CHI 2001 Conference on Human Factors in Computing Systems*, 17–22.

Olson, J. R., & Olson, G. M. (1990). The growth of cognitive modeling in human–computer interaction since GOMS. *Human–Computer Interaction, 5*(2–3), 221–266.

Oviatt, S., DeAngeli, A., & Kuhn, K. (1997). Integration and synchronization of input modes during multimodal human–computer interaction. *Proceedings of the CHI '97*, 415–422.

Poupyrev, I., & Maruyama, S. (2003). Tactile interfaces for small touch screens. *Proceedings of the UIST 2003 Symposium on User Interface Software & Technology*, 217–220.

Putz-Anderson, V. (1988). *Cumulative trauma disorders: A manual for musculoskeletal diseases of the upper limbs*. Bristol, PA: Taylor & Francis.

Raskar, R., Beardsley, P., van Baar, J., Wang, Y., Dietz, P. H., Lee, J., Leigh, D., & Willwacher, T. (2004). "RFIG Lamps: Interacting with a Self-Describing World via Photosensing Wireless Tags and Projectors", *ACM Transactions on Graphics (TOG) SIGGRAPH, 23*(3), 406–415.

Raskin, J. (2000). *The humane interface: New directions for designing interactive systems*, ACM Press, New York.

Rekimoto, J. (1996). Tilting operations for small screen interfaces. *Proceedings of the ACM UIST '96 Symposium on User Interface Software & Technology*, 167–168.

Rekimoto, J. (1998). A multiple device approach for supporting whiteboard-based interactions. *Proceedings of the CHI '98*, 344–351.

Rekimoto, J., Ayatsuka, Y., & Kohno, M. (2003). SyncTap: An interaction technique for mobile networking. *Proceedings of the Mobile HCI 2003*, 104–115.

Rempel, D., Bach, J., Gordon, L., & Tal, R. (1998). Effects of forearm pronation/supination on carpal tunnel pressure. *J Hand Surgery, 23*(1), 38–42.

Rime, B., & Schiaratura, L. (1991). Gesture and speech. In *Fundamentals of Nonverbal Behaviour* (pp. 239–281). New York: Press Syndicate of the University of Cambridge.

Rutledge, J., & Selker, T. (1990). Force-to-motion functions for pointing. *Proceedings of the Interact '90: The IFIP Conference on Human–Computer Interaction*, 701–706.

Saund, E., & Lank, E. (2003). Stylus input and editing without prior selection of mode. *Proceedings of the UIST '03*, 213–216.

Sawhney, N., & Schmandt, C. M. (2000). Nomadic radio: Speech and audio interaction for contextual messaging in nomadic environments. *ACM Transactions on Computer-Human Interaction, 7*(3), 353–383.

Schilit, B. N., Adams, N. I., & Want, R. (1994). Context-aware computing applications. *Proceedings of the IEEE Workshop on Mobile Computing Systems and Applications*, 85–90.

Schilit, B. N., Golovchinsky, G., & Price, M. N. (1998). Beyond paper: Supporting active reading with free form digital ink annotations. *Proceedings of the CHI'98*, 249–256.

Schmandt, C. M., Marmasse, N., Marti, S., Sawhney, N., & Wheeler, S. (2000). Everywhere messaging. *IBM Systems Journal, 39*(3–4), 660–686.

Scott, S. D., Grant, K. D., & Mandryk, R. L. (2003). System guidelines for co-located, collaborative work on a tabletop display. *Proceedings of the ECSCW '03, European Conference Computer-Supported Cooperative Work*, 159–178.

Sears, A., Plaisant, C., & Shneiderman, B. (1992). A new era for high precision touchscreens. In H. Hartson, & D. Hix (Eds.), *Advances in Human–Computer Interaction* (pp. 1–33). Ablex Publishers, Norwood, NJ.

Sears, A., & Shneiderman, B. (1991). High precision touchscreens: Design strategies and comparisons with a mouse. *International Journal of Man-Machine Studies, 34*(4), 593–613.

Sellen, A., Kurtenbach, G., & Buxton, W. (1992). The prevention of mode errors through sensory feedback. *J. Human Computer Interaction, 7*(2), 141–164.

Sellen, A. J., & Harper, H. R. (2002). *The myth of the paperless office.* Cambridge, MA: MIT Press.

Shen, C., Everitt, K., & Ryall, K. (2003). UbiTable: Impromptu face-to-face collaboration on horizontal interactive surfaces. *Proceedings of the UbiComp 2003.*

Smith, M. A., Davenport, D., & Hwa, H. (2003). AURA: A mobile platform for object and location annotation. *Proceedings of the UbiComp 2003..*

Smutz, W., Serina, E., Bloom, T., & Rempel, D. (1994). A system for evaluating the effect of keyboard design on force, posture, comfort, and productivity. *Ergonomics, 37*(10), 1649–1660.

Snibbe, S. S., MacLean, K. E., Shaw, R., Roderick, J., Verplank, W. L., & Scheeff, M. (2001). Haptic techniques for media control. In *Proceedings of the 14th Annual ACM Symposium on User interface Software and Technology* (Orlando, Florida, November 11–14, 2001). *UIST '01* (199–200) ACM Press, New York, NY.

Sommer, R. (1965). Further studies of small group ecology. *Sociometry, 28*, 337–348.

Stifelman, L. (1996). Augmenting real-world objects: A paper-based audio notebook. *CHI'96 Conference Companion.*

Swaminathan, K., & Sato, S. (1997). Interaction design for large displays. *Interactions 4*, 1 (Jan. 1997), 15–24.

Trevor, J., Hilbert, D. M., Schilit, B. N., & Koh, T. K. (2001). From desktop to phonetop: A UI for web interaction on very small devices. *Proceedings of the UIST '01 Symposium on User Interface Software and Technology*, 121–130.

Trimble, J., Wales, R., & Gossweiler, R. (2003). NASA's MERBoard. In K. O'Hara, M. Perry, E. Churchill, & D. Russell (Eds.), *Public and Situated Displays: Social and interactional aspects of shared display technologies.* Kluwer.

Viviani, P., & Flash, T. (1995). Minimum-jerk, two-thirds power law and isochrony: Converging approaches to the study of movement planning. *Journal of Experimental Psychology: Perception and Performance, 21*, 32–53.

Vogel, D., & Balakrishnan, R. (2004). Interactive public ambient displays: Transitioning from implicit to explicit, public to personal, interaction with multiple users. *Proceedings from the UIST 2004.*

Vogel, D., & Balakrishnan, R. (2005). Distant freehand pointing and clicking on very large, high resolution displays. *UIST 2005*, 33–42.

Want, R., & Borriello, G. (2000, May–June). Survey on information appliances. *IEEE Personal Communications*, 24–31.

Want, R., Fishkin, K. P., Gujar, A., & Harrison, B. L. (1999). Bridging physical and virtual worlds with electronic tags. *Proceedings of the ACM CHI '99 Conference on Human Factors in Computing Systems*, 370–377.

Want, R., Hopper, A., Falcao, V., & Gibbons, J. (1992). The active bage location system. *ACM Transactions on Information Systems, 10*(1), 91–102.

Weiser, M. (1991, September). The computer for the 21st century. *Scientific American*, 94–104.

Wherry, E. (2003). Scroll ring performance evaluation. *CHI '03 Extended Abstracts*, ACM, New York, 758–759.

Wickens, C. (1992). *Engineering psychology and human performance.* New York: HarperCollins.

Wigdor, D., & Balakrishnan, R. (2003). TiltText: Using tilt for text input to mobile phones. *ACM UIST '03 Symposium on User Interface Software & Technology*, 81–90.

Wilson, A. (2004). TouchLight: An imaging touch screen and display for gesture-based interaction. *International Conference on Multimodal Interfaces (ICMI).*

Wilson, A. D. (2005). PlayAnywhere: a compact interactive tabletop projection-vision system. In *Proceedings of the 18th Annual ACM Symposium on User Interface Software and Technology* (Seattle, WA, USA, October 23–26, 2005). UIST '05. ACM Press, New York, NY, 83–92.

Wilson, A., & Oliver, N. (2003). GWindows: Towards robust perception-based UI. *Proceedings of CVPR 2003 (Workshop on Computer Vision for HCI).*

Wilson, F. R. (1998). *The hand: How its use shapes the brain, language, and human culture.* New York: Pantheon Books.

Wolf, C., Rhyne, J., & Ellozy, H. (1989). The paper-like interface. In G. Salvendy, & M. Smith (Eds.), *Designing and using human computer interface and knowledge-based systems* (pp. 494–501). Amsterdam, Elsevier.

Wu, M., & Balakrishnan, R. (2003). Multi-finger and whole hand gestural interaction techniques for multi-user tabletop displays. *Proceedings of the UIST 2003*, 193–202.

Yamada, H. (1980). A historical study of typewriters and typing methods: From the position of planning Japanese parallels. *J. Information Processing, 24*(4), 175–202.

Yee, K.-P. (2003). Peephole displays: Pen interaction on spatially aware handheld computers. *Proceedings of the CHI 2003.*

Yee, K.-P., Swearingen, K., Li, K., & Hearst, M. (2003). Faceted metadata for image search and browsing. *Proceedings of the CHI 2003*, 401–408.

Zhai, S. (1993). Human performance evaluation of manipulation schemes in virtual environments. *Proceedings of the IEEE Virtual Reality International Symposium (VRAIS '93)*, 155–161.

Zhai, S. (1998). User performance in relation to 3D input device design. *Computer Graphics, 32*(4), 50–54.

Zhai, S., Hunter, M., & Smith, B. A. (2000). The metropolis keyboard—An exploration of quantitative techniques for virtual keyboard design. *CHI Letters, 2*(2), 119–128.

Zhai, S., & Milgram, P. (1993). Human performance evaluation of isometric and elastic rate controllers in a 6DoF tracking task. *Proceedings of the SPIE Telemanipulator Technology.*

Zhai, S., Milgram, P., & Buxton, W. (1996). The influence of muscle groups on performance of multiple degree-of-freedom input. *Proceedings of CHI '96: ACM Conference on Human Factors in Computing Systems*, 308–315.

Zhai, S., Smith, B. A., & Selker, T. (1997). Improving browsing performance: A study of four input devices for scrolling and pointing tasks. *Proceedings of the INTERACT97: The Sixth IFIP Conference on Human–Computer Interaction*, 286–292.

Zhao, S., & Balakrishnan, R. (2004). Simple vs. compound mark hierarchical marking menus. *Proceedings from the UIST 2004*, 33–42.

·10·

SENSOR- AND RECOGNITION-BASED

INPUT FOR INTERACTION

Andrew D. Wilson
Microsoft Research

INTRODUCTION

Sensors convert a physical signal into an electrical signal that may be manipulated symbolically on a computer. A wide variety of sensors have been developed for aerospace, automotive, and robotics applications (Fraden, 2003). Continual innovations in manufacturing and reductions in cost have allowed many sensing technologies to find application in consumer products. An interesting example is the development of the ubiquitous computer mouse. Douglas Engelbart's original mouse, so named because its wire "tail" came out of its end, used two metal wheels and a pair of potentiometers to sense the wheels rolling over a desk surface. Soon, mice used a ball and a pair of optical encoders to convert the movement of the hand into digital signals indicating precise relative motion. Now, even the most inexpensive mice employ a specialized camera and image processing algorithms to sense motions at the scale of one one-thousandth of an inch several thousand times per second. *Accelerometers*, devices that sense acceleration due to motion and the constant acceleration due to gravity, are another interesting example. Today's tiny accelerometers were originally developed for application in automotive air-bag systems. Digital cameras now incorporate accelerometers to sense whether a picture is taken in landscape or portrait mode, and save the digital photo appropriately. Many laptops with built-in hard disks also include accelerometers to detect when the laptop has been dropped, and park the hard drive before impact. Meanwhile, mobile phone manufacturers are experimenting with phones that use accelerometers to sense motion for use in interaction, such as in-the-air dialing, scrolling, and detecting the user's walking pattern.

Research in human–computer interaction (HCI) explores the application of sensing to enhance interaction. The motivation of this work is varied. Some researchers seek to expand the array of desktop input options, or to build completely new computing form factors such as mobile devices that know where they are pointed and intelligent environments that are aware of their inhabitants. Other researchers are interested in using sensors to make our machines behave more like we do, or alternatively to make them complement human abilities. Entertainment, surveillance, safety, productivity, mobile computing, and affective computing are all active areas in which researchers are applying sensors in interesting ways.

While a wide array of sensors is available to researchers, rarely does a sensor address exactly the needs of a given application. Consider building into a computer the capability to sense when its user is frustrated. Detection of user frustration would allow a computer to respond by adopting a new strategy of interaction, playing soothing music, or even calling technical support; however, today, no "frustration meter" may be purchased at the local electronics store. What are the alternatives? A microphone could be used to sense when the user mutters or yells at the machine. A pressure sensor in the mouse and keyboard could detect whether the user is typing harder or squeezing the mouse in frustration (Klein, Moon, & Picard, 2002; Reynolds, 2001). A webcam might detect scowling or furrowing of the eyebrows. Sensors in the chair could detect user agitation

(Tan, Slivovsky, & Pentland, 2001). Ultimately, the system chosen should probably exploit a consistent, predictable relationship between the output of one of these sensors and the user's frustration level; for example, if the mouse is squeezed at a level exceeding some set threshold, the computer may conclude that the user is frustrated.

In our effort to build a frustration detector, we may find a number of issues confounding the relationship between the sensors and the state to be detected:

- There is no easy a priori mapping between the output of the sensors and the presumed state of frustration in the user. Implementation of a pressure sensor on the mouse requires observation of the user over time to determine how much pressure reliably indicates frustration. Implementation of the more complex approach of detecting furrowed brows by computer vision requires an elaborate image processing algorithm.
- The output of the sensors is noisy and often accompanied by a degree of uncertainty.
- Initial experimentation reveals that while no single sensor seems satisfactory, it sometimes may suffice to combine the output of multiple sensors.
- Our preconceived notions of frustration may not correspond to what the sensors observe. This may cause us to revisit our understanding of how people express frustration, which, in turn, may lead us to a different choice of sensors.
- The manner in which the user expresses frustration depends greatly on the user's current task and other contextual factors, such as the time of day and level of arousal. Exploiting knowledge of the user's current application may address many cases where our algorithm for detecting frustration fails.
- After realizing that our frustration detector does not perform flawlessly, we struggle to balance the cost of our system making mistakes with the benefit the system provides.

These are just some of the considerations that are typical in a nontrivial application of sensors to recognition in HCI. While this article does not propose to solve the problem of detecting and responding to user frustration, it will survey aspects of sensor-based recognition highlighted by this example. In particular, this article presents the variety of available sensors and how they are often used in interactive systems. Signal processing, recognition techniques, and further considerations in designing sensor and recognition-based interactive systems are briefly addressed.

SENSORS AND SENSING MODES

This article focuses only on those sensors relevant to interactive applications and their typical modes of use. Experimenting with such sensors has never been easier. Microcontrollers such as the Microchip PIC and BasicStamp can interface sensors to PCs and other devices, and can be programmed using high-level languages such as C and BASIC. The Phidgets hardware toolkit (Greenberg & Fitchett, 2001) enables effortless "plug and play" prototyping with modular sensors and actuators that are plugged into a base interface board, and provides software APIs. The Berkeley and Intel Mote projects also offer wireless sensor packages useful in data collection and sensor networks (Kahn, Katz, & Pister, 2000; Nachman, Kling, Huang, & Hummel, 2005). Another source for inexpensive sensors and sensor interface kits is the hobbyist robotics community.

Occupancy and Motion

Probably owing to the importance of sensing technology in security applications, many devices and techniques exist to sense either motion or a person's presence (occupancy). Among these are

- Air pressure sensors that detect changes in air pressure resulting from the opening of doors and windows.
- Capacitive sensors that detect capacitance changes induced by the body.
- Acoustic sensors.
- Photoelectric and laser-based sensors that detect disruption of light.
- Optoelectric sensors that detect variations in illumination.
- Pressure mat switches and strain gauges.
- Contact and noncontact (magnetic) switches.
- Vibration detectors.
- Infrared motion detectors.
- Active microwave and ultrasonic detectors.
- Triboelectric detectors that detect the static electric charge of a moving object (Fraden, 2003).

Perhaps one of the most familiar motion detectors is the *passive infrared* (PIR) detector, which is sensitive to small changes in the pattern of infrared radiation within the spectral range of 4 to 20 micrometers ("far infrared"). PIR detectors sense heat changes over a small duration of time, indicating the presence of a person moving through the room. These devices often control lights in office buildings, and can be useful in office-awareness applications when combined with other sensors.

More selective motion and occupancy detection can be obtained with video cameras and simple computer vision techniques. For example, a computer vision system allows for the definition of multiple *regions of interest* (ROI) that allow fine distinctions regarding the location of motion. Such a system may thus be able to ignore distracting motion.

Range Sensing

Range sensors calculate the distance to a given object. Such detectors can be used as occupancy detectors, and are also useful in motion- and gesture-driven interfaces.

Many range and proximity sensors triangulate the position of the nearest object. For example, the Sharp IR Ranger emits a controlled burst of near infrared light from a light emitting diode (LED). This light is reflected by any object within a few feet and is focused onto a small linear *charge-coupled devices* (CCD) array that is displaced slightly from the emitter. The position of the reflection on the sensor can be related the distance to the object by trigonometry. Similar approaches can be used over a longer effective distance with the use of lasers rather than LEDs.

Stereo computer vision systems similarly use triangulation to calculate depth. If the same object is detected in two displaced views, the difference in their sensed 2D positions, called "disparity," can be related to the depth of the object (Forsyth & Ponce, 2002; Horn, 1986). Stereo vision techniques may be used to determine the depth of a discrete object in the scene, or to compute depth at each point in the image to arrive at a full range image.

A second approach to calculating range is based on measuring the time-of-flight of an emitted signal. The Polaroid ultrasonic ranging device, for example, was originally developed for auto-focus cameras, and subsequently became popular in robotics. Such sensors emit a narrow ultrasonic "chirp" and later detect the chirp's reflection. The duration in time between the chirp and the detection of the reflection is used to calculate the distance to the object. Ultrasonic range finders can sometimes be confused by multiple reflections of the same chirp; such difficulties are eliminated by measuring the time of flight of emitted light rather than sound, but such sensors are still comparatively exotic.

Position

Designers of sensing-based interactive systems would probably most like a low-power, wireless, inexpensive 3D position sensor that does not rely on the installation of complicated infrastructure. Originally designed for military application, *Global Positioning Satellite* (GPS) devices are useful for sensing street-level movement but are limited to outdoor application. Unfortunately, no indoor tracking standard has gained the popularity of GPS.

The position of a wireless RF receiver can be determined by measuring signal strengths to RF transmitters of known position using Wi-Fi, Bluetooth, and GSM standards (LaMarca et al., 2005). Under the assumption that signal strength approximates distance, position may determined by triangulation, but often interference from buildings, walls, furniture, and even people can be troublesome. Another approach is to treat the pattern of signal strengths as a "signature" which, when recognized later, indicates the position associated with the signature (Krumm & Horvitz, 2004). Using Wi-Fi transceivers and a number of Wi-Fi access points, position can be calculated within several feet of accuracy under ideal laboratory conditions using a combination of approaches (Letchner, Fox, & LaMarca, 2005). Finally, the Ubisense location system achieves accuracy on the order of 15 cm indoors by using arrival time and angle of RF signals in ultrawideband (UWB) frequencies.

Commercially available motion capture systems employ a variety of strategies to track the position and orientation of multiple points. These systems are generally used to record human motions for applications such as video game character animation. Most require the performer to wear several small tracking devices. For example, when precisely calibrated, the Polhemus and Ascension magnetic tracking devices achieve millimeter accurate, six degree of freedom (position and orientation) tracking of multiple points, but rely on technology that connects each tracking device to a base station with wires. Such products have been very useful in prototyping gesture-based interactions that require accurate 3D position and orientation (Fitzmaurice, Ishii, & Buxton, 1995; Hinckley, Pausch, Proffitt, & Kassel, 1998; Ware, 1990; Ware & Jessome, 1988), but often are too expensive for widespread use.

Much research in computer vision focuses on accurate and reliable object tracking. Where multiple cameras are available, it is possible to compute the 3D position of a tracked object using triangulation. To recover 3D position useful for interactive applications in a typical room setting, such cameras require careful calibration. Several prototype interactive systems use vision techniques to track the hands, head, and body of a user. Often, a model of shape and appearance that includes typical skin color is used to track the head and hands. The ALIVE system, for example, determines the 2D position of the head and hands of a user by first extracting the user's silhouette against a controlled (static) background. Point of high curvature along this contour are then extracted and tracked as hands or head. Later variants exploit color information as well. Depth is computed by assuming the user is standing on a flat floor (Maes, Darrell, Blumberg, & Pentland, 1995; Wren, Azarbayejani, Darrell, & Pentland, 1995).

Computer vision-based tracking systems often suffer from poor tracking reliability and sensitivity to variations in background illumination. Tracking reliability can be enhanced by controlling the appearance of the object so that it can be tracked unambiguously. A number of professional motion capture systems, such as the Vicon Peak system, rely on passive, wireless, infrared-reflective pieces, but also require a powerful infrared light source and multiple, redundant sensors (cameras) to minimize missing data resulting from occlusion. Alternatively, cameras sensitive in the infrared domain can be used to track an infrared LED (IR-LED). The *position sensitive device* (PSD), for example, is an inexpensive, camera-like device that reports the brightest spot on its imaging array and is thus suitable for inexpensive IR-LED-based tracking systems. Multiple IR-LEDs can be tracked using a PSD by carefully controlling when each is illuminated. Gross room-level location can be determined using the simplest infrared detectors and IR-LEDs that transmit the identify of the user by blinking specific patterns over time, much like a television remote control (Want, Hopper, Falcao, & Gibbons, 1992).

Acoustic tracking systems are able to triangulate position using time-of-flight measurements. One approach is to equip the room with multiple detectors that are able to hear a mobile tracking device equipped to emit a sound at a known frequency. This configuration can also be inverted, with the detector on the tracked device and the emitters in the environment (Smith, Balakrishnan, Goraczko, & Priyantha, 2004; Ward, Jones, & Hopper, 1997). Related signal processing algorithms can combine the output of two or more microphones to triangulate the position of an arbitrary sound source (Rui & Florencio, 2003). This approach can be particularly effective when combined with other techniques such as computer vision-based face tracking (Zhang & Hershey, 2005).

Movement and Orientation

Unlike most tracking technologies, a number of movement and orientation sensors do not rely on external infrastructure. *Inertial* sensors, for example, sense spatial and angular motion (translation and rotation). They can be used for activity recognition as well as gesture and body motion-based interactive applications where it is acceptable to wear or hold a small wireless sensor package (Bao & Intille, 2004; Hinckley, Pierce, Sinclair, & Horvitz, 2000; Lee & Mase, 2002).

Very simple tilt sensors such as mercury switches have been used for years to sense gross orientation. More recently, inexpensive accelerometer devices using *micro-electro-mechanical systems* (MEMS) technology were developed for application in automotive airbag systems (Kovacs, 1998). MEMS accelerometers feature a tiny proof mass or cantilever beam and deflection sensing circuitry to sense both varying accelerations due to movement as well as the constant acceleration due to gravity. Two-axis MEMS accelerometers can be applied to sensing tilt (pitch and roll) and have been used in gaming controllers. But nonmilitary accelerometers are not sufficiently precise to support the double integration of acceleration necessary to calculate position information for more than a few seconds. In such applications, it may suffice to add a coarse position sensor to combat the effects of drift.

MEMS technology has also allowed the development of *gyroscope* devices that sense angular acceleration rather than absolute orientation. These devices have been used in stabilizing handheld cameras and in the Gyromouse product, which maps relative change in gyroscope orientation to the relative motion of the mouse cursor.

Magnetometers are compact, solid-state devices able to detect the strength of the earth's magnetic field along its principle axis, and so are useful in determining absolute orientation information. The output of a pair of orthogonally mounted magnetometers held level may be combined to find magnetic north. It is common to combine a two-axis accelerator with a two-axis magnetometer to "correct" the output of the magnetometer when it is not held level (Caruso, 1997). Three axis magnetometers are available, but alone do not give a true 3D orientation (e.g., a magnetometer's reading does not change when it is rotated about magnetic north).

Touch

The microswitch typical of today's mouse requires a certain amount of force to activate, thus allowing a user to comfortably rest their forefinger on the button without accidentally clicking. Pressure sensors, on the other hand, sense a continuous range of pressure states. Historically, these have been useful in robotics, where they play an important role in designing control systems for manipulators. *Polyvinylidene fluoride* (PVDF) films and *force sensitive resistors* (FSRs) are two inexpensive types of pressure sensors with good dynamic range and form factors useful for small devices and interactive systems. Flexible *strain gauges* utilizing the piezoresistive effect have a resistance related to the amount of deformation ("bend") applied to the sensor. Such gauges have been used as the basis for inexpensive glove devices that sense the deflection of each of the fingers of the hand.

Capacitive sensing is based on the property that nearly any object is capable of storing electric charge, and that charge will flow between two objects when touching or in close proximity. A zero-force touch sensor can be implemented with a *charge transfer* technique, whereby the capacitance of an electrode is estimated by measuring the time taken for an electrode to discharge a small applied charge (Hinckley & Sinclair, 1999). This time drops dramatically when the user places a finger on the electrode, since the user's body takes on much of the charge. Other capacitive sensing techniques can sense an object before it touches the electrode, making them suitable as proximity sensors for a wide variety of interactive applications (Baxter, 1996; Smith, 1999; Vranish, McConnell, & Mahalingam, 1991). Multiple electrodes can be used to implement position sensitive sliders, such as the wheel on the Apple iPod.

Most common touch screens report the single 2D position of the user's finger touching or pressing the screen (Sears, Plaisant, & Shneiderman, 1992). *Resistive* touch screens use two large transparent conductive overlays that vary in resistance over their length. When the user presses on the screen, the overlays are brought into contact, and a voltage applied to one or the other overlay is used to detect the horizontal or vertical position of the finger. *Capacitive* touch screens use capacitive sensing to sense touch, and the relative difference in the charge sensed at each corner of the screen to determine position. Recently, the ability to more precisely sense the touch location and also sense the area and shape of the touching object has been enabled by embedding multiple capacitive sensors in the display surface (Dietz & Leigh, 2001; Rekimoto, 2002). Finally, *surface acoustic wave* systems rely on sensing the finger's disruption of surface acoustic waves applied to a screen surface (Pickering, 1986).

Finally, computer vision techniques have been applied to sense touch (Fails & Olsen, 2002; Han, 2005; Matsushita & Rekimoto, 1997; Smart Technologies, Inc., 2007; Tomasi, Rafii, & Torunoglu, 2003; Wellner, 1993; Wilson, 2005). Using computer vision to sense touch over an area has a number of advantages: first, these techniques usually do not require a special instrumented surface as do most touch screens. Secondly, vision techniques naturally support multiple touch points. Finally, vision techniques enable the ability to detect and recognize a variety of objects besides fingers. For example, barcode-like visual codes may be applied to uniquely identify objects such as game pieces placed on a surface.

Gaze and Eyetracking

Gaze detection refers to determining where a person is looking and is principally the domain of computer vision. It is possible to very coarsely determine head orientation using techniques related to face detection (Wu, Toyama, & Huang, 2000), but head orientation is often a poor indicator of where someone is looking. The goal of *eyetracking* systems is precisely determine where the user is looking, or *foveating*. Usually, these techniques are based on precise tracking of multiple reflections of an infrared illuminant off the eye's cornea. For good performance,

however, eyetracking systems require careful per-user calibration, and so have seen limited general application in interactive systems (Beymer & Flickner, 2003; Jacob, 1993; Tobii, 2005; Zhai, Morimoto, & Ihde, 1999).

Rather than determining gaze in a general fashion only to later match the gaze direction to one of several known objects, an alternative is to determine only whether user is looking at the object. The detector can then be embedded in the object itself. Furthermore, the reflection of an infrared illuminant by the cornea and retina can be detected by simple image processing techniques when the camera and infrared illuminant are colocated (Haro, Flicker, & Essa, 2000; Shell et al., 2004).

Speech

The long history of research on speech recognition techniques has resulted in commodity systems that bring modern speech recognition to anyone with a PC and an inexpensive microphone (Rabiner & Juang, 1993). New interactive systems, however, highlight the need for further work. Current systems, for example, function poorly without a "close-talk" microphone and in noisy environments, and so are unsuited for use in such contexts as intelligent rooms and mobile scenarios.

The *array microphone* combines audio from multiple microphones to address the problems of multiple sound sources and noisy environments. Through the process of *beamforming*, the outputs of the multiple microphones of an array is combined to form a single audio signal in which all but the dominant speaker's signal has been removed. Beamforming can also reveal information about the position of the speaker (Tashev & Malvar, 2005).

To achieve robustness, speech may also be combined with other input modalities such as pen gestures (Oviatt, 2002). Such approaches usually require a sophisticated model of the user's interaction. Perhaps inspired by HAL in *2001: A Space Odyssey*, some researchers have proposed incorporating computer vision-based lip-reading techniques into the speech interpretation process (Stork, 1998). Finally, information such as intonation, prosody and conversational turn taking can be valuable in interactive systems (Bilmes et al., 2005; Choudhury & Pentland, 2003; Pentland, 2004).

Gesture

Many notions of *gesture* exist in interactive systems, and thus many sensor systems are applicable. A gesture can be thought of as a specific hand pose, a spatial trajectory of the hands or stylus, pointing or other motion to indicate an object, or the quality of a motion of almost any body part as it relates to a given application context (McNeill, 1992).

Many of the previously mentioned tracking and movement sensing technologies have been applied to sense and recognize gestures. For example, a wireless sensor package with multiple accelerometers or gyros can capture motion information useful in recognizing many gestures. Computer vision techniques also can be used to track body parts such as the hands and head, as well as overall motion qualities that can be interpreted as gesture. Often such systems ease the sensing task by requiring the user to wear brightly colored gloves, or by training precise models of skin color (Brashear, Starner, Lukowicz, & Junker, 2003).

Pen gestures are often studied on the tablet computer, a form factor that often uses the Wacom electromagnetic positioning technology. This system uses coils embedded in the pen and under the display to find pen position, limited height (hover) above the surface, tilt, pressure, and button state. It can also support multiple simultaneous inputs. The gestures themselves are usually modeled as simple "flick" gestures or spatial trajectories.

While computer vision techniques have been explored to recover detailed hand pose information, gloves with built-in sensors are more commonly worn for this purpose (Baudel & Beaudouin-Lafon, 1993). Early *virtual reality* (VR) systems, for example, used magnetic trackers attached to gloves equipped with bend sensors to recover the position, orientation, and pose of the hands. More recently, vision-based professional motion capture systems that track infrared retro-reflective balls have been used in a similar fashion. With such precise hand shape information, it is possible to point at an object with the index finger, and then make a motion similar to pulling the trigger of a gun to effect an action (Vogel & Balakrishnan, 2005).

Identity

In interactive systems, it is often useful to know the identity of an object or user, and a variety of sensing systems are designed to recognize known objects. Object recognition is an active research area in computer vision. There are practical techniques for quickly recognizing one of many known flat objects such as photos or book covers, for example (Lowe, 2004). Computer vision-based face recognition techniques have also been shown to work in fairly controlled settings (Li & Jain, 2005). General object recognition and face recognition in uncontrolled settings is still difficult, however.

Beyond face recognition, biometrics uses a variety of sensing technologies. Fingerprint recognition hardware, for example, uses optical scanning technology, or an array of tiny capacitive sensors, to construct an overall picture of the fingerprint. Since Johansen's early experiments demonstrating an ability to recognize human motion from point-light displays (Johansson, 1973), researchers have worked on gait recognition techniques from video (Boyd & Little, 2005). Iris, retina, hand geometry, vascular pattern, handwritten signature, and voice dynamics are other biometric techniques employing sensing technology (Sugiura & Koseki, 1997; Wayman, Jain, Maltoni, & Maio, 2004).

In the absence of reliable techniques to recognize an object by its natural properties, it is often useful to "tag" an object with a standard, easily recognizable marker that reveals the object's identity. Visual codes such as the ubiquitous UPC bar code symbols, for example, are read by laser scanning systems, which are now small enough to be incorporated into mobile devices. Two-dimensional "matrix codes" such as the QR code pack more bits into the same space, and have been used in a variety of interactive systems that recognize them by image analysis (Kato,

Billinghurst, Poupyrev, Imamoto, & Tachibana, 2000; Rekimoto & Ayatsuka, 2000).

Recently, *radio frequency identification* (RFID) tags have gained in popularity. RFID tags themselves are usually passive, can be made small and unobtrusive, are cheap to manufacture, and can be read at a distance. A scanning antenna that emits an RF signal reads the tags; this signal, in turn, powers the tags with enough energy to respond with an identification code. RFID systems vary in terms of range, power requirements, antenna and tag form factors, bit depth, and so on (Garfinkel & Rosenberg, 2005). They are thus particularly attractive for commercial inventory management applications and have been applied to interactive systems (Want, Fishkin, Gujar, & Harrison, 1999).

Context

Sensors can provide important information about the *context* of the user or device. For example, a computer may listen in on an office to determine whether a meeting is in progress, and if so withhold noncritical notifications (Oliver & Horvitz, 2005). A later section explores the role of context in interpreting the output of sensors.

Simple context sensors are especially useful in mobile applications. Environmental sensors that detect such information as air temperature, lighting quality, and air pressure may be more directly relevant to the application than the absolute location given by a GPS sensor (Lester et al., 2006; Schmidt, Beigl, & Gellersen, 1999). Context sensors may be used to determine the user's *activity*, or what the user is currently doing. In mobile applications, an inertial sensor may be used to determine the current transportation mode of the user, while a microphone may be used to conclude that the user is engaged in a conversation. An array of simple switches placed throughout a household environment, such as on kitchen cabinets and drawers, may be all that is needed to reliably determine the activities of its inhabitants (Tapia, Intille, & Larson, 2004; Wilson & Atkeson, 2005).

Affect

In psychology, *affect* refers to an emotion or subjective feeling. Recently there has been interest in applying sensing technology to allow interactive systems to respond appropriately to (and perhaps influence) the user's affect (Picard, 2000). A system might respond to the user's boredom, interest, pleasure, stress, or frustration (as in the example in the introduction) by changing aspects of the interaction.

Like other multimodal systems, an *affective computing* system is likely to integrate a variety of conventional sensors. There is an emphasis, however, on the use of *physiological sensors* to recover physical data that may be related to the user's affective state. For example, the *galvanic skin response* (GSR) sensor measures the skin's conductivity, which increases quickly when the user is startled or experiences anxiety. The *blood volume pulse* (BVP) sensor measures blood pressure over a local region by measuring the reflectance of a bright infrared light, and can detect certain states of arousal when applied to the fingertips. Respiration rate can be sensed by measuring the amount of stretch in an elastic band worn around the chest. The *electromyogram* (EMG) sensor measures the amount of electrical activity produced when the muscle it is placed over contracts, and is useful in detecting jaw clenching, and contraction of various muscles related to facial expressions. Finally, the *electrocardiogram* (ECG or EKG) measures heart rate.

In designing affective computing systems, it is often difficult to determining the mapping of sensor outputs to application specific quantities, such as emotional state. Particularly challenging is the task of identifying specific physical correlates for broadly defined emotional states such as "frustration." Finally, physiological sensors are unsuitable for many applications because of the difficulty in deploying them: many must be placed on particular locations on the body, may require good contact with the skin, and are susceptible to differences among individual users or even the same user from day to day.

Brain Interfaces

Advances in cognitive neuroscience and brain imaging technology have spurred initial explorations into interfacing computers directly with a user's brain activity. Much of the work is motivated by a desire to help individuals who have lost the motor skills necessary to use traditional interfaces. Thus, the goal of *brain-computer interfaces* (BCI) is often to enable users to explicitly manipulate brain activity in order to provide input to a system. Such interfaces typically emulate traditional interfaces by triggering keystrokes and cursor control. However, future applications will likely take advantage of the unique abilities of BCI systems to enable completely new styles of interaction (Hjelm & Browall, 2000).

BCI is generally limited to brain imaging techniques that are noninvasive and do not require bulky, expensive equipment. The *electroencephalograph* (EEG) measures electrical activity at local parts of the brain using electrodes placed carefully on the scalp. EEG has low spatial resolution compared to other brain imaging techniques, but has relatively good temporal resolution. *Functional near infrared* (fNIR) imaging measures blood flow in local regions of the brain by calculating the absorption of infrared light directed into the scalp. The technology suffers, however, from low temporal resolution, but obtains higher spatial resolution than EEG and generates results that are similar to more impractical blood-flow related imaging techniques such as *functional magnetic resonance imaging* (fMRI).

With today's BCI systems, users must learn how to manipulate their brain activity effectively for a given application, either through operant conditioning or by executing certain predetermined cognitive tasks that are distinguishable to the sensors. Imagining the performance of a motor skill, for example, exercises specific parts of the brain (Curran & Stokes, 2003). This specific activity may be detectable by an imaging technique of coarse resolution. Alternatively, applications can be specifically designed to take advantage of naturally occurring brain activity, such as that associated with a flashing light. Under carefully controlled conditions, it is possible to classify the user's engagement in cognitive tasks, such as rest, mental arithmetic, and mental rotation (Kiern & Aunon, 1990).

SIGNAL PROCESSING

It is rare to find a sensor precisely suited to a given sensing task. Often, sensor output must be manipulated or combined with other sensors to fit the needs of the application. This section surveys *signal processing* techniques useful in applying sensors to input and recognition tasks.

Preprocessing

Preprocessing refers to the earliest stage of processing sensor signals. It is at this stage that noise may be removed from raw sensor signals, or signals may be reduced to make them more compact and otherwise easier to use in later processing stages.

The performance of a sensor relevant to preprocessing can be characterized in several ways. *Accuracy* refers to the degree to which the sensor readings represent the true value of what is measured. *Precision*, or *resolution*, by contrast, refers to the extent to which successive readings of the same physical phenomenon agree in value. It is important to realize that while a device's resolution is often measured in bits, this number is often distinct from the number of bits used to represent or store the sensor's readings.

In general, accuracy and precision can be estimated by collecting several successive measurements (*samples* or *observations*) of the same input, and computing the resultant mean and scatter (covariance). An accurate sensor will put the mean near the true value, and a precise sensor will have a small amount of scatter about the mean. An accurate but *noisy* sensor will have low precision (high scatter), but can still be useful by the Central Limit Theorem from statistics: If we make some assumptions about the noise, and average a sufficient number of successive values we will derive a good estimate of the true value (Hoel, Port, & Stone, 1971).

Averaging of successive sensors readings is but one simple way to *smooth* noisy data to obtain a noise-free estimate of what is measured. Of course, the input in a real application is likely to be changing over time, and the manner in which this average is computed can vary. For example, the *boxcar filter* is simply the average of the last *n* samples, and is thus easy to implement; however, the boxcar filter suffers because it requires a buffer of samples over which the average is computed, and the resulting estimate will *lag* the true value in the case of a changing signal. Related to the boxcar filter is a technique whereby the estimate is obtained as a weighted average of new observation and previous estimate. This filter is even easier to implement and requires no buffer of previous samples. In this technique, however, each estimate depends on previous estimates as the Poisson distribution over time, such that a very quickly moving signal or a signal with many outliers will result in erratic changes in the smoothed signal.

The *Kalman filter* is a popular technique for filtering time-varying signals and can be used to both smooth and predict a signal. It is the optimal linear filter in the sense that it minimizes the difference between the estimate and the true value, assuming a linear model of the input's changing signal and Gaussian noise (Welch & Bishop, 2004). The most common Kalman filter for estimating the position of a moving object models the object's state as a linear function of both velocity and the position estimate in the previous time step. The Kalman filter models uncertainty in two ways. First, there is some uncertainty in the linear model (how much do we believe that the linear model is correct?). Secondly, there is uncertainty resulting from instantaneous noise that corrupts the observation (how precise is the sensor?). A properly tuned Kalman filter balances these uncertainties appropriately and suffers fewer problems with lag than, for example, a boxcar filter. When the changing signal is actually linear, it can completely defeat lag due to filtering. An improperly tuned Kalman filter, however, can impart unnecessary lag and *overshoot*, such that the estimate runs past the input before correcting itself.

Often, rather than obtaining a continuous estimate of a sensed quantity, we are interested only in obtaining a binary result: Is the switch on or off? When a user throws on a switch in a real system, the output state of the switch can change very rapidly from off to on and back several times before settling to a single stable state. *Debouncing* techniques combat this effect; one simple technique is to ignore the switch for some small, fixed time after seeing the first change in switch state (e.g., 40 milliseconds).

More difficult is the situation in which a truly continuous quantity must be transformed into a binary signal. This is commonly done by choosing a *threshold* below which we report the output as "zero," and otherwise, "one." For example, in using a continuous-valued tilt sensor such as an accelerometer to determine whether a tablet PC is being used in "portrait" or "landscape" mode, it is necessary to transform the tilt information into a binary quantity indicating "portrait" or "landscape." An alternative to a single threshold is a scheme with two thresholds and a region between (a *deadband*) in which no change to the output is made. Similar to debouncing, this approach can prevent fluctuation of the output around a single threshold.

Choosing threshold values is generally challenging, and poorly designed complex systems are frequently awash in thresholds that require modification to achieve acceptable performance. Ultimately, the thresholding process destroys information; depending on the nature of subsequent processing, this loss can be detrimental to a system's overall performance. This is particularly true for borderline cases where a system is likely to make erroneous decisions as the result of either an improperly chosen threshold or a noisy input. Such concerns may be eased by adopting a "soft" threshold that reports intermediate results around the "hard" threshold; the *logistic function* (Bishop, 1995) can be useful in this approach.

The signal's effective range of output, or *dynamic range*, must be often be considered both in thresholding and subsequent processing. The relevant range of the property to be sensed must of course lie within the dynamic range of the sensor. If the dynamic range changes—as a consequence, for example, of temperature change, lighting change, or even variations in installation—it may be necessary to calibrate the signal to achieve a normative range. One strategy is to find the sensor's minimum and maximum output during normal use, and to map these to some canonical range. For example, the output of a photodetector may be mapped to a range from zero to one by recording the value of the sensor in the darkest and brightest conditions of regular use. Another strategy is to calibrate the sensor to *ground truth* values that are collected by some other more trusted sensor. Both of these approaches require care if the sensor is not linear in its response; it may then be necessary to fit a curve (e.g., polynomial) to map the sensor to normal values. Characteristics such as dynamic range, linearity of the response, and variation due to temperature are often detailed in a sensor's "data sheet," available from the manufacturer.

In time-varying systems, we are often concerned with the frequency with which we receive new samples from the sensor. An overly high *sampling rate* can result in too much data to process, and can be reduced by *downsampling*. By contrast, many interactive systems will seem to lose their responsiveness if the overall latency is greater than 100 milliseconds. *Latency* or *lag* refers to any delay present in the sensor's response to a change in the sensed property of the world, and can limit the responsiveness of an interactive system built on the sensor (MacKenize & Ware, 1993). A low sampling rate imparts latency, which may be remedied by predictive techniques such as the Kalman filter.

Finally, it is important to consider the true distribution of any noise in filtering and many subsequent processing techniques. Many techniques—including the simplest averaging, Kalman filters, and many probabilistic approaches—assume a Gaussian or uniform distribution of noise. Outliers violating this assumption can be troublesome and should be removed by ad hoc means or techniques from the field of *robust statistics* (Fischler & Bolles, 1981; Huber, 1981). For example, the *median filter*, in which values of a sequence are replaced by the median value, is easy to implement, yet more robust than simple averaging.

Feature Selection

In the context of recognition, a *feature* can refer to a particular sensor or a piece of information derived from one or more sensors, or even derived from other features. Often thought of as a preprocessing step, *feature selection* refers to the process of determining which features are to be computed from the raw inputs and passed to the next level of processing. Appropriate feature selection can sometimes make difficult recognition problems easy. For example, one somewhat unusual approach to detecting faces in video is to detect eye-blinking patterns. Blinking provides a signal that is easily detected by simple image processing operations, and is further supported by the fact that both eyes blink together and are arranged in a symmetric spatial configuration on the face. Blinking thus may be highly *diagnostic* for faces (Crowley & Berard, 1997).

Feature selection begins by determining a set of sensors relevant to the task at hand often with knowledge of the task or domain. In the course of development of a new sensing-based system, it can be beneficial to incorporate as many physical sensors as possible, with the idea that subsequent feature selection processes will indicate which sensors are necessary and sufficient. Furthermore, a number of sensors taken in combination may provide the best overall performance.

Having selected a number of sensors, often the next step is to compute derived features from the raw sensor inputs. For example, when an unimportant and unpredictable offset is present in the sensor inputs raw levels, it may be easier to work with its derivative instead. Again, these features are determined in an ad hoc fashion, in light of special domain or application knowledge. For example, early stylus gesture recognition algorithms relied on simple derived features such as the (a) initial angle of the stroke, (b) maximum speed obtained, (c) size of the containing bounding box, (d) duration of the stroke, (e) amount of change in curvature along the gesture, and so on (Rubine, 1991). Early face recognition approaches relied on features such as the distances between facial features such as the eyes, nose, and mouth (Zhao, Chellappa, Phillips, & Rosenfeld, 2003). In the domain of audio, the Linear Predictive Coding (LPC) and the Fourier transform are useful derived feature spaces. The Fourier transform in particular has general applicability to signals with periodicity. For example, the Fourier transform of a body-worn accelerometer may reveal patterns of the user's walking (Hinckley et al., 2000).

Often, it is desirable to put spatial features in a local coordinate system. For example, a gesture recognition system may begin with the position of the head and hands of the user. To remove the effects of the person moving about the room, the head position may be subtracted from the position of each hand, yielding a "head-centric" coordinate system. In the spirit of asymmetric bimanual models of gesture, we might also consider a coordinate system centered on the nondominant hand (Guiard, 1987; Hinckley et al., 1998). In many cases switching to a local coordinate system eliminates a large source of the irrelevant variation present in the raw signal, thus easing subsequent modeling, and can be superior to using only derivative information.

If there is a large number of sensors, or if each sensor is of high dimension (e.g., images taken from video cameras), each sensor's value is unlikely to be statistically independent from one another. To remove redundancy and make the input a more manageable size, some form of *dimensionality reduction* may be used to transform each observation into one of lower dimension. One broad class of techniques involves approximating each sample as the linear combination of a small number of basis functions; the coefficients in this linear combination form the corresponding sample in the new, smaller feature space. *Principle Components Analysis* (PCA) is a popular technique and is the optimal linear technique in the mean-square error sense. PCA finds orthogonal vectors ("principle components," or *eigenvectors*) in the input space as basis vectors, each vector reducing variance (scatter) in the data set. PCA has been used in a wide variety of recognition systems, such as face recognition from images, where often less than 50 components are necessary to perform recognition (Pentland, Moghaddam, & Starner, 1994). Today, there are numerous techniques related to PCA, many of which are more suited to classification (Fodor, 2002).

Where the number of input features is not large, automatic feature selection techniques may be used to determine the subset of features that matter. While the topic is an active area of research, one technique of general applicability is *cross validation* (Bishop, 1995; Mitchell, 1997). The simplest form of cross validation is the *holdout method,* which begins by dividing the data set into two halves. Several variations of the model are then trained on one half of the data and tested on the other. The variation with the best performance on the test set is selected as the best model. In the case of feature selection, each variation employs a particular subset of the original input features; after trying all such subsets, we are left with the best performing subset of features. For more than a handful of original features this approach will be impractical, so various greedy approximations are often used, such as starting with the full set and eliminating one at a time, or successively adding features from a small set.

Classification and Modeling

Classification refers to the process of determining which of several known classes a given sample or observation is drawn from, and is typically the means by which a novel input is *recognized*. A classifier can be used, for example, to recognize which of several known gestures the user has performed by the motion of the pen on a tablet. *Detection* refers to determining the presence of an observation drawn from a known class against a background of many other observations. The distinction between classification and detection is often rather semantic. For example, a face detection system will determine if there is any face present in an image, while a face recognition system will determine the identity of the detected face. While both operations can be thought of as classification, often they call for different techniques.

When simple thresholding or feature selection operations are not enough to transform a group of sensor readings into a signal that is readily consumed by the application, it is often necessary to exploit more sophisticated classification and modeling techniques. These techniques are particularly useful in cases where it is necessary to use many sensors together, and when there are dependencies among them that are difficult to untangle by simple inspection. *Modeling* refers to the choices in representation of sensor values, their dependencies, and the computations performed on them.

There are many ways to classify a new sensor observation as belonging to one of several known classes. Approaches in which a model is trained automatically from a set of training examples are the most relevant to sensor-based systems. These techniques are typically the domain of *machine learning*. The canonical introductory technique is *Fisher's linear discriminant* (Bishop, 1995) in which a closed-form training procedure determines a line in the feature space that optimally divides two classes of training data. A new, unseen sample may then be classified by determining which side of the line the sample lies. Beyond the two-class case, samples are often classified by computing the *likelihood* that the sample was drawn from each class, and choosing the class with the largest likelihood. Assuming a new observation $\mathbf{x}$, and classes C_i we choose C^* as the maximum value $P(C_i)P(\mathbf{x}|C_i)$. The *prior* $P(C_i)$ indicates our belief that a sample is drawn from a class before we even record it, and is often ignored. There are a variety of techniques to derive such probabilistic models from a set of examples.

Common to all these approaches is the ability to characterize the quality of a recognition result. A sample that is correctly classified as belonging to a given class is a *true positive*. A sample that is incorrectly classified as belonging to the class is a *false positive*. A sample that is correctly classified as *not* belonging to a given class is *true negative,* while a sample that is incorrectly classified as *not* belonging is a *false negative.* In the context of interactive systems, a false negative might correspond to when the user provides an input and the system fails to recognize it. A high false-negative rate can lead to an overall impression of unresponsiveness, or a sense on the part of the user that they are doing something wrong. False positives, on the other hand, may correspond to when the system takes an action when the user had no such intention, and can lead to an impression that the system is erratic or overly sensitive (Zhai & Bellotti, 2005).

In most situations, a clear trade-off exists between the rate of true positives and false positives. Lower the bar for acceptance to increase the true positive rate, and the rate of false positives is likely to increase. In the context of interactive systems, this tradeoff is especially important to consider when developing criteria for when the system takes action as the result of a recognition process. The *receiver operator characteristic curve* (ROC curve) plots true positive rate against false positive rate and best characterizes this trade-off (see Fig. 10.1). The ROC curve is also an established method to compare the performance of recognition techniques, without regard to any application-specific choice on how tolerant we are to false positives.

In a given application, it is also instructive to break out classification performance by each class. The *confusion matrix* summarizes how a labeled test set is classified by each class, and may reveal that much of the overall classification error can be traced to errors classifying observations from a small number of classes. This can thus inform design of the classifier or the set of application-relevant categories. *Boosting* is one technique in which misclassified samples are emphasized in subsequent training of the model to reduce the overall error rate (Schapire, 2003).

The *naïve Bayes* classifier assumes that the value of a given feature is independent of all the others. This property of *conditional independence* may not actually apply to the data set,

but its assumption simplifies computation and often may not matter in practice (hence the label "naïve"). Assuming observations of the form $\mathbf{x} = \langle x_1, x_2, \ldots, x_n \rangle$, the *posterior probability* of a class C is $P(C|\mathbf{x}) = P(C)\, P(\mathbf{x}|C)$ by the Bayes rule. Naïve Bayes treats each feature as independent: $P(C|\mathbf{x}) = P(C)\prod_i P(x_i|C)$. Because each feature is modeled independently, naïve Bayes is particularly suited to high dimensional feature spaces and large data sets. Each feature can be continuous or discrete. Discrete variables are often modeled as a histogram (or *probability mass function*), while continuous variables can be *quantized* or *binned* to discrete values, or modeled as a Gaussian or other parametric distribution.

A number of other popular classification techniques do not have obvious probabilistic interpretations. The *neural network*, for example, is best thought of as a function approximation technique. Often, as applied to classification, the input of the approximated function is the observation itself, and the output is a vector whose ith component indicates belief that the observation belongs to the ith class.

Decision trees can be a powerful classification technique that leads to very compact representations for some problems. Each node of a decision tree corresponds to an assertion about the value of a feature in the observation, and yields a split in the data set. The leaves of the tree then indicate the class to which the observation belongs. Classification of a new sample is then rather like the children's game of "twenty questions." Training the model involves determining how to make the splits to optimize classification performance, and possibly, the size of the tree (Breiman, Freidman, Olsen, & Stone, 1984).

The *Support Vector Machine* (SVM) is a powerful modern alternative to the Fisher linear discriminant (Cristianini & Shawe-Taylor, 2000). SVMs determine the split between the two classes to maximize performance on unseen data. Furthermore, SVMs gain much of their power by allowing nonlinear splits of the feature space, but are often thought of as being computational intensive (though, see (Platt, 1999)).

Where the conditional independence assumption of naïve Bayes is too strong, other techniques that directly model the joint probabilities are applicable. For example, a *mixture of Gaussians* uses a sum of multiple Gaussian distributions to model arbitrarily complex joint distributions: $P(\mathbf{x}|C) = \sum_i P(\omega_i) P(\mathbf{x}|\omega_i)$, where $P(\mathbf{x}|\omega_i)$ is Gaussian with mean μ_i and covariance Σ_i. Such mixture models may be trained by the *expectation maximization* (EM) algorithm (Mitchell, 1997; Neal & Hinton, 1999). The EM algorithm is very similar to clustering approaches such as *k-means*, in which *k* points in the feature space are chosen as representative of the overall set of samples.

Often, there are advantages in treating some subset of the variables as conditionally independent from others. For example, a full joint probability distribution can require a lot of data to train; there may be clear constraints from the application that imply conditional independence, and there may be some subset of the variables that are most effectively modeled with one technique while the rest are best modeled with another. In this case, it may be helpful to selectively apply conditional independence to break the problem into smaller pieces. For example, we might take $P(\mathbf{x}|C) = P(x_1, x_2|C)P(x_3|C)$ for a three-dimensional feature space, model $P(x_1, x_2|C)$ with a mixture of Gaussians, and $P(x_3|C)$ as a histogram. This overall model amounts to an

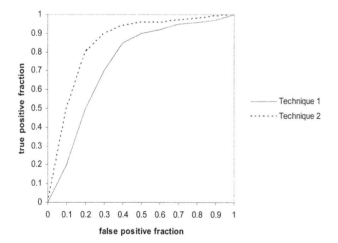

FIGURE 10.1. ROC curves illustrate the trade-off between the rate of true positives and false positives, and can be useful in comparing recognition techniques. Here we see that for a given tolerable rate of false positives, Technique 2 yields better recognition performance than Technique 1.

assertion of condition independence between x_3 and the joint space of x_1 and x_2.

This modularity afforded by assumptions of conditional independence is taken to its logical conclusion in the *Bayesian network,* which is commonly represented as a directed acyclic graph, where each node corresponds to a random variable x_1 with probability distribution $P(x_i | parents(x_i))$, and each variable is conditionally independent of all variables except its parents (Jensen, 2001). Nodes in a Bayesian network for which we have observations are called *evidence* nodes, while others are considered *hidden.* Observations may be entered into network, and through an *inference* procedure, the likelihood of the observation may be calculated, as well as posterior distributions over any hidden nodes. With Bayesian networks, designers may craft complex probability models without becoming mired in mathematical notation, and software packages allow graphical manipulation of the networks directly (Kadie, Hovel, & Horvitz, 2001).

The *dynamic Bayesian network* (DBN) models time-varying sequences, and thus is relevant to systems in which interactions take place over durations of time. A DBN can be thought of as a Bayesian network where certain nodes depend on the same Bayesian network instantiated on the previous time slice. Dependencies can be within a time slice, or across the time slice. For example, a state variable may depend on its past value as $P(x_t | x_{t-1})$. Such relationships with past values of the same variable can encode a probabilistic finite state machine or *Markov model,* where the distribution $P(x_t | x_{t-1})$ is considered a transition matrix. With this dependency on random variables in the past DBNs can effectively encode a time-varying state or "memory" that may be relevant for an interactive application. For example, it can be useful in modeling interactions composed of a sequence of steps, or where application state itself can be modeled as a finite state machine. Finally, by making a strong dependency on the immediate past, the model can be given some inertia or "smoothed."

One popular special case of the DBN is the *Hidden Markov Model* (HMM), often used to model time-varying signals such as speech, gesture, pen strokes, and so on. HMMs model observations y_t conditioned on a state variable x_t, which evolves over time as $P(x_t | x_{t-1})$. As with many probabilistic models, HMMs are *generative* in nature, meaning one of the ways we can understand them is to consider "running them forward" to generate new observations: an HMM can be thought of as a stochastic finite state machine (Markov model) that generates observations drawn from a distribution associated with each state. With each time step, the Markov model takes a transition to a new state. In the inference (recognition) process, the posterior distribution over the hidden state variable x_i is computed from the observation sequence y_t (Rabiner, 1989). HMMs have been applied many types of observation sequences, including hand gestures, handwriting recognition, speech recognition, and so on. In the simplest application paradigm, a separate HMM is trained for each class.

Much of the attraction of Bayesian networks is due to their flexibility to implement complex probabilistic dependencies. Many probability models may be thought of as particular Bayesian networks. Naïve Bayes, mixture of Gaussians, Hidden Markov Models, and Kalman filters, are among the models that have been shown to be special cases of the Bayesian network (Jordan, 1999). The structure of the Bayesian network is often determined by hand using application-specific knowledge, while the various distributions may be tuned by training from data. Parts of a Bayesian network may be completely handcrafted, derived from domain knowledge in the same manner as expert systems. Other sections of the network may in fact be Hidden Markov Models, Kalman filters, mixtures of Gaussians and various hybrids, such as a mixture of Kalman filters. The automatic learning of the structure of the network itself is active area of research (Heckerman, Geiger, & Chickering, 1994).

Many of the techniques outlined above can be applied to the more generic task of *modeling,* where we are interested in more than classification results. For example, a Bayesian network that fully models the user's interactions with a mobile device might include a variable representing the user's location. The value of this variable will be hidden (unknown) if there is no sensor to directly observe the user's location. We may, however, compute the *posterior distribution* of the variable after several other kinds of observations are entered in the network and find that the user's location is sometimes known with some precision (e.g., the device recognizes the nearest wireless access point with a known location). Not only is this model useful in deducing the user's location, it also enables other parts of the model to exploit this knowledge even if we are not ultimately interested in location information.

Finally, sometimes the signal processing task for an application is better thought of as approximating a function that directly maps a set of inputs to outputs. Techniques such as neural networks, *Radial Basis Function networks,* and other *manifold learning* techniques can be useful in learning mappings from sensor inputs to application-specific quantities. Such techniques can be particularly useful in transforming raw, high-dimensional, nonlinear sensor readings into simple calibrated outputs useful in an application. For example, in carefully controlled circumstances, it is possible to map images of a face to gaze angle by providing a number of face image and gaze-angle pairs. A function approximation approach can interpolate over these examples to map new images to gaze angle (Beymer & Poggio, 1996).

AN EXAMPLE SYSTEM

The following example demonstrates a number of the techniques described above in a working, interactive, sensor-based system. After motivating the overall design of the system, a number of aspects of hardware design are illustrated. Also described are subsequent signal processing steps such as sensor fusion, the application of Bayesian networks for modeling, and gesture and speech recognition.

In the design of intelligent rooms, the issue of how the room's inhabitants might best interact with the room often arises. The traditional notions of desktop computing or the multiple, incompatible, button-laden remote controls typical of consumer electronics are perhaps antithetical to the seamless and untethered experience that is a main feature of the vision of intelligent environments. One popular notion of how users could control an intelligent room is borrowed directly from Star Trek:

The user of the room merely speaks to it, as in, "Computer, turn on the lights."

In the development of one intelligent room (Brumitt, Meyers, Krumm, Kern, & Shafer, 2000), a user study was conducted to determine how real users might want to control multiple lights throughout the space (Brumitt & Cadiz, 2001). A Wizard of Oz paradigm was adopted so that the study would not be limited to designs already implemented. The experimenter, seated behind one-way mirrored glass, operated the lighting controls manually in response to user actions. The users were then exposed to multiple ways of controlling the lights: (a) a traditional GUI list box, (b) a graphical touch screen display depicting a plan view of the room with lights, (c) two speech only-based systems, and (d) a speech and gesture-based system. The study concluded that, like Captain Kirk, users preferred to use speech to control the lights, but that the vocabulary used to indicate which light to control was highly unpredictable. This variance in speech chosen poses a problem for the pure speech-based interface.

Interestingly, the majority of subjects looked at the light they were trying to control while speaking. This observation suggests that an intelligent room could resolve ambiguity in spoken commands (e.g., which light to control) by using computer vision techniques to determine the user's gaze, at least where the device under control is within sight. There are a number of general approaches to computing gaze, but each has serious drawbacks. For example, it is possible to roughly compute gaze from a small number of cameras throughout the room (Wu et al., 2000), but such systems presently lack accuracy and reliability, or require a large number of cameras to cover a useful space. Wearing a special device such as glasses solves some problems, but may not be acceptable to casual users. Another approach is to embed a camera in the device to determine whether the user is looking at it, rather than computing general gaze (Shell et al., 2004). This technique can be effective, but presently scales poorly to a large number of devices.

In light of the difficulties of determining gaze reliably, we reasoned that pointing gestures may play a similar role as gaze in indicating objects. While few subjects in the lighting study spontaneously used gestures, this may be partially explained by the near perfect performance of the Wizard of Oz speech recognizer (the experimenter). Furthermore, pointing may have certain advantages over gaze. For example, pointing is typically the result of a conscious decision to take action, while changes in eye gaze direction may be more involuntary (Zhai et al., 1999). On the other hand, pointing may be no easier to detect by computer vision techniques than gaze (Jojic, Brumitt, Meyers, Harris, & Huang, 2000).

To demonstrate the utility of the combination of pointing and speech as an interface modality in an intelligent environment, we built a hardware device to sense pointing gestures and developed associated signal processing algorithms to combine speech and gesture (Wilson & Shafer, 2003). At the center of the XWand system is a handheld device that may be used to select objects in the room by pointing, and a speech recognition system for a simple command and control grammar. To turn on a light in the room, the user may point the wand at a light and say, "Turn on." Because the pointing gesture serves to limit the context of the interaction (the light), the speech recognition task is reduced to recognizing the few operations available on lights: "Turn on" or "Turn off." Alternatively, the user may perform a simple gesture in place of speech to effect the same command. The user may, for example, point at a media player device, hold the button down, and roll the device to adjust the volume. The XWand system illustrates a number of points related to sensor and recognition-based input, including the hardware design of a composite inertial sensor, sensor fusion, dynamic Bayesian networks, and a host of design considerations.

The original XWand system is based on a 3D model of a room and the controllable devices within it. Using onboard sensors, the XWand device can determine its own absolute orientation, while a computer vision system mounted in the environment finds the position of the wand. Given the size and 3D position of an object in the room, it is a simple trigonometric calculation to determine whether the XWand is currently pointing at the object.

The original XWand hardware device contains an Analog Devices ADXL202 two-axis MEMS accelerometer, a Honeywell HMC1023 three-axis magnetometer, a Murata ENC-03 one-axis piezoelectric gyroscope, a 418MHz FM transceiver, a PIC 16F873 microcontroller, an IR-LED, and a pushbutton mounted on a custom printed circuit board (PCB) (see Fig. 10.2). While the accelerometer is useful in detecting pitch and roll (recall that gravity is an acceleration), it cannot detect the yaw attitude of the device. The three-axis magnetometer reports direction cosines against magnetic north, from which yaw can be determined only if the device is held flat (some GPS devices are equipped with two-axis magnetometers that give heading when the device is held flat). Fortunately, pitch and roll information from the accelerometers may be used to "correct" the output of the three-axis magnetometer to yield a full 3D orientation with respect to magnetic north.

To compute the 3D position of the wand, the XWand system uses a pair of FireWire cameras mounted in the corners of the room, which are used to track the IR-LED on the device. Each camera uses an IR-pass filter so that in a typical office environment, only the IR-LED is visible in the image. Furthermore, the IR-LED is programmed to flash at 15Hz. When the host takes the video output of each camera at 30Hz, consecutive images may be subtracted pixelwise so that only objects blinking at 15Hz remain. The IR-LED can thus be located easily in both views. Furthermore, the cameras are calibrated to the geometry of the room so that the 3D position of the IR-LED is obtained from its 2D position in both views. Note that this arrangement assumes a line of sight to the IR-LED from both cameras.

To support speech recognition, an open microphone (low impedance) is placed in the environment. Ultimately, this microphone should be placed on the device, perhaps with the audio encoded and relayed off-board for recognition. The speech recognition engine is programmed with simple command and control grammar based on a simple *command-referent* pattern, where a *referent* can be a device in the environment (e.g., a light) and the *command* refers to one of a number of permitted actions on the device (e.g., "turn on").

Simple gestures made with the wand—such as flicking left, right, up, down, and roll—are recognized by simple routines that measure the change in attitude of the wand from the attitude

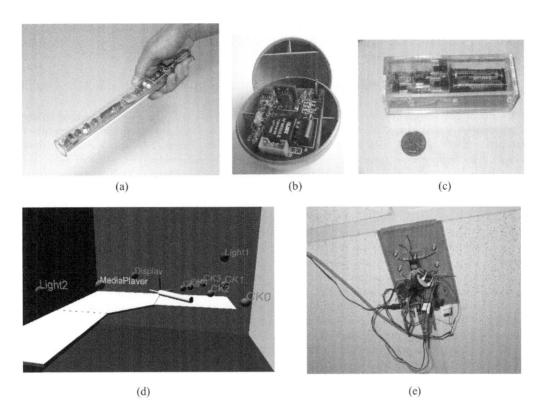

(a)   (b)   (c)

(d)   (e)

FIGURE 10.2.   (a) The first XWand prototype includes accelerometers, magnetometers, gyro, radio, etc. (image © ACM, 2003) (b) The Orb device repackaged much of the XWand. (c) The Warp-Pointer updates most of the components and uses Bluetooth. (d) The XWand 3D geometry model includes the 3D position of all interactive devices in a room. (image © ACM, 2003) (e) The World-Cursor teleoperated laser pointer is driven by the XWand.

recorded when the button on the device is first pressed. We have also experimented with using Hidden Markov Models to recognize more complex gestures, but have chosen instead to exploit a small set of simple, memorable, and reusable gestures in conjunction with other contextual information such as pointing and possibly speech information. This approach allows for a very simple and robust gesture recognition process, and avoids training users on a gesture set of greater complexity.

A dynamic Bayesian network fuses the various quantities to arrive at a multimodal interpretation of the user's interaction. It models the combination of the output of the speech recognition, the object at which the wand is currently pointing, any gesture performed, the known state of the devices under control, and the state of the interpretation in the previous time steps (see Fig. 10.3). The network bases this combination on the *command-referent* pattern outlined above, where the referent may be determined by speech or pointing gesture, and the command may be determined by speech, gesture, button click, or any combination thereof. The ultimate action to be taken (e.g., "turn on light #2") depends on the command and referent, as well as the state of the device itself (e.g., "turn on light #2" is only permitted if the light is off). Finally, both the command and referent at the current time step depend heavily on the command

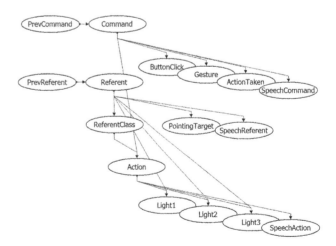

FIGURE 10.3.   The XWand Dynamic Bayesian Network (DBN) models multimodal interaction. It combines wand input (Pointing Target, Gesture, ButtonClick), speech input (SpeechReferent, SpeechCommand, SpeechAction), and world state (Light1, Light2, Light3) to determine the next action (Action) as a combination of command (Command) and referent (Referent) and past beliefs (PrevCommand, PrevReferent). (image © ACM, 2003)

and referent from the previous time step, such that either quantity can be specified at slightly different moments in time.

The multimodal fusion Bayes network offers a number of interesting capabilities for the XWand system. For example, when the state of each controlled device is represented in the network itself, distributions over related quantities change in appropriate ways. For example, if the user points at a device that is currently turned off, speech and gesture recognition results inconsistent with that state are ignored, such that the phrase "turn off" is removed from the speech recognition grammar. In future work, it may also be possible to infer vocabulary by training the network dynamically (e.g., point at a light and label it as a "light," or point several times at a point in space to train the position of a known light while saying, "Light").

We have continued to develop the XWand prototype in various ways. A typical objection to the camera-based tracking system is the lengthy setup and calibration procedures such systems require. We therefore explored an alternative configuration that eliminates cameras in favor of a single teleoperated laser pointer mounted in the ceiling (Wilson & Pham, 2003). The WorldCursor laser (see Fig. 10.2) is programmed to match the motion of the wand, in a manner similar to how the standard mouse controls a cursor with relative motion. Because the set of objects the laser can reach is limited by line of sight, the original 3D model of the system is eliminated in favor of a simpler spherical coordinate system with an arbitrary origin, thus simplifying setup.

The most recent iteration includes a three-axis accelerometer that can be combined with a magnetometer to arrive at a true 3D orientation by a simple cross-product calculation (see Fig. 10.2). The device has also been applied to cursor control with very large displays, where the mixture of sensors enables a variety of cursor control mechanisms including absolute, orientation based pointing (position and orientation), relative angular motion similar to the Gyromouse, and pure absolute position only. This flexibility allows exploration of several new modes of interaction. The relative, gyro-based mode of pointing allows very fine control with clutching over a small area. With very large (wall-sized) displays, however, it is easy to lose the cursor among the many onscreen objects. With the current device, it is possible to momentarily adopt one of the more absolute pointing modes to "warp" the cursor to a point directly in front of the user.

CONSIDERATIONS IN DESIGNING RECOGNITION-BASED SYSTEMS

The previous section touches on issues common to complex sensing-based interactive systems, from hardware design to modeling and sensor fusion. Overall, the strategy in the example is to avoid complex recognition problems whenever possible through thoughtful choice of hardware sensors, design of the interaction, and strong modeling of context. In the following section, we expand upon the motivations behind these design choices. Many of these considerations must be taken into account in the design of recognition-based interactive systems. While many involve difficult problems with no easy answer, it is best to be aware of these issues when designing a sensing-based interactive system.

Computational Cost, Utility, and the Cost of Failure

A number of challenges exist in designing an interactive system that uses sensors and recognition techniques. First, although Moore's Law continually pushes forward the frontier of practical signal processing techniques, many of these algorithms are still computationally intensive. Only in the last six years or so have real-time computer vision techniques become practical on commodity hardware. Furthermore, many of the machine learning and pattern recognition techniques are data-driven, thus requiring large amounts of storage, memory, and training. In the case of mobile devices, where computational power lags desktop computing power by several years, many of these algorithms are impractical, and developers are often forced to make hard choices and take shortcuts. These concerns are often magnified in interactive systems that entirely remake the user interface and employ multiple recognition-based techniques.

Second, recognition-based systems often face serious challenges achieving and guaranteeing the level of robustness required in real applications. In the case of developing consumer products, for example, it is one thing to demonstrate a technique in a laboratory setting; it is quite another to show that the same technique will work in the variety of circumstances in which customers will expect it to work. Unit testing even simple recognition-based interactive systems can be daunting. Computer vision techniques, for example, are often susceptible to variations in lighting, while audio-based techniques may fail in the presence of background noise. Some effort has been devoted to developing signal processing techniques that adapt to both the current circumstances and user. Adaptive speech recognition and handwriting recognition techniques, for example, have become commonplace. Even in these cases, however, it is important that the system have good functionality out of the box or else the user may not use the system long enough for an adaptive algorithm to improve performance.

In the development of a recognition-based interactive system, it may become impractical to seek more improvement in recognition performance. At this point, it is important to consider the cost of recognition failure: often the cost of repairing a false positive recognition can overwhelm any advantage in the use of the system. In speech recognition, for example, the repair of the errors can be awkward, slow, and disruptive to the task (C.-M. Karat, Halverson, C. Karat, & Horn, 1999). Only users that are unable to use a regular keyboard may accept a dictation system that fails three times out of one hundred words, for example (Feng, C.-M. Karat, & Sears, 2005). Another consideration is whether the system returns a result similar to the desired result when it fails ("graceful degradation"), in which case repair is likely to be easier (Horvitz, 1999). If a recognition failure is too costly to consider repair (for example, control of an air-lock on a spacecraft, or more mundanely, closing a window on a desktop GUI), the cost of making a mistake may be incorporated directly in the model so that false positives are more likely to be avoided. This can be done either by seeking some kind of deliberate confirmation from the user, or more simply by moving thresholds up the ROC curve. In the latter approach, it is important to be aware that often users modify their behavior on successive attempts such that their input is no longer modeled by the usual training corpus; in speech this is known as the Lombard effect (Junqua, 1993), but this phenomenon can be observed in other modalities as well.

Considerations of computational cost, robustness, and cost of errors should play prominent roles in the end-to-end design of recognition-based systems. Ultimately, the designer may be forced to recast the interaction and possibly the application altogether. An example of a system in which one can find many of these issues played out is the Sony EyeToy, an add-on camera for the Sony Playstation 2. The EyeToy allows the player to interact with a game through gesture and body motion, rather than through the usual game controller. It also appears to use very simple image processing techniques to determine the player's motion. These techniques are computationally cheap and generally quite robust to varying light levels and other factors that are likely to be encountered in a residential application. Furthermore, the EyeToy works with a small number of games written specifically for use with this device. These games take advantage of the strengths of the EyeToy, rather than risk providing a poor emulation of the regular game controller.

The Role of Feedback

Feedback is important in any interactive application, but may be even more so where sensing and recognition are used (Bellotti et al., 2002). We can characterize some kinds of feedback as "tight" or not, where a "tight" feedback loop provides feedback in a frequent, timely, and informative fashion.

Usually, if there is any chance of recognition failure, the system should provide feedback on all recognition results, errors or otherwise. If possible, the system should provide some indication of the nature of any failure so that the user can modify his or her own behavior to meet the system's expectations. If the system can provide some indication of the quality of the interpretation as it is happening, then it may be possible to allow the user to cancel or modify the interaction on the fly, so as to avoid costly errors or cumbersome confirmation routines (Vogel & Balakrishnan, 2004).

The right feedback can often influence the design of the sensing and recognition algorithm itself. For example, because the onscreen cursor is updated so quickly (a tight feedback loop), a naïve user might not ever realize or care that the mouse provides rapid, small, successive bits of relative movement information rather than true position information, which would be much harder to sense. This trick is used in the WorldCursor system to avoid the use of cameras required by the original XWand system.

Considering the EyeToy again, it is interesting to note that the image of the player is often incorporated into the onscreen presentation. By watching themselves onscreen, players are able to interact with onscreen elements without relying on sophisticated, yet more failure prone and computationally intense, hand-tracking algorithms. This feedback also cleverly ensures that the player stays in the camera's field of view; the flow of the game does not have to be broken to alert a player who has left the field of view.

Implicit and Explicit Interaction

In the previous discussion regarding feedback and cost of failure, we assume that interaction is structured in such a way that the user takes action and expects a timely response from the system—that is, the user's actions and the system's responses are *explicit*. Most interactive systems can be characterized in this way.

In contrast to *explicit* interactions, *implicit* interactions are based not on explicit action by the user, but more commonly on users' existing patterns of behavior (Schmidt, 2000). For example, with the frustration-sensing system outlined in the introduction, the state of frustration is not explicitly entered by the user in order to elicit some behavior from the system. Instead, the state arises naturally and perhaps involuntarily, and upon detection, the system should take appropriate action.

Implicit interactions may take place over a long duration, and may not exhibit an obvious pattern of cause and effect. For example, systems that adapt to perceived user preferences, as indicated by the history of user behavior, might eventually make a recommendation to the user, or even take actions so subtle that the user may not notice them. Such systems can be complex in terms of sensing and modeling, and often tend towards automating or refining aspects of the user's original task (Horvitz, Breese, Heckerman, Hovel, & Rommelse, 1998). For example, a smart home may observe its inhabitants' daily patterns of coming and going to determine an optimal schedule to control the thermostat automatically, balancing comfort and economy (Mozer, 2005). One potential difficulty is that, unless the output of the system is designed very carefully, users may feel unable to correct a mistake made by the system or exert more explicit control in the face of exceptional circumstances (e.g., a party). Designers should consider providing functionality that allows the user to query the system for why it took a given action, provide simple mechanisms to redress errors, and finally, revert to a manual control mode.

Implicit interaction systems driven by patterns of ongoing user behavior are often not as critically dependent on sensing and recognition reliability and the nature of feedback. Rather, it is more important that interpretation processes are more correct than incorrect over time; accordingly, modeling techniques that integrate noisy sensor values over time are often appropriate. The automatic thermostat, for example, should probably incorporate many observations of the users' behavior—including presence and manual thermostat control—and model weekends, weekdays, and holidays differently. Sophisticated modeling techniques in hand, designers of such systems have the opportunity to exploit powerful, though sometimes unreliable, sensing techniques.

The Importance of Context

Notions of *context* can play an important role in sensing-based interaction (Dey, Kortuem, Morse, & Schmidt, 2001). "Context" refers to the overall environment the interaction or device finds itself in, rather than the objects obviously and directly relevant to the task at hand. What constitutes context often depends on point of view; after studying the application in detail, factors that once may have seemed external and tangentially relevant (context) may be central to the model after all.

Note the purposeful vagueness in the definition of "context;" with respect to sensing, "environment" can refer to the actual physical surroundings. The addition of sensors to a system may give it awareness of its environment, and thereby enable interesting "context-dependent" behavior. For example, a mobile device equipped with GPS or other localization technology might bring up web search results corresponding to the nearest points of interest (Abowd et al., 1997; Davies, Ceverst, Mitchell, & Efrat, 2001; Hariharan, Krumm, & Horvitz, 2005). Beyond the physical environment, context can refer to more abstract states, such the user's current activity (e.g., working or not), the history of their interactions, and their preferences. It may also refer to momentary information, such as the action of the nondominant hand, or anaphoric references that provide scope for an ongoing interaction.

Often our activities follow preestablished patterns, either by design or by accident. These patterns can provide strong contextual information for interpreting the user's interactions. For example, entertainment-based scenarios may have a narrative structure that constrains the interaction (Galyean, 1995). The KidsRoom interactive experience, for example, was structured according to a narrative progression involving the fantastical transformation of a children's room (Bobick et al., 1999; Bobick et al.,2000; see Fig. 10.4). This structure in turn guided the selection of recognition at any given moment.

Modeling context explicitly can be a powerful way to solve difficult recognition problems. The more context can be brought to bear, often the easier and more robust the recognition. Although counterintuitive at first, by exploiting context, the combination of multiple sensors with simple signal processing techniques may result in better performance than the use of fewer sensors with more complex signal processing techniques. Recall how in the XWand system, the speech recognition process is constrained by the knowledge of what the user is currently

FIGURE 10.4. The KidsRoom engaged children to participate in an interactive journey. Computer vision and projection technologies were used to transform an ordinary room into a variety of settings including a forest and river. Image © the Massachusetts Institute of Technology, 1999.

pointing at with the wand, the current state of the device indicated, and so forth. In fact, the interaction can be so constrained that, often, the system only needs some indication that the user said *anything*. Similarly, it suffices to use simple, learnable and reusable gestures when the interaction has been contextualized by pointing, speech, or both. Our frustration detection system may be more robust if it incorporates the knowledge of the application currently in focus. For example, it may be easier to write one frustration detector for office applications and a separate one for use while playing video games, rather than one detector that works in both contexts. In the end, the two detectors may only differ in terms of how some threshold is set.

The Importance of A Priori Knowledge

The development of a model for a sensing, interactive, system can benefit from specific knowledge of the domain in which the interaction is situated. Such *a priori* knowledge can lead to insights as to how to determine meaningful categories from raw sensor data. Higher level rules taken from the domain can then be brought to bear. For example, in the case of a pen-based system that automatically parses and manipulates mathematical equations, the rules of how mathematical expressions are combined can be a powerful constraint that drives the correct interpretation of the sloppiest drawings (LaViola & Zeleznik, 2004). Similarly, knowledge of chemistry can guide the transformation of a sketches of molecules to a full 3D model (Tenneson & Becker, 2005). Providing a kind of context, strong assumptions taken from the domain can limit the applicability of a model but can often dramatically improve performance.

Many domains of human behavior have been categorized and described in terms of detailed taxonomies and ontologies. For example, music, gesture, dance, and spoken language each have detailed ontologies, notation schemes, and so on. It can be beneficial to draw from such knowledge when developing a model, but some aspects of a categorization scheme may not be fully supported by the available sensors. For example, there may be some aspect of the domain not covered by the model, or a single category may "alias" to several distinct classes as perceived by the sensors. A detailed analysis of the sensing system output may lead to insights on the original domain model.

One of the advantages of incorporating bits of domain knowledge representation directly into the model itself is that it becomes more transparent to its designers and users, and thus, more reusable. If there is a problem with the system, the designer may directly inspect semantically relevant quantities from the model. Approaches that do not rely on such informed representations, such as neural networks, are often so difficult to inspect that upon discovery of a problem, it may be easier to retrain the model from scratch than to troubleshoot what the network really learned. A good compromise is to use the more data-driven techniques (such as neural networks or probabilistic modeling techniques) to map the raw input signals onto semantically meaningful mid-level primitives. Such choices in representation can support modularity and explorative research.

It is interesting to note, however, that initial research on a given complex sensing problem often draws heavily from domain knowledge, only to be eclipsed later by more purely data-driven approaches. Early approaches to speech recognition, for example, transformed the audio signal into a string of symbols representing phoneme categories developed by linguists; today, one is more likely to see approaches in which *subword acoustic unit* categories are trained from audio signals directly in a purely data-driven approach. Early face detection and recognition approaches similarly relied on prescriptive or ad hoc features, while more recent approaches are more purely data-driven (Li & Jain, 2005; Zhao et al., 2003).

Generalize or Specialize?

The incorporation of a priori and context information leads to potentially complex, yet powerful, modeling techniques. One of the drawbacks of adding more and more detail into a model, however, is that the resulting system may be so tailored to a particular domain or set of contextual circumstances that it fails to generalize to new applications. It seems as if engineering best practices, including modularity and device-independence, run counter to models that are optimized for a given situation.

For example, consider the problem of delivering various location-based services to inhabitants of an indoor space. A good, familiar choice for the representation of each person's location might be two-dimensional Cartesian coordinates on a full map of the environment. A person-tracking system is installed in the environment; it is charged with determining where everyone is on the map. This information is then passed onto another part of the system that delivers services to each user based on the knowledge of where everyone is at any given moment.

Such a choice of representation has many desirable features. We can consider a multitude of person-tracking technologies—as long as each reports 2D Cartesian coordinates we can incorporate its output into our map ("device independence"), and we can consider merging the results of multiple such systems in a simple probabilistic framework where we model the location of each person as a distribution over space rather than simple 2D coordinates. Furthermore, we can exploit familiar rules of geometry to calculate interesting properties such as whether a display is within sight of a particular user. Most importantly, we need not concern ourselves with the particular application services that our ultimate system will provide: as long as each service expects only our generic representation we can expect the service to work properly. The system is *general* in its applicability.

Contrast this approach to one in which we have no unifying geometric model or map, but instead, install proximity sensors at each of the devices of interest, each sensor detecting some one as they approach the device. Perhaps we install special sensors on some of the doors to detect when someone walks through (Abowd, Battestini, & O'Connell, 2003), and maybe a sensor on seats to detect when someone sits down (Brumitt et al., 2000). Often, proximity information is enough and requires no calibration step (Krumm & Hinckley, 2004). For our location-based services, we note that in some cases, it is important to know *who* is at a given location, but in many cases, this information is not necessary. We place various simple sensors throughout the environment in locations that we think will provide the information needed to support the specific location-based

services we currently have in mind. The solution we come up with is quite *specialized*.

It is unclear which approach is superior. The geometric model-based approach depends heavily on the performance of the person-tracking system. When it fails, it may return no useful position information, even when the given circumstances do not require precise position information, and even while the sensor and the underlying signal processing algorithms may produce relevant intermediate results that could be used at a subsequent level of processing. The assumption of device-independence ignores the fact that different sensing technologies typically have very different failure modes. The resulting application may perform poorly due its design around modeling choices that follow the lowest common denominator. On the other hand, it could work very well and provide the utmost in flexibility if the person-tracking is very reliable.

The specialized approach of course suffers in all the ways that the general approach excels; each installation of the system may require significant innovation on the part of the designer. On the other hand, because it is tailored to the application, the system is more likely to gracefully handle sensing failure modes, and furthermore, is more likely to be less wasteful of resources, both physical and computational. Not surprisingly, the specialized approach is likely to exhibit better performance for a given application than the generalized approach.

The consideration of generalized versus specialized designs is a common engineering problem that is especially relevant in the realm of sensor and recognition-based systems. As our example illustrates, the two approaches may demand completely different sensors, representations, and modeling choices.

Traditional vs. Nontraditional Interfaces

A question that naturally arises in the application of sensing and recognition to interactive systems is whether the design should emulate, augment, or completely replace the interfaces we already have.

It is probably not surprising that so many interactive sensor-based systems emulate the mouse. After all, once this functionality is achieved, the system is now relevant to the vast majority of the world's software. It is interesting to note, however, the degree to which the original development of the GUI hinged on the development of the mouse itself, and how to this day, the mouse is still the favored input device. This suggests that unless the new interactive system offers significant new functionality over the mouse, it will not be adopted, and that instead of forcing the new techniques on today's interfaces, designers should think about what changes in the interface are implied by new sensing systems.

Another approach is to augment or complement today's interfaces. For example, it is relatively simple to add an accelerometer to a mobile device that allows the user to position the cursor or scroll by tilting the device, and it is not hard to imagine how users could easily pick up this interaction (Hinckley et al., 2000).

Still another approach is to completely reinvent the interface. The risks are that the user is required to learn a completely new way of working, the designer is faced with developing the entire system, and the utility of the system will inevitably be compared more established techniques. Again, unless there is a significant improvement to be had, users are unlikely to adopt the new approach.

Many of these issues are illustrated by recent research into interactive table systems. In its simplest form, an interactive table system might be no more than a large flat-panel touch screen turned horizontal to function as a table surface. This simple change in orientation enables new applications that exploit the collaborative nature of multiple users gathering around a table (Shen, Vernier, Forlines, & Ringel, 2004). Furthermore, sensing techniques have been developed to support multiple points of contact on the surface, and to support bimanual interaction, or the multiple hands of multiple users. Distinct from wall displays, tables are able hold objects, and accordingly, some interactive tables use sensing techniques to recognize objects placed up on them. This ability can be used to support various *tangible user interface* scenarios. For example, a puck placed on the table can enable a feature of the application; the user might rotate the puck to adjust an associated parameter, or it might call up a group of photos (Fitzmaurice et al., 1995; Ullmer, Ishii, & Glas, 1998; see also Ishii, chapter , this volume.).

Clearly, in order to exploit the unique affordances and sensing capabilities of an interactive table, we must be willing to let go of many of today's GUI interfaces. For example, the multiple-touch and multiple-user aspect cannot be supported by the single-focus model of the GUI, and even the assumption that the display has a natural reading orientation (an "up" direction) may no longer be valid (Kruger, Carpendale, Scott, & Greenberg, 2004; Shen, Lesh, & Vernier, 2003). Finally, today's GUI model has no role for tangible UIs.

The designers of interactive table systems tend to follow a few guiding principles to innovate without alienating the user. First, there is a desire to leave the overburdened widgets of the modern GUI behind and instead rely on more of a direct manipulation style of interaction. For example, to rotate a virtual object on the screen, it may suffice to place two fingers anywhere on the object and move one finger about the other, much in the same way that one might rotate a piece of paper sitting on a desk. By using the multiple touch capability, we avoid the clumsy widget-heavy and mode-heavy rotation techniques typical of many drawing packages. The addition of pucks and other tangle UIs extends this approach; for example, while the puck is on the table, it may behave as knob (Patten, Ishii, & Pangaro, 2001). Secondly, there is a trend to add widget-based interaction back into the direct manipulation framework, perhaps mainly to address the need to perform a variety of actions on the digital representation of the object. For example, e-mailing, printing, and contrast and color adjustment are just a few things that the user might want to do with their photos; these operations are outside of the scope of direct manipulation. These widget-based interactions can draw upon the advantages a multiple touch table interface provides. For example, putting two fingers down near an object may trigger a scrollbar-like widget in which bringing the two fingers closer or further apart adjusts a parameter (Wu & Balakrishnan, 2003).

Evaluating Novel Interfaces

Determining whether the novel sensing and interaction model actually works is the domain of usability testing. No research into sensor and recognition-based input is complete without an evaluation showing its effectiveness. Unfortunately, the technical difficulty of getting such systems to work at all often leaves little time to quantify performance on a standard task. Furthermore, when the new work is in a preliminary state, it may not be instructive to compare the new technique against one that has had decades to evolve.

Many times, user study subjects are often so impressed by the "magic" of sensor-based systems that the sheer novelty of the interface can skew study results. Subjective surveys are likely to show bias in favor of the novel design (Nielsen, 1994). This is a very difficult problem without an easy solution, particularly in the case of tangible UIs, tabletop interactive systems, and perceptual user interface systems. *Longitudinal studies,* which involve repeated sessions spread out over multiple days, can be used to minimize this effect, but such studies are expensive and time-consuming. As a result, many interactive sensing-based systems come with few convincing quantitative user studies that prove their utility. In the case of a system that uses tracking or detection to select an object, Fitts' Law studies can be an effective technique to compare pointing performance across very different systems (MacKenzie, 1992).

Of course, often the point of the work is not to show a decrease in task completion time, a reduction in errors, or other more conventional metrics from the field of HCI. Rather, the goal may be to highlight completely new ways of conceptualizing the relationship between users and their machines, and to demonstrate that such innovations are technically feasible. There is often more of an emphasis on invention and design than on evaluation. The novelty that can sabotage a user study may even be a desirable effect, compelling users to engaging experiences they may not have otherwise had. Furthermore, traditional evaluation methods seem at odds with the goals of surprising, delighting, and entertaining the user. The field has only recently begun to recognize the need to develop ways to objectively evaluate interactive systems along these dimensions that are basic to the quality of life (Blythe, Overbeeke, Monk, & Wright, 2004; Norman, 2003).

A fantastic example where simple sensing techniques were used to great effect to surprise and delight the user is the Ping-PongPlus system (Ishii, Wisneski, Orbanes, Chun, & Paradiso, 1999). In this research prototype, the usual ping-pong table was augmented with sound effects and a top-down video projection onto the table surface, and electronics to sense where the ping pong ball hits the surface during game play (see Fig. 10.5). This sensing system used eight microphones mounted under the table, and custom electronics to triangulate where the ball hit the surface. The sound effects and video presentation reacted to each strike of the ball, in ways calculated to amuse the players and augment the game play in dramatic ways. For example, in one mode, the ball produces ripples on the table, while in another, thunderstorm audio and video effects build up as the length of the volley increases.

FIGURE 10.5. The PingPongPlus system uses a series of microphones to triangulate where the ping pong ball strikes the surface of the table. This position information is used to drive a variety of interactive graphics displayed on the table by an overhead projector. Image © 2006 Tangible Media Group, MIT Media Laboratory

CONCLUSION

In this chapter, we explored a variety of sensing technologies available today, outlined a number of signal processing techniques common to using these sensing technologies in sensor and recognition-based input for interactive systems, and discussed further issues related to designing such systems.

While much of the discussion highlighted difficulties in designing systems that rely on sensors, the future of interactive sensing systems is bright. Advances in MEMS and nanotechnology will continue to drive innovations in the sensors themselves, while the relentless increase in commodity CPU power and storage capabilities will continue to enable more sophisticated modeling techniques used in interpreting sensor outputs.

Another powerful driver in the development of sensing-based interactive systems is the growth of the computing form factor beyond the traditional desktop computer. The proliferation of cell phones, personal digital assistants, portable gaming devices, music players, tablet PCs, and living room-centric PCs shows a trend towards the vision of *ubiquitous computing,* in which computing is situated throughout our environment and daily life. Individual computing devices will be tailored in deep ways to the task at hand, away from the "one size fits all" mentality of desktop computing. The use of sensors and recognition techniques will play an important role in enabling this diversity, and will naturally support and demand a variety of interaction styles. As interactive systems become more tailored to a given activity, the opportunity to leverage the techniques described in this chapter increases, in turn enabling the application of the sensors themselves. Such a virtuous cycle may speed the development and adoption of sensing-based systems in ways that are hard to imagine today.

ACKNOWLEDGMENTS

The XWand and Orb prototypes were developed with the assistance of Mike Sinclair and Steven Bathiche. Thanks to Ken Hinckley, John Krumm, Steven Bathiche, and Ewa Davison for comments.

References

Abowd, G. A., Atkeson, C. G., Hong, J., Long, S., Kooper, R., & Pinkerton, M. (1997). Cyberguide: A mobile context-aware tour guide. *ACM Wireless Networks, 3,* 421–433.

Abowd, G. A., Battestini, A., & O'Connell, T. (2003). *The location service: a framework for handling multiple location sensing technologies* (GVU Technical Report GIT-GVU-03-07). Atlanta, Georgia: Georgia Institute of Technology.

Bao, L., & Intille, S. (2004). Activity recognition from user-annotated acceleration data. *Proceedings of the Second International Conference in Pervasive Computing,* 3001/2004, 1–17.

Baudel, T., & Beaudouin-Lafon, M. (1993). Charade: Remote control of objects using free-hand gestures. *Communications of the ACM, 36*(7), 28–35.

Baxter, L. (1996). *Capacitive sensors: Design and applications.* Hoboken, NJ: Wiley-IEEE Press.

Bellotti, V., Back, M., Edwards, W. K., Grinter, R. E., Henderson, A., & Lopes, C. (2002). Making sense of sensing systems: Five questions for designers and researchers. *Proceedings of the SIGCHI Conference on Human Factors in Computing Systems,* 415–422.

Beymer, D., & Flickner, M. (2003). Eye gaze tracking using an active stereo head. *IEEE Conference on Computer Vision and Pattern Recognition,* 451.

Beymer, D., & Poggio, T. (1996). Image representations for visual learning. *Science, 272,* 1905–1909.

Bilmes, J. A., Li, X., Malkin, J., Kilanski, K., Wright, R., Kirchoff, K., et al. (2005). The vocal joystick: A voice-based human-computer interface for individuals with motor impairments. *Human Language Technology Conference and Conference on Empirical Methods in Natural Language Processing,* 995–1002.

Bishop, C. M. (1995). *Neural networks for pattern recognition.* Oxford, UK: Oxford University Press.

Blythe, M. A., Overbeeke, K., Monk, A. F., & Wright, P. C. (Eds.). (2004). *Funology: From usability to enjoyment.* Human–Computer Interaction Series. Dordrecht, The Netherlands: Kluwer.

Bobick, A., Intille, S., Davis, J., Baird, F., Pinhanez, C., Campbell, L., et al. (1999). The KidsRoom. A perceptually-based interactive and immersive story environment. *Pressure Teleoperators and Virtual Environments, 8*(4), 367–391.

Bobick, A., Intille, S., Davis, J., Baird, F., Pinhanez, C., Campbell, L., et al. (2000). The KidsRoom. *Communications of the ACM, 43*(3), .

Boyd, J., & Little, J. (2005). Biometric gait recognition. In M. Tistarelli, J. Bigun, & E. Grosso (Eds.), *Advanced Studies in Biometrics: Summer School on Biometrics, Alghero, Italy, June 2–6, 2003.* Revised Selected Lectures and Papers, Vol. 3161/2005 (pp. 19–42), New York, NY: Springer.

Brashear, H., Starner, T., Lukowicz, P., & Junker, H. (2003). Using multiple sensors for mobile sign language recognition. *Proceedings of the 5th International Symposium on Wearable Computing,* 45–52.

Breiman, L., Freidman, J. H., Olsen, R. A., & Stone, C. J. (1984). *Classification and regression trees.* Boca Raton, Florida: Chapman & Hall/CRC.

Brumitt, B., & Cadiz, J. (2001). Let there be light: Examining interfaces for homes of the future. *Proceedings of Interact'01,* 375–382.

Brumitt, B. L., Meyers, B., Krumm, J., Kern A., & Shafer, S. (2000). EasyLiving: Technologies for intelligent environments. *Proceedings of the Handheld and Ubiquitous Computing, 2nd International Symposium,* 12–27.

Caruso, M. J. (1997). Applications of magnetoresistive sensors in navigation systems. *Sensors and Actuators 1997,* 15–21.

Choudhury, T., & Pentland, A. (2003). Sensing and modeling human networks using the Sociometer. *Proceeding of the International Conference on Wearable Computing,* 216–222.

Cristianini, N., & Shawe-Taylor, J. (2000). *An introduction to Support Vector Machines and other kernel-based learning methods.* Cambridge University Press.

Crowley, J. L., & Berard, F. (1997). Multi-modal tracking of faces for video communications. *Proceedings of the IEEE Conference on Computer Vision and Pattern Recognition,* 640–645.

Curran, E., & Stokes, M. J. (2003). Learning to control brain activity: A review of the production and control of EEG components for driving brain-computer interface (BCI) systems. *Brain and Cognition, 51,* 326–336.

Davies, N., Ceverst, K., Mitchell, K., & Efrat, A. (2001). Using and determining location in a context-sensitive tour guide. *IEEE Computer, 34*(8), 35–41.

Dey, A. K., Kortuem, G., Morse, D. R., & Schmidt, A. (Eds.) (2001). Special Issue on Situated Interaction and Context-Aware Computing. *Personal and Ubiquitous Computing, 5*(1).

Dietz, P., & Leigh, D. (2001). DiamondTouch: A multi-user touch technology. *Proceedings of the 14th Annual ACM Symposium on User Interface Software and Technology,* 219–226.

Fails, J. A., & Olsen, D. (2002). Light widgets: Interacting in everyday spaces. *Proceedings of the 7th International Conference on Intelligent User Interfaces,* 63–69.

Feng, J., Karat, C.-M., & Sears, A. (2005). How productivity improves in hands-free continuous dictation tasks: lessons learned from a longitudinal study. *Interacting with Computers, 17*(3), 265–289.

Fischler, M. A., & Bolles, R. C. (1981). Random sample consensus for model fitting with applications to image analysis and automated cartography. *Communications of the ACM, 24,* 381–395.

Fitzmaurice, G. W., Ishii, H., & Buxton, W. (1995). Bricks: Laying the foundations for graspable user interfaces. *Proceedings of the SIGCHI Conference on Human Factors in Computing Systems,* 442–449.

Fodor, I. K. (2002). *A survey of dimension reduction techniques.* Livermore, CA: Center for Applied Scientific Computing, Lawrence Livermore National Laboratory.

Forsyth, D. A., & Ponce, J. (2002). *Computer vision: A modern approach.* Upper Saddle River, NJ: Prentice Hall.

Fraden, J. (2003). *Handbook of modern sensors: Physics, designs, and applications.* New York, NY: Springer.

Galyean, T. (1995). *Narrative guidance of interactivity.* Unpublished doctoral thesis, Media Laboratory, Massachusetts Institute of Technology, Cambridge.

Garfinkel, S., & Rosenberg, B. (Eds.). (2005). *RFID: Applications, security, and privacy.* Boston, MA: Addison-Wesley.

Greenberg, S., & Fitchett, C. (2001). Phidgets: Easy development of physical interfaces through physical widgets. *Proceedings of the UIST 2001 14th Annual ACM Symposium on User Interface Software,* 209–218.

Guiard, Y. (1987). Asymmetric division of labor in human skilled bimanual action: The kinematic chain as a model. *Journal of Motor Behavior, 19,* 486–517.

Han, J. Y. (2005). Low-cost multi-touch sensing through frustrated total internal reflection. *Proceedings of the 18th Annual ACM Symposium on User Interface Software and Technology,* 115–118.

Hariharan, R., Krumm, J., & Horvitz, E. (2005). Web-enhanced GPS. In T. Strang and C. Linhoff-Popien (Eds.), *Location- and Context-Awareness: First international workshop, LoCA 2005, Oberpfaffenhofen, Germany, May 12–13, 2005, Proceedings* (pp. 95–104). New York, NY: Springer.

Haro, A., Flicker, M., & Essa, I. (2000). Detecting and tracking eyes by using their physiological properties, dynamics and appearance. *Proceedings of the IEEE CVPR 2000,* 163–168.

Heckerman, D., Geiger, D., & Chickering, D. M. (1994). Learning Bayesian networks: The combination of knowledge and statistical data. *10th Conference on Uncertainty in Artificial Intelligence*, 293–301.

Hinckley, K., Pausch, R., Proffitt, D., & Kassel, N. (1998). Two-handed virtual manipulation. *ACM Transactions on Computer-Human Interaction (TOCHI), 5*(3), 260–302.

Hinckley, K., Pierce, J., Sinclair, M., & Horvitz, E. (2000). Sensing techniques for mobile interaction. *Proceedings of the ACM UIST 2000 Symposium on User Interface Software and Technology*, 91–100.

Hinckley, K., & Sinclair, M. (1999). Touch-sensing input devices. *Proceedings of the ACM Conference on Human Factors in Computing Systems (SIGGCHI '99)*, 223–230.

Hjelm, S. I., & Browall, C. (2000). Brainball—using brain activity for cool competition. Paper presented at NordiCHI 2000, Stockholm, Sweden.

Hoel, P. G., Port, S. C., & Stone, C. J. (1971). *Introduction to probability theory.* Boston, MA: Houghton Mifflin.

Horn, B. K. P. (1986). *Robot vision.* Cambridge, MA: MIT Press.

Horvitz, E. (1999). Principles of mixed-initiative user interfaces. *Proceedings of the ACM SIGCHI Conference on Human Factors in Computing Systems*, 159–166.

Horvitz, E., Breese, J., Heckerman, D., Hovel, D., & Rommelse, K. (1998). The Lumiere project: Baysian user modeling for inferring the goals and needs of software users. *Proceedings of the fourteenth conference on uncertainty in artificial intelligence*, 256–265.

Huber, P. J. (1981). *Robust statistics.* New York: John Wiley & Sons.

Ishii, H., Wisneski, C., Orbanes, J., Chun, B., & Paradiso, J. (1999). Ping-PongPlus: Design of an athletic-tangible interface for computer-supported cooperative play. *Proceedings of the Conference on Human Factors in Computing Systems (SIGCHI)*, 327–328.

Jacob, R. J. K. (1993). Eye-gaze computer interfaces: what you look at is what you get. *IEEE Computer, 26*(7), 65–67.

Jensen, F. V. (2001). *Bayesian networks and decision graphs.* New York, NY: Springer.

Johansson, G. (1973). Visual perception of biological motion and a model for its analysis. *Perception & Psychophysics, 14*, 201–211.

Jojic, N., Brumitt, B., Meyers, B., Harris, S., & Huang, T. (2000). Detection and estimation of pointing gestures in dense disparity maps. *Proceedings of the Fourth International Conference on Automatic Face and Gesture Recognition*, 468–475.

Jordan, M. (Ed.). (1999). *Learning in graphical models.* Cambridge, MA: MIT Press.

Junqua, J. C. (1993). The Lombard reflex and its role on human listeners and automatic speech recognizers. *Journal of the Acoustical Society of America, 93*(1), 510–524.

Kadie, C. M., Hovel, D., & Horvitz, E. (2001). *MSBNx: A component-centric toolkit for modeling and inference with Bayesian networks* (Microsoft Research Technical Report MSR-TR-2001-67). Redmond, WA: Microsoft Corporation.

Kahn, J. M., Katz, R. H., & Pister, K. S. J. (2000). Emerging challenges: Mobile networking for 'smart dust.' *Journal of Communications and Networks, 2*(3), 188–196.

Karat, C.-M., Halverson, C., Karat, J., & Horn, D. (1999). Patterns of entry and correction in large vocabulary continuous speech recognition systems. *Proceedings of CHI 99*, 568–575.

Kato, H., Billinghurst, M., Poupyrev, I., Imamoto, K., & Tachibana, K. (2000). Virtual object manipulation on a table-top AR environment. *Proceedings of ISAR 2000*, 111–119.

Kiern, Z. A., & Aunon, J. I. (1990). A new mode of communication between man and his surroundings. *IEEE Transactions on Biomedical Engineering, 37*(12), 1209–1214.

Klein, J., Moon, Y., & Picard, R. W. (2002). This computer responds to user frustration: Theory, design and results. *Interacting with Computers, 14*(2002), 119–140.

Kovacs, G. T. (1998). *Micromachined Transducers Sourcebook.* New York: McGraw-Hill.

Kruger, R., Carpendale, S., Scott, S. D., & Greenberg, S. (2004). Roles of orientation in tabletop collaboration: Comprehension, coordination and communication. *Computer Supported Cooperative Work, 13*(5–6), 501–537.

Krumm, J., & Hinckley, K. (2004). The NearMe wireless proximity server. *Sixth International Conference on Ubiquitous Computing (Ubicomp 2004)*, 283–300.

Krumm, J., & Horvitz, E. (2004). Locadio: Inferring motion and location from Wi-Fi Signal Strengths. *First Annual International Conference on Mobile and Ubiquitous Systems: Networking and Services (Mobiquitous 2004)*, 4–13.

LaMarca, A., Chawathe, Y., Consolvo, S., Hightower, J., Smith, I., Scott, J., et al. (2005). Place lab: Device positioning using radio beacons in the wild. *Pervasive Computing, 3468/2005*, 116–133.

LaViola, J., & Zeleznik, R. (2004). Mathpad2: A system for the creation and exploration of mathematical sketches. *ACM Transactions on Graphics (Proceedings of SIGGRAPH 2004)*, 432–440.

Lee, S. W., & Mase, K. (2002). Activity and location recognition using wearable sensors. *IEEE Pervasive Computing, 1*(3), 24–32.

Lester, J., Choudhury, T., & Borriello, G. (2006). A practical approach to recognizing physical activities. *Proceedings Pervasive Computing 2006*, 1–16.

Letchner, J., Fox, D., & LaMarca, A. (2005). Large-scale localization from wireless signal strength. *AAAI*, 15–20.

Li, S. Z., & Jain, A. K. (2005). *Handbook of Face Recognition.* New York: Springer.

Lowe, D. (2004). Distinctive Image Features from Scale-Invariant Keypoints. *International Journal of Computer Vision, 60*(2), 91–110.

MacKenzie, I. S. (1992). Fitts' Law as research and design tool in human computer interaction. *Human–Computer Interaction, 7*, 91–139.

MacKenzie, I. S., & Ware, C. (1993). Lag as a determinant of human performance in interactive systems. *Proceedings of the SIGCHI conference on human factors in computing systems*, 488–493.

Maes, P., Darrell, T., Blumberg, B., & Pentland, A. (1995). The ALIVE system: Full-body interaction with autonomous agents. *Proceedings of the Computer Animation*, 11.

Matsushita, N., & Rekimoto, J. (1997). HoloWall: Designing a finger, hand, body and object sensitive wall. *Proceedings of the ACM Symposium on User Interface Software and Technology (UIST)*, 209–210.

McNeill, D. (1992). *What gestures reveal about thought.* Chicago: The University of Chicago Press.

Mitchell, T. M. (1997). *Machine learning.* Boston: McGraw-Hill.

Mozer, M. C. (2005). Lessons from an adaptive house. In D. Cook & R. Das (Eds.), *Smart environments: Technologies, protocols and applications* (pp. 273–294). Hoboken, NJ: J. Wiley & Sons.

Nachman, L., Kling, R., Huang, J., & Hummel, V. (2005). The Intel mote platform: A Bluetooth-based sensor network for industrial monitoring. *Fourth International Symposium on Information Processing in Sensor Networks*, 437–442.

Neal, R., & Hinton, G. (1999). A view of the EM algorithm that justifies incremental, sparse, and other variants. In M. I. Jordan (Ed.), *Learning in graphical models* (pp. 355–368). Cambridge, MA: MIT Press.

Nielsen, J. (1994). *Usability engineering.* San Francisco: Morgan Kaufmann.

Norman, D. A. (2003). *Emotional design: Why we love (or hate) everyday things.* New York: Basic Books.

Oliver, N., & Horvitz, E. (2005). Selective perception policies for guiding sensing and computation in multimodal systems: A comparative analysis. *Computer Vision and Image Understanding, 100*(1–2), 198–224.

Oviatt, S. L. (2002). Breaking the robustness barrier: Recent progress on the design of robust multimodal systems. In M. Zelkowtiz, *Advances in Computers, Volume 56* (pp. 305–341). : Academic Press.

Patten, J., Ishii, H., & Pangaro, G. (2001). Sensetable: A wireless object tracking platform for tangible user interfaces. *Proceedings of the*

Conference on Human Factors in Computing Systems (SIGCHI), 253–260.

Pentland, A. (2004). Social dynamics: Signals and behavior. *Third International Conference on Development and Learning*, 263–267.

Pentland, A., Moghaddam, B., & Starner, T. (1994). View-based and modular eigenspaces for face recognition. *Proceedings of the IEEE Conference on Computer Vision and Pattern Recognition*, 84–91.

Picard, R. W. (2000). *Affective computing*. Cambridge, MA: The MIT Press.

Pickering, J. (1986). Touch-sensitive screens: The technologies and their application. *International Journal of Man-Machine Studies, 25*(3), 249–269.

Platt, J. (1999). Using analytic QP and sparseness to speed training of support vector machines. *Advances in Neural Information Processing Systems, 11*, 557–563.

Rabiner, L., & Juang, B. (1993). *Fundamentals of speech recognition*. Englewood Cliffs, NJ: Prentice Hall.

Rabiner, L. R. (1989). A tutorial in hidden Markov models and selected applications in speech recognition. *Proceedings of the IEEE, 77*(2), 257–386.

Rekimoto, J. (2002). SmartSkin: An infrastructure for freehand manipulation on interactive surfaces. *Proceedings of the SIGCHI conference on human factors in computing systems*, 113–120.

Rekimoto, J., & Ayatsuka, Y. (2000). CyberCode: Designing augmented reality environments with visual tags. *Proceedings of the Designing Augmented Reality Environments (DARE 2000)*, 1–10.

Reynolds, C. (2001). *The sensing and measurement of frustration with computers*. Unpublished master's thesis. Media Laboratory, Massachusetts Institute of Technology, Cambridge.

Rubine, D. (1991). Specifying gestures by example. *Computer Graphics, 25*(4), 329–337.

Rui, Y., & Florencio, D. (2003). New direct approaches to robust sound source localization. *Proceedings of IEEE International Conference on Multimedia Expo*, 737–740.

Schapire, R. E. (2003). The boosting approach to machine learning: An overview. In D. D. Denison, M. H. Hansen, C. Holmes, B. Mallick, & B. Yu, (Eds.), *Nonlinear estimation and classification*, New York: Springer.

Schmidt, A. (2000). Implicit human interaction through context. *Personal Technologies, 4*(2&3), 191–199.

Schmidt, A., Beigl, M., & Gellersen, H. W. (1999). There is more to context than location. *Computers & Graphics Journal, 23*(6), 893–902.

Sears, A., Plaisant, C., & Shneiderman, B. (Eds.). (1992). A new era for high precision touchscreens. In (Eds.), *Advances in human computer interaction*. Norwood, NJ: Ablex.

Shell, J., Vertegaal, R., Cheng, D., Skaburskis, A. W., Sohn, C., Stewart, A. J., et al. (2004). ECSGlasses and EyePliances: Using attention to open sociable windows of interaction. *Proceedings of the ACM Eye Tracking Research and Applications Symposium*, 93–100.

Shen, C., Lesh, N., & Vernier, F. (2003). Personal digital historian: Story sharing around the table. *ACM Interactions, 10*(2), 15–22.

Shen, C., Vernier, F. D., Forlines, C., & Ringel, M. (2004). DiamondSpin: An extensible toolkit for around-the-table interaction. *Proceedings of the ACM Conference on Human Factors in Computing Systems (SIGCHI)*, 167–174.

Smith, A., Balakrishnan, H., Goraczko, M., & Priyantha, N. (2004). Tracking moving devices with the Cricket location system. *Proceedings of the 2nd USENIX/ACM MOBISYS Conference*, 190–202.

Smith, J. R. (1999). *Electric field imaging*. Unpublished doctoral dissertation. Media Laboratory, Massachusetts Institute of Technology, Cambridge.

Stork, D. G. (1998). *HAL's legacy: 2001's computer as dream and reality*. Cambridge, MA: MIT Press.

Sugiura, A., & Koseki, Y. (1997). A user interface using fingerprint recognition—holding commands and data objects on fingers. *Proceedings of the Symposium on User Interface Software and Technology*, 71–79.

Tan, H. Z., Slivovsky, L. A., & Pentland, A. (2001). A sensing chair using pressure distribution sensors. *IEEE/ASME Transactions on Mechatronics, 6*(3), 261–268.

Tapia, E. M., Intille, S., & Larson, K. (2004). Activity recognition in the home setting using simple and ubiquitous sensors. *Proceedings of Pervasive 2004*, 158–175.

Tashev, I., & Malvar, H. S. (2005). A new beamformer design algorithm for microphone arrays. *Proceedings of International Conference of Acoustic, Speech and Signal Processing*, 101–104.

Tenneson, D., & Becker, S. (2005). ChemPad: Generating 3D molecules from 2D sketches. *SIGGRAPH 2005 Extended Abstracts*.

Tobii Technology. http://www.tobii.com. Retrieved February 10, 2007.

Tomasi, C., Rafii, A., & Torunoglu, I. (2003). Full-size Projection Keyboard for Handheld Devices. *Communications of the ACM, 46*(7), 70–75.

Ullmer, B., Ishii, H., & Glas, D. (1998). MediaBlocks: Physical containers, transports, and controls for online media. *Computer Graphics, 32*, 379–386.

Vogel, D., & Balakrishnan, R. (2004). Interactive public ambient displays: Transitioning from implicit to explicit, public to personal, interaction with multiple users. *Proceedings of the ACM Symposium on User Interface Software and Technology (UIST 2004)*, 137–146.

Vogel, D., & Balakrishnan, R. (2005). Distant freehand pointing and clicking on very large high resolution displays. *Proceedings of the ACM Symposium on User Interface Software and Technology*, 33–42.

Vranish, J. M., McConnell, R. L., & Mahalingam, S. (1991). "Capaciflector" collision avoidance sensors for robots. *Robotics research at the NASA/Goddard Space Flight Center, 17*(3), 173–179.

Want, R., Fishkin, K. P., Gujar, A., & Harrison, B. L. (1999). Bridging physical and virtual worlds with electronic tags. *SIGCHI '99*, 370–377.

Want, R., Hopper, A., Falcao, V., & Gibbons, J. (1992). The active badge location system. *ACM Transactions on Information Systems (TOIS), 10*(1), 91–102.

Ward, A., Jones, A., & Hopper, A. (1997). A new location technique for the active office. *IEEE Personal Communications, 4*(5), 42–47.

Ware, C. (1990). Using hand position for virtual object placement. *Visual Computer, 6*(5), 245–253.

Ware, C., & Jessome, D. R. (1988). Using the Bat: A six-dimensional mouse for object placement. *Proceedings of the IEEE Computer Graphics and Applications*, 65–70.

Wayman, J., Jain, A., Maltoni, D., & Maio, D. (Eds.). (2004). Biometric systems: Technology, design and performance evaluation. New York: Springer.

Welch, G., & Bishop, G. (2004). *An introduction to the Kalman Filter*. (Dept. Computer Science Technical Report TR-95-041). Chapel Hill, NC: University of North Carolina at Chapel Hill.

Wellner, P. (1993). Interacting with paper on the DigitalDesk. *Communications of the ACM, 36*(7), 87–96.

Wilson, A. D. (2005). PlayAnywhere: A compact tabletop computer vision system. *Proceedings of the 18th Annual ACM Symposium on User Interface Software Technology*, 83–92.

Wilson, A. D., & Pham, H. (2003). Pointing in intelligent environments with the WorldCursor. *Proceedings of Interact*, 495–502.

Wilson, A. D., & Shafer, S. (2003). XWand: UI for intelligent spaces. *Proceedings of SIGCHI*, 545–552.

Wilson, D. H., & Atkeson, C. G. (2005). Simultaneous tracking and activity recognition (STAR) using many anonymous, binary sensors. *Proceedings of Pervasive 2005*, 62.

Wren, C., Azarbayejani, A., Darrell, T., & Pentland, A. (1995). Pfinder: Real-time tracking of the human body. *Proceedings of SPIE Photonics East*, 89–98.

Wu, M., & Balakrishnan, R. (2003). Multi-finger and whole hand gestural interaction techniques for multi-user tabletop displays. *Proceedings of the 16th Annual Symposium on User Interface Software and Technology*, 193–202.

Wu, Y., Toyama, K., & Huang, T. (2000). Wide-range, person- and illumination-insensitive head orientation estimation. *Proceedings of the International Conference on Face and Gesture Recognition*, 183.

Zhai, S., & Bellotti, V. (Eds.). (2005). Special Issue on Sensing-Based Interaction. *ACM Transactions on Computer-Human Interaction*.

Zhai, S., Morimoto, C., & Ihde, S. (1999). Manual and gaze input cascaded (MAGIC) pointing. *Proceedings of the SIGCHI conference on human factors in computing systems*, 246–253.

Zhang, Z., & Hershey, J. (2005). *Fusing array microphone and stereo vision for improved computer interfaces* (Microsoft Research Technical Report MSR-TR-2005-174) Redmond, WA: Microsoft Corporation. Smart Technologies, Incorporated. http://www.smarttech.com. Retrieved February 10, 2007.

Zhao, W., Chellappa, R., Phillips, R. J., & Rosenfeld, A. (2003). Face recognition: A literature survey. *ACM Computing Surveys, 35*(4), 399–458.

·11·

VISUAL DISPLAYS

*Christopher Schlick, Martina Ziefle,
Milda Park, and Holger Luczak*
RWTH Aachen University

INTRODUCTION

Since the existence of mankind, humans have been inventing and building tools to make life easier and more comfortable. One of mankind's first tools with a visual display was the sundial, which was invented more than 3,000 years ago in Babylon. Due to its basic physical principle, one of its significant disadvantages was that it could not show the time during the night or when the weather was cloudy. Moreover, the first sundials were in public places, so that people had to make the effort to go there in order to check the time. Later developments of clocks (also in public places) eliminated the disadvantages of the sundial. Church clocks could not only show the time under any weather condition and at any time of day or night, but they could also display the time acoustically and therefore bridge distances up to a few miles. The following developments are well known–from the grandfather clock and the fob watch continuing with the first analog and digital wristwatches to today's high performance wrist computers with high-resolution wrist-worn visual displays (Luczak & Oehme, 2002). They do not only display the time accurately, but they are truly multifunctional. For instance, some can visualize individual heart-rate profiles, and others can be used to chart one's precise location on earth with the help of the global-positioning system. A few can even replace personal digital assistants.

BASIC PRINCIPLES AND CONCEPTS

Image Generation

Different visual displays rely on different physical principles to generate an image: light can be emitted, transmitted, or reflected by the display. Examples from noncomputer displays might be helpful to explain the basic principles and point out relevant advantages and restrictions.

Writing on a piece of paper alters the reflective properties of the paper from a highly reflective white to a less reflective blue or black. Ambient light is needed to read what is written on the paper, but the contrast ratio between text and background remains the same under different lighting conditions. The same principle holds for reflectance-based displays such as liquid crystal displays (LCDs) without backlighting or electronic ink and paper (see sections "Nonemitter/Liquid-Crystal Display" and "Electronic Ink and Electronic Paper.")

Transparency film used with an overhead projector and slides used with a projector are examples of transmission. Different parts of the transparency or slide transmit light of different wavelengths (i.e. color) in different amounts. Although transmission is used as a basic principle in many electronic information displays, it is also often combined with other physical principles. For instance, a front-projection display (see section "Classification of Electronic Information Display Technologies") consists of a light source, an LCD generates the image and transmits the rays of light, and finally the projection surface reflects the light into the eye of the observer.

A lighthouse is an example of emission. Whereas its light can be easily seen at night, it is barely visible in bright daylight. Examples of emission-based displays are CRTs (see section "Cathode Ray Tube") electroluminescent displays (see section "Electroluminscent Displays"), and cathodoluminescent displays (see section "Cathodoluminescent Displays"). Similar to a lighthouse, these displays need to be brighter than the ambient light to be perceived properly.

Electronic Information Displays

The development of advanced display technologies began with the cathode ray tube or CRT, which was first discovered in the nineteenth century, although the observation of a glow from the electronic excitation of gas in an evacuated tube may go back as early as the 17th century. The invention of the device itself is generally attributed to Karl Ferdinand Braun. The "Braun tube" reportedly first built in Strasbourg, Germany, in 1896 or 1897, used both deflection and a fluorescent material for the screen. It was probably the first application of an electronic information display in natural sciences (Castellano, 1992). CRT displays have dominated the market for many years since their invention. However, the increasing requirements of consumers have led to rapid developments of alternative concepts for electronic information displays. Today, a large variety of display technologies are competing with CRTs in terms of image quality, and new technologies, which are currently under development, may soon cause more major changes in display demand (see section "Classification of Electronic Information Display Technologies").

There are two basic methods for displaying information visually: (a) digital and (b) analog. A digital display uses binary numbers of arbitrary length for displaying symbols such as characters or icons, whereas an analog system uses a continuous signal spectrum for information presentation. If an instant impression is required, analog displays often present information better. Many people glance quickly at their analog watch and know roughly what the time is or at an automobile dashboard and know that they are driving too fast. Analog displays translate a value of a continuous variable into an angle or a length. Analog displays used for control devices consist of a scale and an indicator or hand. Either the scale or the hand moves. There are a number of guidelines for the design of analog displays (Woodson, 1987; Baumann & Lanz 1998).

When accuracy is a critical issue, however, digital displays are preferred. Reading analog meters accurately requires time and cognitive skills, whereas writing down the value on a digital display is merely a case of copying down the numbers. In cases where both accuracy and quick reckoning are required, hybrid displays are often used.

Because a computer is a digital device, all data displayed are binary numbers, and therefore all commonly used electronic information displays are digital displays. Nonetheless, some application programs mimic analog display devices because of their ergonomic advantages. If the spatial resolution and the refresh rate of the digital display are sufficiently high and the covered color space is sufficiently large, no significant differences in visual performance from an analog display occur (see section "Resolution").

Display Segmentation

Common personal computer (PC) displays have a sufficiently high resolution to display a virtually unlimited set of characters, symbols, and graphics. Conversely, a wide variety of displays with lower spatial resolution is found on other kinds of computerized technical devices, such as mobile phones or music players. Here, the set of displayable tokens is often far more restricted.

Basic display elements are binary, whether they are on or off. A ternary display element can be built from a two-color light-emitting diode displays (LED). To transmit more information, display elements with more discrete states are necessary using i.e., different colors or brightness levels to distinguish between states.

A more common way to increase the amount of displayable information is the grouping of a number of binary display elements into a unit. The classical seven-segment display (Fig. 11.1) and many custom displays are examples of this approach.

If multiple binary display elements are arranged in matrix form and the elements are enhanced to display multiple colors and different levels of brightness, the basic concept of the common CRT and flat-panel displays can be derived. The shape of picture elements (pixels) in current electronic information displays is typically square or rectangular. The pixels are arranged in horizontal rows and vertical columns, which are called "pixel matrices" or "pixel formats."

Display Dimensionality

Visual displays can be distinguished by means of the spatial dimensionality of the image generated. Clearly, as the price of the display and the complexity of the display technology increase, the more dimensions are used.

The simplest way to display information is to rely on only one spatial dimension. Although the notion of one dimensional (1D) display is not strictly correct since human perception requires a display unit to have two spatial dimensions, the secondary dimension of such displays does not provide information and is simply a function of the primary dimension. 1D display represents the minimal approach to presenting information, which is encoded by either discrete or continuous variables in one dimensional states space. Nevertheless, a simple source of light, such as a bulb or a candle, can convey a binary message of arbitrary length. For most current applications, for example, to show the status of a device as "on" or "off," LEDs are used. LEDs give off light radiation when biased in the forward direction. Most light emitting diodes function in the near infrared and visible ranges, with the exception of ultra violet LEDs. When displaying a continuous variable in one dimension, a control window usually indicates the actual level of the

parameter, which is measured or controlled. The volume level of a speaker or the temperature in a room can be indicated using a row of pixels, LEDs, or a slider on a scale.

In order to display more complex information, which can only be shown using two independent spatial dimensions (length and width), two-dimensional (2D) displays are used. This category of displays has by far the biggest segment in the market. They are suitable for displaying all kinds of graphical information. Images are formed by activating pixels, each of which has a unique location (x, y) within a two-dimensional plane, and a color and gray scale value may be assigned to each. Although these displays have only two dimensions, stereoscopic perception is also feasible using depth cues such as relative size, height relative to the horizon, interposition or occlusion, shadows and shading, spatial perspective, linear perspective, and texture gradients. Even more realistic three-dimensional (3D) scenes can be viewed on 2D screens by applying special viewing devices, such as shutter glasses or polarizing glasses. In order to simulate depth perception, each of the eyes must see a slightly different image (binocular parallax). For instance, the so-called shutter glasses alternately block each eye's view so that each perceives only the image intended for it. The alternation of images occurs many times per second and these separate images are fused into one true 3D image at higher cortical levels. The speed of the shutters is directly proportional to the refresh rate of the monitor. In the case of polarising glasses, different light polarisation modes are used to encode the information for the left and the right eye, respectively. The polarizing glasses filter out the unwanted photon polarization from each view and two separate images are projected onto the visual cortex. These techniques are used with immersive desks or a Computer Animated Virtual Environment (CAVE).

One of the limitations of such stereoscopic displays is that motion parallax, as one of the depth cues, is not provided automatically. Display devices that are able to provide the correct stereoscopic perspective to each of the viewer's eyes over a range of viewing positions can be defined as auto-stereoscopic. Volumetric 3D displays fall into this category. There is a growing need to present complex and voluminous 3D information in such way that it may be interpreted rapidly, naturally, and accurately. Volumetric displays (also called direct volume display devices) permit 3D-image data to be presented in a transparent volume. Innate depth cues inherent in three-dimensional objects are then automatically present, and, in principle, the three-dimensional images may be viewed from an arbitrary direction. Volumetric display systems provide a transparent volume in space where animated images may be depicted. Since these images have three physical dimensions, depth cues are inherent without the need to compute complex transformations, which is a considerable benefit of this type of display. Furthermore, images may be observed naturally without the need to wear special glasses. Common to all volumetric display systems is a volume or region occupying three physical dimensions within which image components may be positioned, depicted, and possibly manipulated. Volumetric display systems employ a method for image construction similar to 2D displays. Elements, which are called "voxels," are activated at the appropriate locations within the 3D space. Each voxel has a unique position (x, y, z) and can be assigned a color and gray-scale value. Depending on the

FIGURE 11.1. Seven-segment display.

technology used to activate the voxels, the volumetric displays can be assigned to several categories, such as emissive volume display, varifocal mirror display, laser displays, or holographic displays. A more detailed description of volumetric display technology can be found in Blundel and Schwarz (2000).

Human-Related Display Functionality and Other Criteria

The paradigm of wearable computing requires the visual displays to be worn on the human body (Azuma, 1997; Feiner, 2002). A popular approach for mobile applications is to mount the display on the user's head. Thus, the user can work hands-free and is always able to perceive information in his or her field of vision. These displays are commonly referred to as head-mounted displays (HMDs). From the technological point of view, HMDs can roughly be split into three main categories, according to the way the image is provided to the user (Fig. 11.2).

The first category, referred to as "screen-based HMDs," comprises all HMDs whose picture elements are created in a spatially adjacent way. The ocular image forming displays use technologies such as CRT, LCD, digital mirror devices (DMD), or organic light-emitting diodes (OLEDs). Most of today's HMDs on the market, however, are based on transmissive or reflective liquid-crystal minidisplays (LCD) (Holzel, 1999). Miniature LCDs are available for relatively low prices and provide an appropriate resolution (SXGA resolution of up to 1280 × 1024 pixels), while being lightweight (von Waldkirch, 2005).

In the retinal-projection method, the display image is projected directly onto the retina in the same way as a slide is projected onto the screen. Projection displays are normally designed in the form of a Maxwellian-view optical system (Bass, 1995), where the screen plane is optically conjugated to the retina and the illumination source is conjugated to the eye's pupil plane. Consequently, such displays can only be implemented on the basis of illuminated screens (like LCD and DMD), and not with self-emitting technologies like OLED and CRT (von Waldkirch, 2005).

Scanning displays, where the image is scanned pixel by pixel directly onto the retina, are an alternative to screen-based displays. Retinal scanning displays (RSDs), also referred to as virtual retinal displays (VRDs), are most important in this category. VRD technology was first proposed in 1992 by Sony. Since 1992, researchers at the Human Interface Technology Lab (Washington, DC) have been developing this technology to obtain a commercial product. In 2003, they presented the first commercial VRD, called "Nomad," together with the U.S.-based company Microvision. With this technology, an image is scanned directly onto a viewer's retina using low-power red, green, and blue light sources, such as lasers or LEDs (Urey, Wine, & Lewis, 2002). The VRD system has superior brightness and contrast compared to LCDs and CRTs, as it typically uses spectrally pure lasers as its light source (Stanney & Zyda, 2002). Finally, a combination of a scanning technology and a screen-based system is possible (see Pausch, Dwivedi, & Long, 1991; Aye, Yu, Kostrzewski, Savant, & Jannson, 1998; Fruehauf, Aye, Yua, Zou, & Savant, 2000).

Beside technological categorization, the various HMD concepts can be classified according to their functionality or intended purpose. Fig. 11.3 shows a rough functional categorization of HMDs (von Waldkirch, 2005). Here, HMDs can be divided into monocular, biocular, and binocular displays. In addition, all these types can provide the image in either a closed-view or see-through mode.

Monocular displays have one display source and thus provide the image to one eye only. Therefore, in comparison with biocular or binocular HMDs, monocular displays are lighter and

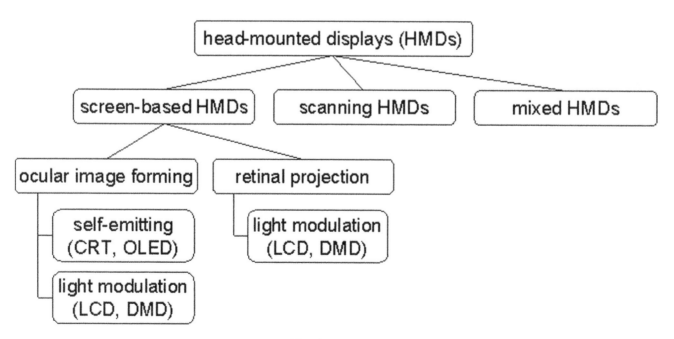

FIGURE 11.2. Systematics of head-mounted displays (von Waldkirch, 2005).

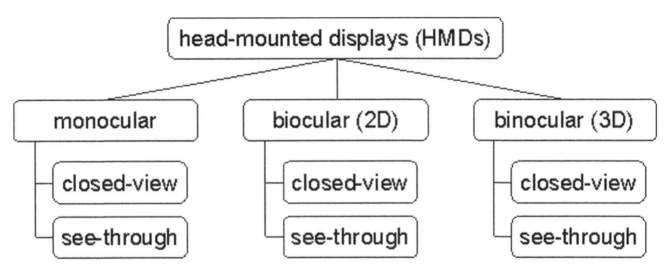

FIGURE 11.3. Head-mounted displays categorized according to functions (von Waldkirch, 2005).

cheaper. However, in a monocular HMD, the view seen by the eye, which is not blocked by a screen, may produce binocular rivalry with the image seen through the HMD. Biocular HMDs have two displays with separate screens and optical paths, allowing both eyes to see exactly the same images simultaneously. Thus, the user perceives a 2D image only, similar to a computer screen. Binocular displays allow stereoscopic viewing with 3D-depth perception. To produce the stereoscopic view, two spatially slightly incongruent images are provided to the left and to the right eye (binocular parallax, see section "Human Related Display Functionality and Other Criteria").

Beside the HMDs, important categories of displays are wrist-worn and handheld displays. These displays are very widespread, because they are integrated in wristwatches, mobile phones, personal digital assistants (PDAs), media players, and other portable devices (see section "Mobile Phones and Handheld Devices").

A truly "wearable" visual display was introduced recently by Philips Research and Textile Institute TITV Greiz. The company has been developing photonic textiles–fabrics that contain lighting systems and therefore can serve as electronic information displays. Multicolored light-emitting diodes (LEDs) were successfully integrated into fabrics without compromising the softness of the cloth. Passive matrixes of 10 10 red, green, and blue LED packages were placed on interconnecting substrates, made entirely of cloth. Applied in soft fabric, the light from the smallest pixels diffuses, resulting in a more or less continuous light-emitting surface. Photonic textiles can also be made interactive. Philips has achieved interactivity by incorporating sensors (such as orientation and pressure sensors) and communication devices (such as Bluetooth, GSM) into the fabric. Photonic textiles open up a wide range of possible applications in the fields of communication and personal health care (Philips Research Press Information, 2005).

QUALITY CRITERIA

Display Colors

Different displays can present differing numbers of colors. The simplest displays have two pixel states, either on or off and, therefore, are monochrome. The greater the difference between these two states, expressed as a contrast ratio or the weighted wavelength differences of the light falling upon the observer's eye, the more easily the viewer can distinguish between them. More complex displays, called gray-scale displays, are also able to present different luminance levels of pixels.

Clearly, all modern electronic information displays reproduce color as well (Hanson, 2003). The Commission International de l'Eclairage (CIE) chromaticity diagram is often used to represent a display's color capabilities. The diagram is based on the CIE standard XYZ system. X, Y, and Z are hypothetical primary colors. Fig. 11.4 depicts the equal energy-matching functions of the standard XYZ system. The XYZ primaries were chosen so that whenever equal amounts of them are combined, they match white light.

Independently of the absolute value of X, Y, and Z, relative amounts—denoted by lowercase letters—can be used to describe a color. Because x, y, and z are relative amounts, $x + y + z = 1$. Consequently, when x and y are known, z is known as well, because $z = 1 - (x + y)$. Therefore, a curve can be used to depict all the possible combinations of the XYZ primaries, as shown here in Fig. 11.5. At the intersection of $x = 0.33$ and $y = 0.33$ pure white (i.e. achromatic light) can be found. The curved line, the locus of spectral colors, represents the x, y, and z values for all of the spectral colors. The hue of any color within the curved line can be determined by drawing a line from pure white

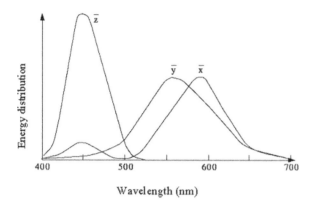

FIGURE 11.4. Equal energy-matching functions of the standard XYZ system. The curves $\overline{X}$, $\overline{Y}$, and $\overline{Z}$ show the relative amounts of the X, Y and Z primaries needed to match the color of the wavelength of light (cf. Kaufman, 1974).

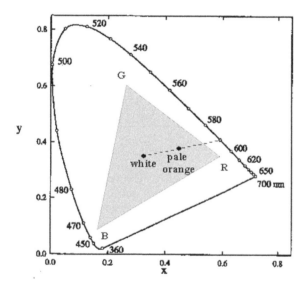

FIGURE 11.5. The CIE chromaticity diagram. The curve, the locus of spectral colors, represents the x, y and z values for all of the spectral colors between 380 nm and 770 nm. The triangle depicts the colors that can be presented by a display that uses the three primary colors at the corners of the triangle. The hue of any color within the curved line can be determined by drawing a line from pure white ($x = 0.33$ and $y = 0.33$) through the color of interest to the locus line. A pale orange ($x = 0.45$ and $y = 0.35$) is marked by a black dot as an example. The saturation is given by the relative length of the dashed line between white and the color of interest and white and the intersection of this line with the locus (cf. Kaufman, 1974).

through the color of interest to the locus line. The saturation is given by the relative length of the line between white and the color of interest, and white and the intersection of this line with the locus (see Fig. 11.5).

The representation of this information in the computer's hardware and software places a further restriction on the variety

of colors. Popular configurations devote 4 to 32 bits to the representation of color information, yielding 16 to 16,777,216 colors with different hues and saturation.

Because x and y are relative values, an additional parameter, not shown in the Fig. 11.5, is needed to specify the luminance. It is denoted by Y, and hence this system of color characterization is referred to as the "xyY-system."

In order to depict a great variety of different colors, common displays use the three primary colors: red, green, and blue. The actual scope of colors is restricted by the position of each of the three primaries in the CIE chromaticity diagram and lies within the triangle defined by these primaries (see Fig. 11.5).

Whereas the CIE chromaticity diagram is useful in depicting the color space, additional systems are used to describe color in the context of human–computer interaction (HCI). The specification of *hue, saturation, and brightness* (HSB; sometimes instead of brightness the terms *luminance* or *lightness* are used) provides a device-independent way to describe color. For this model, the color space is an upside-down cone (see Fig. 11.6). On the edge of the cone base, the visible light spectrum is arranged in a circle by joining red and violet. Hue is the actual color and can be specified in angular degrees around the cone starting and ending at red $= 0°$ or $360°$, in percent or eight-bit values (0 to 255). Saturation is the purity of the color, measured in percent or eight-bit values from the center of the cone (min.) to the surface (max.). At 0% saturation, hue is meaningless. Brightness is measured in percent or eight-bit values from black (min.) to white (max.). At 0% brightness, both hue and saturation are meaningless.

RGB stands for the three basic colors—red, green and blue—that are produced by the visual display. A number of other colors can be produced by *additive color mixing*. If any two of the color channels are mixed in equal proportions, new colors are

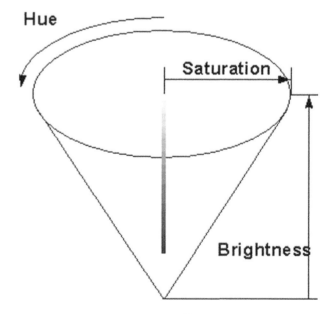

FIGURE 11.6. Hue, saturation and brightness system for specifying color.

created: blue and green combine to create cyan (bright, light blue); red and blue make magenta (a bright pink); and red and green make yellow. If all three colors are mixed together equally at full power, the result is white light. Colors can be specified by giving the relative power of each color in percent or as an 8-bit number (0 to 255 in decimal or 00 to FF in hexadecimal). So, for example, (0, 0, 0) specifies black; (255, 0, 0), a pure red; and (255, 255, 255), white.

CMYK model defines cyan, magenta, yellow, and black as main colors. This system is used to specify colors on the monitor for printing. In printing, colors are mixed subtractively and, using red, green, and blue, it would not be possible to produce many colors. By choosing cyan, magenta, and yellow as basic colors, many other colors, including red, green, and blue, can be produced. Theoretically, when all three basic colors are printed over each other, the resulting color should be black. In practice, this is not the case, and a fourth printing process with black ink is also used. Values for CMYK are often specified as percentages.

Brightness and Glare

Brightness can be defined as the perceived amount of light that comes from an object. It is a physiological interpretation of luminance (L), an important photometric quantity. It is a measure of the luminous intensity of light (I) emitted from a light source per unit surface area (A) normal to the direction of the light flux (Çakir, Hart, & Stewart, 1979). In this context, θ is the angle between the perpendicular from the surface and the direction of measurement (Boff & Lincoln, 1988):

$$L = \frac{I}{A}\cos\theta \qquad (11.1)$$

The unit of luminance is lumen per steradian per square meter or candela per square meter (cd/m²). In some cases the unit foot-lambert (fL) is used to describe luminance (1 fL = 3,426 cd/m²).

The variable I is the luminous intensity (also called "radiant intensity"), which represents the emitted flux-per-unit solid angle Ω:

$$I = \frac{d\Phi}{d\Omega} \qquad (11.2)$$

Luminous intensity is measured in candela (cd); the unit is lumen per steradian.

Brightness of a display depends not only on the optical power generated, transmitted, or reflected by the display but also on the response of the human eye at certain wavelengths. The measure for this relation is photopic luminosity, $K(\lambda)$, which relates optical power in watts at a given wavelength to its effect on the human visual system. This luminosity can be considered as the optical spectral response of the eye of a "standard observer" (Nelson & Wullert, 1997). Mathematical relation of luminance to the spectral distribution of optical power $P(\lambda)$

can be expressed using the following equation, in which the integration is performed over the range of visible wavelengths (Macadam, 1982):

$$\Phi = \int_{\lambda=380}^{\lambda=770} K(l)P(l)dl \qquad (11.3)$$

The human eye cannot collect all of the light that is radiated or reflected from the source. Brightness also depends on the size of the surface spot, which the light is emanating from. Figure 11.7 shows how the brightness of a surface is actually given by the luminous flux per unit of the projected area of the emitting surface per unit solid angle depending on the viewing angle.

Some displays, such as LCDs, appear dimmer from an oblique angle than from the normal viewing angle; whereas most emissive displays, such as CRTs, emit light in such a way that the angular luminous intensity approximately follows Lambert's cosine law (Lambertian surface), resulting in approximately constant luminance across all viewing directions.

High luminance levels in the field of view cause glare discomfort. Glare caused by light sources in the field of view is called "direct glare"; glare caused by light being reflected by a surface in the field of view is called "reflected glare" (Fig. 11.8).

Reflected glare can occur from specular (smooth, polished, or mirror-like) surfaces, spread (brushed, etched, or pebbled) surfaces, diffuse (flat or matt) surfaces or as a combination of the above three (compound) (Sanders & McCormick, 1993). Glare sources are more disturbing when they have higher luminance and when they are closer to the fixation point (Sheedy, 2005). Experiments also show that visibility is decreased by glare, and the decrease is greatest when the source of the glare is in the line of vision (Boff & Lincoln, 1988).

In order to avoid glare, it is advisable to position the display right-angled to the window (so that the line of vision is parallel to the window). The display can be protected with curtains, blinds, or movable walls. Lamps that can be reflected in the monitor must not have a mean luminance of more than 200 cd/m², and the maximum luminance must be less than 400 cd/m² according to German standard DIN 5035-7 (Fig. 11.9).

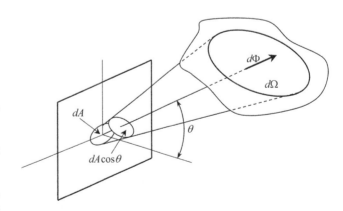

FIGURE 11.7. Definition of the brightness of a surface as a function of the direction from which the surface is observed (Nelson & Wullert, 1997).

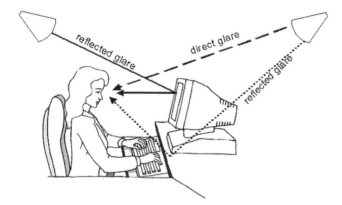

FIGURE 11.8. Direct and reflected glare.

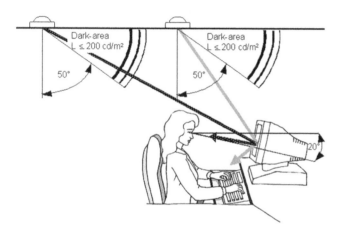

FIGURE 11.9. Glare can be avoided by limiting lamp luminance and by positioning the display correctly.

Resolution

The arrangement of pixels into horizontal rows and vertical columns is defined as the *pixel format*. It is often referred to as a "resolution." However, this is not a resolution parameter by itself. The resolution of a monitor mainly depends on its screen diagonal and its dot pitch (stripe pitch, SP). When set to lower resolutions, a pixel encompasses multiple dots. Thus, defining the resolution pixel density, or the number of pixels per linear distance (pixels per inch or pixels per centimeter) plays an important role. This parameter indicates how close the pixels are.

As a picture element, the pixel is the basic unit of programmable color on a display or in a digital image. In color CRT, the pixel is formed from a number of phosphor dots and may consist of a number of triads that are composed of red, green, and blue phosphor dots. The dot pitch in a CRT display with a shadow mask is defined as the distance between the holes in the shadow mask, measured in millimeters (mm). In desktop monitors, common dot pitches are 0.31 mm, 0.28 mm, 0.27 mm, 0.26 mm, and 0.25 mm. In matrix-driven, flat, panel displays, every single pixel is composed of a red, a green, and a blue phosphor

dot or filter element, and dot pitch is defined as the center-to-center distance between adjacent green phosphor or filter element dots. Thus, the pixel density is the reciprocal of pixel pitch, which is equal to the dot pitch.

There is a simple relationship between resolution, in pixel/inch, and viewable diagonal screen size in inches for different monitor pixel formats. The resolution quality levels are defined as follows (Castellano, 1992):

- Low resolution: <50 pixels/inch
- Medium resolution: 51 to 70 pixels/inch
- High resolution: 71 to 120 pixels/inch
- Ultrahigh resolution: >120 pixels/inch

The highest resolution (x_{max} y_{max}) that can be represented on a CRT monitor can be calculated as follows:

$$x_{max} = \frac{D_{eff}\cos\alpha}{SP} \qquad (11.4)$$

D_{eff} is the effective screen diagonal, where the picture can be viewed. The angle α (angle between the horizontal and the screen diagonal) can be calculated using knowledge about the aspect ratio between the vertical (V) and horizontal (H):

$$\alpha = \arctan\left(\frac{V}{H}\right) \qquad (11.5)$$

As an example, a 17″ monitor (D_{eff} = 15,8 inches) with a horizontal stripe pitch of SP = 0.26 mm and an aspect ratio $H:V$ = 3:4 (common computer monitor) a maximum number of dots presentable in the horizontal of 1,235 or 926 dots in the vertical is possible. For our example, this means that with a resolution of 1,024 × 768 on the graphic card, each dot that has to be presented needs to be provided with at least one phosphorus wire. However, with a resolution of 1,280 × 1,024, not all dots can be represented clearly.

Contrast and Gray Scales

Strictly speaking, contrast is not a physiologic unit, but is nevertheless one of the most important photometry quantities. It is related to a lot of visual performance criteria, such as visual acuity, contrast sensitivity, speed of recognition, and so forth. There are different definitions of contrast.

The *Weberscher Contrast* C_w (luminance contrast) in accordance with the International Lighting Commission is defined as follows:

$$C_w = \frac{L_{object} - L_{background}}{L_{background}} \qquad (11.6)$$

The definition implies that a negative luminance contrast is created by presenting a dark object on a bright background and a positive contrast is created by presenting a bright object on a dark background (Ziefle, 2002). This formula is not symmetric; if the luminance of the object is much greater than the background luminance, the numerical values of contrast are large and increase rapidly. If the background luminance is much greater than the object luminance, the luminance contrast tends to move asymptotically to the value $C_w = -1$.

The *Michelson Contrast* C_m (also called "modulation contrast," "depth of modulation," or "relative contrast about the mean luminance") is generally used for periodic stimuli that deviate symmetrically above and below a mean luminance value, for example, gratings or bar patterns, and is computed as follows (Boff & Lincoln, 1988):

$$C_m = \frac{L_{max} - L_{min}}{L_{max} + L_{min}} \qquad (11.7)$$

L_{max} and L_{min} are maximum and minimum luminance in the pattern. The Michelson Contrast will take on a value between 0 and 1. The minimum modulation threshold, which means the smallest brightness modulation that can be detected, occurs at a brightness modulation of approximately 0.003 at about 3 cycles per degree of the visual field (Nelson & Wullert, 1997).

Sanders and McCormick (1993) specified another popular possibility for defining contrast (called *luminous contrast*) as the difference between maximum and minimum luminance in relationship to maximum luminance:

$$C_l = \frac{L_{max} - L_{min}}{L_{max}} \qquad (11.8)$$

For a simple luminance increment or decrement relative to background luminance (such as a single point), the *contrast ratio* is used (Boff & Lincoln, 1988). This simple ratio of two levels of luminances is widely used in engineering (Bosman, 1989):

$$C_r = \frac{L_{object}}{L_{background}} \qquad (11.9)$$

These contrast measures can be converted from one to another. For example, given the contrast ratio C_r, and knowledge of positive contrast conditions ($L_{max} = L_{object}$; $L_{min} = L_{background}$), the other visual contrasts can be calculated as follows:

Weberscher Contrast:

$$C_w = C_r - 1 \qquad (11.10)$$

Michelson Contrast:

$$C_m = \frac{C_r - 1}{C_r + 1} \qquad (11.11)$$

Luminous Contrast:

$$C_l = \frac{C_r - 1}{C_r} \qquad (11.12)$$

In a computer display, the contrast ratio of symbols to background has to be more than 3:1. The signs have to be represented sharply up to the edges of the screen. Modern LCDs have a contrast ratio of 500:1 or more. Since LCDs do not emit light, the luminance in the previous equation for contrast refers to the luminance of light either passing through the display (for a backlit transmissive type) or the luminance of the light re-

flected off the display's surface (for a reflective type LCD). In multiplexed LCDs, the contrast is affected by the viewing angle. Therefore, the contrast should be indicated by referring to the solid angle, known as the "viewing cone" (Castellano, 1992).

Large contrast ratios are also necessary in order to satisfy gray-scale requirements. Based on the idea of the brightest areas being white and the darkest areas being black, levels of brightness in between the two extremes are referred to as gray levels or gray shades, and the ability to display them as gray scale (Nelson & Wullert, 1997). Technically, gray scale is a term that should be applied only to monochrome or "gray" displays. The term is now often applied to color displays where intermediate brightness controls are required by the system (Castellano, 1992). The number of gray scales is determined both by the contrast level and the ability of the human-visual system to distinguish between the different brightness levels. Our visual system reacts to the changes in brightness level as a logarithmic function; therefore very small differences in brightness might be not perceived (Theis, 1999). The acceptable difference in brightness levels between scales is 1.414 (the square root of 2). In order to obtain five levels of gray scale above background, a display must have a contrast ratio of at least 5.6:1 (1, 1.41, 2, 2.82, 4 and 5.65:1) (Castellano). Full-color displays have about 128 to 256 linear gray levels. The number of gray shades (G) that can be displayed can be defined as a logarithmic function based on contrast ratio (Nelson & Wullert):

$$G = 1 + \frac{1}{\log \sqrt{2}} \log \left(\frac{L_{max}}{L_{min}} \right) \qquad (11.13)$$

Refresh Rates

The number of times that the image on the display is drawn per second is called the *display refresh rate* (also called *frame rate* or *vertical scanning rate*). The unit of measurement is Hertz (Hz). A high-display refresh rate prevents a flickering image, because there is only a small amount of time between two successive stimulations of a single dot. A refresh rate of between 70 and 80 Hz is needed to ensure a flicker-free image on the display (the flicker is not perceivable) (Bosman, 1989).

In CRT technology, the phosphorus dots (phosphorus stripes) are stimulated by the electron beam. Because they light up for only a fraction of a second, stimulation must occur several times a second. It passes line by line while the electron beam is writing from left to right and then returns to the beginning of the following line. Line build begins in the upper left corner, line by line, until each dot has been stimulated once. Then the electron beam goes back to the upper-left corner and begins to build up the picture again.

The line frequency (or horizontal frequency 2 measured in kHz) is another characteristic of CRT displays. It measures the number of lines the electron beam can draw per second. In this case, line frequency, display-refresh rate, and resolution are directly connected with each other. A monitor with a display-refresh rate of 70 Hz and a resolution of 1,024 × 768 needs to have an electron beam that is capable of drawing 70 × 768 lines = 53,760 lines per second. This means that the monitor has to process a line frequency of at least 53.76 kHz. If one wants to use a higher resolution, it is necessary to check whether the monitor is capable of processing such a line frequency; otherwise damage might occur.

In addition to the frequencies described, the image representation mode can also lead to poor-quality representation. There are two image representation modes: interlaced and non-interlaced mode. With the interlaced mode, the image formation is divided into two half images. Here, the electron beam first builds all uneven lines and then all even lines.

This is the common process for televisions (for Phase Alternation Line, PAL); 50 half-images per second are normal, which means a vertical scanning frequency of 50 Hz but a full screen frequency of only 25 Hz. Nowadays, monitors with a non-interlaced mode are commonly used. In this case, the electron beam builds all of the lines one after another, without exception.

In general, active or light emitting displays, such as CRTs, can be written more quickly than passive or reflecting displays, such as LCDs. The higher speed is usually related to increased power consumption. Multiplexed LCDs have refresh rates of about 40 to 50 frames per second. The physics of the display is such that it takes some time to turn on any pixel or line. Recent advantages in LCD technology have reduced response time to about 8–20 ms.

TECHNOLOGIES

Classification of Electronic Information Display Technologies

Direct-view displays. These are video displays in which the light produced by a display device is viewed directly without first bouncing off a screen. All CRT, LCD, plasma televisions, and computer monitors are direct-view displays. These displays tend to work best in bright-light conditions and have greater light output than projection displays.

Projection displays. Unlike a direct-view system, a projection display relies on the projection of an image onto a screen. There are front- and rear-projection systems, which mainly differ with regard to screen technology. The front projection utilizes a reflective screen surface while a rear projection uses a transmissive screen surface. Projection displays work best in dimly lit environments. In particular, a front projection set up requires a darkened room for optimum viewing quality.

Off-screen display systems. These display systems do not utilize any projection screen. Instead, a natural medium, like simple glass or even the retina, can be used for image projection. Off-screen systems are based either on coherent or non-coherent light emission. Coherence is a property of waves that measures the ability of the waves to interfere with each other. Usually, laser light has much higher coherence than nonlaser light. VRDs (see section "Laser Displays") and 3D holographic head-up displays are examples of off-screen display systems.

Cathode Ray Tube

The CRT is a cathodoluminescent display: light is generated by exciting a luminescent material with energetic electrons. A CRT consists of a glass bulb, a cathode, an electron gun, deflection yoke, a mask, and a phosphor coating (Fig. 11.11).

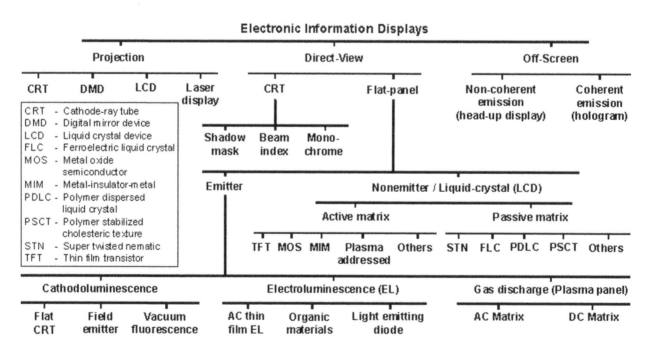

FIGURE 11.10. Classification of electronic information technologies with high information content (Theis, 1999).

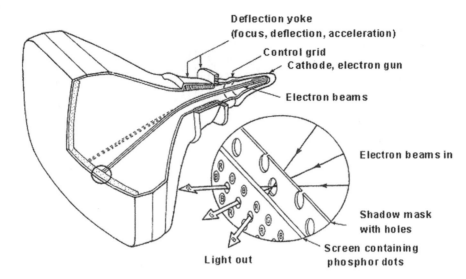

FIGURE 11.11.   Major parts and components of a cathode ray tube. In this example shadow mask technology is depicted (cf. Hanson, 2003).

An electron gun located in the back of the device emits negatively charged electrons, which are attracted and accelerated by an anode that is located in front of the screen. The electron beam is diverted by an electromagnetic field, built up by the deflection coils, and thus directed toward the screen. Electrons are extracted from the cathode by thermal emission from low-surface-potential materials (typically metallic oxides). The electron beam generated at the cathode is then accelerated, deflected, and focused by a series of electrostatic lenses and deflection coils.

A screen mask is attached in front of the ground-glass plate so that the electron beam is focused and then steered on the phosphorus layer and deposited on the front surface. As a consequence, the outside areas of the beam are prevented from mistakenly hitting adjoining phosphorus dots, which would lead to a blurred representation and chromatic distortion. Display-screen masks are either dot-mask screens, Trinitrons, or slot-mask screens.

Dot-mask screen.   As the name suggests, a thin metal or ceramic screen with multiple holes is used in a dot screen (Fig. 11.12 (a)). The shadow-mask technique is applied, in which three electron beams pass through the holes and focus on a single point on the tube's phosphor surface. Thus, the other display screen dots are shaded. The electron guns are arranged in the form of a triangle (delta gun), which requires a high amount of adjustment as insufficient adjustment will cause color defects. The measurement of the dot pitches of dot-type screens is taken diagonally between adjoining dots of the same color. Because of their horizontal measurement, they cannot be easily compared with the distance between the dots of a Trinitron or a slot mask screen (Blankenbach, 1999).

Slit mask (Trinitron technology).   Monitors based on the Trinitron technology use an aperture grid instead of a shadow mask (Fig. 11.12 (b)). The phosphorus surface does not consist of colored dots but instead consists of a multitude of tiny vertical phosphorus wires. Their arrangement is alternating, similar to the dot mask. The electron guns are on a single line, which makes adjustment easier. Instead of a dot-type screen, the Trinitron has vertically taut wires. In contrast to the dot-type screen, the equivalent of the dot pitch, the SP of these monitors is measured by the horizontal distance between wires. Slit masks are relatively insensitive to the warmth that develops during use, because they do not bend, but merely change their length. They are very sensitive to mechanical load (i.e., vibration). Depending on the size of the monitor, either one or two horizontal holding wires are used for stabilization. These wires are recognizable (i.e., on the Trinitron they are visible as small gray lines in the top and bottom third).

Slot-mask screen.   This screen technology (Fig. 11.12 (c)) also has tiny phosphor wires, which are similar to monitors based on Trinitron technology. The openings in the shadow mask are executed as slots (Precht, Meier, & Kleinlein, 1997).

CRT is a mature, well-understood technology, which is found in color and monochrome TV screens, projection TVs, vehicle displays, aircraft cockpit displays, marine instruments, VDT terminals, communication equipment, medical devices, military systems, etc. (Castellano, 1992). It offers high-information content at a low cost and is capable of displaying a large color gamut (>256 colors readily available). It is capable of high resolution and high-pixel-count displays are readily available. Direct-view displays can be made in diagonal screen sizes of up to 40 inches; projection systems using smaller tubes can be made for much larger screen viewing (Castellano). Other advantages of CRTs are an intrinsically high gray scale, good temporal dynamics, and consistent brightness for nearly all viewing angles (Candry, 2003). On the negative side, the color CRT is large and heavy, power consumption and operating voltage are high, and vibration robustness is low. Detrimental imaging characteristics include poor ambient contrast due to diffuse reflection from the phosphor surface, a propensity toward flickering, geometric distortion, and a correlation between spot size and image luminance

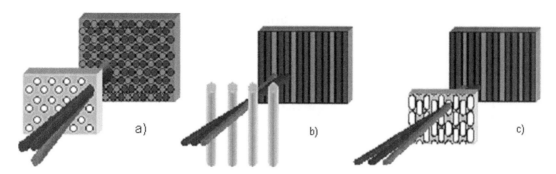

FIGURE 11.12. Dot-mask (a), slit mask (b) and slot mask (c) arrangement.

(Candry). The operation time of CRT displays is limited to about 10,000 hours of continuous operation. The main representation errors that occur in CRT displays are linearity error, convergence error, and moiré.

Linearity error. Geometric elements and letters should be represented at the same size everywhere on the screen. If displays are only adjustable vertically, problems can arise. A deflection error can cause the marginal, equispaced raster elements to disperse. Therefore, the figures' widths change constantly as they move from the middle of the screen.

Convergence error. Color representation is obtained by three electron beams and three phosphorus layers, one of each for the colors red, green, and blue. Each electron beam hits one layer. Congruent dots of the layers build a pixel that can be seen in any color, depending on the intensity of the beams. If the electron beams are not adjusted exactly, misconvergence is the consequence. "Ghostings" of the removed color arise, causing a blurred and distorted image.

Moiré. Certain color combinations and pixel arrangements can cause interference called moiré. The interference can result when a mask is deposited imprecisely, and electrons pass through the mask imprecisely. As a result, streaks are seen on the screen.

Nonemitter/Liquid-Crystal Display

Liquid Crystal Displays (LCD) have become increasingly viable in recent years and now tend to surpass CRTs. Early commercial developments of LCDs concentrated on small numeric and alphanumeric displays, which rapidly replaced LEDs and other technologies in applications such as digital watches and calculators (Bosman, 1989). Now there are even more complex displays for use in many applications, such as laptops, handheld computers, flat-panel displays (FPDs), HMDs, and miniature televisions (i.e. those in airplane seats). LCDs have the following advantages:

- Power consumption is low.
- They operate at low voltages.
- Lifetime in normal environments is very long.

- Displays may be viewed either directly in transmission or reflection or may be projected onto large screens.

LCDs consist of two glass plates with microscopic lines or grooves on their inner surfaces and a liquid crystal layer between them. Liquid crystal materials do not emit light, so external or back illumination must be provided. The physical principle is based on the anisotropic material qualities of liquid crystals. When substances are in an odd state that is somewhat like a liquid and somewhat like a solid, their molecules tend to point in the same direction, like the molecules in a solid, but can also move around to different positions, like the molecules in a liquid. This means that liquid crystals are neither a solid nor a liquid but are closer to a liquid state. LCDs operate by electrically modulating the anisotropy between optical states in order to produce visible contrast. In an electric field, liquid crystals change their alignment and therefore their translucence. If no voltage is applied, light can pass through and the pixels appear bright. When voltage is applied, the pixels become dark (Matschulat, 1999). The light to be modulated may either originate from ambient or an additional bright-light source placed behind the LCD (Theis, 1999).

The two principal flat panel technologies are the passive matrix LCD and active matrix LCD. Passive matrix addressing is used in twisted nematic (TN) LCDs. In TN LCD displays the microscopic lines of the glass plates are arranged orthogonally to each other, and the glass plates serve as polarizers (Precht et al., 1997). Their directions of translucence lie at right angles on top of one another, so that no light can pass through. Because of the fine grooves on the inner surface of the two glass panels (arranged vertically on one panel and horizontally on the other), the liquid crystal is held between them and can be encouraged to form neat spiral chains. These chains can alter the polarity of light. In the so-called nematic phase, the major axes of the crystal's molecules tend to be parallel to each other.

Nonpolarized light from background illumination can pass the polarization filter with just one plane of polarization. It is twisted about 90° along the helix and can thus pass through the second polarization layer. The display appears to be bright when there is no electric current. By applying an electric current to twisted nematics, they untwist and straighten, changing the angle of the light passing through them so that it no longer matches the angle of the top polarizing filter. Consequently, no light passes through that portion of the LCD, which becomes darker than the surrounding areas, and the pixel appears black.

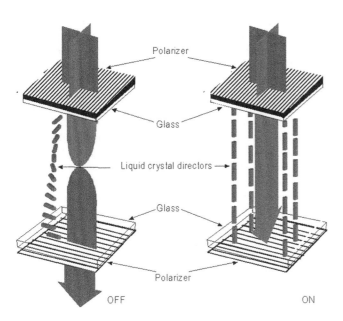

FIGURE 11.13. Principle of operation of a twisted nematic LCD.

By applying different electric currents, gray scales can be produced with LCD technology. One disadvantage with this is the relatively low contrast, but this can be improved by applying a steep electro-optic characteristic line of the liquid crystals. Low voltage is then sufficient to change the translucence, causing the liquid crystals to twist by more than 90 degrees. Such displays are called super-twisted nematic (STN), for example, double super-twisted nematic (DSTN) or triple super-twisted nematic (TSTN) (Schadt, 1996).

In passive matrix LCDs, the electric field expands over the pixel to be addressed to the entire horizontal and vertical electrodes, which results in disturbing stripes, or "ghosting." Other disadvantages are slow-response times and a prelightening of the pixels.

Active matrix (AM) addressing places an electronic switch at each pixel of an LCD, thus controlling the charging of the pixel capacitor up to the voltage corresponding to the desired gray shade, and then holding this voltage unit the next video information is written in. The available switches are thin-film transistors (TFTs), which act like diodes (Lüder, 2003). In order to display a full range of colors, each element on a LCD is assigned to a primary color by a special filter, which is deposited on the faceplate of the LCD. To guarantee high-display contrast quality, brightness and high quality of color representation, many transistors are needed. For a resolution of 1,024 × 768 pixel, 2.36 million transistors are required, one per primary color and subpixel.

The traditional representation errors caused by an incorrect deflection do not occur because of the fixed matrix, but defective transistors can cause errors, leading to permanently bright or dark pixels. These pixels are distracting, particularly when all three transistors of an elementary cell are switched to bright, which causes a white pixel to appear on the screen. Furthermore, the representation of moving pictures is restricted. The response latency of the liquid crystals is about 20 to 30 ms. Fast sequences (i.e. scrolling) are blurred.

LCD monitors with a standard video graphic array (VGA) plug have to convert the analog VGA signal back into a digital signal. If an incorrect A/D changer is used (i.e., an 18-bit-A/D changer), the color representation can be affected, although the LCD monitor is actually capable of representing true color (24 Bit).

CRTs easily represent black; the cathode simply does not emit electrons. LCDs, however, have to completely block out the light from the back light. Technically, this is impossible, and as a consequence the contrast is reduced.

When an LCD display is viewed from an angled position, it appears darker and color representation is distorted (see section "Brightness and Glare"). Recently developed technologies, such as in-plane switching (IPS), multidomain vertical alignment (MVA) and thin-film transistors (TFT), have improved the width of the viewing angle. The welcome effect of the restricted viewing angle for privacy purposes, as in the case of automated teller machines (ATMs), is worth mentioning here.

The nominal screen diagonal of a LCD is equivalent to the effective screen diagonal. In contrast, CRT nominal screen diagonals are smaller than the effective screens.

Plasma Displays

The oldest electro-optical phenomenon able to produce light is an electrical discharge in gas (Bosman, 1989). Millions of years elapsed before this effect was identified, analyzed, and mastered by humans. The first attempts to produce a matrix display panel were made in 1954. Since then, research has continued and a host of approaches have evolved (Bosman). The beauty of this technique is that, unlike front-view projection screens, one does not have to turn off the lights to see the image clearly and easily. Therefore, plasmas are excellent for video conferencing and other presentation needs (Pioneer, 2001).

In plasma technology, two glass plates are laid with their parallel thin conducting paths at right angles to one another (Precht et al., 1997). The gap between the plates is evacuated and filled with a gas mixture (see Fig. 11.14). If sufficiently high voltage is applied to the cross point of two orthogonal conducting paths, the gas ionizes and begins to shine (like a lot of small gas-discharge lamps). The inside of one glass plate is coated with a phosphorus layer, which is, according to the fundamental colors, composed of three different kinds of phosphorus. There is a wired matrix below the phosphorus-layer to trigger the PDP. The space between the glass plates is divided into gas-filled chambers. When voltage is applied to the wired matrix on the bottom of the display, the gas is transformed into a plasmatic state and emits ultraviolet radiation, causing the phosphorus to glow.

The advantage of this technology is the flat construction of large screens that perform extraordinarily well under most ambient light conditions. For example, even very bright light does not wash out the image on the screen. Another characteristic of a plasma panel is the extreme viewing angles both vertically and horizontally. With a 160°-viewing angle, people sitting to the side of the screen will still be able to see the image without losing any information (Pioneer, 2001).

Plasma screens do have disadvantages. First, they consume a large quantity of power, making this technology unsuitable for battery-operated and portable computers. Further, high voltage

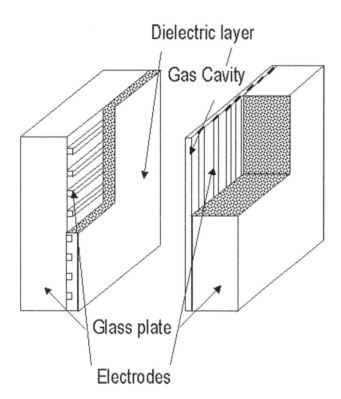

Dielectric layer

Gas Cavity

Glass plate

Electrodes

FIGURE 11.14. Major parts and components of a plasma display panel.

is required to ignite the plasma, and the phosphorus layers degrade over time. Finally, the pixels are oversized, so the user must be situated at a distance from the display.

Electroluminescent Displays

Electroluminescence (EL) refers to the generation of light by nonthermal means in response to an applied electric field that produces light (Nelson & Wullert, 1997). In early EL displays, the phosphors were used in powder form as in CRTs. Current ELDs use a thin film of phosphorescent substance (yellow-emitting ZnS: Mn), which is sandwiched between two conducting electrodes. The top electrode is transparent. One of the electrodes is coated with vertical wires and the other with horizontal wires, forming a grid. When electrical current is passed though a horizontal and a vertical wire, electrons are accelerated and pass through the phosphorous layer. Light is emitted by the excited activator atoms in the form of a pulse. The activators are transition metal or rare-earth atoms. Once excited, these centers can decay to the ground state through a radiating transition, and thus emit light with their specific emission spectrum (Budin, 2003). Electroluminescence can also be obtained from organics. Organic EL devices generally consist of a hole-transport layer (HTL) and an electron-transport layer (ETL) between electrodes. The radiating recombination can be further optimized by introducing fluorescent centers near the interface of the two layers. Such structures are particularly useful in tuning EL colors (Theis, 1999).

High-quality color presentation is the main concern with regard to ELDs. A large number of laboratories have experimented with many materials, activators, and full-color solutions (Ono, 1993). Color ELDs can be manufactured in different ways. One of the approaches, which is quite common for emitters, is that of additive-color synthesis, using three juxtaposed patterned phosphors. This approach, however, suffers from a reduced spatial-fill factor for each monochromatic emitter. The second approach to full color ELDs consists of using a single white emitting structure with patterned color filters. It results in a much simpler manufacturing process. A hybrid solution, which is being developed into a full-color commercial product at Planar (Fig. 11.15), consists in stacking and registering several plates (King, 1996). The first plate is a glass plate with the active structure on its far side, and transparent electrodes on both sides of a patterned ZnS:Mn (filtered to red) and ZnS:Tb (green) structure similar to that described above. The second glass plate has a fully transparent blue-emitting Ce:Ca thiogallate structure on top of the plate. On both plates, row electrodes are reinforced by a thin metal bus (Theis, 1999; Budin, 2003).

There are many different types of EL technology, among which the alternating current thin film electroluminescent displays (ACTFEL) are the most commonly used ELDs. Currently, ELDs are manufactured in sizes ranging from 1 to 18 inches with resolutions from 50 to 1,000 lines per inch (Rack et al., 1996). ELD qualities of note are the sharp edges of the pixels, good contrast (10:1 at 400 lux), the wide viewing angle, fast response time, durability, shock resistance, operation at high and low temperatures, and the smallest thickness and weight compared to the other flat panel displays (Budin, 2003). The bigger disadvantage of ELDs is their limited ability to be used in full-color or large-area applications (Nelson & Wullert, 1997). ELDs are primarily used for medical, industrial, and instrumentation applications (Budin, 2003).

EL devices also include LEDs. LED technology is used in almost every consumer electronic product on the market. The operational principle of LEDs can be briefly described as follows: with no voltage, or reversed voltage applied across the pn junctions, an energy barrier prevents the flow of electrons and holes. When a forward bias voltage (1.5 to 2 V) is applied across the junction, the potential barrier height is reduced by allowing electrons to be injected into the p region and holes into the n region. The injected minority carriers recombine with carriers of opposite sign, resulting in the emission of photons (Castellano, 1992).

LEDs as well as some types of ELD can also be used as a backlight for LCD applications.

Cathodoluminescent Displays

The working principles of these displays are comparable to the CRT; therefore, they are also referred to as flat CRTs. One of the most successful flat CRT configurations is called vacuum fluorescent display (VFD). Vacuum fluorescent displays are broadly used in consumer electronic products and are recognizable by their green or blue-green glow, which is due to the ZnO:Zn phosphor (Nelson & Wullert, 1997). It is essentially a triode with arrays of grids and phosphor-coated anodes used for coincident

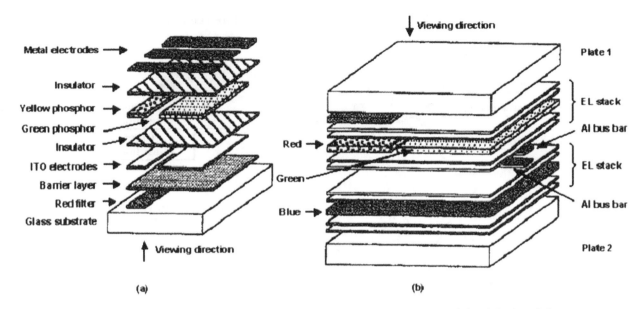

FIGURE 11.15. Schematic of (a) a Planar multicolor EL prototype, and (b) of the first full-(saturated) color dual substrate EL panel (cf. King, 1994).

addressing. The grid structure is positioned over the anode and the cathode filaments are stretched and held above the grid. The entire structure is then sealed in an evacuated cell. The cathode filament is a very fine wire that is heated to a temperature just below incandescence (about 600°C). At that temperature it remains virtually invisible but it emits electrons. The anode is supplied with a voltage of between 10 V and 50 V and at the same time the grid voltage is applied to the grid of that selected segment. The electrons, which are emitted by cathode filaments, are controlled by grids. When a grid is supplied by a positive voltage, it attracts the negative electrons, whilst positive electrons are attracted when the grid is supplied with the negative voltage. The electrons collide with the phosphor-coated anode, resulting in photon emission. The VFD display can be driven in a dynamic or static system. The dynamic system is a type of multiplexing where a pulse is applied to the grid of each digit or row and to the anode of each selected digit. The static system uses a DC signal for each anode of each digit (Castellano, 1992; Nelson & Wullert, 1997).

One of the disadvantages of VFDs is that they cannot be scaled to arbitrarily large sizes because the front and back plates are only supported at the edges. Another problem is equalization of the luminous efficiencies. The blue phosphors tend to be very low in luminous efficiency and make the realization of a full-color matrix VFD difficult (Castellano, 1992).

Because the thermionic emission used in VFDs seems to be incompatible with an internal array of spacers, field emission has been proposed as a source of electrons. Internal spacers can be incorporated between emitters to allow field-emission displays (FEDs) to be thinner, lighter, and potentially much larger than VFDs (Nelson & Wullert, 1997). The basic principle of VFDs is similar to that of FEDs. In FEDs, electrons are not emitted by a cathode consisting of an oxide-coated heated filament, compared to the working principle of VFDs, but by field emission from a cold cathode.

Electrons emitted from a cathode are accelerated toward a screen, where their energy is transformed into light (Theis, 1999) (Fig. 11.16).

In real devices, efficient electron emission is obtained from a few thousand microtips per pixel formed from Mo or Si (Spindt-type, with tip radii below 1 mm for high field strength), or by replacing the microtip with planar diamond-like carbon films, which are also suitable for electrons.

The gate voltage in Spindt-type devices is about 50 V and the dielectric thickness separating gate and cathode is about 1 μm. The anode-cathode distance is 200 μm. A spacer technique is used to avoid glass bending due to atmospheric pressure (Theis, 1999; Budin, 2003).

The greatest FED challenges include finding low-voltage color phosphors, simplifying processing, and increasing efficiency (Nelson & Wullert, 1997). However, the technology is attractive since it is relatively thin and it is possible to create a Lambertian light with 180°-viewing cones and good video-response times.

Laser Displays

LASER is the acronym for Light Amplification by Stimulated Emission of Radiation. Laser light is generated in the active medium of the laser. Energy is pumped into the active medium in an appropriate form and is partially transformed into radiation energy (ILT, 2001). In contrast to thermal emitters, a laser emits a concentrated and monochromatic light with high local and temporal coherence (see Fig. 11.17).

In contrast to conventional light sources, laser light has rays that are nearly parallel with each other (it has a small divergence), is focusable down to a wave length, has a high power density, a monochromatic character (light of one wavelength), and high local and temporal coherence (the same phase). These

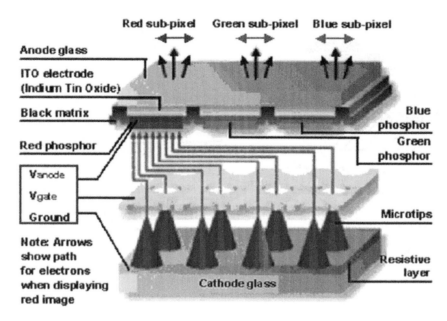

FIGURE 11.16. Principal structure of field emission displays (Theis, 1997).

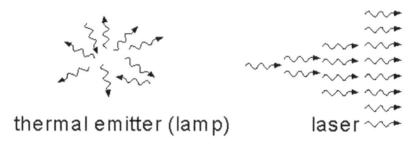

FIGURE 11.17. Dispersion of light between thermal emitters and laser light.

characteristics can be used to generate an image analogous to a conventional CRT system: the laser is directed line by line across the display area. The line diversion, horizontal and vertical, is achieved by the use of rotary or oscillating mirrors. Light output from the diode-pumped solid-state lasers is modulated according to the input signal, and the red, green, and blue light is combined. This combined light is then raster-scanned onto the screen to create an image. The laser projects the image either onto a flat area or directly onto the retina. Figure 11.18 shows the principle of a retinal-scanning laser display.

A collimated low-power laser diode is normally used as a modulated light source. The tiny laser beam is subsequently deflected in u- and v-direction by two uni-axial scanners. The horizontal scanner (u-direction) operates at several kHz. The vertical scanner frequency (v-direction) defines the image refresh rate, which must exceed the critical fusion frequency (CFF) of about 60 Hz to prevent flickering effects. Finally, a viewing optics projects the laser beam through the center of the eye pupil onto the retina. In the retinal-scanning display system the pixels are projected serially in time onto the retina. As the image refresh rate is above the temporal resolution limit of the human eye, the user does not perceive any flickering effects (von Waldkirch, 2005).

Electronic Ink and Electronic Paper

For nearly 2,000 years, ink on paper was the most popular way to display words and images. It still has many advantages compared to computer displays with regard to text readability and price. The biggest limitation of paper displays is that the printed symbols cannot be changed or removed without leaving noticeable marks. The invention of electronic ink has made it possible to overcome this disadvantage. Electronic ink displays promise to have paper-like properties. For example, electronic ink displays are viewed in reflective light, have a wide viewing angle, and are thin, flexible, and relatively inexpensive. Unlike paper, they are electrically writeable and erasable. A big advantage that they have over other displays is their very low power consumption, possibly extending the battery life of devices with such displays into months or even years. The principal components of electronic

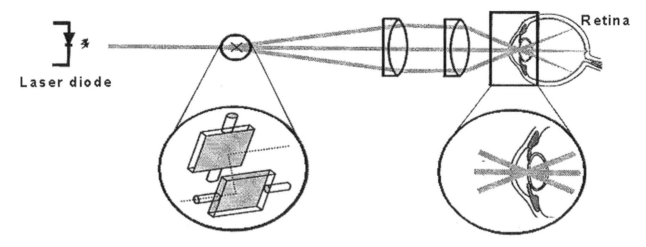

FIGURE 11.18. Schematical illustration of the RSD principle (according to Menozzi et al., 2001).

ink (see E Ink Corporation, 2001) are tiny capsules, about 100 mm diameter, filled with a clear fluid. Suspended in the fluid are positively charged white particles and negatively charged black particles. The particles will move toward an inversely charged electrode. Therefore, when a top transparent electrode is negatively charged, the white particles move to the top and become visible to the user. The black particles will, at the same time, move to a positively charged bottom electrode and the microcapsule appears white to the user. Similarly, the microcapsule will appear black when an electric field of the opposite direction is applied. The microcapsules are suspended in a "carrier medium" that can be printed onto a sheet of plastic film (or onto many other surfaces). The film is again laminated to a layer of circuitry that forms a pattern of pixels that can be controlled similarly to other displays.

An invention made 20 years ago at Xerox Palo Alto Research Center forms the basis of Gyricon reusable paper, which has similar advantages to electronic ink. It consists of an array of small balls of 0.03 to 0.1 mm in diameter embedded in oil-filled pockets in transparent silicone rubber. Every ball is half white and half black and can rotate freely. Each ball has an electric charge, and when an electric field is applied to the surface of the sheet, the balls rotate showing either their black or white side and attach to the wall of the sheet. They remain there displaying the intended image as long as no new electric field is applied (Gibbs, 1998).

Unlike the first prototypes of electronic ink and electronic paper, which could only display two colors, usually black and white, photonic ink can display any color value in the spectrum. Photonics (from photon) is an area of study based on the utilization of radiant energy, such as light, for various applications. Photonic ink, or p-ink, consists of planar arrays of silica microspheres—an opal film-embedded in a matrix of cross-linked polyferrocenylsilane (PFS). Photonic ink is able to display color through controlled Bragg diffraction of light. Varying the size of the spaces between the particles creates different colors. A polymer gel is filled between the stacked spheres. The gel swells when soaked in a solvent and shrinks when it dries out, chang-

ing the distance between the tiny spheres. Because the size of the spaces determines the wavelength that will be reflected, the swelling and drying of the gel results in a continuously tunable display of color across the entire spectral range visible. The amount of solvent absorbed by the gel is controlled by applying an electrical voltage. The optical response of the film to a change in solvent is less than half a second (American Chemical Society, 2003).

Applications for electronic paper are seen as a substitute for paper products such as books and newspapers, as rewriteable paper in the office, as a material for price tags and retail signs, as wall sized displays, foldable displays, and as an alternative display for PDAs and mobile phones.

VISUAL PERFORMANCE

Trouble-free usage and user acceptance of any new technology depend substantially on the quality of the visual display as a central communication unit as well as the ease with which the displays allow visual information to be processed. Thus, a careful visual evaluation is indispensable in order to assess the efficiency of visual performance and to identify existing shortcomings of the displays.

Looking back, there is quite a long history of studies concerned with the evaluation of visual displays (Dillon, 1992; Schlick, Daude, Luczak, Weck, & Springer, 1997; Luczak & Oehme, 2002; Pfendler & Schlick, in press). In order to estimate the costs and benefits of electronic displaying in terms of human performance and visual load, the main interest was to learn which factors affect visual performance to what extent. One research approach was to compare display types with respect to the efficiency of visual processing. In most cases a basic comparison of the traditional hard copy with different electronic displays was performed (Heppner, Anderson, Farstrup, & Weidenmann, 1985; Gould, Alfaro, Barnes, Finn, Grischkowsky, & Minuto, 1987). Another approach was to study the effects of specific display factors on performance (Pfendler, Widdel, & Schlick, 2005; Sheedy, Subbaram, & Hayes, 2003; Plainis, & Murray, 2000; Ziefle, 1998). This procedure helps to distinguish different sources that account for performance shifts. Moreover, user characteristics (Bergqvist, Wolgast, Nilsson, & Voss, 1995; Kothiya & Tettey, 2001; Ziefle, 2003a) and workplace settings (Sommerich, Joines, & Psihoisos, 2001; Ziefle, 2003b) as well as individual body postures received attention (Aarås, Horgen, Bjorset, Ro, Thoresen, & Larsen, 1997; Aarås, Horgen, Bjorset, Ro, & Thoresen, 1998; Helander, Little, & Drury, 2000) in the context of visual performance. Beyond display characteristics, any visual evaluation should also consider the diverse task demands, because they can strongly influence performance outcomes. Their influence can be either positive, by compensating other suboptimal visual settings, or negative, when several negative factors accumulate, possibly leading to visual complaints.

Task Demands

Task demands are of crucial importance for visual evaluations because they represent a complex entity of different elements. The combination of (a) the task type (what the user is requested to do), (b) user characteristics (level of expertise, visual abilities or working motivation), (c) text factors (font size, line pitch), (d) display factors (i.e. contrast, refresh rates, resolution) and (e) surrounding factors (i.e. ambient lighting or time on task) has a considerable impact on performance outcomes. As these factors were found to interplay, it is necessary to examine the particular conditions and settings that were used in the different studies concerned with visual evaluations.

In order to evaluate visual displays, simple detection tasks, memory and recognition tasks, visual search tasks, and (proof) reading were used. Overall, two basic task forms can be differentiated between in this context: tasks in which a semantic context is present (i.e., (proof) reading) and tasks with no semantic context (i.e., visual search). This distinction has implications with respect to the outcomes' ecological validity and generalizability as well as to the sensitivity with which shortcomings of visual displays are reflected. Tasks that have a semantic context have the general advantage of simulating what users mostly do when using displays, thus the evaluation process is ecologically valid. However, these tasks were found not to be very sensitive to visual degradation effects (Stone, Clark, & Slater, 1980; Ziefle 1998), because of two main reasons. First, the encoding and processing of text material represents behavior gained through intensive training: representing a top-down process, comprehension guides the reader through the text and possibly masks degradation effects. The second objection is concerned with reading strategies involving different combinations of cognitive and visual processes: when proofreading, for example, participants may read for comprehension or scan for unfamiliar letter clusters and word shapes. Thus, it is possible that performance outcomes do not reflect the degradation properly even though the display quality is suboptimal: deteriorations might then be overlooked, especially in short test periods. Even though ecological validity is lower, visual search and detection tasks showed a higher sensitivity to visual degradation effects, because the visual encoding process predominantly relies on visual properties of the display (bottom-up process) without being masked by compensating cognitive strategies.

However, text factors were also found to considerably affect performance: among those, display size, the amount of information to be processed at a time, as well as font size (Duncan & Humphreys, 1989; Oehme, Weidenmaier, Schmidt, & Luczak, 2001, Ziefle, Oehme, & Luczak, 2005) were revealed to be crucial in this context. The smaller the font size and the larger and denser the amount of (text) material to be processed, the stronger the performance decrements were.

Whenever a critical visual evaluation procedure is needed that reflects effects of visual displays on performance, a two-step procedure is to be recommended. As a first step, a benchmark procedure is advised. This procedure includes young and well-sighted participants and a task that is visually rather than cognitively strenuous. The second step includes a broadened scope. Older users (as they represent a major part of the workforce) and tasks with different visual and cognitive demands have to be addressed. In addition, extended periods of on-screen viewing are to be examined in order to realistically assess the long-term effects of computer displays.

Measures Used for Visual Evaluations

The measures used to quantify the effects of visual display quality also differ regarding their sensitivity to visual degradation. Apart from effectiveness and efficiency of performance, judgments of visual strain symptoms and the presence of visual complaints were also assessed.

Among performance measures, global and local parameters can be distinguished. Global performance measures are the speed and accuracy of task performance; they were widely used across visual evaluation studies. Reading and search times reflect the basic velocity of the encoding and visual processing. However, system speed can only then be meaningfully interpreted when the accuracy of task completion is also taken into account. In this context, the speed-accuracy trade off has a considerable impact. Whenever visual conditions, for example in computer displays, are suboptimal (under low resolution or contrast conditions), the speed and accuracy of information processing cannot be kept at a constantly high level simultaneously. Rather, one of the two components deteriorates while the other is kept constant. When both components are considered, the overall costs for the information processing can be assessed. Nevertheless, reliable, global measures do not provide an understanding of the processes that cause the performance decrement. Therefore, the oculomotor behavior was consulted in order to gain deeper insights into the nature of the deteriorated encoding process (Owens, & Wolf-Kelley, 1987; Iwasaki & Kurimoto, 1988; Jaschinski, Bonacker, & Alshuth, 1996; Best, Littleton, Gramopadhye, & Tyrell, 1996). The spatial and temporal characteristics of saccades were predominantly analyzed. Under visual degradation conditions, more saccades are executed, smaller in size, and accompanied by increased fixation times (Ziefle, 1998; Ziefle 2001a, 2001b). Also, accommodation states, pupil size, vergence efforts, and visual scan path complexity (Schlick, Winkelholz, Motz, & Luczak, 2006) were taken as measures to quantify the effects of display quality. The higher effort required for the visual system observed when reading electronic texts is assumed to form the physiological basis of visual fatiguing (Wilkins, Nimmo-Smith, Tait, McManus, Della Sala, Tilley, et al., 1984). Beyond oculomotor measures, stress indicators such as heart-rate variability and eye blinks were used to determine the effects of display quality (Oehme, Wiedenmaier, Schmidt, & Luczak, 2003).

Another approach to quantifying effects of display quality is to collect user judgments with respect to what is called visual fatiguing, visual stress, strain, visual load, or discomfort. Visual fatiguing is considered as a complex subjective measure based on the awareness of several symptoms: burning, dry, aching, and watering eyes, difficulties in reading (text becomes blurred and fades away) as well as increased blinking and eye pressure (Stone et al., 1980; Hung, Ciuffreda, & Semmlow, 1986). User judgments represent an important aspect of display quality, as perceived visual comfort is the most direct source of users' satisfaction. Even though ratings are easy to obtain, their validity is not without controversy. First, users differ considerably with respect to their responsiveness to visual strain symptoms. Furthermore, the sensitivity to visual stress is not constant, but changes with the time spent on the task and age (Wolf & Schaffra, 1964). Second, factors like fear of failing and the misinterpretation of one's own performance may contaminate judgments and make it necessary to prove that ratings match performance outcomes. Third, the emergence of visual fatiguing does not necessarily follow the same time course as performance shifts. Visual fatiguing is therefore not necessarily accompanied by performance decrements and vice versa (Yeh & Wickens, 1984; Howarth & Istance, 1985). In order to obtain a valid picture, it is therefore advisable to include both subjective and objective measurements.

Effects of Specific Display Factors on Visual Performance

As a first factor, effects of display resolution (see section "Resolution") are considered. Technically speaking, display resolution is defined as the number of pixels per inch out of which objects and letters are displayed on the screen surface. The psychophysical correlate of display resolution, and this is of visual ergonomic interest, is the sharpness of contours and the clarity with which objects can be identified. High-display resolutions allow more objects to be displayed on a given screen space, as object sizes decrease with increasing resolutions. Even if this may be advantageous when screen space restrictions are considered, it might be counter-productive: the benefit of higher resolutions in terms of contour sharpness may be negated by the smaller object sizes, and this has to be counterbalanced. The central question is whether higher resolutions lead to quantifiable improvements. The relevant literature shows that this is indeed the case (Gould et al., 1987; Miyao, Hacisalihzade, Allen, & Stark,1989; Ziefle, 1998). Due to unequal technical standards over time, resolution levels differ considerably between the studies (40–90 dpi). In conclusion, it can be said that up to 90 dpi performance, the higher resolution, is better. Young adults' search performance was increased by 20% in the 90 dpi condition compared to the 60 dpi condition. This greater effectiveness was caused by a more efficient oculomotor control: fixation times decreased by 11%

and, in addition, less saccades (5%) were executed for processing the visual information. Furthermore, performance in the low-resolution condition was not only found to be interrelated with the emergence and strength of visual fatiguing symptoms, but also, the probability of fatiguing symptoms was significantly higher when two suboptimal viewing conditions were coincident: the longer participants worked in the low-resolution condition the stronger performance decrements were (Ziefle, 1998).

A second factor to be examined extensively was intermittent light stimulation, which is characteristically present in CRT screens, referred to as the "refresh rate" (see section "Refresh Rates"). The perceptual component here is a flicker sensation, which predominately occurs when low refresh rates are present. Reading on a screen with low refresh rates (50 Hz) is extremely hard work for users and leads to considerable eyestrain symptoms even after short reading periods. With increasing refresh rates (>70 Hz) the flicker sensation decreases, but it should be noted that physically, the intermittent stimulation is nevertheless still present and may affect performance. Many ergonomic studies have concerned themselves with screen flicker and found that it is a major source of performance decrements and visual fatiguing, involving the emergence of migraines (Boschmann & Roufs, 1994; Jaschinski et al., 1996; Küller & Laike, 1998; Lindner & Kropf, 1993). The nature of the flicker effect is referred to as "a basic disturbance of eye movement control" caused by the intermittent light (Kennedy & Murray, 1993; Baccino, 1999). Regarding the impact of refresh rates on performance, it was shown that performance with the 100 Hz screen was 14% better compared to 50 Hz, and ocular efficiency was 16% better (fixation times and number of fixations per line). Apart from that, effects of the time spent on the task were found (comparing the beginning and the end of the task completion). Already in a test period of only 30 minutes, visual and ocular performance decreased by 8% (Ziefle, 2001a). However, effects of refresh rates on performance do not follow a linear ("the higher the better"), but rather a curvelinear relationship: when a 140 Hz screen was compared to a screen driven with 100 Hz, search performance and ocular efficiency were found to deteriorate in the 140 Hz condition (by 5–10% across measures). This confirms that the visual system is sensitive to high-frequent intermittent light, even though users are no longer able to detect flicker in this high refresh-rate condition. Overall, we can say in summary that refresh rates of 90–100 Hz in CRTs facilitate reasonably good performance.

Third, effects of luminance (see section "Brightness and Glare") and contrast (see section "Resolution") were also reported to play a major role for visual performance when computer displays are used (Plainis & Murray, 2000; van Schaik & Ling, 2001; Ziefle, Gröger, & Sommer, 2003; Oetjen, Ziefle, & Gröger, 2005). However, results are not easy to integrate. The contrast itself is a complex factor, but it is also escorted by several other lighting characteristics, which are highly interdependent and affect visual performance outcomes separately as well as in combination. The contrast, always a ratio out of two luminance levels (background and object) was proven to markedly affect visual performance. Low-contrast ratios lead to performance decrements and oculomotor efficiency in a range of 10–20%. But contrast ratios do not specify which absolute luminance levels constitute the respective contrast and—depending on which contrast definition is used—they mostly do not differentiate between whether information is displayed in negative or positive polarity (bright letters on dark vs. dark letters on a bright background). In addition, ambient lighting considerably interferes with the display contrast. Thus, knowing the contrast level without detailing polarity, absolute luminance levels and the level of ambient lighting does not allow the interpretation of performance outcomes. Generally, it was found that positive polarity displays result in a better visual performance than negative ones (Bauer & Cavonious, 1980). The positive effect of positive polarity occurs due to the (simple) fact that light is essential for visual encoding (the absolute light level is higher with positive displays), and because it is less demanding when the computer display has the same polarity as hard copy (especially because both media are often used simultaneously in the same work setting), and does not require the user to alternately readapt. Ambient lighting strongly reduces the contrast levels on the screen, because the two lighting sources interfere. Thus, it is advantageous for performance if illumination in the room is sparse (Kokoschka & Haubner, 1986). Regarding display luminance it is recommended that luminance levels should not be too bright, preventing interfering glare, especially when ambient lighting is low and the probability of glare rises (Schenkmann, Fukunda, & Persson, 1999).

The importance of contrast and luminance has received renewed interest in recent studies (Hollands, Cassidy, McFadden, & Boothby, 2001; Hollands, Parker, McFadden, & Boothby, 2002; Gröger, Ziefle, & Sommer, 2003; Ziefle et al., 2003; Oetjen & Ziefle, 2004). Among the visual factors, TFT screens (see section "Nonemitter/Liquid-Crystal Display") have the basic advantage of being flicker free. However, they also have one major disadvantage: the displayed information is "perfectly" visible if users

work in front of the TFT screen. Whenever this "optimal" position is not present, visibility is distinctly worse. This specific property of TFT screens is called "anisotropy." A display is called anisotropic if it shows a deviation of more than 10% of its luminance subject to the target location or viewing angle (ISO 13406-2 2001). The nature of anisotropy is such that photometric measures (contrast and luminance) are not constant over the screen surface, but rather decrease with increasing viewing angle. In order to quantify the change in photometric measures at different viewing angles, a measurement set up was developed that made it possible to exactly correlate photometric measures and visual performance (Gröger et al., 2003; Ziefle et al., 2003). The screen was virtually cut into 63 (9 × 7) fields. The luminance of bright/dark areas was individually measured by a photometer and contrasts were determined. Then different measuring positions were adopted: first, the "standard view" was applied, commonly used by the industry. The photometer was set in front of the screen and displaced gradually from field to field, with the photometer always set at right angles to the screen (Fig. 11.19, left).

From an ergonomic point of view, this procedure (as shown in Fig. 11.19, left) is highly artificial, as users do not displace themselves, but rather change their view: viewing angles change remarkably depending on where users are looking. This is entirely disregarded in this measurement procedure. In order to simulate real viewing conditions, the "user view" (Fig. 11.19, center) as well as the "bystander view" (Fig. 11.19, right) were realized. For the user view, the photometer was positioned centrally in front of the screen. As the position of the photometer did not change, viewing angles increased with distance from the center of the screen, thereby emulating the user's head movements when looking toward the screen edges. For the "bystander

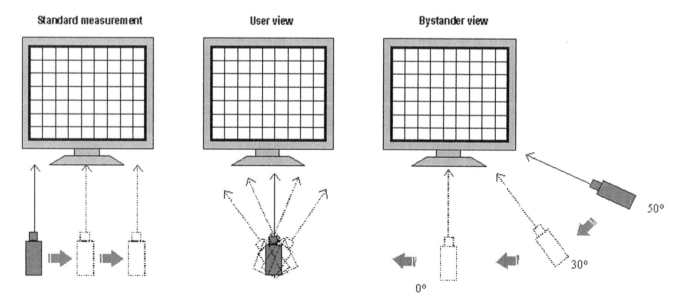

FIGURE 11.19. Quantifying anisotropy in TFT screens: left: "standard view" with the photometer displaced at right angles; center: "user view" with the photometer emulating the user's head movements; right: "bystander view" with the photometer positioned off-axis (Ziefle et al., 2003).

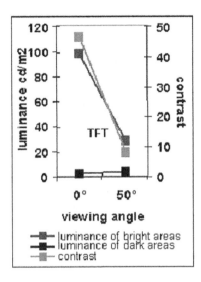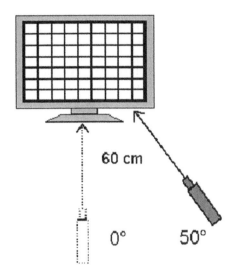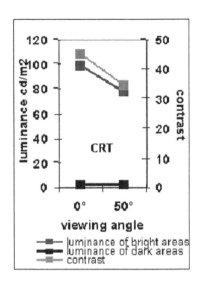

FIGURE 11.20.   Photometric measures in the "bystander view" for a TFT (left side) display and a CRT (right sight) display (Oetjen & Ziefle, 2004)

view," the photometer was set to a central point of the display and turned to the different measuring fields (30° and 50°). For extended-viewing conditions, the photometer was set off axis, and its view pointed to the screen from the side (left and right side, respectively).

In Fig. 11.20 it becomes evident that photometric measures change dramatically as a function of viewing angle in TFT Displays (Figure 11.20, left), and to a much lesser extent in CRTs (Fig., 11.20, right).

Performance outcomes showed that anisotropic effects have to be taken seriously. Visual performance with TFT screens considerably deteriorates by about 10% for young adults when they had to look at the screen off axis. Anisotropy plays a role in many real-life work settings: (air, rail) traffic controlling environments or stock exchanges use several TFT screens simultaneously (placed parallel to and/or upon another) that have to be surveyed by one operator. Another example is the schooling and training context where it is quite usual for several users to work with only one screen. Thus, from an ergonomic point of view, TFT's anisotropy must be regarded as a visual limitation of the TFT Display technology, at least when a fast and accurate visual detection performance is of importance.

Comparison of Different Display Types

During the last 30 years, since the first evaluation studies of different displays were published, a huge number of studies have dealt with the fundamental question as to which display type assures the highest reading comfort and the best visual performance. Typically, and this reflects the chronological development, hard copy, CRT, and TFT displays were compared with respect to visual performance. As electronic display quality has improved continuously over time, the technical standards on which the evaluations were based differ greatly. This is a factor that should be borne in mind. Recent studies include new developments (augmented reality, see section "Augmented Reality") in visual-display technology.

However, independently of the time and technical standard of computer displays, the one and only display that outperformed all others with respect to visual performance and comfort is the traditional hard copy. Its development covers more than 200 years (since the invention of the industrial production of hard copy in France) and has been continuously amended with respect to readability and visibility, by the expertise of typesetters and typographers. Thus, hard copy can be regarded as an outstandingly suitable display with regard to visual ergonomic demands. It provides high contrast and resolution without disturbance by glare, screen reflections, or flicker. Accordingly, the majority of studies show that performance in CRT screens is significantly lower compared to paper (Gould et al., 1987; Heppner et al., 1985; Ziefle, 1998). Hardcopy was also found to outperform modern TFT displays (Ziefle, 2001a).

As the private and public need for electronically displayed information continuously increases, the evaluation should focus on the quality of different electronic displays. Here, the classical comparison of CRT and TFT displays is of central interest. The CRT, the hitherto widespread screen type, seems to be being phased out due to its specific lighting characteristics (flicker). The development of TFT technology was therefore highly welcome: TFT screens, lightweight and flat, are flicker free and display information at much higher levels of luminance and contrast.

At first sight, the outcomes of studies that compare CRTs and TFTs reveal an inconsistent picture: There are studies in which TFT displays led to a higher visual performance compared to CRT screens (Menozzi, Lang, Näpflin, Zeller, & Krueger, 2001; Ziefle 2001a,b), but there are also studies in which CRT displays outperform TFT screens (Oetjen & Ziefle, 2004; Oetjen et al., 2005; Oetjen & Ziefle, in press). This inconsistency can be resolved when taking the experimental purposes and evaluation settings into account. Whenever the impact of screen flicker was of main interest, the methodological focus was to compare both displays with respect to refresh rates (present in the CRT and absent in the TFT). This procedure ruled out any other differences between both displays that may confound or compensate flicker effects (anisotropy). Thus, participants were usually seated in front of the screen and had to work on tasks that were displayed centrally (otherwise, anisotropic effects would have been mixed up). In these cases, visual performance clearly favors the TFT and shows that screen flicker in CRTs is disadvantageous. In addition, it was found that the relative benefit of TFT technology is disproportionately higher for older users: while young adults (20–25 years) showed performance superiority of 10% for the TFT compared to the CRT (100 Hz), the benefit from the TFT was 16% and 27%, respectively, when older users (40–65 years) were examined (Ziefle, 2001b). Note, however, that the older adults generally showed nearly 40% lower visual performance.

However, the benefit of the TFT thus determined has not yet taken anisotropic effects into account. As soon as anisotropy is considered, the picture changes. Studies concerned with the two display types from the anisotropy perspective showed that the TFT superiority over the CRT disappears when extended viewing angles were taken into account (Gröger et al., 2003; Hollands, et al., 2001, 2002; Oetjen & Ziefle, 2004; Oetjen & Ziefle, in press). When considering all screen positions, visual performance decreased by 8% when a TFT was used instead of a CRT screen. When a central view was applied, detection times were 14% faster than when viewing 50° off axis. A further aggravating factor was font size. Deterioration when detecting small targets (1.5 mm vs. 2.4 mm) rose to almost 30% with the TFT, in contrast to "only" 20% with the CRT. Whenever all suboptimal factors (small font, TFT screen, off-axis viewing) occurred simultaneously, performance decrements were found to mount up to as much as 38% (Oetjen & Ziefle, 2004).

STANDARDS, RULES, AND LEGAL REGULATIONS AND HEALTH PROTECTION ASPECTS

There are a large number of national and international standards, regulations, and seals of approval that regulate diverse aspects of the design and use of visual displays. Among the areas covered are ergonomics, emissions, energy consumption, electrical safety, and documentation. Due to very fast technological development, only selected standards are listed in Table 11.1. Up-to-date information can be found on the websites listed in Table 11.2.

DISPLAYS FOR SELECTED APPLICATIONS

Virtual Reality

Immersive virtual reality (VR) is a technology that enables users to "enter into" computer-generated 3D environments and in-

TABLE 11.1. A Selected List of National and International Standards, Regulations, and Seals of Approval Regulating Design and Use of Visual Displays

Ergonomics	
Image quality	DIN EN 29241-3, ISO 13406-2, TCO'99, GS-Mark, Ergonomics Approved Mark
Reflection characteristics	DIN EN ISO 9241-7, ISO 13406-2, TCO'99, GS-Mark, Ergonomics Approved Mark
Color requirements	DIN EN ISO 9241-8, ISO 13406-2, TCO'99, GS-Mark, Ergonomics Approved Mark
Brightness and contrast adjustable	European VDU directive 90/270 EEC, German ordinance for work with visual display units, TCO'99, GS-Mark, Ergonomics Approved Mark
Tilt and swivel	TCO'99, GS-Mark, German ordinance for work with visual display units
Gloss of housing	GS-Mark, German ordinance for work with visual display units
Emissions	
Noise	European VDU directive 90/270 EEC, ISO 7779 (ISO 9296) German ordinance for work with visual display units, TCO'99, GS-Mark
Electrostatic potential, electrical and magnetic fields	PrEN 50279, TCO'99, Ergonomics Approved Mark
X-ray radiation	TCO'99
Energy consumption	
	EPA Energy Star, TCO'99, VESA DPMS
Electrical safety	
	TCO'99, GS-Mark
Documentation	
Technical documentation and user manual	German Equipment Safety Law, GS-Mark

TABLE 11.2. Websites that Provide Information about National and International Standards, Regulations, and Seals of Approval for Visual Displays

Websites	
European directives	European directives regarding health and safety are available via the website of the European Agency for Safety and Health at Work. Internet: http://europe.osha.eu.int/legislation/
EPA	The US Environmental Protection Agency (EPA) promotes the manufacturing and marketing of energy efficient office automation equipment with its Energy Star Program. Internet: http://www.energystar.gov/
National laws and ordinances	National laws, directives and regulations regarding health and safety for many European and some other countries are available via the website of the European Agency for Safety and Health at Work. Internet: http://europe.osha.eu.int/legislation/
ISO	The International Organization for Standardization (ISO) is a worldwide federation of national standards bodies from 140 countries. Internet: http://www.iso.ch
EN	European standards are available via the website of the European Agency for Safety and Health at Work. Internet: http://europe.osha.eu.int/legislation/standards
DIN	DIN is the German Institute for Standardization. Internet: http://www.din.de
TCO	TCO (The Swedish Confederation of Professional Employees) developed requirements for PCs. The TCO'99 label specifies ergonomic, ecological, energy consumption and emission requirements. Internet: http://www.tco.se
VESA	The Video Electronics Standards Association (VESA) creates standards for transmissions between computers and video monitors that signal inactivity. Internet: http://www.vesa.org
GS-Mark	The German safety approval mark shows conformity with the German Equipment Safety Law and signals that the product, as well as the user manual and production process, has been tested by an authorized institution such as TÜV Rheinland. Internet: http://www.tuv.com
Ergonomics Approved Mark	The Ergonomics Approved Mark demonstrates that a Visual Display Terminal complies with numerous ergonomic standards. The test mark is devised by TÜV Rheinland. Internet: http://www.tuv.com

teract with them. VR technology involves the additional monitoring of body movements using tracking devices, enabling intuitive participation with and within the virtual world. Additional tracked peripheral devices permit virtual navigation, pick-and-place manipulation of virtual objects (Schlick, Reuth, & Luczak, 2000), interaction with humanoids and avators by using data gloves, space joysticks, or 3D-tracked balls (Holmes, 2003).

A 3D-view is generated using different concepts and technologies (see sections "Human-Related Display Functionality and other Criteria"). The HMD is a commonly used display device. For VR, a closed-view HMD in non-see-through mode is used (see section "Augmented Reality"). Visual displays, especially HMDs, have decreased substantially in weight since the first invention of an immersive head-worn display, but are still hindered by cumbersome designs, obstructive tethers, suboptimal resolution, and an insufficient field of view. Recent advantages in wearable computer displays, which can incorporate miniature LCDs directly into conventional eyeglasses or helmets, should simplify ergonomic design and further reduce weight (Lieberman, 1999). Most of the advanced closed-view HMDs have adjustable inter pupillary distance (IPD) in order to avoid mismatches in depth perception. They provide a horizontal field of view of 30 to 35 degrees per eye and a resolution of at least 800 600, and therefore outperform predecessor systems (Stanney & Zyda, 2002). Large images of VR scenes can be reproduced by a projector on the front or back of one or more screens. Stereoscopic images for both eyes are projected either alternately or with different polarization at the same time (see section "Human-Related Display Functionality and Other Criteria").

The "CAVE" (Computer Animated Virtual Environment) is a further development in projection technology. It consists of a cube with several panels onto which the images are projected from behind. Depending on the construction they are C3 (two walls and the floor), C4, C5, or C6, respectively. A CAVE provides space for small groups, but can track and optimize the stereoscopic view for only one person. The others perceive distortions, especially on corners and edges.

Augmented Reality

Augmented Reality (AR) characterizes the visual fusion of 3D virtual objects into a 3D real environment in real time. Unlike virtual reality or environments, AR supplements reality, rather than completely replacing it (Azuma, 2001). AR can be used in many applications, such as production (Schlick et al., 1997), assembly and service, medical (Park, Schmidt, & Luczak, 2005), architecture, entertainment and edutainment, military training, design, robotics, and telerobotics.

One approach to overlaying the real world with virtual information is to use an HMD. Superimposition can occur in two ways: using an HMD with see-through mode (optical-see-through) and an HMD with non-see-through mode, called video-see-through (feed-through). The HMD in the non-see-through mode, which is also used in virtual reality, visually isolates the user completely from the surrounding environment, and the system must use video cameras to obtain a view of the real world. The optical see-through HMD eliminates the video channel; the user is directly looking at the real scene. Instead, the merging of real world and virtual augmentation is performed optically in front of the user using a mini projector and half-silvered mirror. Both systems have advantages and disadvantages in usability and technology. In an optical see-through system there is a time lag between the real-world information and the virtual information that is blended into the field of view, which is caused by the computing time for image generation. In addition, the calibration of such a system is rather complicated. In video-see-through systems the quality of the video depends on the technology of the cameras and the display. There are several factors (i.e. limited refresh rate, displacement of the cameras from eye level, and time delay) that have an adverse effect on human perception and hand-eye coordination (Biocca & Rolland, 1998; Oehme et al., 2001; Luczak, Park, Balazs, Wiedenmaier, & Schmidt, 2003; Park et al., 2005; Ziefle et al., 2005). Another disadvantage is the severe loss of information because the perception of the real-work environment is limited to the maximum resolution and FOV of the displays and cameras.

Another approach to overlaying the visual information is to use a VRD. In contrast to the screen-based HMDs just described, the VRD reaches the retina directly with a single stream of pixels, thus guaranteeing a clear, sharp projection of different kinds of information (Microvision, 2000). Because of the higher laser-light intensity, the half-silvered mirrors commonly used for see-through HMDs can be optimized to a maximum translucence, improving the see-through quality. Furthermore, the display's maximum resolution is no longer determined by the tolerances of manufacturing of the display pixels, but by the control logic and quality of the deflection mirror.

A clip-on display is a combination of the optical and video-see-through modus. A tiny LCD screen with video-based real-time images can be clipped onto eyeglasses or safety glasses. The user can easily move back and forth between the images on the monitor and the real world beyond it.

Mobile Phones and Handheld Devices

Mobile phones and handheld electronic devices represent one of the fastest-growing technological fields ever. Already in 1999, 500 million mobile phones were sold worldwide, and it is forecasted that about 2.6 billion mobile devices will be in use by the end of the decade. Mobile communication technologies, such as the Internet, universal mobile telecommunications systems (UMTS), wireless local area networks (WLAN) and wireless application services, have changed and will change social, economic, and communicative pathways in private and business sectors still more. Mobile computing has already expanded into many service fields, including sales, health care, administration, and journalism. Different mobile or handheld devices can be found on the market: mobile phones, smart phones, communicators, and PDAs. The main reason for their topicality is that the devices provide on-the-go lookup and entry of information, quick communication and instant messaging (Weiss, 2002). Handheld devices may differ with respect to physical dimensions, (color) resolution, contrast, luminance, and touch/stylus sensitivity, but they are all characterized by small screen size, which has a considerable impact on information processing. Displays of mobile phones usually vary between 3 and 4 cm, and

have fewer than 6 lines with up to 20 characters per line. PDAs and smart phones provide somewhat more display space (about 5×7 cm), which can result in up to 20 lines of 20 characters each.

The limited screen space is very problematic for providing optimized information access and the question of how to "best" present the information on the small display is challenging. At first glance, the challenge seems to be mainly related to visibility. If so, visually ergonomic principles are to be considered predominately: in order to provide fast and accurate information access, objects and letters should be big enough, text lines should not be too close together and information density should be low. This is especially important for older adults, who often have problems with sight (Brodie, Chattratichart, Perry, & Scane, 2003; Omori, Watanabe, Takai, & Miyao, 2002). However, visibility concerns are not the only point at issue. There is also the cognitive aspect of information visualization, the requirement that the presentation of information should help users to orientate themselves properly. Disorientation in handheld devices' menus is a rather frequent problem (Ziefle & Bay, 2005, 2006). Users have to navigate through a complex menu of functions, which is mostly hidden from sight, as the small window allows only few functions or small text fragments to be displayed at a time. Considering visual and cognitive demands concurrently, two alternatives can be contrasted: one alternative is to display only a little information per screen at a time, which helps avoid visibility problems due to high information density. The other alternative is to display as much information per screen as possible; this allows users to have maximal foresight (cognitive preview) of other functions on the menu, which should benefit information access from a cognitive point of view and minimize disorientation. Apparently, a sensitive cutoff between visual and cognitive impacts has to be met and it is to be determined whether the impact of the cognitive preview, or rather the impact of visual density is decisive for efficient information access. This was examined in a recent study (Bay & Ziefle, 2004). Young adults processed tasks on a simulated mobile phone where one, three, or seven menu items were presented at a time. Results corroborated the meaningful impact of information presentation on small screens for efficiency when using the device: intermediate foresight (three functions) was found to lead to the best performance. If only one menu item was shown at a time—as is the case in a number of devices on the market—the users needed 40% more steps to process tasks than when three items were shown, confirming the cognitive facet to be crucial. But also when information density was high (seven functions), performance declined by more than 30% compared to the presentation of three menu functions per screen, confirming the visual facet as also playing an important role.

In order to accommodate the problem of displaying a lot of information on a small screen, several other techniques were proposed. One technique is rapid serial visual presentation (RSVP), which is based on the idea of presenting information temporally instead of spatially (Rahman & Muter, 1999; Goldstein, Öqvist, Bayat, Ljungstrand, & Björk, 2001). With RSVP, one or more words are presented at a time at a fixed location on the screen and users have to integrate the single text fragments bit by bit, by scrolling the text forwards and (if necessary) backwards. A similar technique is the times square method (TSM) also known as "leading." Here, the information is not static, but moves across the screen (word by word or sentence by sentence), with the text scrolling autonomously from left to right. Even if trained readers may reach a reasonable level of efficiency in both presentation modes (RSVP and TSM), users report disliking both presentation forms. The low acceptance may be due to a high cognitive and visual demand imposed by the information presentation: either memory load is high, as it is difficult for users not to lose the plot while integrating the words and sentences, which are displayed one after another (RSVP), or, visual and attentional demands are high, as it is crucial to catch the text contents in moving sentences. Accordingly, when users fail to catch the contents on the first attempt, they have to wait until the scrolling information appears again (TSM).

References

Aarås, A., Fostervold, K., Ro, O., Thoresen, M., & Larsen, S. (1997). Postural load during VDU work: a comparison between various work postures. *Ergonomics, 40*(11), 1255–1268.

Aarås, A., Horgen, G., Bjorset, H.-H., Ro, O., & Thoresen, M. (1998). Musculosceletal, visual and psychosocial stress in VDU operators before and after multidisciplinary ergonomic interventions. *Applied Ergonomics, 29*(5), 335–354.

American Chemical Society (2003). *Multicolored Ink.* Retrieved October 1, 2005, from http://pubs.acs.org/cen/topstory/8112/8112notw8.html

Aye, T., Yu, K., Kostrzewski, A., Savant, G., & Jannson, J. (1998). Line-image-scanning head-mounted display. *In SID International Symposium (SID '98), Digest of Technical Papers, 29*, San Jose, CA, USA, 564–567.

Azuma, R. (1997). A survey of augmented reality. *Presence, 6*(4), 355–385.

Azuma, R. T. (2001). Augmented Reality: Approaches and technical challenges. In W. Barfield & T. Caudell (Eds.), *Fundamentals of wearable computers and augmented reality* (pp. 27–63). Mahwah, NJ: LEA.

Baccino, T. (1999). Exploring the flicker effect: The influence of in-flight pulsations on saccadic control. *Ophthalmologic and Physiological Optics, 19*(3), 266–273.

Bass, M. (1995). *Handbook of optics,* Vol. 1. New York: McGraw-Hill.

Bauer, D., & Cavonius, C. R. (1980). Improving the legibility of visual display units through contrast reversal. In E. Grandjean & E. Vigliani (Eds.), *Ergonomic aspects of visual display terminals* (pp. 137–142). London: Taylor & Francis.

Baumann, K., & Lanz, H. (1998). *Mensch-Maschine-Schnittstellen elektronischer Geräte. Leitfaden für Design und Schaltungstechnik* [Human-machine-interface for electronic appliances: Guideline for design and circuitry]. Berlin, Germany: Springer.

Bay, S., & Ziefle, M. (2004). Effects of menu foresight on information access in small screen devices. *48th annual meeting of the Human Factors and Ergonomic Society* (pp. 1841–1845). Santa Monica, CA: Human Factors Society.

Bergqvist, U., Wolgast, E., Nilsson, B., & Voss, M. (1995). The influence of VDT work on musculoskeletal disorders. *Ergonomics, 38*(4), 754–762.

Best, P., Littleton, M., Gramopadhye, A., & Tyrell, R. (1996). Relations between individual differences in oculomotor resting states and visual inspecting performance. *Ergonomics, 39*, 35–40.

Biocca, F. A., & Rolland, J. P. (1998). Virtual eyes can rearrange your body: Adaptation to visual displacement in see-through, head-mounted displays. *Presence, 7*(3), 262–277.

Blankenbach, K. (1999). Multimedia-Displays 2 Von der Physik zur Technik [Multimedia displays 2 from physics to technology]. *Physikalische Blätter, 55*(5), 33–38.

Blundel, B. G., & Schwarz, A. J. (2000). *Volumetric three-dimensional display systems.* New York: John Wiley & Sons.

Boff, K. R., & Lincoln, J. E. (1988). *Engineering data compendium—human perception and performance.* AAMRL, WPAFB, Ohio, Volumes 1–3. New York: John Wiley and Sons.

Boschmann, M., & Roufs, J. (1994). Reading and screen flicker. *Nature, 372*, 137–142.

Bosman, D. (1989). *Display engineering: Conditioning, technologies, applications.* Amsterdam: Elesevier Science.

Brodie, J., Chattratichart, J., Perry, M., & Scane, R. (2003). How age can inform the future design of the mobile phone experience. In C. Stephanidis (Ed.), *Universal Access in HCI: Inclusive design in the information society* (pp. 822–826). Mahwah, NJ: LEA.

Budin J.-P. (2003). Emissive displays: the relative merits of ACTFEL. In L. MacDonald & A. Lowe (Eds.), *Display systems–design and applications* (2nd ed., pp. 191–219). New York: John Wiley & Sons.

Çakir, A., Hart, D. J., & Stewart, T. F. M. (1979). *The VDT manual–ergonomics, workplace design, health and safety, task organization.* Darmstadt, Germany: IFRA–Inca-Fiej Research Association.

Candry, P. (2003). Projection systems. In L. MacDonald & A. Lowe (Eds.), *Display systems—design and applications* (2nd ed., pp. 237–256). New York: John Wiley & Sons.

Castellano, J. A. (1992). *Handbook of display technology.* San Diego, CA: Academic.

Dillon, A. (1992). Reading from paper versus screens: a critical review of the empirical literature. *Ergonomics, 35*(10), 1297–1326.

Duncan, J., & Humphreys, G. (1989). Visual search and stimulus similarity. *Psychological Review, 96*(3), 433–458.

E Ink Corporation (2001). *What is Electronic Ink?* Retrieved September 30, 2005, from http://www.eink.com/technology/index.htm

Feiner, S. K. (2002). Augmented reality–A new way of seeing. *Scientific American, 286*(4), 48–55.

Fruehauf, N., Aye, T., Yua, K., Zou, Y., & Savant, G. (2000). Liquid crystal digital scanner-based HMD. *Proceedings of SPIE Helmet- and Head-Mounted Displays V, 4021*, 2–10, Orlando, FL, USA.

Gibbs, W. W. (1998). The reinvention of paper. *Scientific America Online.* Retrieved September 30, 2005, from http://www.sciam.com/1998/0998issue/0998techbus1.html

Goldstein, M., Öqvist, G., Bayat, M., Ljungstrand, P., & Björk, S. (2001). Enhancing the reading experience: Using adaptive and sonified RSVP for reading on small displays. *Proceedings of the Mobile HCI, 2001* (pp. 1–9). Berlin, Germany: Springer.

Gould, J., Alfaro, L., Barnes, V., Finn, R., Grischkowsky, N., & Minuto, A. (1987). Reading is slower from CRT displays than from paper. *Human Factors, 29*(3), 269–299.

Gröger, T., Ziefle, M., & Sommer, D. (2003). Anisotropic characteristics of LCD TFTs and their impact on visual performance. In D. Harris, V. Duffy, M. Smith, & C. Stephanidis (Eds.), *Human-centred computing: Cognitive, social and ergonomic aspects* (pp. 33–37). Mahwah, NJ: LEA.

Hanson, A. (2003). Measurements and standardization in the colorimetry of CRT displays. In L. MacDonald & A. Lowe (Eds.), *Display systems—Design and applications* (2nd ed., pp. 309–327). New York: John Wiley & Sons.

Helander, M., Little, S., & Drury, C. (2000). Adaptation and sensitivity to postural changes in sitting. *Human Factors, 42*(4), 617–629.

Heppner, F., Anderson, J., Farstrup, A., & Weidenman, N. (1985). Reading performance on standardized test is better from print than from computer display. *Journal of Reading, 28*, 321–325.

Hollands, J., Cassidy, H., McFadden, S., & Boothby, R. (2001). LCD versus CRT Displays: Visual search for colored symbols. *Proceedings of the Human Factors and Ergonomics Society 45th Annual Meeting* (pp. 1353–1355). Santa Monica, CA: Human Factors Society.

Hollands, J., Parker, H., McFadden, S., & Boothby, R. (2002). LCD versus CRT Displays: A Comparison of visual search performance for colored symbols. *Human Factors, 44*(2), 210–221.

Holmes, R. (2003). Head-mounted display technology in virtual reality systems. In L. MacDonald & A. Lowe (Eds.), *Display systems—Design and applications*. (2nd., pp. 61–82). New York: John Wiley & Sons.

Holzel, T. (1999). Are head-mounted displays going anywhere? *Information Display, 15*(10), 16–18.

Howarth, P., & Istance, H. (1985). The association between visual discomfort and the use of visual display units. *Behaviour & Information Technology, 4*, 131–149.

Hung, G., Ciuffreda, K., & Semmlow J. (1986). Static vergence and accomodation: population norms and orthopic effects. *Documenta Ophthalmologica, 62*, 165–179.

ILT (2001). *Fraunhofer Institut für Lasertechnik (ILT)*. Retrieved October 15, 2005, from http://www.ilt.fhg.de

Iwasaki, T., & Kurimoto, S. (1988). Eye-strain and changes in accommodation of the eye and in visual evoked potential following quantified visual load. *Ergonomics, 31*(12), 1743–1751.

Jaschinski, W., Bonacker, M., & Alshuth, E. (1996). Accommodation, convergence, pupil and eye blinks at a CRT-display flickering near fusion limit. *Ergonomics, 19*, 152–164.

Jaschinski, W., Heuer, H., & Kylian, H. (1998). Preferred position of visual displays relative to the eyes: a field study of visual strain and individual differences. *Ergonomics, 41*, 1034–1049.

Kaufman, L. (1974). *Sight and mind 2 An introduction to visual perception*. London: Oxford University.

Kennedy, A., & Murray, W. (1993). 'Flicker' on the VDU screens. *Nature, 365*, 213.

King, C. N. (1996). Electroluminescent displays, SID 96 Seminar Lecture Notes, M-9, 1–36.

Kokoschka, S., & Haubner, P. (1986). Luminance ratios at visual display workstations and visual performance. *Lighting Research & Technology, 17*(3), 138–144.

Kothiyal, K., & Tettey, S. (2001). Anthropometry for design for the elderly. *International Journal of Occupational Safety and Ergonomics, 7*(1), 15–34.

Küller, R., & Laike, T. (1998). The impact of flicker from fluorescent lighting on well-being performance and physiological arousal. *Ergonomics, 4* (4), 433–447.

Lieberman, D. (1999). Computer display clips onto eyeglasses. *Technology News*. Retrieved October 15, 2005, from http://www.techweb.com/wire/story/TWB19990422S0003

Lindner, H., & Kropf, S. (1993). Asthenopic complaints associated with fluorescent lamp illumination (FLI). *Lighting Research and Technology, 25*(2), 59–69.

Luczak, H., & Oehme, O. (2002). Visual Displays 2 Developments of the past, the present and the future. In H. Luczak, A. Çakir, & G. Çakir (Eds.), *Proceedings of the 6th International Scientific Conference on Work with Display Units* (pp. 2–5). Berlin, Germany: Ergonomic Institute.

Luczak, H., Park, M., Balazs, B., Wiedenmaier, S., & Schmidt, L. (2003). Task performance with a wearable augmented reality interface for welding. In D. Harris, V. Duffy, M. Smith, & C. Stephanidis (Eds). *Human–Computer Interaction. Cognitive, Social and Ergonomic Aspects* (pp. 98–102). Mahwah, NJ: LEA.

Lüder, E. (2003). Active matrix addressing of LCDs: merits and shortcomings. In L. MacDonald & A. Lowe (Eds.), *Display systems—Design and applications* (2nd ed., pp. 157–171). New York: John Wiley & Sons.

Macadam, D. L. (1982) Caliometry. *American Institute of Physics handbook* (pp. 6-182–6-197). New York: McGraw-Hill.

Matschulat, H. (1999). *Lexikon der Monitor-Technologie* [Lexicon of display technology]. Aachen, Germany: Elektor-Verlag

Menozzi, M., Lang, F., Näpflin, U., Zeller, C., & Krueger, H. (2001). CRT versus LCD: Effects of refresh rate, display technology and background luminance in visual performance. *Displays, 22*(3), 79–85.

Menozzi, M., Näpflin, U., & Krueger, H. (1999). CRT versus LCD: A pilot study on visual performance and suitability of two display technologies for use in office work. *Displays, 20*, 3–10.

Microvision (2000). *Homepage Microvision Inc*. Retrieved September 15, 2005, from http://www.mvis.com/

Miyao, M., Hacisalihzade, S., Allen, J., & Stark, L. (1989). Effects of VDT resolution on visual fatigue and readability: an eye movement approach. *Ergonomics, 32*(6), 603–614.

Nelson, T. J., & Wullert J. R. II (1997). *Electronic information display technologies* (Series on information display, Vol.3). River Edge, NJ: World Scientific.

Oehme, O., Schmidt, L., & Luczak, H. (2003). Comparison between the strain indicator HRV of a head-based virtual retinal display and LC head mounted displays for augmented reality. *International Journal of Occupational Safety and Ergonomics, 9*(4), 411–422.

Oehme, O., Wiedenmaier, S., Schmidt, L., & Luczak, H. (2001). Empirical studies on an augmented reality user interface for a head based virtual retinal display. In M. Smith, G. Salvendy, D. Harris, & R. Koubek (Eds.), *Usability evaluation and interface design: Cognitive engineering, intelligent agents and virtual reality* (pp. 1026–1030). Mahwah, NJ: LEA.

Oetjen, S., & Ziefle, M. (2004). Effects of anisotropy on visual performance regarding different font sizes. In H. Khalid, M. Helander, & A. Yeo (Eds.), *Work with computing systems* (pp. 442–447). Kuala Lumpur, Malaysia: Damai Sciences.

Oetjen, S., Ziefle, M., & Gröger, T. (2005). Work with visually suboptimal displays—in what ways is the visual performance influenced when CRT and TFT displays are compared? In *Proceedings of the HCI International 2005: Vol. 4: Theories, Models and Processes in Human Computer Interaction*. St. Louis, MO: Mira Digital Publisher.

Oetjen, S., & Ziefle, M. (in press). The effects of LCD anisotropy on the visual performance of users of different ages. *Human Factors, 49*(4).

Omori, M., Watanabe, T., Takai, J., Takada, H., & Miyao, M. (2002). Visibility and characteristics of the mobile phones for elderly people. *Behaviour & Information Technology, 21*(5), 313–316. St. Louis (MO).

Ono, Y. A. (1993) Electroluminescent displays, Seminar Lecture Notes, F-1/1-30, Society for Information Display, Santa Ana: CA.

Owens, D., & Wolf- Kelly, K. (1987). Near work, visual fatigue, and variations of oculomotor tonus. *Investigative Ophthalmology and Visual Science, 28*(4), 743–749.

Park, M., Schmidt, L., & Luczak, H. (2005). Changes in hand-eye-coordination with different levels of camera displacement from natural eye position. In *10th International Conference on Human Aspects of Advanced Manufacturing: Agility and Hybrid Automation 2 HAAMAHA 2005*, San Diego, CA, USA, 191–199.

Pausch, R., Dwivedi, P., & Long, J. A. C. (1991). A practical, low-cost stereo head-mounted display. *Proceedings of SPIE. Stereoscopic Displays and Applications II, 1457*, 198–208.

Pfendler, C., Widdel, H., & Schlick, C. (2005). *Bewertung eines Head-Mounted und eines Hand-Held Displays bei einer Zielerkennungsaufgabe* [Task related evaluation of head-mounted and hand-held displays]. *Zeitschrift für Arbeitswissenschaft, 59*(1), 13–21.

Pfendler, C., & Schlick, C. (in press). A comparative study of mobile map displays in a geographic orientation task. *Behaviour and Information Tecchnology, 26*(2).

Philips Research Press Information (2005). *Philips demonstrates photonic textiles that turn fabric into intelligent displays.* Retrieved October 15, 2005, from http://www.research.philips.com/newscenter/archive/2005/050902-phottext.html

Pioneer (2001). *Why choose Plasma.* Retrieved September 15, 2005, from http://www.pioneerelectronics.com/Pioneer/CDA/Common/ArticleDetails/0,1484,1547,00.html

Plainis, S., & Murray, I. (2000). Neurophysiological interpretation of human visual reaction times: effect of contrast, spatial frequency and luminance. *Neuropsychologia, 38*, 1555–1564.

Precht, M., Meier, N., & Kleinlein, J. (1997). *EDV-Grundwissen: Eine Einführung in Theorie und Praxis der modernen EDV* [Computer basics: An introduction in theory and application of modern computing]. Bonn, Germany: Addison-Wesley-Longman.

Rahman, T., & Muter, P. (1999). Designing an interface to optimize reading with small display windows. *Human Factors, 41*(1), 106–117.

Sanders, M. S., & McCormick, E. J. (1993). Human Factors in Engineering and Design (7th ed.). New York: McGraw-Hill.

Schadt, M. (1996). *Optisch strukturierte Flüssigkeitskristall-Anzeigen mit großem Blickwinkelbereich* [Optically structured liquid crystal displays with wide viewing angles]. *Physikalische Blätter, 52*(7/8), 695–698.

Schenkmann, B., Fukunda, T., & Persson, B. (1999). Glare from monitors measured with subjective scales and eye movements. *Displays, 20*, 11–21.

Schlick, C., Daude, R., Luczak, H., Weck, M., & Springer, J. (1997): Head-mounted display for supervisory control in autonomous production cells. *Displays, 17*(3/4), 199–206.

Schlick, C., Reuth, R., & Luczak, H. (2000). Virtual Reality User Interface for Autonomous Production. In L. M. Camarinha-Matos, H. Afsarmanesh, & H. Erbe (Eds.), *Advances in Networked Enterprises* (pp. 279–286). Dordrecht, The Netherlands: Kluwer Academic Publishers.

Schlick, C., Winkelholz, C., Motz, F., & Luczak, H. (2006). Self-Generated Complexity and Human-Machine Interaction. *IEEE Transactions on Systems, Man, and Cybernetics—Part A: Systems and Humans, 36*(1), 220–232.

Sheedy, E. J. (2005). Office Lighting for Computer Use. In J. Anshel, (Ed.), *Visual ergonomics handbook* (pp. 37–51). Boca Raton, FL: CRC.

Sheedy, J., Subbaram, M., & Hayes, J. (2003). Filters on computer displays-effects on legibility, performance and comfort. *Behaviour and Information Technology, 22*(6), 427–433.

Sommerich, C., Joines, S., & Psihoios, J. (2001). Effects of computer monitor viewing angle and related factors on strain, performance, and preference outcomes. *Human Factors, 43*(1), 39–55.

Stanney, K. M., & Zyda, M. (2002) Virtual environments in the 21st century. In K. M. Stanney, (Ed.) *Handbook of virtual environments* (pp. 1–14). Mahwah, NJ: LEA.

Stone, P., Clarke, A., & Slater, A. (1980). The effect of task contrast on visual performance and visual fatigue at a constant illuminance. *Lighting Research and Technology, 12*, 144–159.

Theis, D. (1999). Display Technologie [Display technology]. In C. Müller-Schloer & B. Schallenberger (Eds.), *Vom Arbeitsplatzrechner zum ubiquitären Computer* [From the desktop PC to the ubiquitous computer]. (pp. 205–238), Berlin, Germany: VDE.

Urey, H., Wine, D. W., & Lewis J. R. (1999). Scanner design and resolution tradeoffs for miniature scanning displays. In B. Gnade & E. F. Kelley (Eds.), *Proceedings SPIE, Flat Panel Display Technology and Display Metrology,3636,* San Jose, CA, USA, 60–68.

van Schaik, P., & Ling, J. (2001). The effects of frame layout and differential background contrast on visual search performance in web pages. *Interacting with Computers, 13*, 513–525.

von Waldkirch, M. (2005). *Retinal projection displays for accommodation-insensitive viewing.* Unpublished doctoral thesis, Zurich, Swiss Federal Institute of Technology.

Weiss, S. (2002). *Handheld Usability.* New York: John Wiley.

Wilkins, A., Nimmo-Smith, I., Tait, A., McManus, C., Della Sala, S., Tilley, A., et al. (1984). A neurological basis for visual discomfort. *Brain, 107*, 989–1017.

Wolf, E., & Schraffa, A. (1964). Relationship between critical flicker frequency and age in flicker perimetry. *Archives of Ophthalmology, 72*, 832–843.

Woodson, W. E. (1987). *Human factor reference guide for electronics and computer professionals.* New York: McGraw-Hill.

Yeh, Y.-Y., & Wickens, C. (1984). Why do performance and subjective workload measures dissociate? *Proceedings of the 28th Annual meeting of the Human Factors Society* (pp. 504–508). Santa Monica, CA: Human Factors and Ergonomics Society.

Ziefle, M., & Bay, S. (2005). How older adults meet cognitive complexity: Aging effects on the usability of different cellular phones. *Behavior & Information Technology, 24*(5), 375–389.

Ziefle, M., & Bay, S. (2006). How to overcome disorientation in mobile phone menus: A comparison of two different types of navigation aids. *Human Computer Interaction, 21*(4), 393–433.

Ziefle, M. (1998). Effects of display resolution on visual performance. *Human Factors, 40*(4), 554–568.

Ziefle, M. (2001a). CRT screens or TFT displays? A detailed analysis of TFT screens for reading efficiency. In M. Smith, G. Salvendy, D. Harris, & R. Koubek (Eds.), *Usability evaluation and interface design* (pp. 549–553). Mahwah, NJ: LEA.

Ziefle, M. (2001b). Aging, visual performance and eyestrain in different screen technologies. *Proceedings of the Human Factors and Ergonomics Society 45th annual meeting* (pp. 262–266). Santa Monica, CA: Human Factors Society.

Ziefle, M. (2002). *Lesen am Bildschirm* [Reading from screens]. Münster, Germany: Waxmann.

Ziefle, M. (2003a). Users with body heights above and below the average: How adequate is the standard VDU setting with respect to visual performance and muscular load? In H. Luczak & K. J. Zink (Eds.), *Human Factors in Organizational Design and Management* (pp. 489–494). Santa Monica, CA: IEA Press.

Ziefle, M. (2003b). Sitting posture, postural discomfort, and visual performance: A critical view on the independence of cognitive and anthropometric factors in the VDU workplace. *International Journal of Occupational Safety and Ergonomics, 9*(4), 495–506.

Ziefle, M., Gröger, T., & Sommer, D. (2003). Visual costs of the inhomogeneity of contrast and luminance by viewing LCD-TFT screens off-axis. *International Journal of Occupational Safety and Ergonomics, 9*(4), 507–517.

Ziefle, M., Oehme, O., & Luczak, H. (2005). Information presentation and visual performance in head-mounted displays with augmented reality. *Zeitschrift für Arbeitswissenschaft, 59*(3–4), 331–344.

·12·

HAPTIC INTERFACE

Hiroo Iwata
University of Tsukuba

It is well known that sense of touch is inevitable for understanding the real world. The use of force feedback to enhance computer-human interaction (CHI) has often been discussed. A Haptic interface is a feedback device that generates sensation to the skin and muscles, including a sense of touch, weight, and rigidity. Compared to ordinary visual and auditory sensations, a haptic is difficult to synthesize. Visual and auditory sensations are gathered by specialized organs, the eyes and ears. On the other hand, a sensation of force can occur in any part of the human body and is therefore inseparable from actual physical contact. These characteristics lead to many difficulties when developing a haptic interface. Visual and auditory media are widely used in everyday life, although little application of haptic interface is used for information media.

In the field of virtual reality, haptic interface is one of the major research areas. The last decade has seen significant advances in the development of haptic interface. High-performance haptic devices have been developed and some of them are commercially available. This chapter presents current methods and issues in haptic interface.

The next section describes the mechanism of haptic sensation and overall view of feedback technologies. This section is followed by three sections that introduce examples of haptic-interface technologies developed by the author. The last section presents application areas and future prospects of haptic interface.

MECHANISM OF HAPTICS AND METHODS FOR HAPTIC FEEDBACK

Somatic Sensation

Haptic interface presents synthetic stimulation to somatic sensation. Somatic sensation is composed of proprioception and skin sensation. Proprioception is complemented by mechanoreceptors of skeletal articulations and muscles. There are three types of joint position receptors: (a) free nerve ending as well as (b) Ruffini and (c) Pacinian corpuscles. Ruffini corpuscle detects static force. On the other hand, Pacinian corpuscle has a function to measure acceleration of the joint angle. Position and motion of the human body is perceived by these receptors. Force sensation is derived from mechanoreceptors of muscles, muscle spindles, and Goldi tendons. These receptors detect contact forces applied by an obstacle in the environment.

Skin sensation is derived from mechanoreceptors and thermorecepters of skin. Sense of touch is evoked by those receptors. Mechanoreceptors of skin are classified into four types: (a) Merkel disks, (b) Ruffini capsules, (c) Meissner corpuscles, and (d) Pacinian corpuscles. These receptors detect edges of objects, skin stretching, velocity, and vibration, respectively.

Proprioception and Force Display

Force display is a mechanical device that generates a reaction force from virtual objects. Recently, much research has been conducted on haptic interfaces, although the technology is still in a state of trial and error. There are several approaches to implementing haptic interfaces:

Exoskeleton-Type of Force Display

An *exoskeleton* is a set of actuators attached to a hand or a body. In the field of robotics research, exoskeletons have often been used as master manipulators for teleoperations. However, most master manipulators entail a large amount of hardware and therefore have a high cost, which restricts their application areas. Compact hardware is needed to use them in HCI. The first example of a compact exoskeleton suitable for desktop use was published in 1990 (Iwata, 1990). The device applies force to the fingertips as well as the palm. Figure 12.1 shows an overall view of the system.

Lightweight and portable exoskeletons have also been developed. Burdea used small pneumatic cylinders to apply the force to the fingertips (Burdea, Zhuang, Roskos, Silver, & Langlana, 1992). Cyber Grasp (Fig. 12.2) is a commercially available exoskeleton, in which cables are used to transmit force [http://www.vti.com].

Tool-Handling Type of Force Display

The tool-handling type of force display is the easiest way to realize force feedback. The configuration of this type is similar to that of a joystick. Unlike the exoskeleton, the tool-handling-type force display is free from the need to be fitted to the user's hand. It cannot generate a force between the fingers, but has practical advantages.

FIGURE 12.1. Desktop force display.

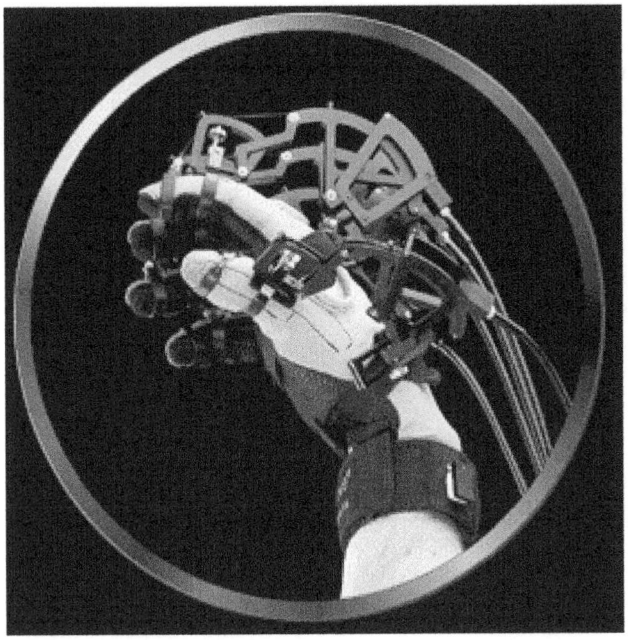

FIGURE 12.2.   Cyber Grasp.

FIGURE 12.3.   PHANToM.

A typical example of this category is the pen-based force display (Iwata, 1993). A pen-shaped grip is supported by two 3DOF pantographs that enable a 6DOF force/torque feedback. Another example of this type is the Haptic Master, which was demonstrated at the edge venue of SIGGRAPH'94. The device has a ball-shaped grip to which 6 DOF force/torque is fed back (Iwata, 1994). This device employs a parallel mechanism in which a top triangular platform and a base triangular platform are connected by three sets of pantographs. This compact hardware has the ability to carry a large payload.

Massie and Salisbury developed the PHANToM, which has a 3 DOF pantograph (Massie & Salisbury, 1994). A thimble with a gimbal is connected to the end of the pantograph, which can then apply a 3 DOF force to the fingertips. The PHANToM became one of the most popular commercially available haptic interfaces (Fig. 12.3).

Object-Oriented Type of Force Display

The object-oriented type of force display is a radical idea for the design of a haptic interface. The device moves or deforms to simulate the shapes of virtual objects. A user can make physical contact the surface of the virtual object.

An example of this type can be found in Tachi and colleagues' work (Tachi et al., 1994). Their device consists of a shape-approximation prop mounted on a manipulator. The position of the fingertip is measured and the prop moves to provide a contact point for the virtual object. McNeely proposed an idea named "Robotic Graphics" (McNeely, 1993), which is similar to Tachi and colleagues' method. Hirose developed a surface display that creates a contact surface using a 4 × 4 linear actuator array (Hirota & Hirose, 1996). The device simulates an edge or a vertex of a virtual object.

Passive Prop

A passive input device equipped with force sensors is a different approach to the haptic interface. Murakami and Nakajima used a flexible prop to manipulate a three dimensional (3D) virtual object (Murakami & Nakajima, 1994). The force applied by the user is measured and the deformation of the virtual object is determined based on the applied force. Sinclair developed a force-sensor array to measure pressure distribution (Sinclair, 1997). These passive devices allow users to interact using their bare fingers. However, these devices have no actuators, so they cannot represent the shape of virtual objects.

Proprioception and Full-Body Haptics

One of the new frontiers of haptic interface is full-body haptics that includes foot haptics. Force applied to a whole body plays a very important role in locomotion. The most intuitive way to move about the real world is walking on foot. *Locomotion interface* is a device that provides a sense of walking while the walker's body is localized in the real world. There are several approaches to realizing locomotion interface:

Sliding Device

A project named "Virtual Perambulator" developed locomotion interface using a specialized sliding device (Iwata & Fujii, 1996). The primary objective of the first stage was to allow the walker's feet to change direction. Controlling steering bars or joysticks are not as intuitive as in locomotion. The first prototype of the Virtual Perambulator was developed in 1989 (Iwata & Matsuda, 1990). Figure 12.4 shows overall view of the apparatus. A

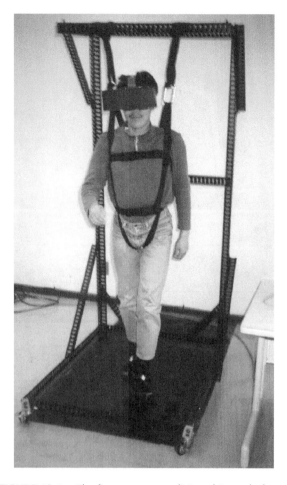

FIGURE 12.4.   The first prototype of Virtual Perambulator.

user wore a parachute-like harness and omni-directional roller skates. The trunk of the walker was fixed to the framework of the system by the harness. An omni-directional sliding device is used for changing direction by foot. Specialized roller skates equipped with four casters were developed, which enabled two-dimensional motion. The walker could freely move his or her feet in any direction. Motion of the feet was measured by ultrasonic range detector. From the result of this measurement, an image of the virtual space was displayed in the head-mounted display corresponding with the motion of the walker. The direction of locomotion in virtual space was determined according to the direction of the walker's step.

Treadmill

A simple device for virtual walking is a treadmill, a device ordinarily used for physical fitness. An application of this device used as a virtual building simulator was developed at UNC (Brooks, 1986). This treadmill has a steering bar similar to that of a bicycle. A treadmill equipped with a series of linear actuators underneath the belt was developed at ATR (Noma, Sugihara, & Miyasato, 2000). The device was named GSS, and it

simulates the slope of virtual terrain. The Treadport developed at the University of Utah is a treadmill combined with a large manipulator connected to a walker (Christensen, 1998). The manipulator provides gravitational force while the walker is passing a slope. Figure 12.5 shows overall view of the Treadport.

The omni-directional treadmill employs two perpendicular treadmills, one inside of the other. Each belt is made from approximately 3,400 separate rollers, which are woven together into a mechanical fabric. Motion of the lower belt is transmitted to the walker with rollers. This mechanism enables omni-directional walking (Darken, Cockayne, & Carmein, 1997).

Foot Pads

A foot pad applied to each foot is an alternative implementation of locomotion interface. Two large manipulators driven by hydraulic actuators were developed at University of Utah and applied to a locomotion interface. These manipulators are attached to a walker's feet. The device is named BiPort [http://www.sarcos.com]. The manipulators can present viscosity of virtual ground. A similar device has been developed at the Cybernet Systems Corporation, which uses two 3 DOF motion platforms for the feet (Poston et al., 1997). These devices, however, have not been evaluated or applied to VE.

Pedaling Device

In the battlefield simulator of NPSNET project, a unicycle-like pedaling device is used for locomotion in the virtual battlefield (Plat et al., 1994). A player changes direction by twisting his or her waist.

The OSIRIS, a simulator of night-vision battle, utilizes a stair-stepper device (Lorenzo et al., 1997). A player changes direction using a joystick or by twisting his or her waist.

FIGURE 12.5.   Treadport.

Gesture Recognition of Walking

Slater et al. proposed locomotion in virtual environments by "walking in place." They recognized the gesture of walking using a position sensor and a neural network (Slater et al., 1994).

Skin Sensation and Tactile Display

The tactile display that stimulates skin sensation is a well-known technology. It has been applied to communication aids for blind persons as well as a master system of tele-operation. A sense of vibration is relatively easy to produce, and a good deal of work has been done using vibration displays (Kontarinis & Howe, 1995; Minsky & Lederman, 1997). The micropin array is also used for tactile displays. Such a device has enabled the provision of a teltaction and communication aid for blind persons (Moy et al., 2000; Kawai & Tomita, 2000). It has the ability to convey texture or 2D-geometry (Burdea, 1996).

The micropin array looks similar to the object-oriented-type force display, but it can only create the sensation of skin. The stroke distance of each pin is short, so the user cannot feel the 3D-shape of a virtual object directly. The major role of tactile display is to convey the sense of fine texture of an object's surface. The latest research activities of tactile display focus on selective stimulation of mechanoreceptors of the skin. As mentioned at the beginning of this section, there are four types of mechanoreceptors in the skin: (a) Merkel disks, (b) Ruffini capsules, (c) Meissner corpuscles, and (d) Pacinian corpuscles. Stimulating these receptors selectively, various tactile sensations such as roughness or slip can be presented. Microair jet (Asanura et al.,1999) and microelectrode array (Kajimoto et al., 1999) are used for selective stimulation.

TECHNOLOGIES IN FINGER/HAND HAPTICS

Exoskeleton

An exoskeleton is one of the typical forms of haptic interface. Figure 12.6 shows detailed view of an exoskeleton (Iwata, 1990).

Force sensation contains six-dimensional information: three-dimensional forces and three-dimensional torque. The core element of the force display is a six-degree-of-freedom parallel manipulator. The typical design feature of parallel manipulators is an octahedron called "Stewart platform." In this mechanism, a top triangular platform and a base triangular platform are connected by six length-controllable cylinders. This compact hardware has the ability to carry a large payload. The structure, however, has some practical disadvantages in its small-working volume and its lack of backdrivability (reduction of friction) of the mechanism. In our system, three sets of parallelogram linkages (pantograph) are employed instead of linear actuators. Each pantograph is driven by two DC motors. Each motor is powered by a PWM (pulse width modulation) amplifier. The top end of the pantograph is connected to the vertex of the top platform with a spherical joint. This mechanical configuration has the same advantages as an octahedron mechanism. The pantograph

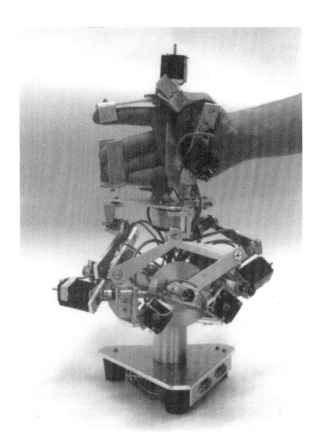

FIGURE 12.6. Desktop force display.

mechanism improves the working volume and backdrivability of the parallel manipulator. The inertia of motion parts of the manipulator are so small that compensation is not needed.

The working space of the center of the top platform is a spherical volume whose diameter is approximately 30 cm. Each joint angle of the manipulator is measured by potentiometers. Linearity of the potentiometers is 1%. The maximum payload of the manipulator is 2.3 kg, which is more than a typical hand.

The top platform of the parallel manipulator is fixed to the palm of the operator with a U-shaped attachment, which enables the operator to move the hand and fingers independently. Three actuators are set coaxially with the first joint of the thumb, forefinger, and middle finger of the operator. The last three fingers work together. DC servo motors are employed for each actuator.

Tool-Handling-Type of Haptic Interface

Exoskeletons are troublesome for users to put on or take off. This disadvantage obstructs practical use of force displays. A tool-handling-type of haptic interface implements force display without a glove-like device. A pen-based force display is proposed as an alternative device (Iwata, 1993). A six-degree-of-freedom force reflective master manipulator, which has pen-shaped grip, was developed. Users are familiar with pens because most human intellectual works are done in pen. In this aspect, the pen-based force display is easily applied to the design of 3D shapes.

The human hand has the ability of six-degree-of-freedom motion in 3D space. If a six-degree-of-freedom master manipulator is built using serial joints, each joint must support the weight of upper joints. Because of this, manipulator hardware must be large; we use parallel mechanism in order to reduce size and weight of the manipulator. The pen-based force display employs two three-degree-of-freedom manipulators. Both ends of the pen are connected to these manipulators. Total degree-of-freedom of the force display is six. Three-degree-of-freedom force and three-degree-of-freedom torque are applied at the pen. An overall view of the force display is shown in Fig. 12.7. Each three-degree-of-freedom manipulator is composed of pantograph link. By this mechanism, the pen is free from the weight of the actuators.

Figure 12.8 shows a diagram of mechanical configuration of the force display. Joints MA1, MA2, MA3, MB1, MB2, and MB3 are equipped with DC motors and potentiometers. Other joints move passively. The position of joint A and B is measured by potentiometers. A three-dimensional force vector is applied at the joint A and B. The joint A determines the position of the pen point, and the joint B determines the orientation of the pen. Working space of the pen point is a part of a spherical volume whose diameter is 44 cm. The rotational angle around the axis of the pen is determined by the distance between joint A and B. A screw-motion mechanism converts rotational motion of the pen into transition of the distance between joint A and B.

Applied force and torque at the pen is generated by combination of forces at the point A and B. In case these forces have the same direction, translational force is applied to the user's hand. If direction of the forces are reversed, torque around the yaw axis or the pitch axis is generated. If two forces are opposite, torque around the roll axis is generated by the screw-motion mechanism.

TECHNOLOGIES IN FINGER/HAND OUTPUT: OBJECT-ORIENTED-TYPE HAPTIC INTERFACE

Basic Idea of FEELEX

The author demonstrated haptic interfaces to a number of people, and found that some of them were unable to fully experience virtual objects through the medium of synthesized haptic sensation. There seem to be two reasons for this phenomenon. First, these haptic interfaces only allow the users to touch the virtual object at a single point or at a group of points. These contact points are not spatially continuous, due to the hardware configuration of the haptic interfaces. The user feels a reaction force through a grip or thimble. Exoskeletons provide more contact points, but these are achieved using Velcro bands attached to a specific part of the user's fingers, which are not continuous. Therefore, these devices cannot recreate a natural interaction sensation when compared to manual manipulation in the real world.

The second reason why users fail to perceive the sensation is related to a combination of the visual and haptic displays.

FIGURE 12.7. Pen-based force display.

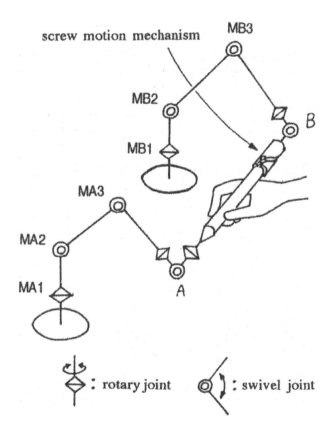

FIGURE 12.8. Mechanical configuration of pen-based force display.

A visual image is usually combined with a haptic interface using a conventional CRT or projection screen. Thus, the user receives visual and haptic sensation through different displays, and therefore has to integrate the visual and haptic images in his or her brain. Some users, especially elderly people, have difficulty in this integration process.

Considering these problems, new interface devices have developed. The project is named "FEELEX." The word FEELEX is derived from a conjunction of "feel" and "flex." The major goals of this project are,

- to provide a spatially continuous surface that enables users to feel virtual objects using any part of the fingers or even the whole palm
- to provide visual and haptic sensations simultaneously using a single device that doesn't oblige the user to wear any extra apparatus

A new configuration of visual/haptic display was designed to achieve these goals. Figure 12.9 illustrates the basic concept of the FEELEX. The device is composed of a flexible screen, an array of actuators, and a projector. The flexible screen is deformed by the actuators in order to simulate the shape of virtual objects. An image of the virtual objects is projected onto the surface of the flexible screen. Deformation of the screen converts the 2D image from the projector into a solid image. This configuration enables users to touch the image directly using any part of their hand. The actuators are equipped with force sensors to measure the force applied by the user. The hardness of the virtual object is determined by the relationship between the measured force and its position on the screen. If the virtual object is soft, a large deformation is caused by a small, applied force.

DESIGN SPECIFICATION AND IMPLEMENTATION OF PROTOTYPES

FEELEX 1

The FEELEX 1, developed in 1997, was designed to enable double-handed interaction using the whole palms. Therefore, the optimum size of the screen was determined to be 24 cm x 24 cm. The screen is connected to a linear actuator array that deforms its shape. Each linear actuator is composed of a screw mechanism driven by a DC motor. The screw mechanism converts the rotation of an axis of the motor to the linear motion of a rod. The motor must generate both motion and a reaction force on the screen. The diameter of the smallest motor that can drive the screen is 4 cm. Therefore, a 6 x 6 linear-actuator array can be set under the screen. The deformable screen is made of a rubber plate and a white nylon cloth. The thickness of the rubber is 3 mm. Figure 12.10 shows an overall view of the device.

The screw mechanism of the linear actuator has a self-lock function that maintains its position while the motor power is off. Hard virtual wall is difficult to simulate using tool-handling-type force displays. Considerable motor power is required to generate the reaction force from the virtual wall, which often leads to uncomfortable vibrations. The screw mechanism is free from this problem. A soft wall can be represented by the computer-

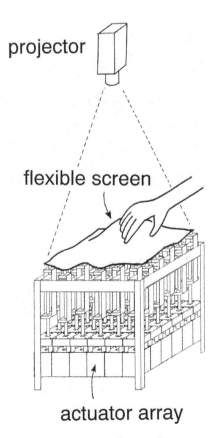

FIGURE 12.9.   Basic design of FEELEX.

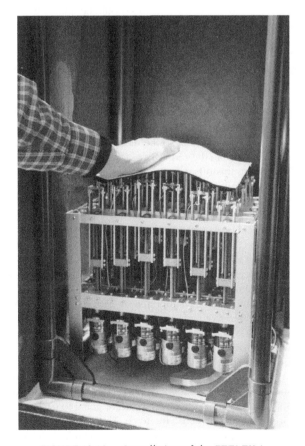

FIGURE 12.10.   Overall view of the FEELEX 1.

controlled motion of the linear actuators based on the data from the force sensors. A force sensor is set at the top of each linear actuator. Two strain gauges are used as a force sensor. The strain gauge detects small displacements of the top end of the linear actuator caused by the force applied by the user. The position of the top end of the linear actuator is measured by an optical encoder connected to the axis of the DC motor. The maximum stroke of the linear actuator is 80 mm, and the maximum speed is 100 mm/s.

The system is controlled via a personal computer (PC). The DC motors are interfaced by a parallel I/O unit, and the force sensors are interfaced by an A/D converter unit. The force sensors provide interaction with the graphics. The position and strength of the force applied by the user are detected by a 6 × 6 sensor array. The graphics projected onto the flexible screen are changed according to the measured force.

FEELEX 2

The FEELEX 2 is designed to improve the resolution of the haptic surface. In order to determine the resolution of the linear actuators, we considered the situation where a medical doctor palpates a patient. After interviewing several medical doctors, it was discovered that they usually recognized a tumor using their index finger, middle finger, and third finger. The size of a tumor is perceived by comparing it to the width of their fingers, i.e. two-fingers large or three-fingers large. Thus, the distance between the axis of the linear actuators should be smaller than the width of a finger. Considering the condition just described, the distance is set to be 8 mm. This 8 mm resolution enables the user to hit at least one actuator when he or she touches any arbitrary position on the screen. The size of the screen is 50 mm × 50 mm, which allows the user to touch the surface using three fingers.

In order to realize 8 mm resolution, a piston-crank mechanism is employed for the linear actuator. The size of the motor is much larger than 8 mm, so the motor should be placed at a position offset from the rod. The piston-crank mechanism can easily achieve this offset position. Figure 12.11 illustrates the mechanical configuration of the linear actuator. A servo-motor from a radio-controlled car is selected as the actuator. The rotation of the axis of the servo-motor is converted to the linear motion of the rod by a crank-shaft and a linkage. The stroke of the rod is 18 mm, and the maximum speed is 250 mm/s. The maximum torque of the servo-motor is 3.2 Kg-cm, which applies a 1.1 Kgf force at the top of each rod. This force is sufficient for palpation using the fingers.

The flexible screen is supported by 23 rods, and the servo-motors are set remotely from the rods. Figure 12.12 shows an overall view of the FEELEX 2. The 23 separate sets of piston-crank mechanisms can be seen in the picture.

Figure 12.13 shows the top end of the rods. The photo was taken while the flexible screen is off. The diameter of each rod is 6 mm. A strain gauge cannot be put on top of the rod because of its small size. Thus, the electric current going to each servo-motor is measured to sense the force. The servo-motor generates a force to maintain the position of the crank shaft. When the user applies force to the rod, the electric current on the mo-

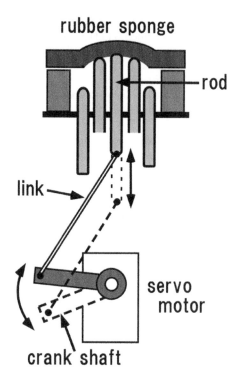

FIGURE 12.11.   Piston-crank mechanism.

FIGURE 12.12.   Overall view of the FEELEX 2.

FIGURE 12.13.   Top end of the rods.

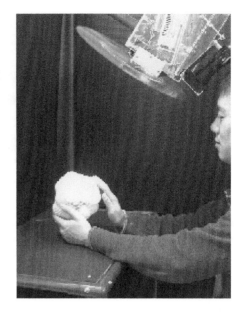

FIGURE 12.14.   Overall view of Volflex.

tor increases to balance the force. The relationship between the applied force and the electric current is measured. The applied force at the top of the rods is calculated using data from the electric current sensor. The resolution of the force sensing capability is 40 gf.

Characteristics of the FEELEX

The performance of existing haptic interfaces is usually represented by the dynamic range of force, impedance, inertia, friction, etc. However, these parameters are only crucial while the device is attached to the finger or hand. In the case of the tool-handling-type of haptic interface or the exoskeleton, the devices move with the hand even though the user doesn't touch the virtual objects. Therefore, inertia or friction degrades the usability and the dynamic range of force determines the quality of the virtual surface. On the other hand, the FEELEX is entirely separate from the user's hand, so its performance is determined by the resolution and speed of the actuators. The resolution of the actuator corresponds to the smoothness of the surface and the speed of the actuator determines the motion of the virtual object. The FEELEX 2 has improved resolution and motion speed compared to the FEELEX 1. Each actuator of the FEELEX 2 has a stroke rate of up to 7 Hz, which can simulate the motion of a very fast virtual object. The rod pushes the rubber sponge so that the user feels as if the object was pulsating. 7 Hz is much faster than the human pulse rate.

The major advantage of the FEELEX is that it allows natural interaction using only the bare hand. In SIGGRAPH'98, 1,992 subjects spontaneously enjoyed the haptic experience. One of the subject contents of the FEELEX 1 system, known as Anomalocaris, was selected as a long-term exhibition at the Ars Electronica Center (Linz, Austria). The exhibition has been popular among visitors, and especially children.

Another advantage of FEELEX is safety. The user of FEELEX doesn't wear any special equipment while the interaction is taking place. The exoskeleton and tool-handling-type force displays have control problems in their contact surface for the virtual objects. Vibration or unwanted forces can be generated back to the user, which is sometimes dangerous. The contact surface of the FEELEX is physically generated, so it is free from such control problems.

The major disadvantage of the FEELEX is the degree of difficulty in its implementation. It requires a large number of actuators that have to be controlled simultaneously. The drive mechanism of the actuator must be robust enough for rough manipulation. Since the FEELEX provides a feeling of natural interaction, some of the users apply large forces. Our exhibit at the Ars Electronica Center suffered from overload of the actuators.

Another disadvantage of the FEELEX is its limitation in the shape of objects that can be displayed. The current prototypes cannot present a sharp edge on a virtual object. Furthermore, the linear actuator array can only simulate the front face of objects. Some of the participants of the Anomalocaris demonstration wanted to touch the rear of the creature, but an entirely new mechanism would be required to simulate the reverse side of the object.

VOLMETRIC OBJECT-ORIENTED-TYPE OF HAPTIC INTERFACE

Basic Design of Volflex

In order to present a side or the back of a virtual object, we designed a volumetric object-oriented haptic interface. Volflex is a new haptic interface that provides the user with a physical 3D surface for interaction. The device is composed of a group of air balloons. The balloons fulfill the interaction surface and are arranged in a body-centered cubic lattice. A tube is connected to each balloon, and the volume of each balloon is controlled by an air cylinder. The tubes are connected to each other by springs. This mechanical flexibility enables an arbitrary shape of the

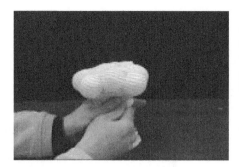

FIGURE 12.15. Examples of deformation.

interaction surface. Each air cylinder is equipped with a pressure sensor that detects force applied by the user. According to the pressure data, the device is programmed to perform like clay. Unlike real clay, Volflex allows the user to "Un do" an operation.

A projector is set above the balloons. An image is projected onto the surface of the device, not on the user's hand. We developed a mechanical rotary shutter that separates the projector and camera. The camera captures the user's hand, which is eliminated from the projected image.

Virtual Clay Volflex

Virtual clay is one of the ultimate goals of interactive technique of 3D graphics. Digital tools for 2D paint are mature technology. On the other hand, tools for 3D-shape manipulation are at only preliminary stage. Shape design of 3D objects is one of the major application areas of haptic interface. Shape design of 3D object requires good sensation of haptics.

The Volflex provides an effective interface device for manipulating virtual clay using a lattice of air balloons. Two-dimensional paint tools have been popular, and digital picture is easy to draw. The Volflex is a new digital tool for making 3D shapes. It has the potential to be a revolution in industrial design. Designers use their palm or the joints of their fingers to deform a clay model when carrying out rough-design tasks. The Volflex has the ability to support such natural manipulation.

The Volflex is not only a tool for 3D-shape design, but is also an interactive artwork. Physical properties of the virtual object can be designed by programming the balloons' controllers. A projected image can be also designed. The combination of haptic and visual display provides a new platform for interactive sculpture.

TECHNOLOGIES IN FULL-BODY HAPTICS

Treadmill-Based Locomotion Interface

Basic Design of the Torus Treadmill

A key principle of treadmill-based locomotion interface is to make the floor move in a direction opposite of that of the walker. The motion of the floor cancels displacement of the walker in the real world. The major problem with treadmill-based locomotion interface is that there is not yet a way to allow the walker to change direction. An omni-directional active floor enables a virtually infinite area. In order to realize an infinite walking area, geometric configuration of an active floor must be chosen. A closed surface driven by actuators has an ability to create an unlimited floor. The following requirements for implementation of the closed surface must be considered:

1. The walker and actuators must be put outside the surface.
2. The walking area must be a plane surface.
3. The surface must be made of a material that stretches very little.

The shape of a closed surface, in general, is a surface with holes. If the number of holes is zero, the surface is a sphere. The sphere is the simplest infinite surface. However, the walking area of the sphere is not a plane surface. A very large diameter is required to make a plane surface on a sphere, which restricts implementation of the locomotion interface.

A closed surface with one hole like a doughnut is called *torus*. A torus can be implemented using a group of belts, which make a plane surface for the user to walk on. A closed surface with more than one hole cannot make a plane walking surface. Thus, the torus is the only form suitable for a locomotion interface.

Mechanism and Performance

The Torus Treadmill operates using a group of belts connected to each other. The Torus Treadmill is realized by these belts (Iwata, 1999). Figures 12.16 and 12.17 illustrate basic structure of the Torus Treadmill, which employs 12 treadmills. These treadmills move the walker along in an "X" direction. Twelve treadmills are connected side by side and driven in a perpendicular direction., a motion that moves the walker along a "Y" direction.

Figure 12.18 shows the apparatus. Twelve treadmills are connected to four chains and mounted on four rails. The chain drives the walker along the Y direction. The rail supports the weight of the treadmills and the walker. An AC motor is employed to drive the chains. The power of the motor is 200 W and is controlled by an inverter. The maximum speed of rotation is 1.2 m/s. The maximum acceleration is 1.0 m/s^2. The deceleration caused by friction is 1.5 m/s^2. Frequency characteristics are

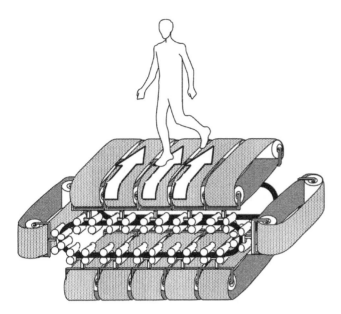

FIGURE 12.16. Structure of Torus Treadmill (X motion).

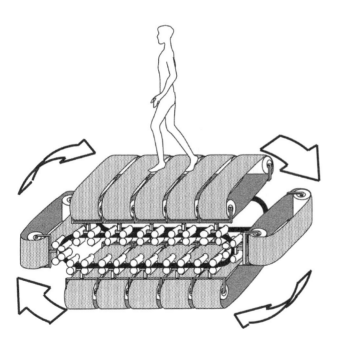

FIGURE 12.17. Structure of Torus Treadmill (Y motion).

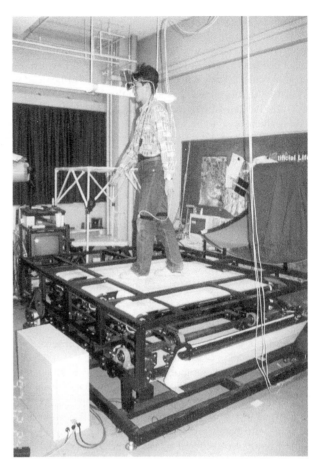

FIGURE 12.18. Torus Treadmill.

A problem with this mechanical configuration is the gap between the belts in the walking area. In order to minimize the gap, we put a driver unit of each treadmill alternatively. The gap is only 2 mm wide in this design.

Control Algorithm of the Torus Treadmill

A scene of the virtual space is generated corresponding with the results of motion tracking of the feet and head. The motion of the feet and head is measured by a Polhemus FASTRACK. The device measures a six-degree-of-freedom motion. The sampling rate of each point is 20 Hz. A receiver is attached to each knee. We cannot put the sensors near the motion floor because a steel frame distorts magnetic field. The length and direction of a step is calculated by the data from those sensors. The user's viewpoint in virtual space moves in correspondence with the length and direction of the steps.

To keep the walker in the center of the walking area, the Torus Treadmill must be driven in correspondence with the walker. A control algorithm is required to achieve safe and natural walking. From our experience in the Virtual Perambulator project, the walker should not be connected to a harness or mechanical linkages, since such devices restrict the motion and inhibit natural walking. The control algorithm of the Torus Tread-

limited by a circuit protector of the motor driver. The maximum switching frequency is 0.8 Hz.

Each treadmill is equipped with an AC motor. In order to shorten the length of the treadmill, the motor is put underneath the belt. The power of each motor is 80 W and is controlled by an inverter. The maximum speed of each treadmill is 1.2 m/s. The maximum acceleration is 0.8 m/s². The deceleration caused by friction is 1.0 m/s². The width of each belt is 250 mm and the overall walkable area is 1m × 1m.

mill must be safe enough to allow removal of the harness from the walker. At the final stage of the Virtual Perambulator Project, we succeeded in removing the harness using a hoop frame. The walker can freely walk and turn around in the hoop, which supports the walker's body while he or she slides his or her feet. We simulated the function of the hoop in the control algorithm of the Torus Treadmill by putting circular deadzone in the center of the walking area. If the walker steps out of the area, the floor moves in the opposite direction, so that the walker is carried back into the deadzone.

Foot-Pad-Based Locomotion Interface

Methods of Presentation of Uneven Surface

One of the major research issues in locomotion interface is the presentation of uneven surface. Locomotion interfaces are often applied for simulation of buildings or urban spaces. Those spaces usually include stairs. A walker should be provided with a sense of climbing up or going down those stairs. In some applications of locomotion interface, such as training simulators or entertainment facilities, rough terrain should be presented.

The presentation of a virtual staircase was tested at the early stage of the Virtual Perambulator project (Iwata & Fujii, 1996). A string was connected to the roller skate of each foot. The string is pulled by a motor. When the walker climbed up a stair, the forward foot was pulled up. When the walker went down the stairs, the backward foot was pulled up. However, this method was not considered successful because of instability.

Later, a 6 DOF motion platform was applied to a final version of the Virtual Perambulator, where a user walks in a hoop frame. The walker stood on the top plate of the motion platform. Pitch and heave motion of the platform were used. When the walker stepped forward to climb a stair, the pitch angle and vertical position of the floor increased. After finishing the climbing motion, the floor went back to the neutral position. When the walker stepped forward to go down a stair, the pitch angle and vertical position of the floor decreases. This inclination of the floor is intended to present height differences between the feet. The heave motion is intended to simulate vertical acceleration. However, this method failed in simulation of stairs, mainly because the floor was flat.

A possible method to create a difference in height between the feet is the application of two large manipulators, such as the BiPort. A 4 DOF manipulator driven by hydraulic actuators is connected to each foot. The major problem with this method is the way in which the manipulators trace the turning motion of the walker. When the walker turns around, two manipulators interfere with each other.

The Torus Treadmill provides a natural turning motion in which the walker can physically turn about on the active floor. This turning motion using the feet is a major contribution to human spatial recognition performance. Vestibular and proprioceptive feedback is essential to the sense of orientation (Iwata & Yoshida, 1999). The Torus Treadmill can be modified for simulation of an uneven surface. If we install an array of linear actuators on each treadmill, an uneven floor can be realized by controlling the length of each linear actuator. However, this method

is almost impossible to implement, because a very large number of linear actuators are required to cover the surface of the torus-shaped treadmills and control of the signal for each actuator must be transmitted wirelessly.

Basic Design of the GaitMaster

A new locomotion interface that simulates an omni-directional uneven surface has been designed, which is called the "GaitMaster." The core elements of the device are two 6 DOF motion-bases mounted on a turntable. Figure 12.19 illustrates the basic configuration of the GaitMaster.

A walker stands on the top plate of the motion base. Each motion base is controlled to trace the position of the foot. The turntable is controlled to trace the orientation of the walker. The motion of the turntable removes interference between the two motion bases.

The X and Y motion of the motion base traces the horizontal position of the feet and cancels its motion by moving in the opposite direction. The rotation around the yaw axis traces the horizontal orientation of the feet. The Z motion traces vertical position of the feet and cancels its motion. The rotation around the roll-and-pitch axis simulates the inclination of a virtual surface.

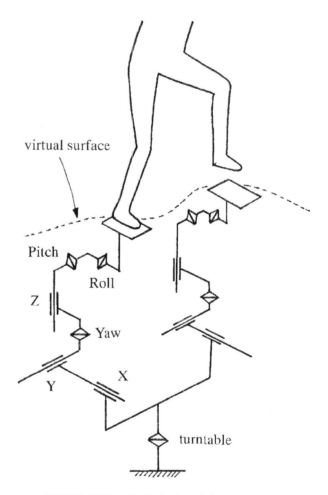

FIGURE 12.19. Basic design of the GaitMaster.

Control Algorithm of the GaitMaster

The control algorithm must keep the position of the walker at the neutral position of the GaitMaster. In order to keep the position maintained, the motion platforms have to cancel the motion of the feet. The principal of the cancellation is as follows:

1. Suppose the right foot is at the forward position and the left foot is at the backward position while walking.
2. When the walker steps forward with the left foot, the weight of the walker is laid on the right foot.
3. The motion platform of the right foot goes backward in accordance with the displacement of the left foot, so that the central position of the walker is maintained.
4. The motion platform of the left foot follows the position of the left foot. When the walker finishes stepping forward, the motion platform supports the left foot.

If the walker climbs up or goes down stairs, a similar procedure can be applied. The vertical motion of the feet is canceled using the same principle. The vertical displacement of the forward foot is canceled in accordance with the motion of the backward foot, so that the central position of the walker is maintained at the neutral height. Figure 12.20 illustrates the method of canceling the climbing-up motion.

The turntable rotates so that the two motion platforms can trace the rotational motion of the walker. If the walker changes the direction of his or her steps, the turntable rotates to trace the walker's orientation. The orientation of the turntable is determined according to direction of the feet. The turntable rotates so that its orientation is at the middle of the feet. The walker can physically turn around on the GaitMaster using this control algorithm of the turntable.

Prototype GaitMaster

Figure 12.21 shows the prototype GaitMaster. In order to simplify the mechanism of the motion platform, the surface of the virtual space was defined as sets of plainer surfaces. Most buildings or urbane spaces can be simulated without inclining of the floor. Thus, we can neglect the roll-and-pitch axis of the motion platforms. Each platform of the prototype GaitMaster is composed of three linear actuators top of which a yaw joint is mounted. We disassembled a 6 DOF Stewart platform and made two XYZ stages. Three linear guides are applied to support the orientation of the top plate of the motion platform. The payload of each motion-platform is approximately 150 Kg. A rotational joint around the yaw axis is mounted on each motion platform. The joint is equipped with a spring that moves the feet in the neutral direction.

A turntable is developed using a large DD motor. The maximum angular velocity is 500 deg/sec. A 3 DOF goniometer is connected to each foot. The goniometer measures back-and-forth and up-and-down motion as well as yaw angle. The control algorithm mentioned in the former section was implemented and succeeded in the presentation of virtual staircases.

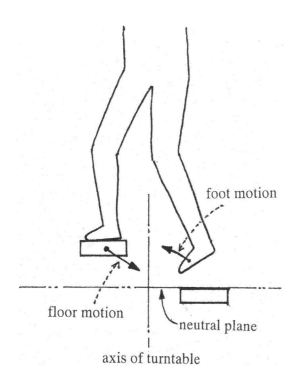

FIGURE 12.20. Canceling the climbing-up motion.

FIGURE 12.21. GaitMaster.

Robot-Tile-Based Locomotion Interface

CirculaFloor Project

Locomotion interfaces often require bulky hardware, since they have to carry the user's whole body. Also, the hardware is not easy to reconfigure to improve its performance or to add new functions. Considering these issues, the goals of the Circula Floor project are

1. to develop compact hardware for the creation of an infinite surface for walking; the major disadvantage of existing locomotion interfaces is their difficult installation. We need to solve this problem for a demonstration at SIGGRAPH.
2. to develop scalable hardware architecture for future improvement of the system; another disadvantage of existing locomotion interfaces is the difficulty of improving the system. We have to design a new hardware architecture that allows us to easily upgrade the actuation mechanism and add new mechanisms for the creation of uneven surfaces.

In order to achieve these goals, we designed a new configuration for a locomotion interface by using a set of omni-directional movable tiles. Each tile is equipped with a holonomic mechanism that achieves omni-directional motion. An infinite surface is simulated by the circulation of the movable tiles. The motion of the feet is measured by position sensors. The tile moves opposite to the measured direction of the walker so that the motion of the step is canceled. The position of the walker is fixed in the real world by this computer-controlled motion of the tiles. The circulation of the tiles has the ability to cancel the displacement of the walker in an arbitrary direction. Thus, the walker can freely change direction while walking. Figure 12.22 shows overall view of CirculaFloor.

The CirculaFloor is a new method that takes advantage of both the treadmill and footpad. It creates an infinite omnidirectional surface using a set of movable tiles. The combination of tiles provides a sufficient area for walking, and thus precision

tracing of the foot position is not required. It has the potential to create an uneven surface by mounting an up-and-down mechanism onto each tile.

Method of Creating Infinite Surface

The current method of circulating the movable tiles is designed to satisfy the following conditions: (a) Two of the movable tiles are used for pulling back the user to the center of the dead zone; (b) The rest of the movable tiles are used to create a new front surface; (c) These tiles are moved the shortest distances to the next destination, while they avoid colliding with other tiles; (d) The control program allocates all destinations to the tiles, when the tiles reach their destination; (e) The tiles don't rotate corresponding to walking direction to simplify the algorithm.

Considering the above conditions, the circulation method is varied corresponding to the walking direction. Three modes, "alternating circulation," "unidirectional circulation," and "cross circulation," are designed corresponding to the direction (Fig. 12.23). Representative motion of each mode is illustrated in Fig. 12.24 and 12.25.

Alternating circulation (Fig. 12.24): This mode is adopted for the directions between ±15 deg and ±75–105 deg. The tiles used for creating a new front surface (white-colored) move around to the front of the tiles for alternatively pulling back

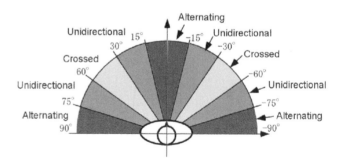

FIGURE 12.23. Pulling-back modes corresponding to walking direction.

FIGURE 12.22. CirculaFloor.

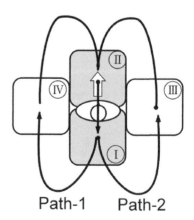

FIGURE 12.24. Circulation of movable tiles in alternating mode.

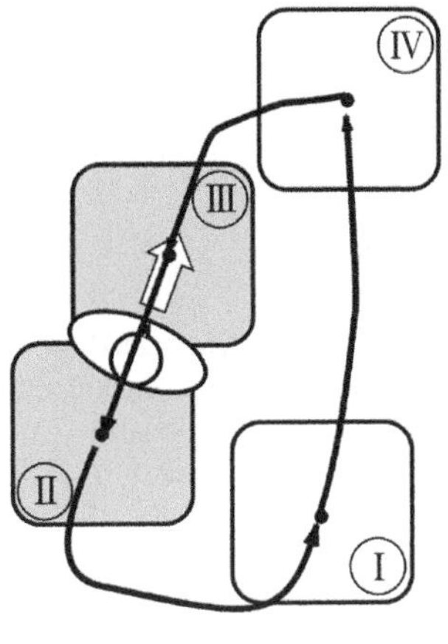

FIGURE 12.25. Circulation of movable tiles in unidirectional mode.

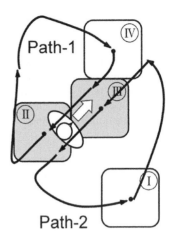

FIGURE 12.26. Circulation of movable tiles in cross mode.

(in Figure 12.24, gray-colored tiles) from left (Path-1)/right (Path-2) sides.

Unidirectional circulation (Fig. 12.25): This mode is adopted for the directions between ±15–30 deg and ±60–75 deg. The tiles used for creating a new front surface move around to the right/left front of the tiles for pulling back with a unidirectional circulation.

Cross-circulation (Fig. 12.26): This mode is adopted for the directions between ±30–60 deg. The tiles used for creating a new front surface move around to the left/right front (Path-1) or the left/right sides (Path-2) of the tiles for pulling back.

When a user of the CirculaFloor switches the walking direction, the control program calculates the nearest phase of each tile by using a template-matching technique corresponding to the new direction. Then the tiles take the shortest way to their destinations.

APPLICATION AREA FOR HAPTIC INTERFACE

Application Area for Finger/Hand Haptics

Medicine. Medical applications for haptic interfaces are currently growing rapidly. Various surgical simulators have been developed using a tool-handling-type force display. We developed a simulator for laparoscopic surgery using the Haptic Master. Simulator software using PHANToM are commercially available.

Palpation is typically used in medical examinations. The FEELEX 2 is designed to be used as a palpation simulator. If we display a virtual tumor based on a CT or an MRI image, a medical doctor can palpate the internal organs before surgery, and this technique can be also applied to telemedicine. Connecting two FEELEXs together via a communication line would allow a doctor to palpate a patient remotely.

3D-shape modeling. The design of 3D shapes definitely requires haptic feedback. A typical application of the tool-handling-type force display is in 3D-shape modeling. One of the most popular applications of the PHANToM system is as a mod-

eling tool. Such a tool-handling-type force display allows a user to point contact, and point-contact manipulation is most suited for precision-modeling tasks. However, it isn't effective when the modeling task requires access to the whole shape. Designers use their palm or the joints of their fingers to deform a clay model when carrying out rough-design tasks. The FEELEX has the ability to support such natural manipulation.

HUI (Haptic User Interface). Today, touchscreens are widely used in automatic teller machines, ticketing machines, information kiosks, and so on. A touchscreen enables an intuitive-user interface, although it lacks haptic feedback. Users can see virtual buttons, but they can't feel them. This is a serious problem for a blind person. The FEELEX provides a barrier-free solution to the touchscreen-based user interface. Figure 12.10 shows an example of a haptic touch-screen using the FEELEX 1.

Art. Interactive art may be one of the best applications of the FEELEX system. As we discussed earlier, the Anomalocaris has been exhibited in a museum in Austria. It succeeded in evoking haptic interaction with many visitors. The FEELEX can be used for interactive sculptures. Visitors are usually prohibited from touching physical sculptures. However, they can not only touch sculptures based around FEELEX, they can also deform them.

Application Areas for Locomotion Interfaces

As a serious application of our locomotion interface, we are working with the Ship Research Laboratory to develop an "evacuation simulator (Yamao et al., 1996)." The Ship Research Laboratory is a national research institute that belongs to the ministry of transportation of Japan. Analysis of evacuation of passengers during maritime accidents is very important for ship safety. However, it is impossible to carry out experiments with human subjects during an actual disaster. Therefore, they introduced virtual-reality tools for simulation of disaster in order to analyze evacuation of passengers. They built a virtual ship that models the generation of smoke and the inclination of the vessel. Experiments of evacuation are carried out for the construction of mathematical models of passengers' behavior in a disaster. The Torus Treadmill will be effective in such experiments.

Locomotion by walking is intuitive and inevitable in the study of human behavior in virtual environments. We are applying the system to research a human model of evacuation in maritime accidents. The GaitMaster can be applied to other areas than the Torus Treadmill. Its application may include rehabilitation of walking or simulation of mountain climbing.

CONCLUSION AND FUTURE PROSPECTS

This chapter describes major topics of haptic interfaces. A number of methods have been proposed implementing haptic interfaces. Future research in this field will include following two issues: (a) safety issues, and (b) psychology in haptics.

Safety Issues

Safety is an important issue in haptic interface. Inadequate control of actuator may injure the user. The exoskeleton and tool-handling-type-force displays have control problems in their contact surface for the virtual objects. Vibration or unwanted forces can be generated back to the user, which is sometimes dangerous. One of the major advantages of FEELEX is safety. The user of FEELEX doesn't wear any special equipment while the interaction is taking place. The contact surface of the FEELEX is physically generated, so it is free from such control problems.

Locomotion interface has much more important safety issues. The system supports the full body of the user, so that inadequate control causes major damage to the user. Specialized hardware for keeping the walker safe must be developed.

Psychology in Haptics

There have been many findings regarding haptic sensation. Most of these are related to skin sensation, and very few research activities include muscle sensation. Among these, Lederman and Klatzky's work was closely related to the design of the force display (Lederman & Klatzky, 1987). Their latest work involved spatially distributed forces (Lederman & Klatzky, 1999). They performed an experiment involving palpation. The subjects were asked to find a steel ball placed underneath a foam-rubber cover. The results showed that steel balls smaller than 8 mm in diameter decreased the score. This finding supports our specification for the FEELEX 2 in which the distance between rods is 8 mm. This kind of psychological study will support future development of haptic interface.

Haptics is indispensable for human interaction in the real world. However, haptics is not commonly used in the field of HCI. Although there are several commercially avalable haptic interfaces, they are expensive and limited in their fuction. Image display has a history of over 100 years. Today, image displays, such as television and movies, are used in everyday life. On the other hand, haptic interface has only 10-year history. There are hazards to overcome for popular use of haptic interface. However, haptic interface is a new frontier of media technology and will definitly contribute to human life.

References

Asanuma, N., Yokoyama, N., & Shinoda, H. (1999). A method of selective stimulation to epidermal skin receptors for realistic touch feedback. *Proceedings of IEEE Virtual Reality '99* (pp. 274–281).

Brooks, F. P., Jr. (1986). A dynamic graphics system for simulating virtual buildings. *Proceedings of the 1986 Workshop on Interactive 3D Graphics* (pp. 9–21). New York: ACM.

Brooks, F. P., et al. (1990). Project GROPE—Haptic displays for scientific visualization. *Computer Graphics, 24*(4).

Burdea, G. C. (1996). *Force and touch feedback for virtual reality.* Wiley-Interscience.

Burdea, G., Zhuang, J., Roskos, E., Silver, D., & Langlana, L. (1992). A Portable Dextrous Master with Force Feedback. *Presence, 1*(1).

Christensen, R., Hollerbach, J. M., Xu, Y., & Meek, S. (1998). Inertial force feedback for a locomotion interface. Proceedings of ASME Dynamic Systems and Control Division (Vol. 64; pp. 119–126).

Darken, R., Cockayne, W., & Carmein, D. (1997). The Omni-directional treadmill: A locomotion device for virtual worlds. *Proceedings of UIST '97.*

Hirota, K., & Hirose, M. (1996). Simulation and presentation of curved surface in virtual reality environment through surface display. *Proceedings of IEEE VRAIS '96.*

Iwata, H. (1990). Artificial reality with force-feedback: Development of desktop virtual space with compact master manipulator. ACM SIGGRAPH *Computer Graphics, 24*(4).

Iwata, H. (1990). Artificial reality for walking about large scale virtual space. *Human Interface News and Report, 5*(1), 49–52.

Iwata, H. (1993). Pen-based Haptic virtual environment. *Proceedings of IEEE VRAIS '93.*

Iwata, H. (1994). Desktop Force Display. *SIGGRAPH 94 Visual Proceedings.*

Iwata, H. (1999). Walking about virtual space on an infinite floor. *Proceedings of IEEE Virtual Reality '99* (pp. 236–293).

Iwata, H., & Fujii, T. (1996). Virtual perambulator: A novel interface device for locomotion in virtual environment. Proceedings of IEEE 1996 Virtual Reality Annual International Symposium (pp. 60–65).

Iwata, H., & Yoshida, Y. (1999). Path reproduction tests using a torus treadmill. *Presence, 8*(6), 587–597.

Kajimoto, H., Kawakami, N., Maeda, T., & Tachi, S. (1999). Tactile Feeling Display using Functional Electrical Stimulation. *Proceedings of ICAT '99* (pp. 107–114).

Kawai, Y., & Tomita, F. (2000). A support system for the visually impaired to recognize three-dimensional objects. *Technology and Disability, 12*(1), 13–20.

Kontarinis, D. A., & Howe, R .D. (1995). Tactile display of vibratory information in teleoperation and virtual environment. *Presence, 4*(4), 387–402.

Lederman, S. J., & Klatzky, R. L. (1987). Hand movements: A window into haptic object recognition. *Cognitive Psychology, 19*(3), 342–368.

Lederman, S. J., & Klatzky, R. L. (1999). Sensing and displaying spatially distributed fingertip forces in haptic interfaces for teleoperators and virtual environment system. *Presence, 8*(1).

Lorenzo, M. et al. (1995). OSIRIS. *SIGGRAPH '95 Visual Proceedings* (p. 129).

Massie, T., & Salisbury, K. (1994). The PHANToM Haptic Interface: A device for probing virtual objects. *ASME Winter Anual Meeting* (DSC-Vol.55-1).

McNeely, W. (1993). Robotic graphics: A new approach to force feedback for virtual reality. *Proceedings of IEEE VRAIS '93.*

Minsky, M., & Lederman, S. J. (1997). Simulated Haptic Textures: Roughness. Symposium on Haptic Interfaces for Virtual Environment and Teleoperator Systems, Proceedings of the ASME Dynamic Systems and Control Division, DSC-Vol. 58.

Moy, G.., Wagner, C., & Fearing, R. S. (2000, April). A compliant tactile display for teletaction. *IEEE International Conference on Robotics and Automation.*

Murakami, T., & Nakajima, N. (1994). Direct and intuitive input device for 3D shape deformation. *ACM CHI, Conference on Human Factors in Computing Systems,* (pp. 465–470).

Noma, H., Sugihara, T., & Miyasato. (2000). Development of ground surface simulator for Tel-E-Merge System. *Proceedings of IEEE Virtual Reality* (pp. 217–224).

Poston, R. et al. (1997). A whole body kinematic display for virtual reality applications. *Proceedings of the IEEE International Conference on Robotics and Automation* (pp. 3006–3011).

Prat, David R. et al. (1994). Insertion of an articulated human into a networked virtual environment. *Proceedings of the A I Simulation, and Planning in High Autonomy Systems Conference* (pp. 7–9).

Sinclair, M. (1997) The Haptic lens. *SIGGRAPH '97 Visual Proceedings* (p. 179).

Slater, M. et al. (1994). Steps and ladders in Virtual Reality. Virtual Reality Technology, World Scientific Publication, pp. 45–54 (1994).

Tachi, S. et al. (1994). A construction method of virtual Haptic space. *Proceedings of ICAT '94.*

Yamao, T., Ishida, S., Ota, S., & Kaneko, F. (1996). Formal safety assessment—Research project on quantification of risk on lives, MSC67/INF.9 IMO information paper, (1996).

·13·

NONSPEECH AUDITORY OUTPUT

Stephen Brewster
University of Glasgow

INTRODUCTION AND A BRIEF HISTORY OF NONSPEECH SOUND AND HCI

Our sense of hearing is very powerful, and we can extract a wealth of information from the pressure waves entering our ears as sound. Sound gives us a continuous, holistic contact with our environments; we hear a rich set of sounds from our interactions with objects close to us, familiar sounds of unseen friends or family nearby, noises of things to avoid, such as traffic, and noises of things to attend to, such as a ringing telephone. Nonspeech sounds (such as music, environmental sounds, and sound effects) give us different types of information than those provided by speech; they can be more general and more ambient where speech is precise and requires more focus. Nonspeech sounds complement speech in the same way that visual icons complement text. For example, icons can present information in a small amount of space as compared to text; nonspeech sounds can present information in a small amount of time as compared to speech. There is less research about nonspeech than speech interfaces, and this chapter will show something of where it has been used and of what it is capable.

The combination of visual and auditory feedback at the user interface is a powerful tool for interaction. In everyday life, these primary senses combine to give complementary information about the world. Our visual system gives us detailed information about a small area of focus, whereas our auditory system provides general information from all around, alerting us to things outside our view. Blattner and Dannenberg (1992) discussed some of the advantages of using this approach in multimedia/multimodal computer systems:

In our interaction with the world around us, we use many senses. Through each sense, we interpret the external world using representations and organizations to accommodate that use. The senses enhance each other in various ways, adding synergies or further informational dimensions. (p. xviii)

These advantages can be brought to the multimodal (or multisensory) human-computer interface by the addition of nonspeech auditory output to standard graphical displays (see chapter 21 for more on multimodal interaction). Whilst directing our visual attention to one task, such as editing a document, we can still monitor the state of others on our machine using sound. Currently, almost all information presented by computers uses the visual sense. This means information can be missed because of visual overload or because the user is not looking in the right place at the right time. A multimodal interface that integrated information output to both senses could capitalize on the interdependence between them and present information in the most efficient way possible.

The classical uses of nonspeech sound can be found in the human factors literature (see McCormick & Sanders, 1982). Here it is used mainly for alarms, warnings, and status information. Buxton (1989) extended these ideas and suggested that encoded messages could be used to present more complex information in sound. This type of auditory feedback will be considered here.

The other main use of nonspeech sound is in music and sound effects for games and other multimedia applications. These kinds of sounds indicate to the user something about what is going on and try to create a mood for the piece (much as music does in film and radio). As Blattner and Dannenberg (1992) said, "Music and other sound in film or drama can be used to communicate aspects of the plot or situation that are not verbalized by the actors. Ancient drama used a chorus and musicians to put the action into its proper setting without interfering with the plot" (p. xix). Work on auditory output for interaction takes this further and uses sound to present information, such as things that the user might not otherwise see or important events that the user might not notice.

The use of sound to convey information in computers is not new. In the early days, programmers used to attach speakers to their computers' busses or program counters (Thimbleby, 1990). The speaker would click each time the program counter was changed. Programmers would get to know the patterns and rhythms of sound and could recognize what the machine was doing. Another everyday example is the sound of a hard disk. Often, users can tell when a save or copy operation has completed by the noise their disks make. This allows them to do other things while waiting for the copy to finish. Nonspeech sound is therefore an important information provider, giving users knowledge about things in their systems that they may not see.

Two important events kick-started the research area of nonspeech auditory output: the first was the special issue of the HCI journal on nonspeech sound, edited by Buxton (1989). This laid the foundations for some of the key work in the area; it included papers by Blattner on earcons (Blattner, Sumikawa, & Greenberg, 1989), Gaver (1989) on auditory icons, and Edwards (1989) on Soundtrack. The second event was the First International Conference on Auditory Display (ICAD'92) held in Santa Fe in 1992 (Kramer, 1994b). For the first time, this meeting brought the main researchers interested in the area together (see http://www.icad.org for the proceedings of the ICAD conferences). Resulting from these two events was a large growth in research in the area during the 1990s that continues today.

The rest of this chapter goes into detail on all aspects of auditory interface design. It presents some of the advantages and disadvantages of using sound at the interface. Then, a brief introduction to psychoacoustics, or the study of the perception of sound, follows. This is followed by an introduction to sound sampling and synthesis techniques needed for auditory interfaces. The next sections describe the main techniques that are used for auditory information presentation and some of the main applications of sound in human–computer interaction. The chapter finishes with some conclusions about the state of research in this area.

WHY USE NONSPEECH SOUND IN HCI?

Some Advantages of Sound

It is advantageous to use sound at the user interface for many reasons.

Vision and hearing are interdependent. Our visual and auditory systems work well together. Our eyes provide high-resolution information around a small area of focus (with peripheral vision extending further). According to Perrott, Sadralobadi, Saberi, and Strybel (1991), humans view the world through a window of 80° laterally and 60° vertically. Within this visual field, focusing capacity is not uniform. The foveal area of the retina (the part with the greatest acuity) subtends an angle of only 2° around the point of fixation. Sounds, on the other hand, can be heard from above, below, in front, or behind the user, but with a much lower resolution. Therefore, our ears tell our eyes where to look: if there is an interesting sound from outside our view we will turn to look at it to get more detailed information.

Superior temporal resolution. As Kramer (1994a) said, "Acute temporal resolution is one of the greatest strengths of the auditory system" (p. 2). In certain cases, reactions to auditory stimuli have been shown to be faster than reactions to visual stimuli (Bly, 1982).

Reduce the overload from large displays. Modern, large, or multiple monitor graphical interfaces use the human visual sense very intensively. This means that we may miss important information because our visual system is overloaded—we have just too much to look at. To stop this overload, information could be displayed in sound so that the load could be shared between senses.

Reduce the amount of information needed on screen. Related to the point above is the problem with information presentation on devices with small visual displays, such as mobile telephones and personal digital assistants (PDAs). These have very small screens that can easily become cluttered. To solve this, some information could be presented in sound to release screen space.

Reduce demands on visual attention. Another issue with mobile devices is that users who are using them on the move cannot devote all of their visual attention to the device—they must look where they are going to avoid uneven surfaces, traffic, pedestrians, and so forth. In this case, visual information may be missed because the user is not looking at the device. If this were played in sound, then the information would be delivered while the user was looking at something else.

The auditory sense is underutilized. The auditory system is very powerful; we can listen to (and some can compose) highly complex musical structures. As Alty (1995) said, "The information contained in a large musical work (say a symphony) is very large . . . The information is highly organized into complex structures and substructures. The potential therefore exists for using music to successfully transmit complex information to a user" (p. 409).

Sound is attention grabbing. Users can choose not to look at something, but it is harder to avoid hearing it. This makes sound very useful for delivering important information.

Some objects or actions within an interface may have a more natural representation in sound. Bly (1982) suggested, "Perception of sound is different to visual perception, sound can offer a different intuitive view of the information

it presents . . ." (p. 14). Therefore, sound could allow us to understand information in different ways.

To make computers more usable by visually disabled people. With the development of graphical displays, user interfaces became much harder for visually impaired people to operate. A screen reader (see the section on sounds for users with visual impairments below) cannot easily read this kind of graphical information. Providing information in an auditory form can help solve this problem and allow visually disabled persons to use the facilities available on modern computers.

Some Problems with Sound

Kramer (1994a) suggested some general difficulties with using sound to present information.

Low resolution: Many auditory parameters are not suitable for high-resolution display of quantitative information. For example, when using sound volume, only very few different values can be unambiguously presented (Buxton, Gaver, & Bly, 1991). Vision has a much higher resolution. The same also applies to spatial precision in sound. Under optimal conditions, differences of about 1° can be detected in front of a listener (see section on three dimensional sound; Blauert, 1997). In vision, differences of an angle of two seconds can be detected in the area of greatest acuity in the central visual field.

Presenting absolute data is difficult: Many interfaces that use nonspeech sound to present data do it in a relative way. Users tell if a value is going up or down through the difference between two sounds. It is difficult to present absolute data unless the listener has perfect pitch, which is rare. In vision, a user needs only to look at a number to get an absolute value.

Lack of orthogonality: Changing one attribute of a sound may affect the others. For example, changing the pitch of a note may affect its perceived loudness and vice versa (see the section on perception of sound).

Transience of information: Sound disappears after it has been presented; users must remember the information that the sound contained or use some method of replaying. In vision, the user can easily look back at the display and see the information again. (This is not always the case—think, for example, of an air conditioning system: its sounds continue for long periods and become habituated. Sounds often continue in the background and only become apparent when they change in some way.)

Annoyance due to auditory feedback: There is one problem with sound that has not yet been mentioned, but it is the one most commonly brought up against the use of sound in user interfaces: annoyance. As this is an important topic, it will be discussed in detail in a later section.

Comparing Speech and Nonspeech Sounds for Interface Design

One obvious question is why not just use speech for output? Why do we need to use nonspeech sounds? Many of the advantages presented above apply to speech as well as nonspeech sounds. However, there are some advantages to nonspeech sounds. If we think of a visual analogy, speech output is like the text on a visual display and nonspeech sounds are like the icons. Presenting information in speech is slow because of its serial nature; to assimilate information, the user must typically hear it from beginning to end, and many words may have to be comprehended before a message can be understood. With nonspeech sounds, the messages are shorter and therefore can be heard more rapidly (although the user might have to learn the meaning of the nonspeech sound whereas the meaning is contained within the speech and so requires no learning—just like the visual case). Speech suffers from many of the same problems as text in text-based computer systems, as this is also a serial medium. Barker and Manji (1989) claimed that an important limitation of text is its lack of expressive capability: It may take many words to describe something fairly simple. Graphical displays that speeded up interactions were introduced, as users could see a picture of the application they wanted instead of having to read its name from a list (Barker & Manji, 1989). In the same way, an encoded sound message can communicate its information in fewer sounds. The user hears the sound then recalls its meaning rather than having the meaning described in words. Increasing the presentation rate of synthetic speech decreases the accuracy of recognition of what is being said. Pictorial icons can also be universal: they can mean the same thing in many different languages and cultures, and the nonspeech sounds have similar universality (given some of the different musical cultures).

An important ability of the auditory system is habituation, where continuous sounds can fade into the background of consciousness after a short period. If the sound changes (or stops), then it would come to the foreground of attention because the auditory system is sensitive to change. Habituation is difficult to achieve with speech because of the large dynamic range it uses. According to Patterson (1982), "The vowels of speech are often 30 dB more intense than the consonants, and so, if a voice warning were attenuated to produce a background version with the correct vowel level the consonants would be near or below masked threshold" (p. 11). It is easier to habituate certain types of nonspeech sounds (think of an air conditioner where you only notice that it was on when it switches off) and sounds can be designed to facilitate habituation if required.

Baddeley (1997) gave evidence to show that background speech, even at low intensities, is much more disruptive than nonspeech sound when recalling information. He reports the unattended speech effect. Unattended speech (e.g., in the background) causes information to be knocked out of short-term memory, whereas noise or nonspeech sound does not. This problem is unaffected by the intensity of the speech, provided that it is audible. This shows a problem for speech at the interface, as it is likely to prove disruptive for other users in the same environment unless it is kept at a very low intensity and, as we saw above, this can cause problems with the ability to hear consonants.

Nonspeech sounds are also good for presenting continuous information, such as a graph of stock market data. In speech, particular values can be spoken out at particular times, but there would be no way to monitor the overall trend of the data. Methods

for doing this in nonspeech sounds were developed over 20 years ago (Mansur, Blattner, & Joy, 1985) and have proved to be very effective (see Sound Graphs in perception of sound section below for more on this).

This discussion has shown that there are many reasons to think of using nonspeech sounds in addition to speech in HCIs. Few interfaces make good use of both. Speech in general is good for giving instructions and absolute values; nonspeech sounds are good for giving rapid feedback on actions, quickly presenting highly structured information, and presenting continuous data. Together, they make a very effective means of presenting information nonvisually.

Avoiding Annoyance

The main concern potential users of all auditory interfaces have is annoyance due to sound pollution. There are two aspects to annoyance: A sound may be annoying to the user whose machine is making the noise (the primary user), and it may be annoying to others in the same environment who overhear it (secondary users). Buxton (1989) discussed some of the problems of sound and suggested that some sounds help us (information) and some impede us (noise). We therefore need to design sounds that are more informative and less noise. Of course, one person's informative sounds are another's noise, so it is important to make sure that the sounds on one computer are not annoying for colleagues working nearby.

Few studies focus on the problems of annoyance due to nonspeech sounds in computers. There are, however, many studies of annoyance from speech (e.g., Berglund, Harder, & Preis, 1994), from the sounds of aircraft, traffic, and other environmental noises, and most of these suggest that the primary reason for the annoyance of sound is excessive volume. In a different context, Patterson (1989) investigated some of the problems with auditory warnings in aircraft cockpits. Many of the warnings were added in a "better-safe-than-sorry" manner that led to them being so loud that the pilot's first response was to try to turn them off rather than deal with the problem being indicated. One of Patterson's main recommendations was that the volume of the warnings should be reduced.

A loud sound grabs the attention of the primary user, even when the sound is communicating an unimportant event. As the sound is loud, it travels from one machine to the ears of other people working nearby, increasing the noise in their environments.

So, how can annoyance be avoided? One key way is to avoid using intensity as a cue in sound design for auditory interfaces. Quiet sounds are less annoying. Listeners are also not good at making absolute intensity judgments (see the next section). Therefore, intensity is not a good cue for differentiating sounds anyway.

Headphones can be used so that sounds are heard only by the primary user. This may be fine for users of mobile telephones and music players, but is not always a good solution for desktop users who do not want to be cut-off from their colleagues. Manipulating sound parameters other than intensity can make sounds attention grabbing (but not annoying).

Rhythm or pitch can be used to make sounds demanding because the human auditory system is very good at detecting changing stimuli (see Edworthy, Loxley, & Dennis, 1991). Therefore, if care is taken with the design of sounds in an interface, specifically avoiding the use of volume changes to cue the user, then many of the problems of annoyance can be avoided.

PERCEPTION OF SOUND

This section provides some basic information about the perception of sound that is applicable to nonspeech auditory output. The auditory interface designer must be conscious of the effects of psychoacoustics, or the perception of sound, when designing sounds for the interface. As Frysinger (1990) said, "The characterization of human hearing is essential to auditory data representation because it defines the limits within which auditory display designs must operate if they are to be effective" (p. 31). Using sounds without regard for psychoacoustics may lead to the user being unable to differentiate one sound from another, unable to hear the sounds, or unable to remember them (see Moore, 2003).

What are sounds? Sounds are pressure variations that propagate in an elastic medium (in this case, the air). The pressure variations originate from the motion or vibration of objects. These pressure variations hit the listener's ear and start the process of perceiving the sound. The pressure variations plotted against time can be seen in Fig. 13.1. This shows the simplest form of sound: a sine wave (which might be produced by a tuning fork). A sound is made up from three basic components. Frequency is the number of times per second the wave repeats itself (Fig. 13.1 shows three cycles). It is normally measured in Hertz (Hz). Amplitude is the deviation away from the mean pressure level, or force per unit area of a sound. It is normally measured in decibels (dB). Phase is the position of the start of the wave on the time axis (measured in milliseconds).

Sounds from the real world are normally much more complex than Fig. 13.1 and tend to be made up of many sine waves with different frequencies, amplitudes, and phases. Figure 13.2 shows a more complex sound made of three sine wave components (or partials) and the resulting waveform. Fourier analysis allows a sound to be broken down into its component sine waves (Gelfand, 1981).

The sounds in Fig. 13.1 and Fig. 13.2 are periodic—they repeat regularly over time. This is very common for many types of musical instruments that might be used in an auditory interface. Many natural, everyday sounds (such as impact sounds) are not periodic and do not repeat. The sound in Fig. 13.2 is also harmonic—its partials are integer multiples of the lowest (or fundamental) frequency. This is again common for musical instruments but not for everyday sounds. Periodic harmonic sounds have a recognizable pitch, where as nonperiodic, inharmonic sounds tend to have no clear pitch.

The attributes of sound described above are the physical aspects. There is a corresponding set of perceptual attributes. Pitch is the perceived frequency of a sound. Pitch is roughly a logarithmic function of frequency. It can be defined as the

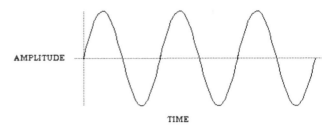

FIGURE 13.1.   A sine wave.

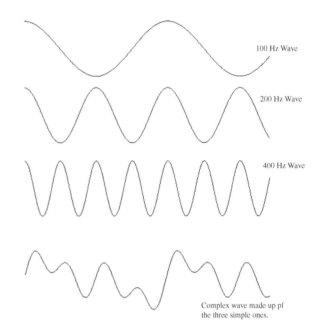

100 Hz Wave

200 Hz Wave

400 Hz Wave

Complex wave made up pf the three simple ones.

FIGURE 13.2.   A complex wave made up of three components with its fundamental at 100 Hz.

attribute of auditory sensation in terms of which sounds may be ordered on a musical scale (Moore, 2003). In the western musical system, there are 96 different pitches arranged into eight octaves of 12 notes. Tones separated by an octave have the frequency ratio two to one. For example, middle C is 261.63 Hz, the octave above this is at 523.25 Hz, and the octave below at 130.81 Hz. It is one of the most useful and easily controlled aspects of sound and is very useful for auditory interface designers. However, as Buxton et al. (1991) said, "It is important to be aware of the myriad interactions between pitch and other attributes of sound . . ." (p. 2.10). For example, pitch is affected by sound intensity: at less than 2 kHz an increase in intensity increases the perceived pitch, at 3 kHz and over an increase in intensity decreases the perceived pitch (Gelfand, 1981).

Humans can perceive a wide range of frequencies. The maximum range we can hear is from 20 Hz to 20 kHz. This decreases with age, so that at age 70, a listener might only hear a maximum of 10 kHz. It is therefore important to ensure that the sounds in an auditory interface are perceivable by its users (also poor quality loudspeakers may not be able to cope with the highest or lowest frequencies). Listeners are not good at making absolute judgments of pitch (Moore, 2003). Only 1% of the population has perfect pitch. Another important factor is tone deafness. Moore suggested that this is a misnomer and almost everyone is able to tell that two sounds are different; they are not always able to say which is higher or lower. This can often be overcome with practice, but it is important for the auditory interface designer to be aware of the problem. Mansur et al. (1985) gave evidence of one other important effect: "There appears to be a natural tendency, even in infants, to perceive a pitch that is higher in frequency to be coming from a source that is vertically higher in space when compared to some lower tone" (p. 171). This is important when creating an auditory interface as it could be used to give objects a spatial position. If only stereo position is available to provide spatial cues in the horizontal plane, then pitch could provide them in the vertical plane. Guidelines for the use of pitch (and the other parameters below) are described in the section on Earcons.

Loudness is the perceived intensity of a sound. Loudness (L) is related to intensity (I) according to the *Power Law*: $L = k\,I^{0.3}$ (Gelfand, 1981). Therefore, a 10dB increase in intensity doubles the perceived loudness of a sound. Loudness is again affected by the other parameters of sound. For example, sounds of between 1 kHz and 5 kHz sound louder at the same intensity level than those outside that frequency range. Humans can perceive a very wide range of intensities: the most intense sound that a listener can hear is 120 dB louder than the quietest. This equates to a ratio of 1,000,000,000,000:1 (Moore, 2003). Buxton et al. (1991) also reported that listeners are "very bad at making absolute judgments about loudness" (p. 2.10) and also that "our ability to make relative judgments of loudness are limited to a scale of about three different levels" (p. 2.11). It is also a primary cause of annoyance (see the section on avoiding annoyance above), so it should be used sparingly by auditory interface designers.

Timbre is the quality of the sound. This attribute of auditory sensation allows a listener to judge two sounds with the same loudness and pitch as dissimilar. It is what makes a violin sound different from a piano, even if both are playing at the same pitch and at the same loudness. Its structure and dimensions are not yet fully understood. It is based partly on the spectrum and dynamics of a sound. Its structure is not well understood, but it is one of the most important attributes of sound that an interface designer can use. As Blattner et al. (1989) said, "Even though timbre is difficult to describe and notate precisely, it is one of the most immediate and easily recognizable characteristics of sound" (p. 26) (both auditory icons and earcons use timbre as one of their fundamental attributes; see section on nonspeech sound presentation techniques below). Many of the synthesis techniques in the next section make it easy for a designer to create and use different timbres.

Duration is another important attribute of sound. Sounds of different durations are used to form rhythmic structures that are a fundamental part of music. Duration can also affect the other parameters of sound. For example, for sounds of less than one second, loudness increases with duration. This is important in auditory interfaces because short sounds are often needed so that the auditory feedback can keep pace with the interactions

taking place; accordingly, they must be made loud enough for listeners to hear.

Direction is the position of the sound source. This is often overlooked but is an important aspect of sound in our everyday lives. As mentioned above, one of the key differences between sight and hearing is that sounds can be heard from all around the listener. If a sound source is located to one side of the head, then the sound reaching the further ear will be reduced in intensity (Interaural Intensity Difference; IID) and delayed in time (Interaural Time Difference; ITD; Blauert, 1997). These two factors are key in allowing a listener to localize a sound in space. Humans can detect small changes in the position of a sound source. The minimum auditory angle (MAA) is the smallest separation between two sources that can be reliably detected. Strybel, Manligas and Perrott (1992) reported that in the median plane sound sources only 1° apart can be detected. At 90° azimuth (directly opposite one ear), accuracy falls to ± 10°. This has important implications for auditory displays that use sound position as a cue because higher-resolution placement can be used when sounds are presented in front of the user (see section on Earcons for more on sound positioning).

TECHNOLOGY AND SOUND PRODUCTION

Most desktop PCs, handheld computers, and mobile telephones have sophisticated sound hardware available for the auditory interface designer. This is normally used for playing games, but it is sufficient to do most of the things required by an auditory interface. The aim of this section is to describe briefly some of the main technologies that are important for designers to understand when creating interfaces.

There are two main aspects to sound production: the first is sound synthesis, and the second is sound sampling and playback. A basic overview will be given focusing on aspects related to audio interfaces. (For much more detail on sound synthesis and MIDI see Roads, 1996; Miranda, 1998; for more on sampling, see Pohlmann, 2005.)

There are many tools available for synthesis and sample playback, and devices such as desktop PCs and mobile telephones have the necessary processing power. The Java programming language (http://www.java.sun.com), for example, has built-in support for a range of synthesis and sampling techniques for many different platforms. Libraries such as Fmod (http://www.fmod.org) allow standard cross-platform sound and work on many different devices and programming languages. All current desktop operating systems provide support for synthesis and sampling. The basic technologies necessary to make the sounds needed for auditory interfaces are thus readily available, but it is important to know something of how they work to utilize them most effectively.

A Brief Introduction to Sound Synthesis and MIDI

The aim of sound synthesis is to generate a sound from a stored model, often a model of a musical instrument. For auditory interfaces, we need a wide, good-quality range of sounds that we can generate in real time as the user interacts with the interface. Synthesizers come in three main forms: soundcards on PCs, external hardware devices, and software synthesizers.

Most synthesizers are polyphonic, such as able to play multiple notes at the same time (as opposed to monophonic). This is important for auditory interface design, as you might well want to play a chord made up of several notes. Most synthesizers are multitimbral, such as able to play multiple different instruments at the same time. This is again important, as in many situations, a sound composed of two different instruments might be required. The main forms of synthesis will now be briefly reviewed.

Wavetable synthesis is one of the most common and low cost synthesis techniques. Many of the most popular PC soundcards use it (such as the SoundBlaster™ series from Creative Technology—http://www.creative.com). The idea behind wavetable synthesis is to use existing sound recordings (which are often very difficult to synthesize exactly) as the starting point and to create very convincing simulations of acoustical instruments based on them (Miranda, 1998; Roads, 1996). A sample (recording) of a particular sound will be stored in the soundcard. It can then be played back to produce a sound. The sample memory in these systems contains a large number of sampled sound segments, and they can be thought of as a table of sound waveforms that may be looked up and utilized when needed. Wavetable synthesizers employ a variety of different techniques, such as sample looping, pitch shifting, mathematical interpolation, and polyphonic digital filtering, to reduce the amount of memory required to store the sound samples or to get more types of sounds. More sophisticated synthesizers contain more wavetables (perhaps one or more for the initial attack part of a sound and then more for the sustain part of the sound and then more for the final decay and release parts). Generally, the more wavetables that are used the better the quality of the synthesis, but this does require more storage. It is also possible to combine multiple, separately controlled wavetables to create a new instrument.

Wavetable synthesis is not so good if you want to create new timbres, as it lacks some of the flexibility of the other techniques below. Most wavetable synthesizers contain many sounds (often hundreds), so there may not be a great need to create new ones. For most auditory interfaces, the sounds from a good quality wavetable synthesizer will be perfectly acceptable. For desktop computers the storage of large wavetables is no problem, but for mobile telephones with less storage, there may be a much smaller, lower quality set, so care may be needed to design appropriate sounds.

FM (frequency modulation) Synthesis techniques generally use one periodic signal (the modulator) to modulate the frequency of another signal (the carrier; Chowning, 1975). If the modulating signal is in the audible range, then the result will be a significant change in the timbre of the carrier signal. Each FM voice requires a minimum of two signal generators. Sophisticated FM systems may use four or six operators per voice, and the operators may have adjustable envelopes that allow adjustment of the attack and decay rates of the signal. FM synthesis is cheap and easy to implement, and it can be useful for creating expressive new synthesized sounds. However, if the goal is to recreate the sound of an existing instrument, then FM synthesis

is not the best choice, as it can generally be done more easily and accurately with wavetable techniques.

Additive (and Subtractive) Synthesis is the oldest form of synthesis (Roads, 1996). Multiple sine waves are added together to produce a more complex output sound (subtractive synthesis is the opposite: a complex sound has frequencies filtered out to create the sound required). It is theoretically possible to create any sound using this method (as all complex sounds can be decomposed into sets of sine waves by Fourier analysis). However, it can be very difficult to create any particular sound. Computer musicians often use this technique, as it is very flexible and easy to create new and unusual sounds, but it may be less useful for general auditory interface design.

Physical Modeling Synthesis uses mathematical models of the physical acoustic properties of instruments and objects. Equations describe the mechanical and acoustic behavior of an instrument. The better the simulation of the instrument, the more realistic the sound produced. Nonexistent instruments can also be modeled and made to produce sounds. Physical modeling is an extremely good choice for synthesis of many classical instruments, especially those of the woodwind and brass families. The downside is that it can require large amounts of processing power, which limits the polyphony.

The Musical Instrument Digital Interface—MIDI

The Musical Instrument Digital Interface (MIDI) allows the real-time control of electronic instruments (such as synthesizers, samplers, drum machines, etc.) and is now very widely used (http://www.midi.org). It specifies a hardware interconnection scheme, a method for data communications, and a grammar for encoding musical performance information (Roads, 1996). For auditory interface designers, the most important part of MIDI is the performance data, which is a very efficient method of representing sounds. Most soundcards support MIDI with an internal synthesizer and provide a MIDI interface to connect to external devices. Both Apple Mac OS and Microsoft Windows have good support for MIDI. Most programming languages now come with libraries supporting MIDI commands.

MIDI performance information is like a piano roll: notes are set to turn on or off and play different instruments over time. A MIDI message is an instruction that controls some aspect of the performance of an instrument. A MIDI message is made up of a status byte, which indicates the type of the message, followed by up to two data bytes that give the parameters. For example, the Note On command takes two parameters: one value giving the pitch of the note required and the other the volume. This makes it a very compact form of presentation.

Performance data can be created dynamically from program code or by a sequencer. In an auditory interface, the designer might assign a particular note to a particular interface event—for example, a click on a button. When the user clicks on the button, a MIDI Note On event will be fired. When the user releases the button, the corresponding Note Off event will be sent. This is a very simple and straightforward way of adding sounds. With a sequencer, data can be entered using classical music notation by dragging and dropping notes onto a stave or by using an external piano-style keyboard. This could then be saved to a MIDI file for later playback (or could be recorded and played back as a sample—see comparing MIDI synthesis to sampling for auditory interface design section below).

A Brief Introduction to Sampling

In many ways, sampling is simpler than synthesis. The aim is to make a digital recording of an analogue sound and then to be able to play it back later, with the played back sound matching the original as closely as possible. There are two important aspects: sample rate and sample size.

Sample Rate

Sample rate is the number of discrete snapshots of the sound that are taken, often measured per second. The higher the sampling rate is, the higher the quality of the sound when it is played back. With a low sampling rate, few snapshots of the sound are taken, and the recording will not match well the sound being recorded. The Sampling Theorem (Roads, 1996) states that to reconstruct a signal, the sampling frequency must be at least twice the frequency of the signal being sampled. As mentioned above, the limit of human hearing is around 20 kHz, therefore a maximum rate of 40 kHz is required to be able to record any sound that a human can hear. The standard audio CD format uses a sample rate of 44.1 kHz, meaning that it covers all of the frequencies that a human can hear. If a lower sampling rate is used, then higher frequencies are lost. For example, the .au sample format uses a sampling rate of 8 kHz, meaning that only frequencies of less than 4 kHz can be recorded (see Bagwell, 1998).

Higher sampling rates generate much more data than lower ones, so they may not always be suitable if storage is limited (for example, on a mobile computing device). An auditory interface designer must think about the frequency range of the sounds needed in an interface. This might allow the sample rate to be reduced. If the highest quality is required, you must be prepared to deal with large audio files.

Sample Size

The larger the sample size is, the better the quality of the recording, as more information is stored at each snapshot of the sound. Sample size defines the volume (or dynamic) range of the sound. With an 8-bit sample, only 256 discrete amplitude (or quantization) levels can be represented. To fit an analogue sound into one of these levels might cause it to be rounded up or down, and this can add noise to the recording. CD quality sounds use 16-bit samples, giving 65536 different levels, so the effects of quantization are reduced. Many high quality samplers use 24-bit samples to reduce the problems of quantization further.

The two main bit sizes used in most soundcards are 8 and 16 bits. As with sample rates, the main issue is size: 16-bit samples require a lot of storage, especially at high sample rates.

Audio CD quality sound generates around 10 Mbytes of data per minute. Compression techniques such as MP3 can help reduce the amount of storage but keep quality high.

Comparing MIDI Synthesis to Sampling for Auditory Interface Design

MIDI is very flexible, as synthesizers can generate sound in real time, as it is needed. If you do not know all of the sounds you might want in your auditory interface in advance, then this can very effective—as the sound is needed, it is just played by the synthesizer. A system that is based around samples can only play back samples that have been prerecorded and stored.

Another advantage of MIDI is that sounds can be changed. Once a sample has been stored, it is difficult to change it. For example, it is possible to change the speed and pitch of a sound independently with MIDI. If a sample is played back at a different speed its pitch will change, which may cause undesirable effects.

MIDI commands are also very small; each command might only take up two or three bytes. Generating sounds from code in your auditory interface is straightforward. For instance, files containing high quality stereo sampled audio require about 10 Mbytes of data per minute of sound, while a typical MIDI sequence might consume less than 10 Kbytes of data per minute of sound. This is because the MIDI file does not contain the sampled audio data; it contains only the instructions needed by a synthesizer to play the sounds.

Samples have the advantage that you, as the interface designer, know exactly what your sound will sound like. With MIDI, you are at the mercy of the synthesizer on the user's machine. It may be of very poor quality, and the sounds might not sound anything like the ones you designed. With samples, all of the information about the sound is stored so that you can guarantee it will sound like the recording you made (given the possible limitations of the speakers on the user's machine).

It is also not possible to synthesize all sounds. Much of the work in sound synthesis has focused on synthesizing musical instruments. Few synthesizers can do a good job of creating natural, everyday sounds. If you want to use natural sounds in your interface, you are limited to using samples.

Three Dimensional Sound

Much of the recorded sound we hear is in stereo. A stereo recording uses differences in intensity between the ears. From these differences, the listener can gain a sense of movement and position of a sound source in the stereo field. The perceived position is along a line between the two loudspeakers or, if they are wearing headphones, inside the head between the listeners' ears. This simple, inexpensive technique can give useful spatial cues at the auditory interface. This is being taken further to make sounds appear as coming from around a user (in virtual 3D) when only a small number of loudspeakers (or even just a pair of headphones) are used. Spatial sound can be used for a range of things, including giving directional information, spreading sound sources around the head to help users differentiate simultaneous sounds, and creating audio windows in which to present information (see the section on sound for mobile and ubiquitous computing).

As well as the ITD and IID, in the real world we use our pinnae (the outer ear) to filter the sounds coming from different directions so that we know where they are coming from. To simulate sounds as coming from around the user and outside of the head when wearing headphones, sounds entering the ear are recorded by putting microphones into the ear canals of listeners. The differences between the sound at the sound source and at the ear canal are then calculated and the differences, or head-related transfer functions (HRTFs), derived are used to create filters with which stimuli can be synthesized (Blauert, 1997). This research is important, as three-dimensional auditory interfaces that are more natural can be created, with sounds presented around the user, as they would be in real life. Almost all current PC soundcards can generate such 3D sounds, as they are often used in games.

The main problem with providing simulated 3D sound through headphones comes from the general HRTFs used. If your ears are not like the ears of the head (often a dummy head) from which the HRTFs were generated, then you are likely to feel that the sounds are coming from inside your head and not outside. It is also very easy to confuse front and back so that listeners cannot tell if a sound is in front or behind. Vertical positioning is also difficult to do reliably. This means that many designers who use spatial sound in their interfaces often limit themselves to a plane cutting through the head horizontally at the level of the ears, creating a 2.5D space. This reduces the space in which sounds can be presented but avoids many of the problems of users not being able to localize the sounds properly.

To improve quality, head tracking is often used. Once the orientation of the user's head is known, sounds can be respatialized to remain in position when the head turns. Active listening is used to disambiguate the location of a sound—listeners naturally make small head movements and these change the IID and ITD, cueing the listener to the location of the sound. Using such tracking can significantly improve the performance of 3D auditory user interfaces (Marentakis & Brewster, 2004). Marentakis and Brewster also showed that targets should be around ±10° in size when head tracking is used to enable accurate localization and selection.

NONSPEECH SOUND PRESENTATION TECHNIQUES

The two main types of nonspeech audio presentation techniques commonly used are auditory icons and earcons. Substantial research has gone into developing both of these, and the main work is reviewed below.

Auditory Icons

Gaver (1989, 1997) developed the idea of auditory icons. These natural, everyday sounds can be used to represent actions and objects within an interface. Gaver defined them as "everyday sounds mapped to computer events by analogy with everyday

sound-producing events. Auditory icons are like sound effects for computers" (p. 68). Auditory icons rely on an analogy between the everyday world and the model world of the computer (Gaver, 1997; for more examples of the use of earcons see the work on Mercator and Audio Aura described in sections on sounds for users with visual impairments and sound for mobile and ubiquitous computing).

Gaver (1997) used sounds of events that are recorded from the natural environment, such as tapping or smashing sounds. He used an ecological listening approach (Neuhoff, 2004), suggesting that people do not listen to the pitch and timbre of sounds but to the sources that created them. When pouring liquid, a listener hears the fullness of the receptacle, not the increases in pitch. Another important property of everyday sounds is that they can convey multidimensional data. When a door slams, a listener may hear the size and material of the door, the force that was used, and the size of room on which it was slammed. This could be used within an interface so that selection of an object makes a tapping sound, the type of material could represent the type of object, and the size of the tapped object could represent the size of the object within the interface.

Gaver (1989) used these ideas to create auditory icons and from these built the SonicFinder. This ran on the Apple Macintosh and provided auditory representations of some objects and actions within the interface. Files were given a wooden sound, applications a metal sound, and folders a paper sound. The larger the object the deeper the sound it made. Thus, selecting an application meant tapping it—it made a metal sound that confirmed that it was an application and the deepness of the sound indicated its size. Copying used the idea of pouring liquid into a receptacle. The rising of the pitch indicated that the receptacle was getting fuller and the copy progressing.

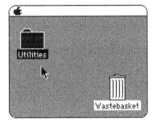

A) Papery tapping sound to show selection of folder.

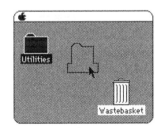

B) Scraping sound to indicate dragging folder.

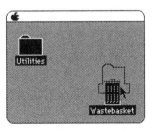

C) Clinking sound to show wastebasket selected

D) Smashing sound to indicate folder deleted.

FIGURE 13.3. An interaction showing the deletion of a folder in the SonicFinder (from Gaver, 1989).

To demonstrate how the SonicFinder worked a simple interaction is provided in Fig. 13.3, showing the deletion of a folder. In (a), a folder is selected by tapping on it; this causes a papery sound, indicating that the target is a folder. In (b), the folder is dragged toward the wastebasket, causing a scraping sound. In (c), the wastebasket becomes highlighted, and a clinking sound occurs when the pointer reaches it. Finally, in (d), the folder is dropped into the wastebasket, and a smashing sound occurs to indicate it has been deleted (the wastebasket becomes fat to indicate there is something in it).

Problems can occur with representational systems, such as auditory icons, because some abstract interface actions and objects have no obvious representation in everyday sound. Gaver (1989) used a pouring sound to indicate copying because there was no natural equivalent; this is more like a sound effect. He suggested the use of movie-like sound effects to create sounds for things with no easy representation. This may cause problems if the sounds are not chosen correctly, as they will become more abstract than representational and the advantages of auditory icons will be lost.

Gaver, Smith, and O'Shea (1991) developed the ideas from the SonicFinder further in the ARKola system, which modeled a soft drink factory. The simulation consisted of a set of nine machines split into two groups: those for input, and those for output. The input machines supplied the raw materials; the output machines capped the bottles and sent them for shipping. Each machine had an on/off switch and a rate control. The aim of the simulation was to run the plant as efficiently as possible, avoid waste of raw materials, and make a profit by shipping bottles. Two users controlled the factory, with each user able to see approximately one third of the whole plant. This form of plant was chosen because it allowed Gaver et al. to investigate how the sounds would affect the way users handled the given task and how people collaborated. It was also an opportunity to investigate how different sounds would combine to form an auditory ecology (integrated set of sounds), or soundscape. Gaver et al. related the way the different sounds in the factory combined to the way a car engine is perceived. Although the sounds are generated by multiple distinct components, these combine to form what is perceived as a unified sound. If something goes wrong, the sound of the engine will change, alerting the listener to the problem, but in addition, to a trained ear, the change in the sound would alert the listener to the nature of the problem. The sounds used to indicate the performance of the individual components of the factory were designed to reflect the semantics of the machine.

Each of the machines had a sound to indicate its status over time; for example, the bottle dispenser made the sound of clinking bottles. The rhythm of the sounds reflected the rate at which the machine was running. If a machine ran out of supplies or broke down, its sound stopped. Sounds were also added to indicate that materials were being wasted. A splashing sound indicated that liquid was being spilled; the sound of smashing bottles indicated that bottles were being lost. The system was designed so that up to 14 different sounds could be played at once. To reduce the chance that all sounds would be playing simultaneously, sounds were pulsed once a second rather than playing continuously.

An informal evaluation was undertaken where pairs of users were observed controlling the plant, either with or without sound. These observations indicated that the sounds were effective in informing the users about the state of the plant and that the users were able to differentiate the different sounds and identify the problem when something went wrong. When the sounds were used, there was much more collaboration between the two users. This was because each could hear the whole plant and therefore help if there were problems with machines that the other was controlling. In the visual only condition, users were not as efficient at diagnosing what was wrong even if they knew there was a problem.

One of the biggest advantages of auditory icons is the ability to communicate meanings which listeners can easily learn and remember, other systems (for example earcons, see the next section) use abstract sounds where the meanings are harder to learn. Problems did occur with some of the warning sounds used, as Gaver et al. (1991) indicated: "The breaking bottle sound was so compelling semantically and acoustically that partners sometimes rushed to stop the sound without understanding its underlying cause or at the expense of ignoring more serious problems" (p. 89). Another problem was that, when a machine ran out of raw materials, its sound just stopped; users sometimes missed this and did not notice that something had gone wrong.

Design Guidelines for Auditory Icons

There have been few detailed studies investigating the best ways to design auditory icons, so there is little guidance for interaction designers. Mynatt (1994) proposed the following basic design methodology: (a) Choose short sounds which have a wide bandwidth, and where length, intensity, and sound quality are roughly equal; (b) evaluate the identifiability of the auditory cues using free-form answers; (c) evaluate the learnability of the auditory cues which are not readily identified; (d) test possible conceptual mappings for the auditory cues using a repeated measures design where the independent variable is the concept that the cue will represent; (e) evaluate possible sets of auditory icons for potential problems with masking, discriminability and conflicting mappings; and (f) conduct usability experiments with interfaces using the auditory icons.

Earcons

Earcons were developed by Blattner et al. (1989). They used abstract, synthetic tones in structured combinations to create auditory messages. Blattner et al. defined earcons as "nonverbal audio messages that are used in the computer/user interface to provide information to the user about some computer object, operation or interaction" (p. 13). Unlike auditory icons, there is no intuitive link between the earcon and what it represents; the link must be learned. They use a more traditional musical approach than auditory icons.

Earcons are constructed from simple building blocks called "motifs" (Blattner et al., 1989). These short, rhythmic sequences can be combined in different ways. Blattner et al. suggested their most important features include the following:

Rhythm: Changing the rhythm of a motif can make it sound very different. Blattner et al. (1989) described this as the most prominent characteristic of a motif.

Pitch: There are 96 different pitches in the western musical system, and these can be combined to produce a large number of different motifs.

Timbre: Motifs can be made to sound different by the use of different timbres, for example playing one motif with the sound of a violin and the other with the sound of a piano.

Register: This is the position of the motif in the musical scale. A high register means a high-pitched note, and a low register means a low-pitched note. The same motif in a different register can convey a different meaning.

Dynamics: This is the volume of the motif. It can be made to increase as the motif plays (crescendo) or decrease (decrescendo).

There are two basic ways in which earcons can be constructed. The first, and simplest, are compound earcons. These simple motifs can be concatenated to create more complex earcons. For example, a set of simple, one-element motifs might represent various system elements such as create, destroy, file, and string (see Fig. 13.4A). These could then be concatenated to form earcons (Blattner et al., 1989). In the figure, the earcon for create is a high-pitched sound that gets louder, and for destroy, it is a low-pitched sound that gets quieter. For file, there are two long notes that fall in pitch and for string, two short notes that rise. In Fig. 13.4B, the compound earcons can be seen. For the create file earcon, the create motif is simply followed by the file motif. This provides a simple and effective method for building up earcons.

Hierarchical earcons are more complex but can represent complex sound structures. Each earcon is a node in a tree and inherits properties from the earcons above it. Figure 13.5 shows a hierarchy of earcons representing a family of errors. The top level of the tree is the family rhythm. This sound just has a rhythm and no pitch; the sounds used are clicks. The rhythmic structure of level one is inherited by level two, but this time, a second motif is added where pitches are put to the rhythm. At this level, Blattner et al. (1989) suggested that the timbre should be a sine wave, which produces a colorless sound. This is done so that at level three, the timbre can be varied. At level three, the pitch is also raised by a semitone to make it easier to differentiate from the pitches inherited from level two. Other levels can be created where register and dynamics are varied.

Blattner et al. (1989) proposed the design of earcons but did not develop or test them. Brewster, Wright, and Edwards (1994) carried out a detailed evaluation of compound and hierarchical earcons based on the design proposed by Blattner et al., which involved simple system beeps and a richer design based on more complex musical timbres using psychoacoustical research (see the section on perception of sound). In these experiments, participants were presented with earcons representing families of icons, menus, and combinations

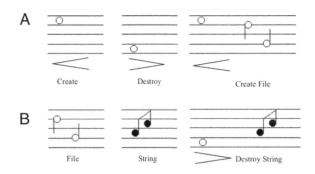

FIGURE 13.4. Compound earcons. A shows the four audio motifs 'create,' 'destroy,' 'file,' and 'string.' B shows the compound earcons 'create file' and 'destroy string' (Blattner et al., 1989).

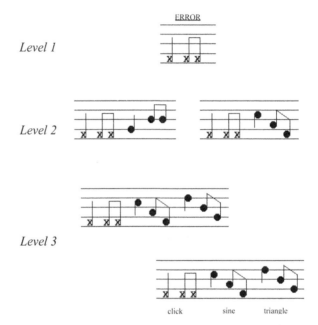

FIGURE 13.5. A hierarchy of earcons representing errors (From Blattner et al., 1989).

of both (examples can be heard at http://www.dcs.gla.ac.uk/~stephen/demos.shtml). They heard each sound three times and then had to identify them when played back. Results showed that the more complex musical earcons were significantly more effective than both the simple beeps and Blattner's proposed design, with over 80% recalled correctly. Brewster et al. (1994) found that timbre was a much more important than previously suggested whereas pitch on its own was difficult to differentiate. The main design features of the earcons used were formalized into a set of design guidelines:

Timbre: This is the most important grouping factor for earcons. Use musical instrument timbres with multiple harmonics, as this helps perception and can avoid masking. These timbres are more recognizable and differentiable.

Pitch and Register: If listeners are to make absolute judgments of earcons, then pitch/register should not be used as a cue on its own. A combination of register and another parameter gives better rates of recall. If register alone must be used, then there should be large differences (two or three octaves) between earcons. Much smaller differences can be used if relative judgments are to be made. The maximum pitch used should be no higher than 5 kHz and no lower than 125 Hz–150 Hz so that the sounds are not easily masked and are within the hearing range of most listeners.

Rhythm, duration and tempo: Make rhythms as different as possible. Putting different numbers of notes in each earcon is very effective. Earcons are likely to be confused if the rhythms are similar, even if there are large spectral differences. Very short note lengths might not be noticed, so do not use very short sounds. Earcons should be kept as short as possible so they can keep up with interactions in the interface being sonified. Two earcons can be played in parallel to speed up presentation.

Intensity: This should not be used as a cue on its own because it is a major cause of annoyance. Earcons should be kept within a narrow dynamic range so that annoyance can be avoided (see section on avoiding annoyance above for more on this issue).

Major/minor mode: Lemmens (2005) showed that by changing from a major to minor key, he could change the affective responses of users to earcons. In western music, the minor mode is broadly thought of as sad and the major mode as happy. This can be used as a further cue to create differentiable earcons.

One aspect that Brewster also investigated was musical ability—as earcons are based on musical structures, is it only musicians who can use them? The results showed that more complex earcons were recalled equally well by nonmusicians as they were by musicians, indicating that they are useful to a more general audience of users.

In a further series of experiments, Brewster (1998b) looked in detail at designing hierarchical earcons to represent larger structures (with over 30 earcons at four levels). These were designed building on the guidelines above. Users were given a short training period and then were presented with sounds, and they had to indicate where the sound was in the hierarchy.

Results were again good, with participants recalling over 80% correctly, even with the larger hierarchy used. The study also looked at the learning and memorability of earcons over time. Results showed that, even with small amounts of training, users could get good recall rates, and the recall rates of the same earcons tested a week later were unchanged.

In recent work, McGookin and Brewster (2004) looked at presenting multiple earcons in parallel. This is problematic unless done carefully, as the structures used to create earcons also cause them to overlap when played in parallel. McGookin and Brewster suggested each earcon should have an onset delay and different spatial location to improve understanding. (For examples of earcons in use, see the sonically enhanced widgets and Palm III work in the section on applications of auditory output).

Comparing Auditory Icons and Earcons

Both earcons and auditory icons are effective at communicating information in sound. There is more formal evidence of this for earcons, as more basic research has looked at their design. There is less basic research into the design of auditory icons, but the systems that have used them in practice have been effective. Detailed research is needed into auditory icons to correct this problem and to provide designers with design guidance on how to create effective sounds. It may be that each has advantages over the other in certain circumstances and that a combination of both is the best. In some situations, the intuitive nature of auditory icons may make them favorable. In other situations, earcons might be best because of the powerful structure they contain, especially if there is no real-world equivalent of what the sounds are representing. Indeed, there may be some middle ground where the natural sounds of auditory icons can be manipulated to give the structure of earcons.

The advantage of auditory icons over earcons is that they are easy to learn and remember because they are based on natural sounds, and the sounds contain a semantic link to the objects they represent. This may make their association to certain, more abstract actions or objects within an interface more difficult. Problems of ambiguity can also occur when natural sounds are taken out of the natural environment and context is lost (people may have their own idiosyncratic mappings). If the meanings of auditory icons must be learned then they lose some of their advantages, and they come closer to earcons.

Earcons are abstract so their meaning must always be learned. This may be a problem, for example, in walk up and use type applications. Research has shown that little training is needed if the sounds are well designed and structured. Leplâtre and Brewster (2000) began to show that it is possible to learn the meanings implicitly while using an interface that generates the sounds as it is being used. However, some form of learning must take place. According to Blattner et al. (1989), earcons may have an advantage when there are many highly structured sounds in an interface. With auditory icons, each one must be remembered as a distinct entity because there is no structure linking them together. With earcons, there is a strong structure linking them that can easily be manipulated. There is not yet any experimental evidence to support this.

Pure auditory icons and earcons make up the two ends of a presentation continuum from representational to abstract (see Fig. 13.6). In reality, things are less clear. Objects or actions within an interface that do not have an auditory equivalent must have an abstract auditory icon made for them. The auditory icon then moves more toward the abstract end of the continuum. When hearing an earcon, the listener may hear and recognize a piano timbre, rhythm, and pitch structure as a kind of catch phrase representing an object in the interface. He or she does not hear all the separate parts of the earcon and work out the meaning from them (listeners may also try and put their own representational meanings on earcons, even if the designer did not intend it as found by Brewster, 1998b). The earcon then moves more toward the representational side of the continuum. Therefore, earcons and icons are closer than they might appear.

There are not yet any systems that use both types of sounds fully, and this would be an interesting area to investigate. Some

FIGURE 13.6. The presentation continuum of auditory icons and earcons.

parts of a system may have natural analogues in sound, and therefore, auditory icons could be used; other parts might be more abstract or structured and earcons would be better. The combination of the two would be the most beneficial. This is an area ripe for further research.

THE APPLICATIONS OF AUDITORY OUTPUT

Auditory output has been used in a wide range of different situations and applications. This section will outline some of the main areas of use and will highlight some of the key papers in each area (for more uses of sound see the ICAD (http://www.icad.org) or ACM CHI (http://www.acm.org/sigchi) series of conferences).

Sonic Enhancement of GUIs

One long-running strand of research in the area of auditory output is in the addition of sound to standard graphical displays to improve usability. One reason for doing this is that users can become overloaded with visual information on large, high-resolution displays. In highly complex graphical displays, users must concentrate on one part of the display to perceive the visual feedback, so that feedback from another part may be missed. This becomes very important in situations where users must notice and deal with large amounts of dynamic data. For example, imagine you are working on your computer writing a report and are monitoring several ongoing tasks, such as compiling, printing, and downloading files over the Internet. The word-processing task will take up your visual attention because you must concentrate on what you are writing. In order to check when your printout is done, the compilation has finished, or the files have downloaded, you must move your visual attention away from the report and look at these other tasks. This causes the interface to intrude into the task you are trying to perform. If information about these other tasks was presented in sound, you could continue looking at the report and hear information in the background about the other tasks. To find out how the file download was progressing, you could just listen to the download sound without moving your visual attention from the writing task.

One of the earliest pieces of work on sonic enhancement of an interface was Gaver's (1989) SonicFinder described above. This used auditory icons to present information about the Macintosh interface redundantly with the graphical display.

Brewster (1998a) investigated the addition of sound to enhance graphical buttons. An analysis of the way buttons are used was undertaken, highlighting some usability problems. It was found that the existing, visual feedback did not indicate when mispresses of a button might have occurred. For example, the

selection of a graphical button is shown in Fig. 13.7 (starting with 1.A and 2.A). The button highlights when it is pressed down (Fig. 13.7 1.B and 2.B). There is no difference in feedback between a correct selection (Fig. 13.7 1.C) and a misselection (Fig. 13.7 2.C), where the user moves the mouse off the graphical button before the selection is complete. The user could therefore slip off the button, fail to press it, and get no feedback. This error can happen when the user is moving away from the button and on to some other task. For example, the user moves to a toolbar to press the "Bold" button and then moves back to the text to position the cursor to start typing. The button press and the mouse move overlap, and the button is not pressed. It is hard for the user to notice this because no feedback is given.

The problems could not easily be solved by adding more graphical feedback: the user is no longer looking at the button's location so any feedback given there will be missed. Feedback could be given at the mouse location but we cannot be sure the user will be looking there either. Brewster (1998a) designed a new button that used auditory feedback to indicate more about the state of the button. This was advantageous, as sound is omnidirectional, and the user does not need to focus attention on any part of the screen to perceive it.

Three earcons were used to improve the effectiveness of graphical buttons. An organ timbre was used for all of the sounds. When the user moved over a button, a continuous tone was played at 130 Hz at a volume just above the background sound level. This informed the user the cursor was over the target (but could easily be habituated). When the mouse was pressed down over the graphical button, a continuous tone was played at 261 Hz. The third sound indicated that the graphical button had been successfully selected. This sound consisted of two short tones with a pitch of 1046 Hz and duration of 40ms. This sound was not played if a slip-off error occurred. If the user pressed the button very quickly, then only the success sound was played to avoid unnecessary feedback.

An experimental evaluation of these sounds was undertaken. Results showed that users recovered from slip-off errors significantly faster and with significantly fewer mouse clicks when sounds were present in the buttons. Users also significantly preferred the buttons with sound when asked to rate subjective preference. They also did not rate the buttons as more annoying than the standard graphical ones. An interesting point to note was that the use of no sound could be attention grabbing when a sound was expected. The participants could easily recognize a slip off due to the demanding nature of the success sound not being played. This is important as reducing the amount of feedback presented is one way to make sure that it is not annoying.

Many other widgets have been successfully sonified. Beaudouin-Lafon and Conversey (1996) showed that nonspeech sounds could improve usability of scrollbars. Maury, Athenes, and Chatty (1999) and Marila (2002) added sounds to improve menu selections in drop-down menus. Ronkainen and Pasanen (2005) did several studies into the design of audio feedback for buttons. Brewster (1998a) investigated a wide range of different widgets including scroll-bars, menus, progress bars, tool palettes, and drag and drop. These widgets have been included in a toolkit (Crease, Gray, & Brewster, 2000) that designers can use to add sound easily to their interfaces.

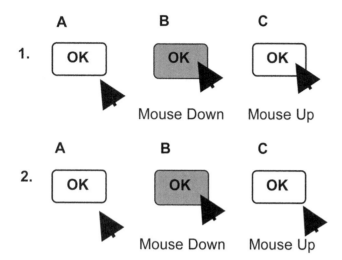

FIGURE 13.7. The visual feedback presented by a graphical button when selected. 1 shows a correct selection and 2 shows a slip-off (From Brewster, 1998a).

Sound for Users with Visual Impairments

One of the most important uses for nonspeech sound is in interfaces for people with visual disabilities. One of the main deprivations caused by blindness is the problem of access to information. A blind person will typically use a screen reader and a voice synthesizer to use a computer. The screen reader extracts textual information from the computer's video memory and sends it to the speech synthesizer to speak it. This works well for text but not well for the graphical components of current user interfaces. It is still surprising to find that many commercial applications used by blind people make little use of nonspeech sound, concentrating on synthetic speech output. This is limiting (as discussed above), as speech is slow, it can overload short-term memory, and it is not good for presenting certain types of information; for example, it is not possible to render many types of images via speech, so these can become inaccessible to blind people. One reason for sound's lack of use has been how to employ it effectively. As Edwards (1995) said, "Currently the greatest obstacle to the exploitation of the variety of communications channels now available is our lack of understanding of how to use them" (p. ii). The combination of speech and nonspeech sounds can increase the amount of information presented to the user. As long as this is done in a way that does not overload the user, then it can improve access to information. Some of the main research into the use of nonspeech auditory in interfaces for blind people will now be described.

Soundtrack was an early attempt to create a word processor designed to be used by blind persons and was developed by Edwards (1989). It used earcons and synthetic speech as output, and it was designed so that the objects a sighted user would see in an interface, for example menus and dialogues, were replaced by auditory equivalents that were analogies of their visual counterparts. Its interface was constructed from auditory objects with which the user could interact. They were defined by a location, a name, a sound, and an action. They were arranged into a grid of two layers, analogous to menus (see Fig. 13.8).

Each auditory object made a sound when the cursor entered it, and these could be used to navigate rapidly around the screen. Soundtrack used sine waves for its audio feedback. Chords were built up for each menu, dependent on the number of menu items. For the edit menu, a chord of four notes was played because there were four menu items within it (cut, copy, paste, and find).

The base sounds increased in pitch from left to right—as in the normal representation of a musical scale (for example on a piano) and the top layer used higher pitches than the bottom. Using these two pieces of information a user could quickly find his or her position on the screen. If any edge of the screen was reached, a warning sound was played. If at any point the user got lost or needed more precise information, he or she could click on an object and it would speak its name.

The approach taken in Soundtrack was to take the visual interface to a word processor and translate it into an equivalent auditory form. The Mercator system (Mynatt & Edwards, 1995; Mynatt & Weber, 1994) took a broader approach. The designers' goal was to model and translate the graphical interfaces of X Windows applications into sound without modifying the applications (and thus create a more general solution than Soundtrack's). Their main motivation was to simulate many of the features of graphical interfaces to make graphical applications accessible to blind users and keep coherence between the audio and visual interfaces so that blind and sighted users could interact and work together on the same applications. This meant that the auditory version of the interface had to facilitate the same mental model as the visual one. This did not mean that they translated every pixel on the screen into an auditory form; instead, they modeled the interaction objects that were present. Modeling the pixels exactly in sound was ineffective due to the very different nature of visual and auditory media and the fact that graphical interfaces had been optimized to work with the visual sense (for example, the authors claim that an audio equivalent of overlapping windows was not needed as overlapping was just an artifact of a small visual display). Nonspeech sound was an important aspect of their design to make the iconic parts of a graphical interface usable.

Mercator used three levels of nonspeech auditory cues to convey symbolic information presented as icons in the visual interface. The first level addressed the question, what is this

File Menu	Edit Menu	Sound Menu	Format Menu
Alert	Dialog	Document 1	Document 2

FIGURE 13.8. The Soundtrack's main screen (From Edwards, 1989).

object? In Mercator, the type of an interface object was conveyed with an auditory icon. For example, touching a window sounded like tapping a piece of glass, container objects sounded like a wooden box with a creaky hinge, and text fields used the sound of a manual typewriter. While the mapping was easy for interface components such as trashcan icons, it was less straightforward components that did not have simple referents in reality (e.g., menus or dialogue boxes, as discussed in the section on sonic enhancement of GUIs). In Mercator, auditory icons were also parameterized to convey detailed information about specific attributes, such as menu length. Global attributes were also mapped into changes in the auditory icons. For example, highlighting and graying out are common to a wide range of different widgets. To represent these, Mynatt and Edwards (1995) used sound filters. A low-pass filter was used to make the sound of a grayed out object duller and more muffled.

Sonification

Building on the work about accessibility emerges the idea of making data accessible. Sonification, or visualization in sound, can be used to present complex data nonvisually. There are many situations where sonification can also be useful for sighted users (if they only have access to a small screen for example), or in combination with graphical feedback in multimodal visualization systems. Sonification is defined as "the transformation of data relations into perceived relations in an acoustic signal for the purposes of facilitating communication or interpretation" (Kramer & Walker, 1999). The range of sonification goes from the clicks of the Geiger counter to multidimensional information presentation of stock market data.

Mansur et al. (1985) performed one of the most significant studies presenting data in sound. Their study, which laid out the research agenda for subsequent research in sound graphs, used sound patterns to represent two-dimensional line graphs. The value on the y-axis of the graph was mapped to pitch and the x-axis to time; this meant that a listener could hear the graph rise and fall over time in a similar way that a sighted person could see the line rising and falling. This is the basic technique used in most sonification systems.

They found that their approach was successful in allowing distinctions to be made between straight and exponential graphs, varying monotonicity in graphs, convergence, and symmetry. However, they did find that there were difficulties in identifying secondary aspects of graphs, such as the slope of the curves. They suggested that a full sound graph system should contain information for secondary aspects of the graph, such as the first derivative. Their suggestion was to encode this information by adding more overtones to the sound to change the timbre. They also suggested utilizing special signal tones to indicate a graph's maxima or minima, inflection points, or discontinuities. Many other studies have been undertaken to develop this presentation technique further (Flowers & Hauer, 1992, 1995; Walker, 2002).

Walker and Cothran (2003) produced the Sonification Sandbox to allow designers to design sound graphs easily and rapidly. The software allows "users to independently map several data sets to timbre, pitch, volume, and pan, with full control over the default, minimum, maximum, and polarity for each attribute." This gives auditory interface designers the chance to prototype sonifications, without having to create their own, custom-made applications. (The software is freely available from http://sonify. psych.gatech.edu/research.)

Sound for Mobile and Ubiquitous Computing

One of the major growth areas in computing at the beginning of the 21st century has been in mobile and ubiquitous computing. People no longer use computers just while at a desk. Mobile telephones, personal digital assistants (PDAs), and handheld computers are now widely used. One problem with these devices is that there is a very limited amount of screen space on which to display information: the screens are small as the devices must be able to fit into the hand or pocket to be easily carried. Small screens can easily become cluttered with information and widgets, and this presents a difficult challenge for interface designers.

The graphical techniques for designing interfaces on desktop interfaces do not apply well to handheld devices. Screen resources are limited; memory and processing power are much reduced from desktop systems. However, in many cases, interface designs and interaction techniques have been taken straight from standard desktop graphical interfaces (where screen space and other resources are not a problem) and applied directly to mobile devices. This has resulted in devices that are hard to use, with small text that is hard to read, cramped graphics and little contextual information. Speech and nonspeech sounds are an important way of solving these problems.

Another reason for using sound is that if users are performing tasks while walking or driving, they cannot devote all of their visual attention to the mobile device. Visual attention must remain with the main task for safety. It is therefore hard to design a visual interface that can work well under these circumstances. An alternative, sonically enhanced interface would require less visual attention, and therefore potentially interfere less in the main activity in which the user is engaged.

Three main pieces of work are surveyed in this section, covering the main approaches taken in this area. The first adds sound to the existing interface of a mobile computer to improve usability, the second creates a purely auditory interface for a mobile, and the third creates an ambient auditory environment.

Brewster (2002) developed the ideas of sonified buttons described in the section on sonic enhancement of GUIs and applied them to buttons on the 3Com Palm series of pen-based handheld computers. Many of the same feedback problems with buttons apply in handhelds as in desktops, but are worse as the screen is smaller (and may be hard to see when the device is moving or the sun is shining). In addition, there is the problem of the stylus (or finger) obscuring the target on the display, which makes it difficult for users to know when they are pressing in the correct place. Simple earcons were used to overcome the problems. One aim of the work was to see the effects when users were on the move and to see if adding audio could reduce the size of the widgets so that screen space could be saved.

In general, the results confirmed those of the previous study. Subjective workload in the sonically enhanced buttons was reduced, as compared to their silent counterparts. The addition of sound allowed the participants to enter significantly more five-digit strings than in the corresponding silent treatment, with smaller sonic buttons as effective as larger silent ones. When walking, there was a 20% drop in performance overall, with the sonic interface still performing better than the standard one. Participants walked further when sound was added, and small buttons with sound allowed as much text to be entered as the large, silent buttons. The suggested reason for this was that users did not have to concentrate so much of their visual attention on the device; much of the feedback needed was in sound, so they could look where they were going. This would therefore allow the size of items on the display to be reduced without a corresponding drop in usability.

Sawhney and Schmandt (1999, 2000) developed a wearable computer-based personal messaging audio system called "nomadic radio" to deliver information and messages to users on the move. One of the aims of this system was to reduce the interruptions to a user caused by messages being delivered at the wrong time (for example mobile telephone calls being received in a meeting, or a PDA beeping to indicate an appointment in the middle of a conversation). In the system, users wore a microphone and shoulder-mounted loudspeakers that provide a basic planar 3D audio environment (see section on three dimensional sound) through which the audio was presented. A clock face metaphor was used with 12:00 in front of the user's nose, 3:00 by the right ear, 6:00 directly behind the head, and so forth. Messages were then presented in the position appropriate to the time that they arrived. The advantage of the 3D audio presentation (as described above) is that it allows users to listen to multiple, simultaneous sound streams and to distinguish and separate each one (the cocktail party effect shows that listeners can attend one stream of sound amongst many, but also monitor the others in case they need attention; Arons, 1992).

The system used a context-based notification strategy that dynamically selected the appropriate notification method based on the user's focus of attention. Seven levels of auditory presentation were used from silent to full speech rendering. If the user was engaged in a task, then the system was silent and no notification of an incoming call or message would be given (so as not to cause an interruption). The next level used ambient cues (based on auditory icons) with sounds like running water indicating that the system was operational. These cues were designed to be easily habituated but to let the user know that the system was working. The next level was a more detailed form of auditory cue, giving information on system events, task completions, and mode transitions. For example, a ringing telephone sound was used to indicate the arrival of voicemail. These were more attention grabbing cues than the ambient cues and would only be played if the user was not fully occupied. The next four levels of cue used speech, expanding from a simple message summary up to the full text of a voicemail message. These might be used if the person wearing nomadic radio was not involved in tasks that required detailed attention. The system attempted to work out the appropriate level to deliver the notifications by listening to the background audio level near the user (using the built-in microphone) and by determining whether the user was speaking. For example, if the user was speaking, the system might use an ambient cue so as not to

interrupt the conversation. Users could also press a button on the device to indicate they were busy and turn it to silent.

Three-dimensional sound has been combined with gestures to create interactions where users can point at sound sources to choose them. An early example was from Cohen (1993) who created audio windows that users could manipulate with gestures, much as windows on a desktop computer could be controlled. Brewster, Lumsden, Bell, Hall, and Tasker (2003) made this idea mobile and created a soundscape of audio sources around a listener's head that presented different types of information. Users nodded at a sound source of interest to select it. A simple study showed that users could walk and nod to select items, but that there were many issues with sound placement and feedback. Further study by Marentakis and Brewster (2004) looked at different types of gestures and feedback to improve the quality of mobile 3D audio interactions.

There has been much work in the area of notification systems using audio for ambient displays. Carefully designed nonspeech audio can grab attention and then fade into the background. An early example was Audio Aura by Mynatt, Back, Want, Baer, and Ellis (1998), which aimed "to provide serendipitous information, via background auditory cues, that is tied to people's physical actions in the workplace" (p. 566). In a similar way to nomadic radio, Audio Aura used auditory icons to provide background information that did not distract users.

The system used active badges so that the location of users could be identified and appropriate audio cues given, along with wireless headphones so that users could hear the sounds without distracting others. The location information from the active badges was combined with other data sources such as online calendars and e-mail. Changes in this information triggered audio cues sent to the user through the headphones.

Here are some examples of how the system might be used. In the first the user might go the office coffee room and as he or she enters the room hear information about the number and type of e-mail messages currently waiting. This would give the user a cue as to whether to stay and talk to colleagues or go back to the office to answer the messages. In the second example, a user goes to a colleague's office but the occupant is not there. Audio Aura would play sounds indicating if the occupant has been in recently or away for a longer period. The authors were keen to make sure the sounds were not distracting and attention grabbing—they were meant to give background information and not to be alarms. To this end, great care was taken with the cue design. They attempted to design sonic ecologies—groups of sounds that fitted together into a coherent whole. For example, one set of cues was based on a beach scene. The amount of new e-mail was mapped to seagull cries: the more mail, the more the gulls cried. Group activity levels were mapped to the sound of surf: the more activity going on within the group, the more active the waves became. These cues were very subtle and did not grab users' attentions, but some learning of the sounds would be needed, as they are quite abstract.

This work was taken on and implemented in a realistic environment by Kilander and Lönnqvist (2001). They created a Weakly Intrusive Ambient Soundscape (WISP), where states in the computational or physical ubiquitous computing environment are presented as subtle, nonintrusive sound cues based on auditory icons, with each cue "sufficiently nonintrusive to be accepted without disturbing the focus of the task at hand, while distinctive enough to be separable from other cues" (Kilander

& Lonnqvist, 2002, p. 1). They describe a meeting room scenario where devices such as handheld computers, public PCs, and clocks might all be able to make sounds and give cues for ambient awareness. The level of intrusiveness of the sounds could be varied. For low intrusiveness, a quiet sound with lots of reverb was played, making the cue sound far away and almost inaudible; for high intrusiveness, sharp sounds with no reverb were played.

One problem with the system was choice of sounds; users could be detected by the environment, their personal sound mappings, and parameters chosen. However, these mappings could conflict with others' choices; for example, two users might use the same sound cue for different events. This could be solved to some extent with local presentation of the sounds, as in nomadic radio, but global cues would be more of a problem for these types of systems.

CONCLUSIONS

Research into the use of nonspeech sounds for information display at the user interface began in the early 1990s, and there has been rapid growth since then. It has shown its benefits in a wide range of different applications from systems for blind people to ubiquitous computing. There are many good examples that designers can look at to see how sounds may be used effectively, and design guidelines are now starting to appear.

Two areas are likely to be important in its future growth. The first is in combining sound with other senses (vision, tactile, force-feedback, etc.) to create multimodal displays that make the most of the all the senses available to users. This is an area ripe for further investigation, and many interesting interaction problems can be tackled when multiple senses are used together. Key questions here are around what sound is best for and how to combine it with the others most effectively. The second area in which nonspeech sound has a large part to play is with mobile and wearable computing devices (again also in a multimodal form). Small screens cause many difficult presentation problems, and this is exactly the situation where sound has many advantages—it does not take up any precious screen space, and users can hear it even if they cannot look at their device. In a ubiquitous setting, there may not even be a screen at all, and sound can provide information on the services available in a particular environment in a nonintrusive way.

References

Alty, J. L. (1995). *Can we use music in human–computer interaction?* Paper presented at British Computer Society Human conference (HCI'95), Huddersfield, UK.

Arons, B. (1992, July). A review of the cocktail party effect. *Journal of the American Voice I/O Society, 12,* 35–50.

Baddeley, A. (1997). *Human memory: Theory and practice.* London: Psychology Press Ltd.

Bagwell, C. (1998). *Audio file formats FAQ.* Retrieved November, 2005, from http://www.cnpbagwell.com/audio.html

Barker, P. G., & Manji, K. A. (1989). Pictorial dialogue methods. *International Journal of Man-Machine Studies, 31,* 323–347.

Beaudouin-Lafon, M., & Conversy, S. (1996). *Auditory illusions for audio feedback.* Paper presented at Association for Computing Machinery Computer Human Interaction conference (CHI'96), Vancouver, Canada.

Berglund, B., Harder, K., & Preis, A. (1994). Annoyance perception of sound and information extraction. *Journal of the Acoustical Society of America, 95*(3), 1501–1509.

Blattner, M., & Dannenberg, R. B. (Eds.). (1992). *Multimedia interface design.* New York: ACM Press, Addison-Wesley.

Blattner, M., Sumikawa, D., & Greenberg, R. (1989). Earcons and icons: Their structure and common design principles. *Human Computer Interaction, 4*(1), 11–44.

Blauert, J. (1997). *Spatial hearing.* Cambridge, MA: MIT Press.

Bly, S. (1982). *Sound and computer information presentation.* Unpublished PhD Thesis No. UCRL53282, Lawrence Livermore National Laboratory, Livermore, CA.

Brewster, S. A. (1998a). The design of sonically-enhanced widgets. *Interacting with Computers, 11*(2), 211–235.

Brewster, S. A. (1998b). Using nonspeech sounds to provide navigation cues. *ACM Transactions on Computer-Human Interaction, 5*(3), 224–259.

Brewster, S. A. (2002). Overcoming the lack of screen space on mobile computers. *Personal and Ubiquitous Computing, 6*(3), 188–205.

Brewster, S. A., Lumsden, J., Bell, M., Hall, M., & Tasker, S. (2003). *Multi-modal 'eyes-free' interaction techniques for wearable devices.* Paper presented at Association for Computing Machinery Computer Human Interaction conference (CHI 2003), Fort Lauderdale, FL.

Brewster, S. A., Wright, P. C., & Edwards, A. D. N. (1994). A detailed investigation into the effectiveness of earcons. In G. Kramer (Ed.), *Auditory display* (pp. 471–498). Reading, MA: Addison-Wesley.

Buxton, W. (1989). Introduction to this special issue on nonspeech audio. *Human Computer Interaction, 4*(1), 1–9.

Buxton, W., Gaver, W., & Bly, S. (1991). *Tutorial number 8: The use of nonspeech audio at the interface.* Paper presented at Association for Computing Machinery Computer Human Interaction conference CHI'91, New Orleans, LA.

Chowning, J. (1975). Synthesis of complex audio spectra by means of frequency modulation. *Journal of the Audio Engineering Society, 21*(7), 526–534.

Cohen, M. (1993). Throwing, pitching and catching sound: Audio windowing models and modes. *International Journal of Man-Machine Studies, 39,* 269–304.

Crease, M. C., Gray, P. D., & Brewster, S. A. (2000). *Caring, sharing widgets.* Paper presented at British Computer Society Human (HCI 2000) conference, Sunderland, UK.

Edwards, A. D. N. (1989). Soundtrack: An auditory interface for blind users. *Human Computer Interaction, 4*(1), 45–66.

Edwards, A. D. N. (Ed.). (1995). *Extra-ordinary human–computer interaction.* Cambridge, UK: Cambridge University Press.

Edworthy, J., Loxley, S., & Dennis, I. (1991). Improving auditory warning design: Relationships between warning sound parameters and perceived urgency. *Human Factors, 33*(2), 205–231.

Flowers, J. H., & Hauer, T. A. (1992). The ear's versus the eye's potential to assess characteristics of numeric data: Are we too visuocentric? *Behavior Research Methods, Instruments, and Computers, 24*, 258–264.

Flowers, J. H., & Hauer, T. A. (1995). Musical versus visual graphs: Cross-modal equivalence in perception of time series data. *Human Factors, 37*, 553–569.

Frysinger, S. P. (1990). *Applied research in auditory data representation.* Extracting meaning from complex data: Processing, display, interaction. Proceedings of the SPIE/SPSE symposium on electronic imaging, Springfield, VA.

Gaver, W. (1989). The sonicfinder: An interface that uses auditory icons. *Human Computer Interaction, 4*(1), 67–94.

Gaver, W. (1997). Auditory interfaces. In M. Helander, T. Landauer, & P. Prabhu (Eds.), *Handbook of human–computer interaction* (2nd ed., pp. 1003–1042). Amsterdam: Elsevier.

Gaver, W., Smith, R., & O'Shea, T. (1991). *Effective sounds in complex systems: The ARKola simulation.* Paper presented at Association for Computing Machinery Computer Human Interaction conference (CHI'91), New Orleans, LA.

Gelfand, S. A. (1981). *Hearing: An introduction to psychological and physiological acoustics.* New York: Marcel Dekker Inc.

Kilander, F., & Lonnqvist, P. (2001). A weakly intrusive ambient soundscape for intuitive state perception. In G. Doherty, M. Massink, & M. Wilson (Eds.), *Continuity in future computing systems* (pp. 70–74). Oxford, UK: CLRC.

Kilander, F., & Lonnqvist, P. (2002). *A whisper in the woods—an ambient soundscape for peripheral awareness of remote processes.* Paper presented at International Conference on Auditory Display (ICAD 2002), Kyoto, Japan.

Kramer, G. (1994a). An introduction to auditory display. In G. Kramer (Ed.), *Auditory display* (pp. 1–77). Reading, MA: Addison-Wesley.

Kramer, G. (Ed.). (1994b). *Auditory display* (Vol. Proceedings volume XVIII). Reading, MA: Addison-Wesley.

Kramer, G., & Walker, B. (Eds.). (1999). *Sonification report: Status of the field and research agenda.* Santa Fe: The International Community for Auditory Display.

Lemmens, P. (2005). *Using the major and minor mode to create affectively-charged earcons.* Paper presented at International Conference on Auditory Display (ICAD 2005), Limerick, Ireland.

Leplâtre, G., & Brewster, S. A. (2000). *Designing nonspeech sounds to support navigation in mobile phone menus.* Paper presented at International Conference on Auditory Display (ICAD 2000), Atlanta, GA.

Mansur, D. L., Blattner, M., & Joy, K. (1985). Sound-graphs: A numerical data analysis method for the blind. *Journal of Medical Systems, 9*, 163–174.

Marentakis, G., & Brewster, S. A. (2004). *A study on gestural interaction with a 3D audio display.* Paper presented at MobileHCI 2004, Glasgow, UK.

Marila, J. (2002). *Experimental comparison of complex and simple sounds in menu and hierarchy sonification.* Paper presented at International Conference on Auditory Display (ICAD 2003), Kyoto, Japan.

Maury, S., Athenes, S., & Chatty, S. (1999). *Rhythmic menus: Toward interaction based on rhythm.* Paper presented at Association for Computing Machinery Computer Human Interaction conference (CHI'99), Pittsburgh, PA.

McCormick, E. J., & Sanders, M. S. (1982). *Human factors in engineering and design* (5th ed.). New York: McGraw-Hill.

McGookin, D. K., & Brewster, S. A. (2004). Understanding concurrent earcons: Applying auditory scene analysis principles to concurrent earcon recognition. *ACM Transactions on Applied Perception, 1*(2), 120–155.

Miranda, E. R. (1998). *Computer sound synthesis for the electronic musician.* Oxford, UK: Focal Press.

Moore, B. C. (2003). *An introduction to the psychology of hearing* (5th ed.). Oxford, UK: Elsevier Science.

Mynatt, E. D. (1994). *Designing with auditory icons: How well do we identify auditory cues?* Paper presented at Association for Computing Machinery Computer Human Interaction conference (CHI '94), Boston, MA.

Mynatt, E. D., Back, M., Want, R., Baer, M., & Ellis, J. B. (1998). *Designing audio aura.* Paper presented at Association for Computing Machinery Computer Human Interaction conference (ACM CHI'98), Los Angeles, CA.

Mynatt, E. D., & Edwards, K. (1995). Chapter 10: Metaphors for nonvisual computing. In A. D. N. Edwards (Ed.), *Extra-ordinary human–computer interaction* (pp. 201–220). Cambridge, UK: Cambridge University Press.

Mynatt, E. D., & Weber, G. (1994). *Nonvisual presentation of graphical user interfaces: Contrasting two approaches.* Paper presented at Association for Computing Machinery Computer Human Interaction conference (CHI'94), Boston, MA.

Neuhoff, J. G. (Ed.). (2004). *Ecological psychoacoustics.* San Diego, CA: Elsevier Academic Press.

Patterson, R. D. (1982). *Guidelines for auditory warning systems on civil aircraft* (CAA Paper No. 82017), London, Civil Aviation Authority.

Patterson, R. D. (1989). Guidelines for the design of auditory warning sounds. *Proceeding of the Institute of Acoustics, Spring Conference, 11*(5), 17–24.

Perrott, D., Sadralobadi, T., Saberi, K., & Strybel, T. (1991). Aurally aided visual search in the central visual field: Effects of visual load and visual enhancement of the target. *Human Factors, 33*(4), 389–400.

Pohlmann, K. (2005). *Principles of digital audio* (5th ed.). New York: McGraw-Hill.

Roads, C. (1996). *The computer music tutorial.* Cambridge, MA: MIT Press.

Ronkainen, S., & Pasanen, L. (2005). *Effect of aesthetics on audio-enhanced graphical buttons.* Paper presented at International Conference on Auditory Display (ICAD 2005), Limerick, Ireland.

Sawhney, N., & Schmandt, C. (1999). *Nomadic radio: Scalable and contextual notification for wearable messaging.* Paper presented at Association for Computing Machinery Computer Human Interaction conference (CHI'99), Pittsburgh, PA.

Sawhney, N., & Schmandt, C. (2000). Nomadic radio: Speech and audio interaction for contextual messaging in nomadic environments. *ACM Transactions on Human–Computer Interaction, 7*(3), 353–383.

Strybel, T., Manligas, C., & Perrott, D. (1992). Minimum audible movement angle as a function of the azimuth and elevation of the source. *Human Factors, 34*(3), 267–275.

Thimbleby, H. (1990). *User interface design.* New York: ACM Press, Addison-Wesley.

Walker, B. N. (2002). Magnitude estimation of conceptual data dimensions for use in sonification. *Journal of Experimental Psychology, Applied, 8*, 211–221.

Walker, B. N., & Cothran, J. (2003). *Sonification sandbox: A graphical toolkit for auditory graphs.* Paper presented at International Conference on Auditory Display (ICAD 2003), Boston, MA.

·14·

NETWORK-BASED INTERACTION

Alan Dix
Lancaster University

INTRODUCTION

In some ways this chapter could be seen as redundant in an HCI book—surely networks are just an implementation mechanism, a detail below the surface; all that matters are the interfaces that are built on them. On the other hand, networked interfaces, especially the Web—and, increasingly, mobile devices—have changed the way we view the world and society. Even those bastions of conservatism, the financial institutions, have found themselves in sea change, and a complete re-structuring of the fundamentals of businesses is just an implementation detail.

Structure

The chapter will begin with a brief overview of types of networks and then deal with network-based interaction under four main headings:

- **Networks as Enablers**—things that are only possible with networks
- **Networks as Mediators**—issues and problems because of networks
- **Networks as Subjects**—understanding and managing networks
- **Networks as Platforms**—algorithms and architectures for distributed interfaces

In addition, there will be a section taking a broader view of the history and future of network interaction and the societal effects and paradigm changes engendered, especially by more recent developments in global and wireless networking.

ABOUT NETWORKS

The word *network* will probably make many think of accessing the Internet and the Web. Others may think of a jumble of Ethernet wires between the PCs in their office. In fact, the range of networking standards, including physical cabling (or lack of cabling) and the protocols that computers use to talk along those cables, is extensive. Although most of the wire-based networks have been around for some time, they are in a state of flux, due to increases in scale and the demands of continuous media. In the wireless world things are changing even more rapidly with two new generations of data service being introduced over the next two years.

As an aid to seeing the broader issues surrounding these changing (and, in some cases, potentially ephemeral) technologies, we can use the following two dimensions to classify them:

- **Global vs. Local**
  How spatially distant are the points connected— ranging from machines in the same room (IrDa, Bluetooth), through those in a building/site (LAN) to global networks (Internet, mobile phone networks)?

	Fixed	Flexible
local	LAN	PAN, IrDA Bluetooth WiFi, UWB
	WAN	
global	Internet	GSM GPRS, 3G
	mobile	

- **Fixed vs. Flexible**
  How permanent are the links between points of the network, from physically fixed machines, to self-reconfiguring devices that recognize other devices in their vicinity?

The fixed vs. flexible dimension is almost, but not quite, terrestrial vs. wireless. The "not quite" is because fixed networks increasingly involve wireless links. Also, it is often possible, when visiting another organization, to plug a portable computer into a (wired) Ethernet network and find you have access to the local printers, Internet connections, and so on—flexible wire-based networking.

Let's look at a few network technologies against these dimensions. Traditional office LANs (local area networks) are squarely in the local–fixed category, whereas the Internet is largely global–fixed. Corporate WANs (wide area networks), connecting offices within the same national or international company, sit somewhere between.

Mobile phones have been placed within the global–fixed category as well. This may seem strange; the phone can go anywhere. However, the interconnections between phones are fixed and location-independent. If two mobile phones are in the same room it is no easier to connect between them than if they were at opposite ends of the earth (bar a shorter lag time perhaps).

Similarly, the Internet, although increasingly accessible through mobile devices and phones, is largely based on fixed domain names, IP numbers, and URLs.

Given the ideas that the placement of mobile phones is a little ambiguous and it is possible to detect the location of phones and thus deliver location-based content, some of the phone technologies have been listed in the global–flexible category. These include GSM (Global System for Mobile Communications), GPRS (General Packet Radio Service), and 3G. There is obviously a steadily increasing data rate, and third-generation services are able to cope with heavy media content including live video; in fact, the "killer app" is live sports highlights! However, the most significant differences are the charging and connectivity model. With GSM you connect when required to the Internet and this is treated like any other telephone call, usually meaning pay-per-minute whilst connected. In contrast, second- and third-generation services are based on sending small packets of data (the P in the GPRS acronym). The connection to the Internet is treated as "always on" and packets of data are sent to or from the phone as required. Charging is also typically by data use or by fixed charge.

In the local–flexible category there is a host of existing and emerging technologies. At the most mundane are the now ubiquitous WiFi networks and hotspots (based on the 802.11 protocol) (IEEE, 2001). These merely treat the machine the same as if it were plugged into the local fixed network. At a more local scale, infrared (IrDa) enabled devices can talk to one another if their infrared sensors are within line-of-sight, and Bluetooth (Bluetooth, 2001) or emerging wireless technologies such as ZigBee or UWB (Ultra Wide Band) (Zigbee, 2006; WiMedia, 2006) allow flexible connections between personal devices. With these, a laptop can use a mobile-phone modem, or a Bluetooth hands-free headset can connect to a phone without having to plug in with a piece of wire.

These same technologies can also be used to establish local connections with printers or other devices or even track people using the unique addresses that are often broadcast continually. Thus they offer both the opportunities of accessing fixed public equipment through personal devices, as well as the threat of surveillance and hacking everywhere!

Finally, research in wearable computers has suggested using the body itself as the connection between worn devices in a personal area network (PAN) (Zimmerman, 1996). The future is networked and we will become the network.

On the whole we have seen in the last 10 years the main focus of network-based interaction has moved anti-clockwise in this picture from fixed/local networks (mainly LAN), through fixed global networks (the Internet and web explosion), through global mobile networks (mostly phone-based, but including WAP (Wireless Application Protocol), i-mode, etc.), and moving towards flexible local connections between devices. In both the local and global spaces there has also been a growth of less centrally controlled networking with peer–peer services establishing decentralized applications over the Internet, and wireless ad-hoc networks allowing machines to establish networks with no fixed infrastructure.

NETWORKS AS ENABLERS

Things That Are Only Possible With Networks

It can be the case that the network is no more than an implementation detail; for example, using a networked disk rather than a local one. However, there are also many applications, like videoconferencing, which are only possible because the network is there. The key feature of networks is the access to remote resources of some kind or other.

Remote Resources

Four kinds of remote things are made accessible by networks:

- People
- Physical things
- Data
- Computation

These things may be remote because they are far away from where you normally are, or because you are yourself on the move and hence away from your own resources (colleagues, databases, etc.). Hence, mobility can create a need for any or all the above.

People

Networks mean we can communicate and work with others in distant places. This is often a direct action, such as e-mailing someone or engaging in a videoconference. These are all the normal study of CSCW (Computer-Supported Cooperative Work) and groupware.

Interaction with remote people may also be indirect. Recommender systems gather information about people's preferences and use this to suggest further information, services, or goods based on their own preferences and those of others who have similar tastes (Resnick & Varian, 1997). Because the people making recommendations are in different locations, the data on who selected what must be stored centrally. If you have been suggested books at Amazon.com, you have experienced a recommender system.

Collaborative virtual environments also offer the ability for remote people to interact, but by embedding them within an apparently local virtual-reality world. Although the person you are dealing with may be half a world away, her avatar (a virtual presence, perhaps a cartoon character, photo, or robot-like creature) may seem only a few yards or meters away in the virtual world.

Physical Things

We can also view and control remote things at a distance. For example, live web cams in public places allow us to see things (and people) there. Similarly, the cameras mounted around rockets as they prepare to take off (and are then usually destroyed during the launch) allow the mission controllers to monitor critical aspects of the physical system, as do the numerous telemetry sensors, which will also be related via some sort of closed network. And of course the launch command itself will be relayed to the rocket by the same closed network as will the ongoing mission, perhaps the Mars robots, via wireless links.

In the rocket example it would be dangerous to be in the actual location; in other circumstances it is merely expensive or inconvenient. Telescopes are frequently mounted in distant parts of the world where skies are clearer than those above the laboratories to which they belong. In order to avoid long international trips to remote places, some of these telescopes now have some form of remote control and monitoring using the Internet (Lavery, Kilgourz, & Sykeso, 1994).

At a more personal level the systems within certain high-end cars are controlled using a within-car network (called CAN). Even an adjustable heated seat may require dozens of control wires, but with a network only one power and one control cable is needed. The engine management system, lighting assemblies, radio, CD player, and wind screen wipers each have a small controller that talks through the network to the driver's console (although critical engine systems will usually have a separate circuit).

Many household appliances are now being made Internet-ready. In some cases this may mean an actual interface; for example, an Internet fridge that can scan the barcodes of items as you put them in and out and then warn you when items are getting out of date, generate a shopping list of items for you, and even order from your favorite store (Electrolux, 1999). Others have instead, or in addition, connectivity for maintenance purposes, sending usage and diagnostic data back to the manufacturer so that technicians can organize service or repair visits before the appliance fails in some way.

In some ways Internet shopping can also be seen in this light. Whilst at one level it is merely a transfer of data, the ultimate end is that you receive physically the ordered goods. This interaction with the remote physical goods is often two-way as you track its progress, and sometimes its physical location, through a web interface.

Data

Anyone using the Web is accessing remote data. Sometimes, data is stored remotely purely for convenience, but often data is necessarily stored remotely

- because it is shared by many remote people;
- because central storage helps maintain control, security, or privacy;
- because it is used by a single user at different locations (e.g., web e-mail);
- because it is too extensive to be stored locally (e.g., large databases and thin client).

In the case of the Web the data is remote because it is accessed by different people at different locations, the author(s) of the material, and all those who want to read it.

Even though the Web is quite complex, we may perceive a web page as a single entity, but in fact it exists in many forms (Fig. 14.1). The author of the page will typically have created it offline on his or her own PC. He or she then uploads the page (which effectively means copying it) onto the web server. Any changes the author makes after uploading the page will not be visible to the world until it is next uploaded. When a user wants to see the page and enters a URL or clicks on a link, his browser asks the web server for the file, which is then copied into the browser's memory and displayed to the user. You can tell the browser has a copy as you can disconnect from the Internet and still scroll within the file. If you access the same page again quite soon, your browser may choose to use the copy it holds rather than going back to the web server, again potentially meaning you see a slightly out-of-date copy of the page. Various other things may keep their own cached copies, including web proxies and firewalls.

This story of copied data in various places is not just about the Web, but is true to some extent or other of all shared networked data. With people or physical things, we do not expect to have the actual person or thing locally, just a representation. This is equally true for shared data except that the representation is so much like the "real thing" it is far less obvious to the user.

For shared networked data, even the "real thing" may be problematic—there may be no single "golden copy," but instead many variants, all with equal right to be called "the real data."

You don't even escape networking issues if you only access data locally on your own PC; networking issues may still arise if your data is backed-up over the network.

Computation

Sometimes remote computational resources are accessed over the network. The most obvious example of this is large supercomputers. These have enormous computational power, and scientists wishing to use them will often pre-book time slots

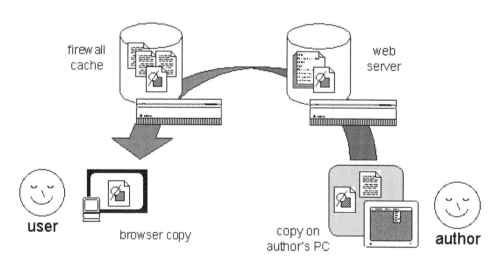

FIGURE 14.1. Copies of a web page in many places.

to perform particularly intensive calculations such as global weather simulations, analysis of chemical structure, stress calculations, and so on. Because these machines are so expensive, programs for them are typically developed on other, less powerful computers and then uploaded over the network when the supercomputer is available.

If the data required as input or output for the calculation is not too great, "fairly simple" means can be used to upload the programs and data. However, some calculations work on large volumes of data; for example, data from microwave readings of the upper atmosphere to probe the ozone hole generate terabytes (millions of millions of bytes) of data per second. High-capacity networks are being created in many countries to enable both high-volume data for this sort of application as well as the expected data required for rich media (Foster, 2000; Foster & Kesselman, 1999; GRID, 2001).

The ease with which data and results can be shipped back and forth across the Internet has enabled the growth of web services, web applications designed to be accessed by other programs supplying services or data. As well as more scientific or heavy commercial uses of these, they have become a standard part of many consumer-oriented applications—for example, "del.icio.us" has a standard API accessible through the Web allowing third-party applications to interact with it.

Sometimes calculations need to be performed centrally, not because the central computer is powerful, but because the local device is a computational lightweight. For example, one may want to create a remote analysis package where engineers in the field enter data into a PDA or phone interface, but where complex stress calculations are carried out on a small server back in the office. The data on materials and calculations involved may not be extensive by supercomputer standards, but may still be too much for a handheld device.

Because transporting large volumes of data is not always practical, calculations are often performed where the data is. (In performing any computation program, data and computational engines must all be in the same place. If they aren't together, then one or the other must be moved or copied to bring them together (Ramduny & Dix, 1997)). For example, when you perform a database access, the request for the data is usually transmitted to the database server as an SQL query, for example: "SELECT name, salary FROM payroll WHERE salary > 70000." In principle the complete contents of the payroll database could be downloaded to your PC and the selection of appropriate records carried out locally, however, it would be more costly to transmit the data hence the calculation is effectively transmitted to the database server. In a similar vein, Alexa allows third parties to run programs on their servers through their Web Search Platform (AlexaWSP, 2006); this allows the programs to access a 100-terabyte web crawl that would be impractical (and commercially unacceptable) to transfer to clients.

Even when the volume of data is not large or the frequency of access would make it cost effective to transmit it, security or privacy reasons may prevent the download of data. For example, some datasets are available to search to a limited degree on the Web, but charge for a download or CD of the complete dataset. My own hcibook.com site (Dix, Finlay, Abowd, & Beale, 1998) is rather like this, allowing searching of the book's contents online and displaying portions of the text, but not allowing a full download as readers are expected to buy the book!

For those who haven't come across it, the SETI (the Search for Extra-Terrestrial Intelligence) project is analyzing radio signals from outer space looking for patterns or regularities that may indicate transmissions from an alien civilization. You can download a SETI screensaver that performs calculations for SETI when you are not using your machine. Each SETI screensaver periodically gets bits of data to analyze from the central SETI servers and then returns results. This means that the SETI project ends up with the combined computational resources of many hundreds of thousands of PCs.

Security considerations may also prohibit the distribution of programs themselves if they contain proprietary algorithms. Also, if the source of the program is not fully trusted, one may not want to run these programs locally. The latter is the reason that Java applets are run in a software "sandbox" confining the ability of the applet to access local files and other potentially vulnerable resources.

The Search for Extra-Terrestrial Intelligence (SETI) is an interesting example of remote computation (SETI@home, 2001). Normally, remote computation involves a device of low-computational power asking a central computer to do work for it. In the case of SETI, large calculations are split up and distributed over large numbers of not particularly powerful computers.

The same technique is used in "PC farms." These occur when large numbers of PCs are networked together to act as a form of super-computer. For example, in CERN (the home of the Web), data from high-energy collisions may consist of many megabytes of data for each event, with perhaps hundreds of significant events per second (CERN, 2001). The data from each event is passed to a different PC, which then performs calculations on the data. When the PC finishes it stores its results and then adds itself back to a pool of available machines.

During coming years we are likely to see both forms of remote computation. As devices become smaller and more numerous, many will become simply sensors or actuators communicating with central computational and data servers (although "central" here may mean one per room, or even one per body). On the other hand, several companies, inspired by SETI, are pursuing commercial ways of harnessing the spare, and usually wasted, computational power of the millions of home and office PCs across the world.

Applications

The existence of networks, particularly the global networks offered by the Internet and mobile phone networks, have made many new applications possible and changed others.

Several of the more major application areas made possible by networks are covered in *The HCI Handbook* (CRC Press): groupware (chapter 29), online communities (chapter 30), mobile systems (chapter 32), e-commerce (chapter 39), telecommunications (chapter 40) and, of course, the web (chapter 37).

In addition, networking impinges on many other areas. Handheld devices (chapter 32) can operate alone, but are increasingly able to interact with one another and with fixed networks via wireless networking. Similarly, wearable computers (chapter 33) are expected to be interacting with one another via short-range networks, possibly carried through our own bodies (which makes mobile phones seem positively safe!) and information appliances (chapter 38) will be Internet connected to allow remote control and maintenance. In the area of government and citizenship (chapter 41), terms such as *e-democracy* and *e-government* are used to denote not just the technological ability to vote or access traditional government publications online, but a broader agenda whereby citizens feel a more intimate connection to the democratic process. Of course, education, entertainment, and game playing are also making use of networks.

Throughout the chapter we will also encounter broader issues of human abilities, especially concerned with time and delays, involving aspects of virtually all of part II (human perception, cognition, motor skills, etc.). Also, we will find that networking raises issues of trust and ethics (chapters 65 and 62) and of course the global network increases the importance of culturally and linguistically accessible information and interfaces (chapter 23).

Networking has already transformed many people's working lives, allowing telecommunication, improving access to corporate information whilst on the move, and enabling the formation of virtual organizations. Networks are also allowing whole new business areas to develop, not just the obvious applications in e-shopping and those concerned with web design.

The Internet has forced many organizations to create parallel structures to handle the more direct connections between primary supplier and consumer (disintermediation). This paradoxically is allowing more personalized (if not personal) services and often a focus on customer–supplier and customer–customer communication (Siegal, 1999; Light & Wakeman, 2001). This restructuring may also allow the more flexible businesses to revolutionize their high street (or mall) presence, allowing you to buy shoes in different sizes, or use next-day fitting services for clothes (Dix, 2001b).

The complexity of installing software and the need to have data available anywhere at any time has driven the nascent application service provider (ASP) sector. You don't install software yourself, but instead use software hosted remotely by providers who charge on a usage rather than once-off basis. By storing the data with third parties, an organization can off-load the majority of its backup and disaster-management requirements.

For the individual user, the ubiquity of Internet access for many has enabled many PIM (personal information management) applications such as e-mail, calendars, bookmark lists, and address books. These things that would once have been seen as personal are being not only accessed via the Web, but in many cases also shared. These web communities are no longer the province of geeks, but have become part of the day-to-day lives of many, engendering whole new ways of finding out and getting to know, including social bookmarks, blogs, and photologs.

The "personal" device has not become redundant though in this web orientation of applications. As well as being an access point to global services, it is also a potential interaction device for things close by. For example, in an installation by .:thePooch:. (thePooch, 2006) in an arts event, the attendees were encouraged to send SMS texts to Andrine, a huge face projected high on the wall. The texts were analyzed using natural language-processing techniques and, depending on the content, the face took on different emotions: happy, sad, or shocked (Lock, Al-

lanson, & Phillips, 2003). The cameras in phones are also being used to enable them to be used as location-finding devices (Sarvas, Herrarte, Wilhelm, & Davis, 2004), to enable the embedding of SpotCodes or other visual codes in paper posters (Toye et al., 2004; Semacode, 2006), and for real time manipulation of large public displays (Rohs, Sheridan, & Ballagas, 2004). Some of these applications use local networking such as Bluetooth; others paradoxically use the "global" connectivity through SMS or WAP to enable local interactions.

NETWORKS AS MEDIATORS

Issues and Problems Because of Networks

This section takes as a starting point that an application is networked and looks at the implications this has for the user interface. This is most apparent in terms of timing problems of various kinds. This section is really about when the network is largely not apparent, except for the unintended effects it has on the user.

We'll begin with a technical introduction to basic properties of networks and then see how these affect the user interface and media delivery.

Network Properties

Bandwidth and Compression

The most commonly cited network property is *bandwidth*, how much data can be sent per second. Those who have used dial-up connections will be familiar with 56 K modems, and those with longer memories or who use mobile phone modems may recall 9.6 K modems or less. The "K" in all of these refers to thousands of bits (0/1 value) per second (strictly Kbps), rather than bytes (single characters) that are more commonly seen in disk and other memory sizes. A byte takes eight bits; taking into account a small amount for overhead, you can divide the bits per second by 10 to get bytes per second.

Faster networks between machines in offices are more typically measured in megabits per second (again, strictly Mbps but often just written "M")—for example, the small "telephone cable" Ethernet is rated at either 10 Mbps or 100 Mbps.

As numbers these don't mean much, but if we think about them in relation to real data the implications for users become apparent.

A small word processor document may be 30 Kb (kilobytes). Down a 9.6 K GSM modem this will take approximately half a minute; on a 56 K modem this is reduced to five seconds; for a 10 Mb Ethernet this is 30 milliseconds. A full-screen, web-quality graphic may be 300 Kb, taking five minutes of 9.6 K modem, less than a minute on a 56 K modem, or one-third

Note that I am using the formula:

$$\text{download time } T = \frac{F \times 10}{M}$$

where:
  F = size of file in bytes
  10 is the number of raw bits per byte
  M = modem speed in bits per second

of a second on 10 Mb Ethernet. (Note that these are theoretical minimum times if there is nothing else using the network.)

Rich media, such as sound or video, puts a greater load on again. Raw, uncompressed HI-FI quality sound needs over 200 kilobits per second, and video tens of megabits per second. Happily there are ways to reduce this; otherwise digital AV would be impossible over normal networks.

Real media data has a lot of redundant information: areas of similar color in a picture, successive similar frames in a video, or sustained notes in music. **Compression** techniques use this similarity to reduce the actual amount of data that needs to be sent (e.g., rather than sending a whole new frame of video, it just sends the differences from the last frame). Also, some forms of compression make use of human perceptual limits; for example, MP3 stores certain pitch ranges with greater fidelity than others, as the human ear's sensitivity is different at different pitches (MPEG, 2001); also, JPEG images give less emphasis to accurate color hue than the darkness/lightness (JPEG, 2001). Between them, these techniques can reduce the amount of information that needs to be transferred significantly, especially for richer media such as video. Thus the actual bandwidth and the effective bandwidth, in terms of the sorts of data that are transmitted, may be very different.

Latency and Start-Up

Bandwidth measures how much data can be transferred; *latency* is how long each bit takes. In terms of a highway, bandwidth would be the number of lanes and latency is the time it takes to travel the length of the highway. The latency is due to two factors. The first is the speed of transmission of electricity through wires or light through optical networks. This may seem insignificant, but for a beam of light to travel across the Atlantic would take 20 ms and in practice this hop takes more like 70 ms. For satellite-based communications, the return trip to and from a geostationary satellite takes nearly a second; think about the typical delay you can hear on a transcontinental telephone call. The second factor contributing to latency is that every electronic switch or computer router has to temporarily store and then decide what to do with the signal before passing it on to the next along the chain. Typically this is a more major factor and in practice transatlantic Internet traffic will take nearer

250 ms from source to final destination, most of which is spent in various computer centers at one end or the other.

Latency is made worse by *setup time*. Every time you establish an Internet connection a conversation is established between your computer and the machine hosting the web-server:

"Hello, are you there?"
"Yes, I'm here. What do you want?"
"I'd like to send you some data."
"Great. I'm waiting."
"Okay, here it is, then."

(This is called handshaking.) Each turn in this conversation involves a round trip, network latency on both outward and return paths, and processing by both computers. And this is before the web server proper even gets to look at your request. Similar patterns happen as you dial a telephone call.

Latency and setup time are critical as they often dominate the delay for the user except for very large files or streaming audio/visual media. Early web design advice (given by those concerned about people with a slow connection, but who clearly had never used one!) used to suggest having only as much text as would fit on a single screen. This was intended to minimize the download time. However, this ignores setup times. A long text page doesn't take long to load even on a slow connection, once the connection to the web server has been established. Then it is far faster to scroll in the browser than to click and wait for another small page to load. A similar problem is the practice of breaking large images up into a jigsaw of small pieces. There are valid reasons for this—allowing rollover interaction, or where parts of the image are of different kinds (picture/text)—however, it is also used without such reasons and each small image requires a separate interaction with the server, encountering latency and setup delays.

Jitter and Buffering

Suppose you send letters to a friend every three days and the postal service typically takes two days to deliver letters (the average latency in network terms). Your friend will receive letters every three days, just delayed from when you sent them. Now

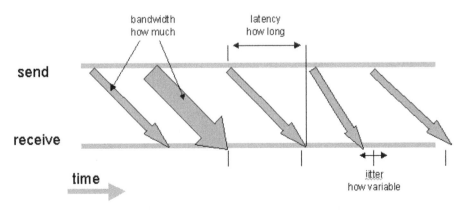

FIGURE 14.2. Bandwidth, latency, and jitter.

imagine that the postal system is a little variable: sometimes letters take two days, but occasionally they are faster and arrive the next day and sometimes they are slower and take three days. You continue to send letters every three days, but if a slow letter is followed by a fast one, your friend will receive them only one day apart; if on the other hand a fast letter is followed by a slow one the gap becomes five days. This variability in the delay is called "jitter." (Note that the fast letters are just as problematic as the slow ones; a fast letter followed by a normal-speed one still gives a four-day gap.)

Jitter doesn't really matter when sending large amounts of data, or when sending one-off messages. However it is critical for continuous media. If you just played video frames or sound when it arrived, jitter would mean that the recording would keep accelerating and slowing down (see Fig. 14.3(a) and (b)).

Jitter can be partially alleviated by *buffering*. Imagine that your friend's postman holds back one letter for three days and then starts giving letters to your friend one every third day. If your mail always arrives in exactly two days, the postman will always hold exactly one letter as mail will arrive as fast as he passes it on. If however a letter arrives quickly he will simply hold two letters for a few days, and if it is slow he will have a spare letter to give. Your friend's mail is now arriving at a regular rate, but the delay has increased to (a predictable) five days.

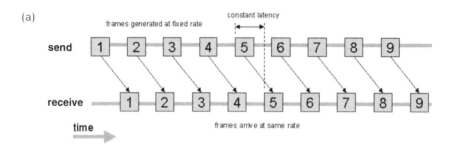

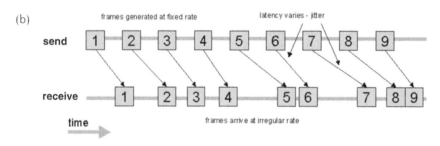

FIGURE 14.3.   (a) No jitter, no problem. (b) Jitter causes irregular perception.

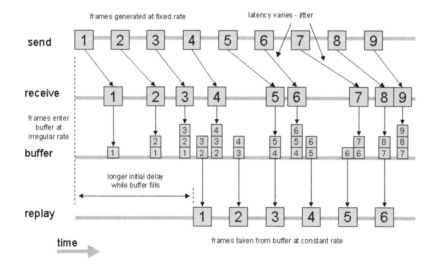

FIGURE 14.4.   Buffering smooths jitter, but adds delay.

Buffering in network multimedia behaves exactly the same, holding back a few seconds of audio/video data and then releasing it at a constant rate.

Reliability and Loss, Datagram, and Connection-Based Services

Virtually all networks are designed on the principle that there will be some loss or damage to data en-route. This arises for various reasons; sometimes there is electrical interference in a wire, or the internal computers and routers in the network may have too much traffic to cope with. This is normal and network software is built to detect damaged data and cope with lost data.

Because of this, the lowest layers of a network are assumed to be "lossy." Any data damaged in transit are discarded and computers and hardware en-route can choose to discard data if they get busy. So when one computer sends a packet of data to another it can assume that if the packet of data arrives it will be intact, but it may never arrive at all.

Some network data, in particular certain forms of real-time multimedia data, are deliberately sent in this unreliable, message-at-a-time, form (called "datagrams").

However, it is usually easier to deal with reliable channels, and so higher levels of the network create what are called *connection-based* services on top of the unreliable lower-level service. Internet users may have come across the term TCP/IP. IP is the name of an unreliable low-level service that routes packets of data between computers. TCP is a higher-level, connection-based service built on top of IP. The way TCP works is that the computer wanting to make a connection contacts the other and they exchange a few (unreliable IP) messages to establish the link. Once the link is established, the sending computer tags messages it sends with sequence data. Whenever the receiving computer has all the data up to a certain point it sends an acknowledgement. If the sending computer doesn't get an acknowledgement after a certain time it resends the data (Stevens, 1998, 1999).

With TCP, the receiving computer cannot send a message back when it notices a gap; it has to wait for the sending computer to resend after the timeout. While it is awaiting the resend it cannot process any of the later data. Notice that this means reliability is bought at the price of potential delays.

Quality of Service (QoS) and Reservation

The above properties are not just determined by a raw network's characteristics, such as the length of wires, types of routers, modems, and so on. They are also affected by other traffic and its volume and nature. If 10 PCs are connected to a single 10 Mbps network connection and require high-volume data transfers (perhaps streaming video), then there is only, on average, 1Mbps available for each. If you are accessing a network service that requires transatlantic connections during peak hours, then intermediate routers and hubs in the network are likely to become occasionally overloaded, leading to intermittent packet loss, longer average delays, and more variability in delay, hence jitter. In the past when the capacity of Internet backbones was lower, it was obvious in the United Kingdom when the United States "woke up" as the Web ground to a crawl!

For certain types of activity, in particular real-time or streaming rich-media, one would like to be able to predict or guarantee a minimum bandwidth, maximum delay, jitter, and so forth. These are collectively called "quality of service (QoS)" issues (Campbell & Nahrstedt, 1997). Some network protocols allow applications to reserve a virtual channel with guaranteed properties; however, the most common large-scale network, the Internet, does not have such guarantees—it operates solely on a best-endeavor basis. Upgrades to the underlying protocol (called by the catchy name IPv6) allow some differentiation of different types of traffic. This may allow routers to make decisions to favor time-critical data, but it will still not be able to reserve guaranteed capacity. However, in practice the increased capacity of Internet backbones is allowing large-scale Voice-over-Internet services such as Skype with acceptable end-to-end service (FCC, 2006; Skype, 2006).

Encryption, Authentication, and Digital Signatures

Some networks, such as closed office networks, offer no greater worries about security of information than talking together (both are capable of being bugged, but with similar levels of difficulty). However, more open networks, such as the Internet or phone networks, mean that data is traveling through a third party and public infrastructure to get to its recipients. Increasing use of wireless devices also means that the data sent between devices is more easily able to be monitored or interfered with by third parties. One option is to only use physically secure networks, but for economic reasons this is often not an option. Furthermore, solutions that do not rely on the network itself being secure are more robust. If you rely on, for example, a private dedicated line between two offices and assume it is secure, then if someone does manage to tap into it, all your interoffice communication is at risk.

The more common approach now is to assume the networks are insecure and live with it. This gives rise to two problems:

Secrecy—how to stop others from seeing your data
Security—how to make sure data is not tampered with

The first problem is managed largely by encryption methods, ensuring that even if someone reads all your communications they cannot understand them (Schneier, 1996). The "https" in some URLs is an example of this, denoting that the communication to the web server is encrypted.

The second problem, security, has various manifestations. Given communications are via a network, how do you know that you are talking to the right person/machine? Authentication mechanisms deal with this. In various ways they allow one machine to verify (usually by secret information that can only be known by the true intended party) that it is talking to the right party.

Even if you know that you are talking to the right person/machine, how do you know that the data you receive hasn't been changed? This is like receiving a signed letter, but, unbeknownst to you, someone has added some lines of text above the signature—although it really comes from the person you think, the message is not as was sent. If data is being encrypted then this may often implicitly solve this problem as any tampered data is uninterpretable by a third party, who therefore cannot alter it in a meaningful way.

If secrecy is not an issue, however, encryption is an unnecessary overhead and instead digital signatures generate a small data block that depends on the whole of the message and secret information known to the sender. It is possible for the recipient to verify that the signature block corresponds to the data sent and the person who is supposed to have sent it. One example of this are "signed applets" in which the Java code is digitally signed so that you can choose to only run Java programs from trusted parties.

UI Properties

Network Transparency

One of the goals of many low-level network systems is to achieve transparency; that is, to make it invisible to the user where on the network a particular resource lies. When you access the Web you use the same kind of URL and same kind of interface whether the web server is in Arizona, Australia, or Armenia. I know that when I'm at home and send an e-mail between two machines less than two meters apart, the message actually goes all the way across the Atlantic and back—but this is only because I have quite a detailed understanding of the computers involved. As a user I press "send mail" on one machine and it arrives near-instantaneously on the other.

Although network transparency has many advantages to the user—you don't care about routes through the network and so on—there are limits to its effectiveness and desirability. Some years ago I was at a Xerox lab in Welwyn Garden City in the United Kingdom. Randy Trigg was demonstrating some new features of Notecards (an early hypertext system; Halasz, Moran, & Trigg, 1987). The version was still under development and every so often would hit a problem and a LISP debugger window would appear. After using it for a while, it suddenly froze—no debugger window, no error message; it just froze. After a few seconds of embarrassment, Randy hit a control key and launched the debugger. A few minutes of frantic scanning through stack dumps, program traces, and so on, and the reason became clear to him. He had demonstrated a feature that he had last used on his workstation at Palo Alto. The feature itself was not at fault, but required an obscure font that he had on his own workstation, but not on the machine there in Welwyn. When Notecards had requested the font the system might have thrown up an error window, or substituted a similar font. However, in the spirit of true network transparency, the location of the font should not matter; having failed to find it on the local machine, it proceeded to interrogate machines on the local network to see if they had it, and then proceeded to scan the Xerox UK network and world network. Eventually, if we had waited long enough, it would have been found on Randy's machine in Palo Alto. Network transparency rarely extends to timing!

Transparency has also been critiqued for CSCW purposes (Mariani & Rodden, 1991). It may well be very important to users where resources and people are. For mobile computing also, an executive takes a laptop on the plane only to discover that the files needed are residing on a network fileserver rather than on the machine itself. If the interface hides location, how can one predict when and where resources will be available?

Later in this chapter, we will discuss recent work where the presence of intermittent connections, limited range, and variable signal strength is being used as a deliberate feature in interfaces.

Delays and Time

As is evident, one of the issues that arises again and again when considering networks is time. How long are the delays, how long to transfer data, and so on? Networking is not the only reason for delays in applications, but is probably one of the most noticeable—the Web has often been renamed the "World-Wide Wait." There is a long-standing literature on time and delays in user interfaces. This is not as extensive as one might think, however, largely because for a long time the prevailing perception in the HCI community was that temporal problems would go away (with some exceptions), leading to what I called the "myth of the infinitely fast machine" (Dix, 1987).

One of the earlier influential papers was Ben Shneiderman's review of research findings on delays (Shneiderman, 1984), mainly based on command-line interfaces. More recently there have been a series of workshops and special journal issues on issues of time, sparked largely by web delays (Johnson & Gray, 1996; Clarke, Dix, Ramduny, & Trepess, 1997; Howard & Fabre, 1998).

Three main timescales are problematic for networked user interfaces:

100 ms	Feedback for hand-eye coordination tasks needs to be less than 100–200 milliseconds to feel fluid. This is probably related to the fact that there are delays of this length in our motor-sensory system anyway. For aural feedback, the timescales are slightly tighter again.
1 second	Timescale for apparent cause–effect links, such as popping a window after pressing a button. If the response is faster than this the effect seems "immediate." This is related to a period of about 1 second that the brain regards as "now."
5–10 seconds	Waits longer than this engender annoyance and make it hard to maintain task focus. This may be related to short-term memory decay.

The 100 ms time is hard to achieve if the interaction involves even local network traffic. The 1 second time is usually achievable for local networks (as are assumed by X-Windows systems), but more problematic for long-haul networks. The 5–10 second time is in principle achievable for even the longest transcontinental connections, but, when combined with bandwidth limitations or overload of remote resources, may become problematic. This is especially evident on web-based services where the delay between hitting a link and retrieving a page (especially a generated page) may well exceed these limits, even for the page to begin to draw.

The lesson for UI designers is to understand the sort of interaction required and to ensure that parts of the user interface are located appropriately. For example, if close hand-eye coordination is required, it must run locally on the user's own machine—in the case of the Web in an applet, in JavaScript code, or so on. If the nature of the application is such that parts of the application cannot reside close enough to the user for the type of interaction required then, of course, one should not simply have a slow version of (for example) dragging an icon around, but instead change the overall interaction style to reflect the available resources.

Two of the factors that alleviate the effects of longer delays are predictability of the delay and progress indicators. Both give the user some sense of control or understanding over the process, especially if users have some indication of expected delays before initiating an action (Johnson, 1997). The many variable factors in networked systems make predicting delays very difficult, increasing the importance of giving users some sense of progress. The psychological effect of progress indicators is exploited (cynically) by those web browsers that have progress bars that effectively lie to the user, moving irrespective of any real activity. (Try unplugging a computer from the network and attempting to access a web page; some browsers will hit 70% on the progress bar before reporting a problem.) Other network applications use recent network activity to predict remaining time for long operations (such as large file downloads). Other solutions include generating some sort of intermediate low quality or partial information while the full information is being generated or downloaded (e.g., progressive image formats or splash pages of Flash movies).

For virtual reality using head-mounted displays, as well as hand-eye coordination tasks, we also have issues of the coordination between head movements and corresponding generated images. The timescales here are even tighter as the sensory paths are faster within our bodies, hence less tolerant of external delays. The brain receives various indications of movement: the position and changes of neck and related muscles, the balance sensors in the inner ear, and visual feedback. Delays between the movement of the generated environment and head movement lead to dissonance between these different senses and have an effect rather like being at sea, with corresponding disorientation and nausea. Also, any delays reduce the sense of immersion—being there within the virtual environment. Early studies of virtual reality (VR) showed that users' sense of immersion was far better when they were given very responsive wire frame images than if they were given fully rendered images at a delayed and lower frame rate (Pausch, 1991).

Coping Strategies

People are very adaptable. When faced with unacceptable delays (or other user interface problems) users develop ways to work around or ameliorate the problem—coping strategies. For example, web users may open multiple windows so that they can view one page whilst reading another (McManus, 1997) and users of "telnet" for remote command line interfaces may type "test" characters to see whether the system has any outstanding input (Dix, 1994).

Coping strategies may hide real problems, so it is important not to assume that just because users do not seem to be complaining or failing that everything is all right. On the other hand, we can also use the fact that users are bright and resourceful by building interface features that allow users to adopt coping strategies where it would be impossible or impractical to produce the interface response we would like. For example, where we expect delays we can ensure that continual interaction is not required (by perhaps amassing issues requiring user attention in a "batch" fashion), thus allowing users to more easily multitask. Unfortunately, this latter behavior is not frequently seen—the "myth" lives on and most networked programs still stop activity and await user interaction whenever problems are encountered.

Timeliness of Feedback/Feedthrough, Pace

Although feedback is one of the most heavily used terms in HCI, we may often ignore the complex levels of feedback when dealing with near-instantaneous responses of GUI interfaces. In networked systems with potentially long delays we need to unpack the concept. We've already discussed some of the critical timescales for feedback. For hand-eye coordination, getting feedback below the 100 ms threshold is far more important than fidelity; quickly moving wire frames or simple representations are better than dragging an exact image with drop shadow.

For longer feedback cycles, such as pressing a button, we need to distinguish

- **Syntactic feedback**—that the system has recognized your action
- **Intermediate feedback**—that the system is dealing with the request implied by your action (and, if possible, progress towards that request)
- **Semantic feedback**—that the system has responded and the results were obtained

The direct-manipulation metaphor has led to identification and hence confusion between these levels, and many systems provide little in the way of syntactic or intermediate feedback, relying solely on semantic feedback. In networked systems where the semantic feedback includes some sort of remote resource, it is crucial to introduce specific mechanisms to supply syntactic and intermediate feedback; otherwise, the system may simply appear to have ignored the user's action (leading to repeated actions with potentially unforeseen consequences) or even frozen or crashed.

This also reminds us of a crucial design rule for slow systems: Wherever possible, make actions idempotent—that is, that invoking the same action twice should, where possible, have the same effect as a single action. This means that the "try again" response to a slow system does not lead to strange results.

For collaborative systems or those involving external or autonomous resources (e.g., remote controlled objects, environmental sensors, software agents), we must also consider feedthrough. Feedback is experiencing the effect of one's own actions; *feedthrough* is the effect of one's own actions on other people and things and experiencing the effects of their actions oneself. For example, in an online chat system, you type a short message, press "send," and your message appears in your transcript (feedback); then, sometime later, it also appears in the transcript of the other chat participants (feedthrough).

Feedback is needed to enable us to work out whether the actions we have performed are appropriate, hence (typically) needs to be much quicker than feedthrough responses. This is fortunate as feedthrough by its very nature usually requires network transmission and ensuing delays. The exception to the rule that feedthrough can afford to be slower is where the users are attempting to perform some close collaborative task (e.g., positioning some items using direct manipulation) or where there is a second, fast communication channel (e.g., on the telephone, User A says to User B, "See the red box," but the relevant item hasn't appeared yet on User B's screen).

Potentially more important to users of collaborative systems than bandwidth or even raw delays is *pace,* the rate at which it is possible to interact with a remote resource or person. This is partly determined by lower level timings, but is also heavily influenced by interface design. For example, you know that someone is sitting at their desk and send them an urgent e-mail. The time that it takes to get a response will be hardly affected by the raw speeds between your machine and your colleague, and more determined by factors such as how often the e-mail client checks the server for new e-mail and whether it sounds an alert when new e-mail arrives, or simply waits there until your colleague chooses to check the in-box.

Race Conditions and Inconsistent Interface States

Alison and Brian are using an online chat program.

Alison writes, "It's a beautiful day. Let's go out after work," and then begins to think about it.
Brian writes, "I agree totally," and then has to leave to go to a meeting.
At almost the same time Alison writes, "Perhaps not; I look awful after the late party."

Unfortunately the messages are so close to simultaneous that both Alison and Brian's machines put their own contribution first, so Alison sees the chat window as in Fig. 14.5.a and Brian sees it as in Fig. 14.5.b. Brian thinks for a few moments, and then writes, "No, you look lovely as ever," but unfortunately Alison never sees this as she took one look at Brian's previous remark and shut down the chat program.

This type of incident where two events happen so close together that their effects overlap is called "a race condition." Race conditions may lead to *inconsistent states* for users as in this example, or may even lead to the software crashing. Although in principle race conditions are possible however fast the underlying network, the likelihood of races occurring gets greater as the network (and other) delays get longer.

Even some of the earliest studies in collaborative systems have shown the disorienting effects of users seeing different views of their shared information space, even when this is simply a matter of seeing different parts of the same space (Stefik, Bobrow, Foster, Lanning, & Tatar, 1987).

Consistency becomes an even greater problem in mobile systems where wireless connections may be temporarily lost, or devices may be unplugged from fixed networks whilst on the move. During these periods of *disconnection,* it is easy for several people to be updating the same information, leading to potential problems when their devices next become network-connected.

Later in this chapter, we will discuss mechanisms and algorithms that can be used to maintain consistency even when delays are long and race conditions are likely to occur.

Awareness

Returning to Alison and Brian, after Brian has typed his response he may not know that Alison hasn't seen his second contribution.

Awareness of who is around and what they are doing is a major issue in CSCW (see Dourish & Bellotti (1992) and McDaniel & Brinck (1997)). It has various forms:

- Being able to tell easily, when you want to know, what other people are doing;
- Being made aware (via alerts, very salient visual cues, etc.) when significant events occur (e.g., new user arrives, someone makes a contribution, etc.);
- Having a peripheral awareness of who is around and what they are up to.

Awareness isn't just about other people. In any circumstance where the environment may change, but not through your own direct action, you may need to know what the current state is and what is happening. This is not confined to networked applications, but applies to any hidden or invisible phenomena, for example, background indexing of your hard-disk contents. In networked applications, anything distant is invisible unless it is made visible (audible) in the interface.

One of the earlier influential experiments to demonstrate the importance of peripheral awareness was ArKola (Gaver, Smith, & O'Shea, 1991). This was a simulated bottling factory

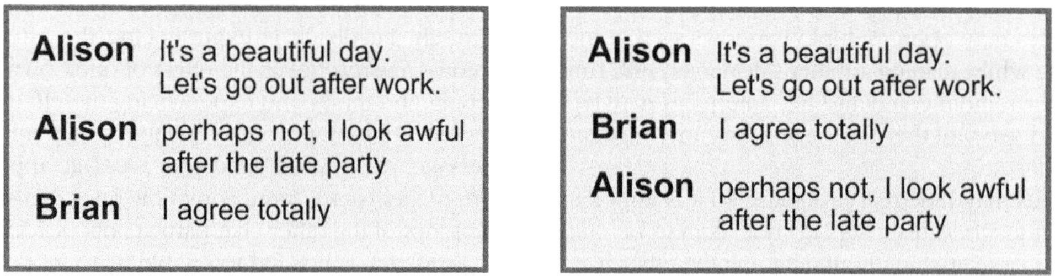

Alison	It's a beautiful day. Let's go out after work.
Alison	perhaps not, I look awful after the late party
Brian	I agree totally

Alison	It's a beautiful day. Let's go out after work.
Brian	I agree totally
Alison	perhaps not, I look awful after the late party

(a) Alison's chat window   (b) Brian's chat window

FIGURE 14.5. Consistency breakdown.

where two people worked together to maintain the factory, supplying, maintaining, and so on. The participants couldn't see the entire factory at once, so relied on the sounds produced to be aware of its smooth running or if there are any problems. For example, the sound of breaking glass might suggest that the end of the production line has run out of crates, but if it immediately stopped, one would assume that the other participant had sorted the problem out.

The numerous forms of shared video and audio spaces are another example of this. Several people, usually in distant offices, establish long-term, always-on audio, video, or audio-video links between their offices (Buxton & Moran, 1990; Olson & Bly, 1991). Sometimes these are used for direct communication, but most of the time they just give a peripheral awareness that the other person is there and the sort of activity he or she is doing. This can be used for functional purposes (e.g., knowing when the other person is interruptable), but also for social purposes—feeling part of a larger virtual office. Other systems have allowed larger numbers of (usually deliberately low quality and so less intrusive) web cam views of colleagues' offices and shared areas (Roussel, 1999, 2001). The aim is the same: to build social cohesion, to allow at-a-glance reading of one another's situation, and to promote "accidental" encounters.

A form of awareness mechanism is now common on the Web with "buddy lists" that tell you when friends are online (ICQ, 2001). I still know of no examples of rich media experiments in a domestic environment, for example a virtual kitchen-share with your elderly mother in Minnesota. However, various forms of domestic sharing via the Internet are becoming more common, including the early Casablanca project at Interval (Hindus, Mainwaring, Leduc, Hagström, & Bayley, 2001) using shared electronic sketchpads in the home, and more recent projects (CASIDE, 2005; Taylor, Izadi, Swan, Harper, & Buxton, 2006). Some of these applications are fairly standard "computer" interfaces; some are soft surveillance, such as monitoring an elderly relative; however, there is also a stream of research looking at more intimate ways of sharing presence, including pads that glow or warm when a loved one touches them far away or Jenny Tillotson's "Scent Whisper," a pair of Internet-connected brooches which emit a pleasant scent when the loved one whispers on the other end (Tillotson, 2005).

Trying to capture all this information within a computer display can be distracting and can use up valuable screen space—and of course assumes that the computer is there and on. For this reason, several projects have looked at ambient interfaces, which in various ways make the physical environment reflect the virtual. These interfaces monitor various events in the electronic worlds and then change things in the physical environment: lights on the wall, moving strings hung from the ceiling, or even a shaking pot plant (Lock, Allanson, & Phillips, 2000). Again, this is not fundamentally limited to networked environments, but is of course not very useful when the relevant activity is close at hand anyway.

The other side of this is finding out what people are doing in order to signal this to others. For computer activity—are you logged on, have you been typing recently, and what web page are you viewing? —this is in principle available, although the various layers of software may make it hard for an awareness service to discover. For non-computer aspects this is more problematic—are you in the room, busy, with other people?—and

may require a range of sensors in the environment, ultrasound, video, and so on, with corresponding privacy issues (see, for example, Bellotti (1993)). Monitoring of everyday objects is another way to achieve this; for example, one experiment used electronic coffee cups with sensors to tell when they were picked up and moved around (Gellersen, Beigl, & Krull, 1999). As more and more devices become networked, it may be that we don't need special sensors, just use the combined information from those available, although the privacy issues remain.

In collaborative virtual-reality environments, knowing that other people are around (as avatars) is as important as in a physical world, but harder due to limited senses (usually just vision and sound) and limited field of view. Furthermore, there are computational costs in passing information such as audio or even detailed positional information around the network when there are tens, hundreds, or thousands of users. Various spatial models have been developed to analyze and implement the idea of proximity in virtual space (Benford, Bowers, Fahlen, Mariani, & Rodden, 1994; Rodden, 1996; Sandor, Bogdan, & Bowers, 1997; Dix et al., 2000). These seek to formalize concepts of (a) where your focus of attention is within the virtual world and thus whether you require full quality audio and visual representation of others, (b) broader areas where you would expect some peripheral awareness where potentially degraded information can be used, and (c) those parts of the space for which you need no awareness information.

Media Issues

When describing the intrinsic network properties, issues for continuous media were mentioned several times. This is because, with the possible exception of close hand-eye coordination tasks, continuous media put some of the tightest requirements on the underlying networks.

Interactive Conversation and Action

Most demanding of all are audio-visual requirements of interactive conversation. Anyone who has had a transcontinental telephone conversation will have some feeling for the problems a delay of a second or two can cause. While actually speaking, the delays are less significant; however, turn-taking becomes very problematic. This is because the speaker in a conversation periodically (and subconsciously) leaves short (200–300 ms) gaps in the flow of speech. These moments of silence act as entry points for the other participant, who is expected to either acknowledge with a "go on" sound, such as "uhm" or perhaps a small nod of the head, or break in with his own conversation. Entries at other points would be seen as butting in and rude, and lack of feedback responses can leave the speaker uncertain as to the listener's understanding. The 200–300 ms is again almost certainly related to the time it takes for the listener's sensory system to get the relevant aural information to the brain and for it to signal the relevant nod, acknowledgement, or start to speak. Clearly our conversational system is finely tuned to the expected intrinsic delays of face-to-face conversation.

When network delays are added it is no longer possible to respond within the expected 200–300 ms window. The speaker therefore gets no responses at the appropriate points and it is very hard for the listener to break into the flow of speech without appearing rude (by the time they hear the gap and speak, the speaker has already restarted). Some telephone systems are half-duplex; that is, they only allow conversation in one direction at a time, which means that the various vocalizations ("uhu," "hmm," and so on) that give the speaker feedback will be lost entirely while the speaker is actually talking. It is not uncommon for the speaker to have to resort to saying, "Are you there?" due to a loss of sense of presence.

These effects are similar whether one is dealing with pure audio stream (as with the telephone), video streams (as with desktop conferencing), or distributed virtual environments. One VR project in the United Kingdom conducted all of its meetings using a virtual environment in which the participants were represented by cuboid robot-like avatars (called "blockies") (Greenhalgh, 1997). The project ended with an online virtual party. As the music played the participants (and their avatars) danced. Although clearly enjoying themselves, the video of the party showed an interesting phenomenon. Everyone danced alone. There are various reasons for this; for example, it was hard to determine the gender of a potential dancing partner. However, one relates directly to the network delays. Although everyone hears the same music they all were hearing it at slightly different time; furthermore, the avatars for other people will be slightly delayed from their actual movements. Given popular music rhythms operate at several beats per second, even modest delays means that your partner appears to dance completely out of time! In more recent work, predictive algorithms have been used to create "ghost" figures showing the "best guess" location of people and things in collaborative virtual environments (Gutwin et al., 2004).

Reliability

As well as delays, we noted previously that network connections may not always be reliable—that is, information may be lost. Video and audio streams behave very differently in the presence of dropped information. Imagine you were watching a film on a long air flight. The break in the sound when the pilot makes an announcement is much more difficult than losing sight of the screen for a moment or two as the passenger in front stands up. At a smaller scale a fraction-of-a-second loss of a few frames of video just makes the movie seem a little jerky; a smaller loss of even a few tens of milliseconds of audio signal would make an intrusive click or distortion. In general, reliability is more important for audio than video streams and where resources are limited it is typically most important to reserve the quality of service for the audio stream.

Sound and Vision

Why is it that audio is more sensitive than video? Vision works (largely) by looking at a single snapshot. Try walking around the room with your eyes shut, but opening them for glances once or twice a second. Apart from the moment or two as your eyes refocus you can cope remarkably well. Now turn a radio on with the sound turned very low and every second turn the sound up for a moment and back to silent—potentially an interesting remixing sound, but not at all meaningful. Sound more than vision is about change in time. Even the most basic sounds, pure tones, are measured in frequencies, how long between peaks and troughs of air pressure. For more complex sounds the shape of the sound through time—how its volume and frequency mix changes—are critical. For musical instruments it is hard to hear the difference between instruments if they are playing a continuous note, but instantly differentiable by their **attack**, how the note starts. (To get some idea of the complexity of sound see the review in Mitsopoulos' thesis (Mitsopoulos, 2000), and for an insight into the way different senses affect interaction see my AVI'96 paper (Dix, 1996).)

Compression

As we discussed earlier, it is possible to produce reliable network connections, but this introduces additional delays. Compression can also help by reducing the overall amount of audio-visual data that needs to be transmitted, but again may introduce additional delays. Furthermore, simple compression algorithms require reliable channels (both kinds of delays). Special algorithms can be designed to cope with dropped data, making sure that the most important parts of the signal are replicated or "spread out" so that dropped data leads to a loss in quality rather than to interruption.

Jitter

As noted previously, jitter is particularly problematic for continuous media. Small variations in delay can lead to jerky video playback, but is again even worse for audio streams. First of all, a longer than normal gap between successive bits of audio data would lead to a gap in the sound, just like dropped data. And perhaps even more problematic, what do you do when subsequent data arrives closer together—play it faster? Changing the rate of playing audio data doesn't just make it jerky, but also changes the frequency of sound, rendering it meaningless.

(*Aside:* Actually, you can do some quite clever things by digitally speeding up sound but not changing its frequency. These are not useful for dealing with jitter, but can be used to quickly overview audio recordings, or catch up on missed audio streams (Arons, 1997; Stifelman, Arons, & Schmandt, 2001).)

For real-time audio streams, such as video conferencing or Voice over IP (VoIP), it is hard to do anything about this and the best one can do is drop late data and do some processing of the audio stream to smooth out the clicks this would otherwise generate.

Broadcast and Prerecorded Media

Where media is prerecorded or being broadcast, but where a few seconds delay are acceptable, it is possible to do far better. Recall that in several places we saw that better quality can be obtained if we are prepared to introduce additional delays.

If you have used streaming audio or video broadcasts, you will know that the quality is quite acceptable and doesn't have many of the problems described above. This is partly because of efficient compression, meaning that video is compressed to a fraction of a percent of its raw bandwidth and so can fit down even a modem line. However, this would not solve the problems of jitter. To deal with this, the player at the receiving end buffers several seconds of audio-visual data before playing it back. The buffering irons out the jitter, giving continuous quality.

Try it out for yourself. Tune onto a radio channel and simultaneously listen to the same broadcast over the Internet with streamed audio. You'll clearly hear up to a minute delay between the two.

Fast and free Internet connections have enabled the sharing and distribution of high-quality media for storing and playing later, initially through illegitimate file-sharing including the pre-lawsuit Napster and many current peer–peer applications, but also increasingly through paid-for services such as the current Napster and Apple iTunes (Wikipedia contributors, 2006). As well as forcing existing media publishers to revisit their business models, the rise of web-distributed MP3 and Podcasting has enabled would-be artists and broadcasters to bypass the traditional distribution channels.

Public Perception: Ownership, Privacy, and Trust

One of the early barriers to consumer e-commerce has been distrust of the transaction mechanisms, especially giving credit card details over the Web. Arguments that web transactions are more secure than phone-based credit-card transactions or even using a credit card at your local restaurant (which gets both card number and signature) did little to alleviate this fear. This was never as major a barrier in the United States as it was, for example, in Europe, but across the world has been a concern, slowing down the growth of e-shopping (or really e-buying, but that is another story (Dix, 2001a)).

It certainly is the case that transactions via secure channels can be far more secure than physical transactions, in which various documents can be stolen or copied en-route and are in a format much more easy to exploit for fraud. However, knowing that a transaction is secure is more than the mechanisms involved, it is about human trust: Do I understand the mechanisms well enough, and the people involved well enough, to trust my money to it?

In fact, with a wider perspective, this distrust is very well founded. Encryption and authentication mechanisms can ensure that I am talking to a particular person or company and that no one else can overhear. But how do I know to trust that person? Being distant means I have few of the normal means available to assess the trustworthiness of my virtual contact. In the real world I may use the location and appearance of a shop to decide whether I believe it will give good service. For mail order goods I may use the size, glossiness, and publication containing an advertisement to assess its expense and again use this to give a sense of quality of the organization. It is not that these physical indicators are foolproof, but that we are more familiar with them. In contrast, virtual space offers few intrinsic affordances. It is easy and quite cheap to produce a very professional web presence, but it may be little more than a facade. This is problematic in all kinds of electronic materials, but is perhaps most obvious when money is involved.

Even if I trust the person at the other end, how do I know whether the network channel I am using is of a secure kind?

Again, the affordances of the physical world are clear. In a closed office, the open street, or in a bar frequented by staff of a rival firm, we will say different things depending on the perceived privacy of the location. In the electronic world we rely on "https" at the beginning of a URL (how many ordinary consumers know what that means?), or an icon inserted by the e-mail program to say a message has been encrypted or signed. We need to trust not only the mechanisms themselves, but also the indicators that tell us what mechanisms are being used (Millett, Friedman, & Felten, 2001; Fogg et al., 2001).

In non-financial transactions, issues of privacy are also critical. We've already seen several examples where privacy issues occur. As more devices become networked, especially via wireless links, and our environment and even our own bodies become filled with interlinked sensors, issues about who can access information about you become more significant. This poses problems at a technical level—ensuring security between devices, at an interface level—being able to know and control what or who can see specific information; and at a perception level, believing in the privacy and security of the systems. There are also legal implications. For example, in the United Kingdom in 2001 it was illegal for mobile telecoms operators to give location information to third parties (Sangha, 2001).

The issue of perception is not just a minor point, but perhaps the dominant one. Networks, and indeed computer systems in general, are by their nature hidden. We do not see the bits traveling down the wires or through the air from device to device, but have to trust the system at even the most basic level. As HCI specialists we believe ourselves a little above the mundane software engineers who merely construct computer systems, as we take a wider view and understand that the interaction between human and electronic systems has additional emergent properties and that it is this complete socio-technical unit which achieves real goals. For networked systems, this view is still far too parochial.

Imagine if the personal e-mail of millions of people was being sucked into the databanks of a transnational computer company, and only being released when accessed through the multinational's own web interface. The public outcry! Imagine Hotmail, Yahoo mail, and so on. How is it that, although stored on distant computers, perhaps half the world away, millions of people feel that it is "their" mailbox and trust the privacy of web mail more than, perhaps, their organization's own mail system. This feeling of ownership of remote resources is more than the technology that protects the security of such systems, it is a cultural phenomenon and a marketing phenomenon. The web mail "product" is not just technology, or interface, but is formed by every word that is written about the product in advertisements, press releases, and media interviews (Dix, 2001a).

NETWORKS AS SUBJECTS

Understanding and Managing Networks

When using a networked application, you don't really care what kind of network is being used, whether the data is sent over copper wires, fiber optic, microwave, or satellite. All you care about is that the two ends manage to communicate and the effects any of the above have on the end-to-end network properties such as bandwidth discussed in the previous section. However, there are times when the network's internals can't be ignored.

Those involved in installing or managing networks need to understand the internal workings of the network in order to optimize performance and find faults. For ordinary users, when things go wrong in a networked application, they effectively become a network manager and so understanding something of the network can help them to deal with the problem. Even when things are working, having some awareness of the current state of the network may help one predict potential delays, avoid problems, and minimize costs. In some cases this can even be used as a positive part of the interactive experience.

We'll start this section by looking at some of the technical issues that are important in understanding networks. This parallels earlier information in this chapter, but is focused on the internal properties of the network. We'll then look at the interface issues for those managing networks and the ways in which interfaces can make users aware of critical network states. Finally, we'll look briefly at the way models of networks can be used as a metaphor for some of the motor and cognitive behaviors of humans.

Network Models

Layers

Networking is dominated by the idea of layers; lower levels of the network offer standard interfaces to higher levels so that it its possible to change the details of the lower level without changing the higher levels. For example, imagine you are using your PDA to access the Internet whilst in a train (see Fig. 14.6).

The web browser establishes a TCP/IP connection to the web server, requests a web page, and then displays it. However, between your PDA and the web server the message may have traveled through an infrared link to your mobile phone, which then used a cell-based radio to send it to a mobile-phone station, then via a microwave link to a larger base station onto a fiber optic telephone backbone, and via various copper wires to your ISP's modem bank. Your ISP is then connected into another fiber optic Internet backbone and eventually via more fiber optic, copper, and microwave links to the web server. To complicate things even further, it may even be that the telephone and Internet backbones may share the same physical cabling at various points. Imagine if your poor PDA had to know about all of this.

In fact, even your PDA will know about at least five layers:

• Infrared—how the PDA talks to the phone
• Modem—how the PDA uses the phone to talk to your ISP
• IP—how your PDA talks to the web-server computer
• TCP—how data is passed as a reliable connection-based channel between the right program on your PDA and the web-server computer
• HTTP—how the browser talks to the web server

Each of these hides most of the lower levels, so your browser needs to know nothing about IP, the modem, or the infrared connection whilst accessing the web page.

The nature of these layers differs both between different types of network (for example, WAP) for sending data over mobile phones and devices; it has five defined layers (Arehart et al., 2000) while the ISO OSI reference model has seven layers (ISO/IEC7498, 1994).

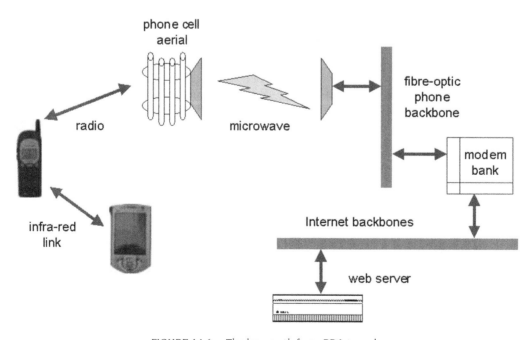

FIGURE 14.6. The long path from PDA to web.

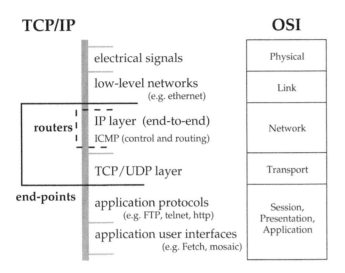

TCP/IP

electrical signals

low-level networks
(e.g. ethernet)

routers

IP layer (end-to-end)

ICMP (control and routing)

TCP/UDP layer

end-points

application protocols
(e.g. FTP, telnet, http)

application user interfaces
(e.g. Fetch, mosaic)

OSI

| Physical |
| Link |
| Network |
| Transport |
| Session, Presentation, Application |

FIGURE 14.7.   OSI 7 layers and TCP/IP.

mail program says	SMTP server replies

220 mail.server.net ESMTP

HELO mypc.mydomain.com

250 mail.server.net Hello mypc.mydomain.com[100.0.1.7],

pleased to meet you

MAIL From:<myself@mydomain.com>

250 <myself@mydomain.com>... Sender ok

RCPT To: <a-friend@theirdomain.com>

250 <a-friend@theirdomain.com>... Recipient ok

DATA

354 Enter mail, end with "." on a line by itself

first line of message

..dotty 2nd line

last line

.

250 KAA24082 Message accepted for delivery

QUIT

221 mail.server.net closing connection

FIGURE 14.8.   Protocol to send e-mail via SMTP.

Protocols

Systems at the same layer typically require some standard language to communicate. This is called a *protocol*; for higher levels this may be quite readable. For example, to send an Internet e-mail message your mail program connects to an SMTP server (a system that relays messages) using TCP/IP and has the exchange shown in Fig. 14.8.

At lower levels data is usually sent in small **packets**, which contain a small amount of data plus header information saying where the data is coming from, where it is going to and other bookkeeping information such as sequence numbers, data length, and so forth.

Even telephone conversations, except those using pre-digital exchanges, are sent by chopping up your speech into short segments at your local telephone exchange, sending each segment as a packet and then reassembling the packets back into a continuous stream at the other end (Stevens, 1998, 1999).

Internetworking and Tunneling

This layering doesn't just operate within a particular network standard, but between different kinds of networks too. The Internet is an example of an internet (notice lowercase "i") that is a network that links together different kinds of low-level networks. For example, many PCs are connected to the Internet via an Ethernet cable. Ethernet sends its own data in packets like those of Fig. 14.9. The Internet protocol (IP) also has packets of a form like Fig. 14.9. When you make an Internet connection via Ethernet, the IP packets are placed in the data portion of the Ethernet packet, so you get something a bit like Fig.14.10.

This placing of one kind of network packet inside the data portion of another kind of network is also used in virtual private networks (VPNs) in a process called **tunneling**. This process is used to allow a secure network to be implemented using a public network like the Internet. Imagine a company has just two offices, one in Australia and the other in Canada. When a computer in the Sydney office sends data to a computer in the Toronto office, the network packet is encrypted, put in the data portion of an Internet packet, and sent via the Internet to a special computer in the Toronto office. When it gets there, the computer at the Toronto office detects it is VPN data, extracts the encrypted data packet, decrypts it, and puts it onto its own local network where the target computer picks it up. As far as both

header				body
to address	from address	info	data length	data .....

FIGURE 14.9.   Typical network packet format (simplified).

ethernet header   ethernet body

			IP header			IP body
ethernet to addr.	ethernet from addr.	IP flag data len. etc.	IP to addr	IP from addr	other header IP info	Internet data .....

FIGURE 14.10.   Internet IP packet inside Ethernet packet.

ends are concerned it looks as if both offices are on the same LAN and any data on the Internet is fully encrypted and secure.

Routing

If two computers are on the same piece of physical network, each can simply "listen" for packets that are destined for them—so sending messages between them is easy. If, however, messages need to be sent between distant machines—for example, if you are dialed into an ISP in the United Kingdom and are accessing a web server in the United States—the message cannot simply be broadcast to every machine on the Internet (Fig. 14.11). Instead, at each stage it needs to be passed in the right direction between different parts of the network. Routers perform this task. They look at the address of each packet and decide where to pass it. A local machine might need to simply put it onto the relevant local network, but if not it may need to pass it on to another intermediate machine.

Routers may be stand-alone boxes in network centers, or may be a normal computer. Often a file server acts as a router between a LAN and the global network.

As well as routers, networks are also linked by hubs and switches that make several different pieces of physical network behave as if they were one local network, and gateways that link different kinds of network. The details of these are not important, but they add more to the sheer complexity of even small networks.

Addresses

In order to send messages on the Internet or any other network, you need to have the address of where they are to go (or at least your computer needs it). In a phone network this is the telephone number and on the Internet it is an IP number. The *IP number* is a 32-bit number, normally represented as a group of four numbers between 0 and 255 (e.g. 212.35.74.132), which you will have probably seen at some stage when using a web browser or other Internet tool. It is these IP numbers that are used by routers to send Internet data to the right places.

However, with any network there is a problem of getting to know the address. With phone numbers you simply look up the person's name in a telephone directory or by phoning the operator. Similarly, most networks have a naming scheme and some way to translate these into addresses. In the case of the

---

The 32-bit IP number space allows for 4 billion addresses. This sounds like quite a lot; however, these have been running out due to the explosive growth in the number of Internet devices and "wasted" IP numbers due to the way ranges of numbers get allocated to sub-networks. The new version of TCP/IP, IPv6, which is being deployed, has 128-bit addresses, which require 16 numbers (IPng, 2001). This allows sufficient unique IP addresses for every phone, PDA, Internet enabled domestic appliance, or even electronic paperclip.

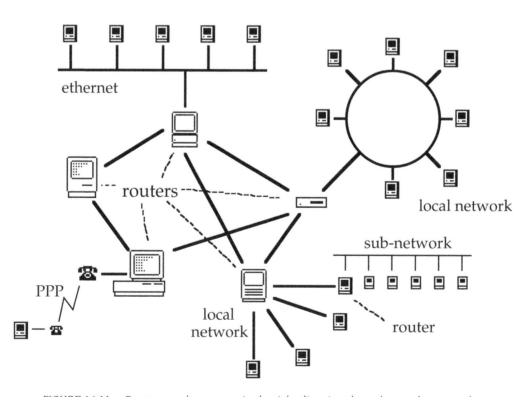

FIGURE 14.11.   Routers send messages in the right direction through complex networks.

Internet, domain names (e.g., acm.org, www.hcibook.com, or magisoft.co.uk) are the naming system. There are many of these and they are changed relatively rapidly, so there is no equivalent of a telephone directory, but instead special computers called "domain name servers (DNS)" act as the equivalent of telephone directory inquiry operators. Every time an application needs to access a network resource using a domain name (e.g., to look up a URL) the computer has to ask a DNS what IP address corresponds to that domain name. Only then can it use the IP address to contact the target computer.

The DNS system is an example of a "white pages" system. You have an exact name and want to find the address for that name. In addition there are so-called "yellow pages" systems where you request, for example, a color postscript printer and are told of addresses of systems supplying the service. Sometimes this may be mediated by brokers who may attempt to find the closest matching resource (e.g., a nonpostscript color printer) or even perform translations (for example, the Java JINI framework; Edwards & Rodden, 2001).

This latter form of resource-discovery system is most important in mobile systems and ubiquitous computing where we are particularly interested in establishing connections with other geographically close devices.

A final piece in the puzzle is how one gets to know the address of the name server, directory service, or brokering service. In some types of network this may be managed by sending broadcast requests to a network ("Is there a name server out there?"). In the case of the Internet this is normally explicitly set for each machine as part of its network settings.

All Together . . .

If you are in your web browser and you try to access the URL http://www.meandeviation.com/qbb/, the following stages take place:

1. Send IP-level request to DNS asking for www.meandeviation.com
2. Wait for reply
3. DNS sends reply 64.39.13.108
4. Establish TCP level connection with 64.39.13.108
5. Send HTTP request ("GET/qbb/HTTP/1.1")
6. Web server sends the page back in reply
7. Close TCP connection

Most of these stages are themselves simplified, all will involve layering on top of lower-level networks, and most stages involve several sub-stages (e.g., establishing a TCP level connection requires several IP-level messages).

The basic message is that network internals are multilevel, multistage, and pretty complicated.

Decentralizing: Peer–Peer and Ad-Hoc Networks

Traditional Internet and web applications tended to work through client–server paradigms in which a user's client application accesses a well-known application such as a web server.

However, there has been a growth of more decentralized models at both higher and lower protocol levels (Androutsellis-Theotokis & Spinellis, 2004).

At an application level, peer–peer file sharing works by having clients running on users' own PCs talk directly to one another, broadcasting requests for particular files by name or type to all connected clients. These then forward the requests to their connected nodes until some maximum hop count is hit. Often semi-centralized services are used to establish initial connections, but thereafter everything is done at the "edges" of the network. These protocols and applications have some technical strengths but their origins are rooted in legal disputes on the distribution of copyright material. Recent research has also shown that the anonymity of peer–peer file sharing also allows "deviant" subgroups sharing illicit material (Hughes, Gibson, Walkerdine, & Coulson, 2006).

At a lower level, ad-hoc networks build more structured higher-level networks for computers that have point-to-point connections, often through wireless connectivity. This can be used to set up a network in a meeting, or at a larger level could even allow everyone in a pop festival to have Internet access through a small number of wireless hot spots. In an ad hoc network, the low-level software works out what machines are connected where and routes messages from end to end even if the two end-point computers do not have direct access. The difficult thing in an ad hoc network is not just setting up this routing information, but the dynamics: dealing with the fact that machines constantly move around, perhaps going out of range of one another, may be turned off, or may join the network late.

Network Management

Those most obviously exposed to this complexity are the engineers managing large national and international networks, both data networks such as the main Internet backbones, and telecoms networks. The technical issues outlined above are compounded by the fact that the different levels of network hardware and network management software are typically supplied by different manufacturers. Furthermore, parts of the network may be owned and managed by a third party and shared with other networks. For example, a transatlantic fiber optic cable may carry telecoms and data traffic from many different carriers.

When a fault occurs in such a network it is hard to know whether it is a software fault or a hardware fault, where it is happening, or who is responsible. If you send engineers out to the wrong location it will cost both their time and the time the service is unavailable. Typically the penalties for inoperative or reduced quality services are high; you need to get it right, fast.

This is a specialized and complex area, but clearly of increasing importance. It poses many fascinating UI challenges: how to visualize complex multilayered structures and how to help operators trace faults in these. Although I know that it is a topic being addressed by individual telecoms companies, published material in the HCI literature is minimal. The exception is visualization of networks, both physical and logical, which is quite extensive (Dodge and Kitchin's Atlas of Cyberspace and associated website (http://www.cybergeography.org/atlas/) is the comprehensive text in this area; Dodge & Kitchin, 2001).

Ordinary systems administrators in organizations face similar problems, albeit on a smaller scale. A near-universal experience and consequently of misconfigured e-mail systems, continual network failures and performance problems certainly suggests this is an area ripe for effective interface solutions, but again there is very little in the current HCI literature.

Finally, it appears that everyone is now a network administrator; even a first-time home PC user must manage modem settings, name server addresses, SMTP and POP servers, and more. It is interesting that the interface used by most such users is identical to that supplied for full systems administrators. Arguably this may ease the path for those who graduate from single machines to administering an office or organization (perhaps less than 5% of users). Unfortunately, it makes life intolerable for the other 95%! The only thing that makes this possible at all is that the "welcome" disks from many ISPs and the Wizards shipping with the OS offer step-by-step instructions or may automatically configure the system. These complications are compounded if the user wishes to allow access through more than one ISP or connect into a fixed network. As many home users now have several PCs and other devices that need to be networked, this is not a minor issue.

In the first edition of this handbook I wrote, "If we look at the current state of the two most popular PC systems, Microsoft Windows and Mac OS, the picture is not rosy." Sadly things have not improved dramatically; you only have to watch a room full of computer scientists at a meeting trying to sort out wireless connections.

On Windows, if you do want to change your network settings you need to set several IP numbers, web proxies, and SMTP settings in several control panels. In fact as I sat here now writing I tried and failed to find the place to change the web proxy setting on my PC. It is possible to set up alternate settings for networks if you move between two locations, but more than that and you are lost! On MacOS the situation has improved in the last few years and now nearly all settings are in one place under "Network Settings." Changing locations with a laptop is simply a matter of selecting "Location . . ." from the main menu. Perversely, the default printer that used to be attached to your location on the older MacOS 9 interface now has to be changed by hand each time!

The problems in both emphasize partly the intrinsic complexity of networking—yes, it does involve multiple logically distinct settings, many of which relate to low-level details. However, it also exposes the apparent view that those involved in network administration are experts who understand the meaning of various internal networking terms. This is not the case even for most office networks, and certainly not at home.

And this is just initially setting up the system to use. For the home user, debugging faults has many of the same problems as large networks. You try to visit a website and get an error box. Is the website down? Are there problems with the wireless LAN, the broadband modem, the phone line, or the ISP's hardware? Are all your configurations settings right? Has a thunderstorm 3,000 miles away knocked out a vital network connection? Trying to understand a multilayered, non-localized and, when things work, largely hidden system is intrinsically difficult and where diagnostic tools for this are provided, they assume an even greater degree of expertise.

The much-heralded promise of devices that connect to one another within our homes and about our bodies is going to throw up many of the same problems. Some old ones may ease as explicit configuration becomes automated by self-discovery between components, but this adds further to the "hiddenness" and thus difficulty in managing faults, security, and so on. You can imagine the scenario: The sound on my portable DVD stops working and produces a continuous noise—why? There are no cables to check of course (wireless networking), but hours of checking and randomly turning devices on and off narrows the problem down to a fault in the washing machine, which is sending continuous "I finished the clothes" alerts to all devices in the vicinity.

Network Awareness

One of the problems noted above is the hiddenness of networks. This causes problems when things go wrong as one hasn't an appropriate model of what is going on, but also sometimes even when things are working properly.

We discussed earlier some of the network properties that may affect usability: bandwidth, delay, jitter, and so forth. These are all affected to some extent by other network loads, the quality of current network connections, and so on. So predicting performance (and knowing whether to panic if things appear to go slow) needs some awareness of the current state of the network.

Personal computers using wireless networks usually offer some indication of signal strength (if one knows where to look and what it means) although this is less common for line quality for modems. As wireless devices and sensors become smaller they will not have suitable displays for this and explicitly making users aware of the low-level signal strength of an intelligent paperclip may not be appropriate. However, as interface designers we do need to think how users will be able to cope and problem-solve in such networks.

Only more sophisticated network management software allows one to probe the current load on a network. This does matter. Consider a small home network with several PCs connected through a single modem. If one person starts a large download they will be aware that this will affect the performance of the rest of their web browsing. Other members of the household will just experience a slowing down of everything and not understand why. If the cause is a file-sharing utility such as Gnutella, even the person whose computer is running it may not realize its network impact. In experiments at MIT the level of network traffic has been used to "jiggle" a string hanging from the ceiling so that heavy traffic leads to a lot of movement (Wisneski et al., 1998). The movements are not intrusive so give a general background awareness of network activity. Although supplying a ceiling-mounted string may not be the ideal solution for every home, other more prosaic interface features are possible.

Cost awareness is also very important. In the United Kingdom, first generation GSM mobile data services are charged by connection time. So knowing how long you have been connected and how long things are likely to take becomes critical. If these charges differ at peak hours of the day, calculating whether to read your e-mail now, or do a quick check and download the big attachments later can become a complex decision. The move to data-volume-based charging means the user needs to estimate the volume of data that is likely to be involved in initiating an action versus the value of that data—do you want to click on that link if the page it links to includes large graphics, perhaps an applet or two? Terrestrial broadband networks have largely shifted to fixed monthly fees and this has dramatically changed the way people perceive the Internet, from smash-and-grab interactions to surf-and-play.

Network Confusion

If the preceding doesn't sound confusing enough, the multilayered nature of networked applications means that it is hard to predict the possible patterns of interference between things implemented at different levels or even at the same level. Again this is often most obvious when things go wrong, but also because unforeseen interactions may mean that two features that work perfectly well in isolation may fail when used together.

This problem, feature interaction, has been studied particularly in standard telecoms (although it is certainly not confined to them). Let's look at an example of feature interaction. Telephone systems universally apply the principle that the caller, who has control over whether and when the call is made, is the person who pays. In the exceptions (free-phone numbers, reverse charges) special efforts are made to ensure that subscribers understand the costs involved. Some telephone systems also have a feature whereby a caller who encounters a busy line can request a callback when the line becomes free. Unfortunately at least one company implemented this callback feature so that the charging system saw the callback as originating from the person who had originally been called. Each feature seemed to be clear on its own, but together meant you could be charged for calls you didn't want to make.

With N features there are $N(N-1)/2$ possible pairs of interactions to consider, $N(N-1)(N-2)/6$ triples, and so on. This is a well-recognized (but not solved) problem with considerable efforts being made using, for example, formal analysis of interactions to automatically detect potential problems. It is worth noting that this is not simply a technical issue; the charging example shows that it is not just who pays that matters, but also the perceptions of who pays. This particular interaction would have been less of a problem if the interface of the phone system had, for example, said (in generated speech), "You have had a call from XXX; press 'callback' and you will be charged for this call." The hybrid-feature interaction research group at Glasgow are attempting to build a comprehensive list of such problems in telecoms (HFIG, 2001), but these sorts of problems are likely to be found increasingly in related areas such as ubiquitous computing and resource discovery.

Exploiting the Limitations: Seamfulness and Virtual Locality

Some companies cynically exploit this user confusion over network charging and in the United Kingdom there have been some high-profile news stories about teenagers running up thousands of pounds of debt after innocently signing up to ringtone delivery services.

Happily there are also more positive uses of the limitations of networks. Given suitable awareness mechanisms for network strength and connectivity, people are very resourceful in exploitation. You will have experienced the way mobile-phone users get to know the sweet spots for their networks when in areas of poor coverage, learning to get out of doors, away from big buildings, or up hills. This can get sophisticated. One mixed-reality experiment, part of the Equator project, consisted of an outside game where real players running in the street were pitted against virtual characters manipulated remotely. The "real" players learned to hide in the GPS shadow of buildings so that their location could not be detected by the virtual participants (Benford et al., 2003). Similar effects have been seen in WiFi-based games where participants get to know the regions of good and bad coverage and may use these to seek out or avoid remote interactions.

In these ways, users do more than cope; they actively exploit the limitations of networks and sensing. The Pirates game did this with RF technology. RF beacons represent islands in an ocean, the locality of the RF signals corresponding to the limited land area of the island (Björk, Falk, Hansson, & Ljungstrand, 2001). This notion of exploiting network limitations has led to the idea of *seamful games*, deliberately designed so that variations in WiFi coverage and connectivity are part of the game play (Chalmers et al., 2005).

Even more strange is a recent game "Hitchers." In this you pick up and drop off hitchers with your phone, just as you might pick up a hitchhiker on the road. You can only pick up a hitcher in the same mobile cell as the last person who dropped the hitcher off. The hitchers are actually stored and managed centrally, but the effect is as if the hitchers were only accessible in a small region. Here, even though there is no real limitation of access, a limitation is constructed to make a more interesting game: *virtual locality*. You can imagine virtual locality being used in synchronous applications—for example, allowing a phone user to broadcast to people in their vicinity apparently as if it were a limited range transmission, whereas in reality it is centrally managed, based on GPS or phone cell location?

The Network Within

So far the story is pretty bleak from a user interface viewpoint: a complex problem, of rapidly growing importance, with relatively little published work in many areas. One good thing as a HCI practitioner about understanding the complexity of networks is that they help us understand better the workings of the human cognitive and motor system.

For at least five decades, computational models have been used to inspire cognitive models. Also of course cognitive and neurological models have been used to inspire computational models in artificial intelligence and neural networks. However, our bodies are not like a single computer, but in various ways more like a networked system.

This is because, first, several things can happen at once. The interacting cognitive subsystems (ICS) model from APU Cambridge (Barnard & May, 1995) takes this into account by looking at various parts of the cognitive system, the conversions between representations between these parts, and the conflicts that arise if the same part is used to perform different tasks simultaneously. Similarly, the very successful PERT-style GOMS analysis used on the NYNEX telephone operators' interface used the fact that the operator could be doing several things simultaneously with no interfering parts of their bodies and brains (John, 1990; Gray, John, & Atwood, 1992).

We are also like a networked system in that signals take an appreciable time to get from our senses to our brains and from our brains to our muscles. The famous homunculus from Card, Moran, and Newell's (1983) Model Human Processor makes this very clear with timings attached to various paths and types of mental processing. In fact, the sorts of delays within our bodies (from 50–200 ms on different paths) are very similar to those found on international networks.

In industrial control one distinguishes between open-loop and closed-loop control systems. *Open-loop control* is where you give the machine an instruction and assume it does it correctly (like a treasure map: "Ten steps North, turn left, three steps forward, and dig"). This assumes the machine is well calibrated and predictable. In contrast, *closed-loop control* uses sensors to constantly feedback and modify future actions based on the outcomes of previous ones (e.g., "Follow the yellow brick road until you come to the Emerald City"). Closed-loop control systems tend to be far more robust, especially in uncontrolled environments, like the real world.

Not surprisingly our bodies are full of closed-loop control systems (e.g., the level of carbon dioxide in your lungs triggers the breathing reflex). However, closed-loop control can become difficult if there are delays (you have not received feedback from the previous action when starting the next one). Delays either mean one has to slow down the task or use some level of prediction to work out what to do next, based on feedback of actions before the last one. This breakdown of closed-loop control

in the face of (especially unexpected) delays is one of the reasons hand-eye coordination tasks, such as mouse movement, break down if delays exceed a couple of hundred milliseconds. The feedback loops in our bodies for these tasks assume normal delays of around 200 ms and are robust to variations around this figure, but adding delays beyond this starts to cause breakdown.

The delays inside our bodies cause other problems too. The path from our visual cortex into our brain is far faster (by 100 ms or so) than that from our touch and muscle tension sensors around our bodies. If we were designing a computer system to use this information, we might consider having a short 100 ms tape loop, so that we could store the video input until we had the appropriate information from all senses. However, the sheer volume of visual information means that our brains do not attempt to do this. Instead there is a part of your brain that predicts where it "thinks" your body is and what it is feeling based on previous nerve feedback and what it knows the muscles have been asked to do. The same bit of the brain then monitors what actually did happen (when the nerve signals have made their way up the spinal column to the brain) and gives an uncomfortable or shocked sensation when a mismatch occurs—for example, if you go to pick something up, but because of poor light or a strange-shaped object you touch it earlier or later than you would expect. Tickling is also connected with this lack of ability to predict the sensations (this is why it is difficult to tickle yourself).

Race conditions also occur within this networked system of our bodies—for example, getting letter inversions whilst typing where signals to the two hands get processed in the wrong order. Brewster and Dix (1994) also used race conditions to understand what goes wrong in certain kinds of mis-hits of on-screen buttons. In certain circumstances two almost simultaneous "commands" from our brains to our hands to release the mouse button and to our arms to move to a new mouse location can get out of order, meaning the mouse moves out of the target before it is released. This analysis allowed us to design an experiment that forced this very infrequent error to occur much more frequently and therefore make it easier to assess potential solutions.

NETWORKS AS PLATFORMS

Algorithms and Architectures for Distributed Interfaces

User interfaces are hard enough to construct on a single machine; concurrent access by users on networked machines is a nightmare!

Happily, appropriate algorithms, architectures, toolkits and frameworks can help—a bit.

Accessing Shared Objects

We saw earlier how race conditions within networked systems can lead to inconsistencies within the user interface and within the underlying data structures. Fortunately, there are a range of techniques for dealing with this.

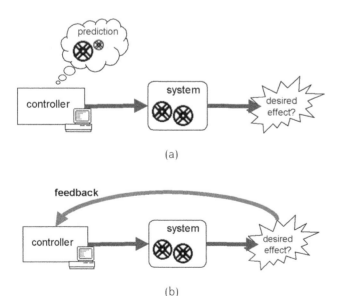

FIGURE 14.12.   (a) Open-loop control. (b) Closed-loop control.

Locking

The standard technique, used in databases and file systems, for dealing with multiple accesses to the same data object is locking. When a user's application wants to update a particular database record, it asks the database manager for a lock on the record. It can then get a copy of the record and send back the update, knowing that nothing else can access it in the meantime. Users are typically unaware that locking is being performed—the act of opening a file or opening an edit form for a database record establishes the lock and later, when the file or the edit form is completed and closed, the lock is released.

Although this is acceptable for more structured domains there are problems in more dynamic domains such as shared editing. Locking a file when one user is editing it is no good as we want several people to edit the same file at the same time. In these cases more lightweight forms of locking can be used at finer granularities: at paragraph, sentence, or even per-character level. For example, the act of clicking over a paragraph to set a text-entry position may implicitly request a paragraph lock, which is released when you go on to edit another paragraph. However, implicit and informal locks, because they are not apparent, can lead to new problems. For example, a user may click on a paragraph and do some changes, but before moving on to another part of the document, get interrupted. No one else can then edit the paragraph. To avoid this, the more informal locks are often time-limited or can be forcibly broken by the server if another user requests a lock on the same object.

Replication

In collaboration systems such as Lotus Notes/Domino or source code systems such as CVS, users do not lock central copies of data, but instead each user (or possibly each site) has their own replica of the complete Notes database/CVS tree. Periodically these replicas are synchronized with central copies or with each other. Updates can happen anywhere by anyone with a replica. Conflicts may of course arise if two people edit the same note between synchronizations. Instead of preventing such conflicts, the system (and software written using it) accepts that such conflicts will occur. When the replicas synchronizes, conflicts are detected and various (configurable) actions may occur: flagging the conflicts to users, adding conflicting copies as versions, and so on.

This view of "replicate and worry later" is essential in many mobile applications, as attempts to lock a file whilst disconnected would first of all require waiting until a network connection could be made, and, worse, if the network connection is lost whilst the lock is still in operation, could lead to files being locked for very long periods. Other examples of replication in research environments include the CODA at CMU, which allows replication of a standard UNIX file system (Kistler & Satyanarayanan, 1992) and Liveware, a contact-information system that replicates and synchronizes when people meet in a manner modeled after the spread of computer viruses (Witten, Thimbleby, Coulouris, & Greenberg, 1991). Although Liveware is now quite a few years old and was "spread" using synchronizing floppy disks, the same principle is now being suggested for PDAs and other devices to exchange information via IrDA or Bluetooth.

Optimistic Concurrency for Synchronous Editing

The "do it now and see if there are conflicts later" approach is called "optimistic concurrency," especially in more synchronous settings. For example, in a shared editing system the likelihood of two users editing the same sentence at the same time is very low. An optimistic algorithm doesn't bother to lock or otherwise check things when the users start to edit in an area, but in the midst or at the end of their edits checks to see if there are any conflicts, and attempts to fix them.

There are three main types of data that may be shared:

- **Orthogonal data**—where the data consists of attributes of individual objects/records that can all be independently updated
- **Sequential data**—particularly text, but any form of list where the order is not determined by an attribute property
- **Complex structural data**—such as directory trees, taxonomic categories, and so on.

In terms of complexity for shared data, these are in increasing difficulty.

Orthogonal data, although by no means trivial, is the simplest case. There is quite a literature on shared graphical editors, which all have this model: independent shapes and objects with independent attributes such as color, size, or position. When merging updates from two users, all one has to do is look at each attribute in turn and see whether it has been changed by only one user, in which case the updated value is used, or if it has been changed by both, in which case either the last update is used or the conflict is flagged.

Structured data is most complicated. What do you do if someone has created a new file in directory D, but at the same time someone else has deleted the directory? I know of no optimistic algorithms for dealing effectively with this in the CSCW literature. CODA deals with directory structures (normal UNIX file system) but takes a very simple view of this as it only flags inconsistencies and doesn't attempt to fix them.

Algorithms for shared text editing sit somewhere between the two and have two slightly different problems, both relating to race conditions when two or more users are updating the same text:

- **Dynamic pointers**—if user A is updating an area of text in front of user B, then the text user B is editing will effectively move in the document.
- **Deep conflict**—what happens if user A and user B's cursors are at the very same location and they perform insertions/deletions

Figure 14.13 shows an example of the first of these problems. The deeper conflict occurs when both cursors are at the same point, say after the "Y" in "XYZ." Adonis types "A" and at the same time Beatrice types "B"—should we have "XYABZ" or "XYBAZ," or perhaps even lose one or other character? Or if Adonis types "A" and Beatrice presses "delete," should we have "XYZ" or "XAZ"?

> Imagine two users, Adonis and Beatrice.
>
> They are working using a shared editor and their current document reads:
>
> Adonis is⌶ and Beatrice is⌶.
>   A   B
>
> The sentence is partial and both users are about to type in their prime personal characteristic in order to complete it.
>
> Adonis' insertion point is denoted by the boxed A and Beatrice's insertion point is the boxed B.
>
> Beatrice types first yielding:
>
> Adonis is⌶ and Beatrice is beautiful⌶.
>   A   B
>
> Adonis then types 'adorable', but unfortunately the implementor of the group editor was not very expert and the resulting display was:
>
> Adonis is adorable⌶ and Beatrice is beautiful.
>   A   B
>
> Beatrice's insertion point followed the 36th character before Adonis' insertion, and followed the 36th character after.
>
> Reasonable but wrong! The actual text should clearly read:
>
> Adonis is adorable⌶ and Beatrice is beautiful⌶.
>   A   B
>
> This correct behaviour is called a dynamic pointer as opposed to the static pointer 'character position 36'.

FIGURE 14.13. Dynamic pointers from Dix (1995).

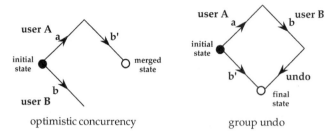

FIGURE 14.14. Multi-user transformations.

A number of algorithms exist for dealing with this (Sun & Ellis, 1998; Mauve, 2000; Vidot, Cart, Ferriz, & Suleiman, 2000), including a retrofit to Microsoft Word called CoWord (Xia, Sun, Sun, Chen, & Shen, 2004). Most of these stem from the dOPT algorithm used in the Grove editor (Ellis & Gibbs, 1989). These algorithms work by having various transformations that allow you to reorder operations. For example, if we have two insertions (labeled "a" and "b") performed at the same time at different locations

a) insert texta at location n
b) insert textb at location m

but decide to give insert a preference, then we have to transform b to b′ as follows:

b′) if (m < n) insert textb at location m   (i)
     if (m = n) insert textb at location m   (ii)
     if (m > n) insert textb at location m+length(texta)   (iii)

Case (i) says that if the location of insert "b" is before insert "a" you don't have to worry. Case (iii) says that if it is after insert "a" you have to shift your insert along accordingly. Case (ii) is the difficult one where a conflict occurs and has to be dealt with carefully to ensure that the algorithm generates the same results no matter where it is. The version above would mean that B's cursor gets left behind by A's edit. The alternative would be to make case (ii) the same as case (iii), which would mean B's cursor would be pushed ahead of A's.

In early work in this area I proposed regarding dynamic pointers as first-class objects and using these in all representations of actions (Dix, 1991, 1995). This means that rules like case (i) and case (iii) happen "for free," but the deep conflict case still needs to be dealt with specially.

Groupware Undo

The reason that undo is complicated in groupware is similar to the problems of race conditions in optimistic concurrency.

In the case of optimistic concurrency user A has performed action **a** and user B has performed action **b**, both on the same initial state. The problem is to transform user B's action into one **b′** that can be applied to the state after action **a** yet still mean the "same things" as the original action **a**.

In the case of groupware undo we may have the situation where user A has performed action **a**, followed by user B performing action **b**, then user A decides to undo action **a**. How do we transform action **b** so that the transformed **b′** means the same before action **a** as **b** did after?

Similar, but slightly different transformation rules can be produced for the case of undo and dynamic pointers can be used for most cases.

As with optimistic concurrency there is slightly more work on group undo in shared graphical editors where the orthogonal data makes conflicts easier (Berlage & Spenke, 1992).

Real Solutions?

Although these various algorithms can ensure there is no internal inconsistency and that all participants see the same thing, they do not necessarily solve all problems. Look again at the case of Alison and Brian's chat earlier in the chapter. Certainly in the case of group undo, when Abowd and Dix (1992) published the first paper on the topic, we proposed various solutions, but also recommended that as well as an explicit undo button, systems ought to provide sufficient history to allow users to recreate what they want to without using the undo button. This is not because it is impossible to find a reasonable meaning for the undo button, but because in the case of group undo, there are several reasonable meanings. Choosing the meaning a user intends is impossible and so it may sometimes be better not to guess.

Architectures for Networked Systems

Software architecture is about choosing what, in terms of code and functionality, goes where. For applications on a single machine these are often logical distinctions between parts of the code. For networked systems, "where" includes physical location, and the choice of location makes an enormous difference to the responsiveness of the system.

The simplest systems are usually centralized client–server architectures in which the majority of computation and data storage happens in the central server. Many of the problems of race conditions and potential inconsistencies disappear. However, this means that every interaction requires a network interaction with the server, meaning that feedback may be very slow. The opposite extreme is replicated peer–peer architectures where all the code is running on users' own PCs, and the PCs communicate directly with one another. Feedback can now be instantaneous, but the complexity of algorithms to maintain consistency, catch-up late joiners, and so on, can be very complex. Most systems operate somewhere between these two extremes with a central "golden copy" of shared data, but with some portion of the data on individual PCs, PDAs, or other devices in order to allow rapid feedback. A notable exception to this is peer–peer sharing networks such as Gnutella where the only central resource is a sort of switchboard to allow client programs to contact one another. The reason for this is largely legal as these are often used to share copyright media! One reason this works is that the data is static; the song "Yesterday" is the same now as when it was first recorded in 1965—although given it has the greatest number of covers of any song this is perhaps not the best example!

In web applications options are constrained by the features allowed in HTML and web browsers. Note that even a web form is allowing some local interaction (filling in the form) as well as some centralized interaction (submitting it). Applets allow more interaction, but the security limitations mean that they can only talk back to the server where they originated. Thus true peer–peer architectures are impossible on the Web, but can be emulated by chat servers and similar programs that relay messages between clients. For Intranets it is easier to configure browsers so that they accept applets as trusted and thus with greater network privileges, or to include special plug-in components to perform more complicated actions.

Mobile systems have yet more issues as they need to be capable of managing disconnected operations (when they have no physical or wireless connection to the network). This certainly means keeping cached copies of central data. For example, the AvantGo browser for PDAs downloads copies of web pages onto a PDA and synchronizes these with their latest versions every time the PDA is docked into a desktop PC (AvantGo, 2001).

Although this is now beginning to change with higher-bandwidth mobile-data services, it is still the case that access whilst mobile is slower and more expensive than whilst using a fixed connection. This pushes one towards replicated architectures with major resynchronization when connected via cheap/fast connections and more on-demand or priority-driven synchronization when on the move.

Supporting Infrastructure

In order to help manage networked applications various types of supporting infrastructure are being developed.

Awareness Servers

These keep track of which users are accessing particular resources (e.g., visiting a particular web page) so that you can be kept informed as to whether others are "near" you in virtual space, or whether friends are online and active. (Palfreyman & Rodden, 1996; ICQ, 2001; SUN Microsystems, 2001)

Notification Servers

These serve a similar role for data, allowing client programs to register an interest in particular pieces of shared data. When the data is modified or accessed the interested parties are "notified." For example, you may be told when a web page has changed or when a new item has been added to a bulletin board. Some notification servers also manage the shared data (Patterson, Day, & Kucan, 1996) whilst others are "pure," just managing the job of notification (Ramduny, Dix, & Rodden, 1998).

Event/Messaging Systems

These allow different objects in a networked environment to send messages to one another in ways that are more convenient than those allowed by the raw network. For example, they may allow messages to be sent to objects based on location-independent names, so that objects don't have to know where each other are.

Resource Discovery

Systems such as the Java JINI framework and Universal Plug-and-Play allow devices to find out about other devices close to them—for example, the local printer—and configure themselves to work with one another. As ubiquitous and mobile devices multiply, this will become increasingly important.

Web Service and Remote Procedure Call

Web service frameworks such as SOAP or XML-RPC, or non-web remote procedure mechanisms such as Java's RMI or CORBA, allow applications on different machines to easily connect to one another even though they may be written in different programming languages and run on different operating systems. In all cases the parameters or arguments have to be serialized into a binary or ASCII test stream to be sent. As Fig. 14.15 shows these are designed to be read by machines, not people!

HISTORY, FUTURES, AND PARADIGM SHIFTS

History

Given the anarchic image of the Web, it is strange that the Internet began its development as a United States military project. The suitability of the Internet for distributed management and independent growth stems not from an egalitarian or anti-centralist political agenda, but from the need to make the network resilient to nuclear attack, with no single point of failure.

In the 1970s when the Internet was first developing, it and other networks were mainly targeted at connecting together large computers. It was during the 1980s, with the rise of personal computing, that local networks began to become popular.

However, even before that point, very local networks at the lab-bench level had been developed to link laboratory equipment—for example, the IEEE488 designed originally to link Hewlett Packard's proprietary equipment and then becoming an international standard. Ethernet too began life in commercial development at Xerox before becoming the de facto standard for local networking.

Timeline—Key Events for the Internet	
1968	First proposal for ARPANET –military and government research contracted to Bolt, Beranek, and Newman
1971	ARPANET enters regular use
1973/4	Redesign of lower level protocols leads to TCP/IP
1983	Berkeley TCP/IP implementation for 4.2BSD— public domain code
1980s	Rapid growth of NSFNET –broad academic use
1990s	WWW and widespread public access to the Internet
2000	WAP on mobile phones Web transcends the Internet

Although it is technical features of the Internet (decentralized, resilient to failures, hardware independent) that have made it possible for it to grow, it is the Web that has made it become part of popular consciousness. Just as strange as the Internet's metamorphosis from military standard to anarchic cult, is the Web's development from medium of exchange of large high-energy physics datasets to e-commerce, home of alternative web pages and online sex!

```
<SOAP-ENV:Envelope

    xmlns:SOAP-ENV="http://schemas.xmlsoap.org/soap/envelope/"

    xmlns:xsi="http://www.w3.org/1999/XMLSchema-instance"

    xmlns:xsd="http://www.w3.org/1999/XMLSchema">

 <SOAP-ENV:Header>

 </SOAP-ENV:Header>

 <SOAP-ENV:Body>

   <ns1:sayHelloTo

       xmlns:ns1="Lookup"

SOAP-ENV:encodingStyle="http://schemas.xmlsoap.org/soap/encoding/">

     <name xsi:type="xsd:string">Hello World</name>

   </ns1:sayHelloTo>

 </SOAP-ENV:Body>

</SOAP-ENV:Envelope>
```

FIGURE 14.15. SOAP XML encoding of Lookup ("Hello World").

Paradigm Shift

During the 1970s through to the mid-1990s, networks were a technical phenomenon, enabling many aspects of business and academic life, but with very little public impact. However, this has changed dramatically over the last five years and now we think of a networked society. The Internet and other network technologies, such as SMS text messages, are not only transforming society but, at a popular and cultural level, defining an era.

International transport, telecommunications, and broadcasting had long before given rise to the term *global village*. However, it seems this was more a phrase waiting for a meaning. Until recently the global village was either parochial (telephoning those you already know) or sanitized (views of distant cultures through the eyes of the travel agent or television camera). It is only now that we see chat rooms and web homepages allowing new contacts and friendship around the world or at least amongst those affluent enough to have Internet access.

Markets too have changed due to global networks. It is not only the transnationals who can trade across the world, and even my father-in-law runs a thriving business selling antiques through eBay.

Marketing has also had to face a different form of cross-cultural issue. Although selling the same product, a hoarding in Karachi may well be different from an advert in a magazine in Kentucky, reflecting the different cultural concerns. Global availability of web pages changes all that. You have to create a message that appeals to all cultures, a tall order. Those who try to replicate the targeting of traditional media by having several country-specific websites may face new problems; the global access to even these country-specific pages means that the residents of Kentucky and Karachi can compare the different adverts prepared for them, and in so doing see how the company views them and their cultures.

Economics drives much of popular, as well as business, development of networked society. One of the most significant changes in the United Kingdom in recent years was due to changes in charging models. In the United States, local calls have long been free and hence so were Internet connections to local points of presence. The costs of Internet access in the United Kingdom (and even more important, the perception of the cost) held back widespread use. The rise of free or fixed-charge unmetered access changed nearly overnight the acceptability and style of use. Internet access used to be like a lightning guerrilla attack into the web territory, quick in and out before the costs mounted up, but is now a full-scale occupation.

The need for telecoms companies across the world to recover large investments in wireless-band franchises, combined with use rather than connection-based charging made possible by GPRS and third-generation mobile services, make it likely that we will see a similar growth in mobile access to global networked information and services.

In an article in 1998, I used the term *PopuNET* to refer to a change in society that was not yet there, but would come. PopuNET is characterized by network access

everywhere, everywhen by everyone

This pervasive, permanent, popular access is similar to the so-called "Martini principle" applied more recently to mobile networking—anytime, anyplace, anywhere. Of course, Martini never pretends to be anything but exclusive, so not surprisingly these differ on the popular dimension! "Anyplace, anywhere" does correspond to the pervasive "everywhere" and "anytime" to the permanent "everywhen." However, there is a subtle difference, especially between "anytime" and "everywhen.". "Anytime" means that at any time you choose you can connect. "Everywhen" means that at all places and all times you are connected. When this happens, one ceases to think of "connectedness" and it simply becomes part of the backdrop of life.

The changes in charging models have brought the United Kingdom closer to "everywhen" and the United States and many other parts of the world are already well down the path. Always-on mobile connectivity will reinforce these changes.

PopuNET will demand new interfaces and products, not just putting web pages on TV screens or spreadsheets on fridge doors. What these new interfaces will be is still uncertain.

Futures (Near)

It is dangerous to predict far into the future in an area as volatile as this. One development that is already underway, which will make a major impact on user interfaces, is short-range networking, which will enable various forms of wearable and ubiquitous networks. Another is the introduction of network appliances, which will make the home "alive" in the network.

We have considered network aspects of continuous media at length. The fact that the existing Internet TCP/IP protocols do not enable guaranteed quality-of-service will put severe limits on its ability to act as an infrastructure for services such as video-on-demand or video-sharing between homes. The update to TCP/IP that has been under development for several years, IPv6, will allow prioritized traffic, but it falls short of real guaranteed QoS (IPng, 2001).

It seems that this is impasse. One of the reasons that IPv6 has taken so long is not the technical difficulty, but backwards compatibility and the problems of uptake on the existing worldwide infrastructure, so evolutionary change is hard. However, revolutionary change is also hard; one cannot easily establish a new parallel international infrastructure overnight. Or perhaps one can.

Mobile phone services have started with an infrastructure designed for continuous voice and are, through a series of quite dramatic changes, moving this towards a fully mixed media/data service—and it is a global network. Furthermore, more and more non-web-based Internet services are using HTTP, the web protocol, to talk to one another in order to be "firewall friendly." This means that mobile phones and PDAs can be web-connected without being Internet-connected. So perhaps the future is not an evolution of Internet protocols, but a gradual replacement of the Internet by nth-generation mobile networks. Instead of the Internet on phones, we may even see mobile phone networking standards being used over wires.

References

All web links below and other related links and material at: http://www.hiraeth.com/alan/hbhci/network/

Abowd, G. D., & Dix, A. J. (1992). Giving undo attention. *Interacting with computers, 4*(3), 317–342.

AlexaWSP (2006). *Alexa Web Search Platform Service.* Retrieved March 13, 2006, from http://pages.alexa.com/prod_serv/web_search_platform.html

Androutsellis-Theotokis, S., & Spinellis, D. (2004). A survey of peer-to-peer content distribution technologies. *ACM Computing Surveys, 36*(4), 335–371.

Arehart, C., Chidambaram, N., Guruprasad, S., Homer, A., Howell, R., Kasippillai, S., et al. (2000). *Professional WAP.* Birmingham, UK: Wrox Press.

Arons, B. (1997). SpeechSkimmer: A system for interactively skimming recorded speech. *ACM Transactions on Computer–Human Interaction (TOCHI), 4*(1), 3–38.

AvantGo (2001). *AvantGo, Inc.* Retrieved March 13, 2006 from http://www.avantgo.com/

Barnard, P., & May, J. (1995). Interactions with advanced graphical interfaces and the deployment of latent human knowledge. In F. Paternò (Ed.), *Eurographics Workshop on the Design, Specification and Verification of Interactive Systems.* Berlin: Springer Verlag, 15–49.

Bellotti, V. (1993). Design for privacy in ubiquitous computing environments. *Proceedings of CSCW'93* (pp. 77–92). New York: ACM Press.

Benford, S., Anastasi, R., Flintham, M., Drozd, A., Crabtree, A., & Greenhalgh, C., et al. (2003). Coping with uncertainty in a location-based game. *IEEE Pervasive Computing, 2*(3), 34–41.

Benford, S., Bowers, J., Fahlen, L., Mariani, J., & Rodden, T. (1994). Supporting cooperative work in virtual environments. *The Computer Journal, 37*(8), 635–668.

Berlage, T., & Spenke, M. (1992). The GINA interaction recorder. *Proceedings of the IFIP WG2.7 Working Conference on Engineering for Human–Computer Interaction,* Amsterdam: North Holland, Ellivuori, Finland, 69–80.

Björk, S., Falk, J., Hansson, R., & Ljungstrand, P. (2001). Pirates!—Using the physical world as a game board. In M. Hirose (Ed.), *Proceedings of Interact 2001* (pp. 9–13). Amsterdam: IOS Press.

Bluetooth. (2001). *Official bluetooth SIG website.* Retrieved March 13, 2006, from http://www.bluetooth.com/

Buxton, W., & Moran, T. (1990). EuroPARC's integrated interactive intermedia facility (IIIF): Early Experiences. In S. Gibbs & A. A. Verrijn-Stuart (Eds.), *Multi-user interfaces and applications, Proceedings of IFIP WG8.4 Conference,* (pp. 11–34). Heraklion, Greece. Amsterdam: North-Holland.

Campbell, A., & Nahrstedt, K. (Eds.). (1997). *Building QoS into distributed systems.* Boston: Kluwer.

Card, S. K., Moran, T. P., & Newell, A. (1983). *The psychology of human computer interaction.* Hillsdale, NJ: Lawrence Erlbaum Associates.

CASIDE. (2005). *CASIDE Project (EP/C005589): Investigating cooperative applications in situated display environments.* Retrieved March 13, 2006, from http://www.caside.lancs.ac.uk/

CERN (2001). *European organisation for nuclear research.* Retrieved March 13, 2006, from http://public.web.cern.ch/Public/

Chalmers, M., Bell, M., Brown, B., Hall, M., Sherwood, S., & Tennent, P. (2005). Gaming on the edge: Using seams in Ubicomp games. In *Proceedings of ACM Advances in Computer Entertainment (ACE05)* (pp. 306–309). New York: ACM Press.

Clarke, D., Dix, A., Ramduny, D., & Trepess, D. (Eds.). (1997, June). *Workshop on time and the web.* Staffordshire University. Retrieved March 13, 2006, from http://www.hiraeth.com/conf/web97/papers/

Dix, A. J. (1987). The myth of the infinitely fast machine. In *People and Computers III—Proceedings of HCI'87* (pp. 215–228). Cambridge, UK: Cambridge University Press.

Dix, A. J. (1991). *Formal methods for interactive systems.* New York: Academic Press.

Dix, A. J. (1994). *Seven years on, the myth continues.* Research Report-RR9405, Huddersfield, U.K.: University of Huddersfield. Retrieved March 13, 2006, from http://www.hcibook.com/alan/papers/myth95/7-years-on.html

Dix, A. J. (1995). Dynamic pointers and threads. *Collaborative Computing, 1*(3), 191–216.

Dix, A. J. (1996). Closing the Loop: modelling action, perception and information. In T. Catarci, M. F. Costabile, S. Levialdi, & G. Santucci (Eds.), *Proceedings of AVI'96—Advanced Visual Interfaces, Gubbio, Italy* (pp. 20–28). New York: ACM Press.

Dix, A. (2001a). artefact + marketing = product. *Interfaces, 48,* 20–21. London: BCS-HCI Group. Retrieved March 13, 2006, from http://www.hiraeth.com/alan/ebulletin/product-and-market/

Dix, A. (2001b). *Cyber-economies and the real world.* Keynote—South African Institute of Computer Scientists and Information Technologists Annual Conference, SAICSIT 2001, Pretoria, South Africa. Retrieved Septmeber 25–20, 2001, from http://www.hcibook.com/alan/papers/SAICSIT2001/

Dix, A., & Brewster, S. A. (1994). Causing trouble with buttons. In D. England (Ed.), *Ancilliary Proceedings of HCI'94, Glasgow, Scotland.* London: BCS-HCI Group. Retrieved March 13, 2006, from http://www.hcibook.com/alan/papers/buttons94/

Dix, A., Finlay, J., Abowd, G., & Beale, R. (1998). *Human–computer interaction* (2nd ed.). Englewood Cliffs, NJ: Prentice-Hall, Inc.

Dix, A., Rodden, T., Davies, N., Trevor, J., Friday, A., & Palfreyman, K. (2000). Exploiting space and location as a design framework for interactive mobile systems. *ACM Transactions on Computer–Human Interaction (TOCHI), 7*(3), 285–321.

Dodge, M., & Kitchin, R. (2001). *Atlas of Cyberspace.* Reading, MA: Addison Wesley. http://www.cybergeography.org/atlas/

Dourish, P., & Bellotti, V. (1992). Awareness and Coordination in Shared Workspaces. In *Proceedings of CSCW'92* (pp. 107–114). New York: ACM Press

Edwards, W. K., & Rodden, T. (2001). *Jini, example by example.* Upper Saddle River, NJ: SUN Microsystems Press.

Electrolux. (1999). *Screenfridge.* Retrieved March 13, 2006, from http://www.electrolux.com/screenfridge/

Ellis, C. A., & Gibbs, S. J. (1989). Concurrency control in groupware systems. In *Proceedings of 1989 ACM SIGMOD International Conference on Management of Data, SIGMOD Record, 18*(2), 399–407.

FCC. (2006). *Voice over Internet Protocol.* Federal Communications Commission. Retrieved March 13, 2006, from http://www.fcc.gov/voip/

Fogg, B. J., Marshall, J., Laraki, O., Osipovich, A., Varma, C., Fang, N., et al. (2001). What makes websites credible? A report on a large quantitative study. *Proceedings of CHI2001,* Seattle, 2001, Also *CHI Letters, 3*(1), 61–68. New York: ACM Press.

Foster, I. (2000). Internet computing and the emerging grid. *Nature, webmatters.* Retrieved March 13, 2006, http://www.nature.com/nature/webmatters/grid/grid.html

Foster, I., & Kesselman, C. (Eds.). (1999). *The grid: Blueprint for a new computing infrastructure.* San Francisco: Morgan-Kaufmann.

Gaver, W. W., Smith, R. B., & O'Shea, T. (1991). Effective sounds in complex situations: The ARKola simulation. In S. P. Robertson, G. M. Olson, & J. S. Olson (Eds.), *Reaching through technology—CHI'91 conference proceedings* (pp. 85–90). New York: ACM Press.

Gellersen, H-W., Beigl, M., & Krull, H. (1999). The MediaCup: Awareness technology embedded in an everyday object. In H.-W. Gellersen (Ed.), *Handheld & ubiquitous computing, Lecture notes in computer science* (Vol. 1707; pp. 308–310). Berlin, Germany: Springer.

Gray, W. D., John, B. E., & Atwood, M. E. (1992). The precis of project ernestine or an overview of a validation of goms. In P. Bauersfeld, J. Bennett, & G. Lynch (Eds.), *Striking a balance, Proceedings of the CHI'92 Conference on human factors in computing systems* (pp. 307–312). New York: ACM Press.

Greenhalgh, C. (1997). Analysing movement and world transitions in virtual reality tele-conferencing. In J. A. Hughes, W. Prinz, T. Rodden, & K. Schmidt (Eds.), *Proceedings of ECSCW 97* (pp. 313–328). Dordrecht, The Netherlands: Kluwer Academic Publishers.

GRID. (2001). *GRID Forum home page*. Retrieved March 13, 2006, from http://www.gridforum.org/

Gutwin, C., Benford, S., Dyck, J., Fraser, M., Vaghi, I., & Greenhalgh, C. (2004). Revealing delay in collaborative environments. In *Proceedings of ACM Conference on Computer–Human Interaction, CHI'04* (pp. 503–510). New York: ACM Press.

Halasz, F., Moran, T., & Trigg, R. (1987). NoteCards in a nutshell. In *Proceedings of the CHI + GI* (pp. 45–52). New York: ACM Press.

HFIG (2001). *Hybrid feature interaction group home page*, Retrieved March 13, 2006, from University of Glasgow, Scotland website http://www.dcs.gla.ac.uk/research/hfig/

Hindus, D., Mainwaring, S. D., Leduc, N., Hagström, A. E., & Bayley, O. (2001). Casablanca: designing social communication devices for the home. In *Proceedings of CHI 2001* (pp. 325–332). New York: ACM Press.

Howard, S., & Fabre, J. (Eds.) (1998). Temporal aspects of usability [Special issue]. *Interacting with Computers, 11*(1), 1–105.

Hughes, D., Gibson, S., Walkerdine, J., & Coulson, G. (2006, February). Is deviant behaviour the norm on P2P file-sharing networks? In *IEEE Distributed Systems Online, 7*(2). Retrieved March 13, 2006, from http://dsonline.computer.org

ICQ (2001). *ICQ home page*. Retrieved March 13, 2006, from http://www.icq.com/products/whatisicq.html

IEEE (2001). *IEEE 802.11 working group*. Retrieved March 13, 2006, from http://www.ieee802.org/11/

IPng (2001). IP Next generation working group home page. Retrieved March 3, 2007, title from web page is: *IP Version 6* (IPv6). Retrieved March 13, 2006, from http://playground.sun.com/pub/ipng/html

ISO/IEC 7498. (1994). Information technology—Open systems interconnection—Basic reference model: The basic model. International standards organisation. Retrieved March 3, 2007, http://isotc.iso.org/livelink/livelink/fetch/2000/2489/Ittf_Home/PubliclyAvailableStandards.htm

John, B. E. (1990). Extensions of GOMS analyses to expert performance requiring perception of dynamic visual and auditory information. In J. C. Chew & J. Whiteside (Eds.), *Empowering people—Proceedings of CHI'90 human factors in computer systems* (pp. 107–115). New York: ACM Press.

Johnson, C. (1997). What's the web worth? The impact of retrieval delays on the value of distributed information. In D. Clarke, A. Dix, D. Ramduny, & D. Trepress (Eds), *Workshop on Time and the web*. Retrieved March 13, 2006, from http://www.hiraeth.com/conf/web97/papers/johnson.html

Johnson, C., & Gray, P. (Eds.) (1996). Temporal Aspects of Usability, report of workshop in Glasgow, June 1995, *SIGCHI Bulletin, 28*(2), 32–61. New York: ACM Press. http://www.acm.org/sigchi/bulletin/1996.2/timeintro.html

JPEG (2001). *Joint Photographic Experts Group home page* [On-line]. http://www.jpeg.org/public/jpeghomepage.htm

Kistler, J. J., & Satyanarayanan, M. (1992). Disconnected operation in the CODA file system. *ACM Transactions on Computer Systems, 10*(1), 3–25.

Lavery, D., Kilgourz, A., & Sykeso, P. (1994). Collaborative use of X-Windows applications in observational astronomy. In G. Cockton, S. Draper, & G. Wier (Eds.), *People and computers* (Vol 9; pp. 383–396). Cambridge, UK: Cambridge University Press.

Light, A., & Wakeman, I. (2001). Beyond the interface: Users' perceptions of interaction and audience on websites. In D. Clarke & A. Dix (Eds.), Interfaces for the active web (Part 1) [Special issue]. *Interacting with Computers, 13*(3), 401–426.

Lock, S., Allanson, J., & Phillips, P. (2000). User-driven design of a tangible awareness landscape. In D. Boyahrski & W. Kellogg, *Proceedings of Symposium on Designing Interactive Systems* (pp. 434–440). New York: ACM Press.

Lock, S., Rayson, P., & Allanson, J. (2003). Personality Engineering for Emotional Interactive Avatars. In *Human Computer Interaction, Theory and practice (Part II), Volume 2 of the Proceedings of the 10th International Conference on Human–Computer Interaction* (pp. 503–507). Mahwah, NJ: Lawrence Erlbaum Associates.

Mariani, J. A., & Rodden, T. (1991). The impact of CSCW on database technology. In *Proceedings of ACM Conference on Computer Supported Cooperative Work (includes critique of 'transparency' in a CSCW setting)*. New York: ACM Press.

Mauve, M. (2000). Consistency in replicated continuous interactive media. In *Proceedings of CSCW'2000* (pp. 181–190). New York: ACM Press.

McDaniel, S. E., & Brinck, T. (1997). Awareness in Collaborative Systems: A CHI 97 Workshop (report). In *SIGCHI Bulletin, 29*(4). New York: ACM Press. Workshop report: http://www.acm.org/sigchi/bulletin/1997.4/mcdaniel.html Workshop web pages: http://www.usabilityfirst.com/groupware/awareness/workshop/

McManus, B. (1997*). Compensatory actions for time delays*. In Clarke, http://www.hiraeth.com/conf/web97/papers/barbara.html

Millett, L. I., Friedman, B., & Felten, E. (2001). Cookies and web browser design: toward informed consent online. In *Proceedings of CHI2001, CHI Letters, 3*(1), 46–52.

Mitsopoulos, E. (2000). A Principled Approach to the Design of Auditory Interaction in the Non-Visual User Interface. *DPhil Thesis*, University of York, UK. http://www.cs.york.ac.uk/ftpdir/reports/YCST-2000 07.zip

MPEG. (2001). *Moving Picture Experts Group home page* [On-line]. http://www.cselt.it/mpeg/

Olson, M., & Bly, S. (1991). The Portland experience: A report on a distributed research group. *International Journal of Man-Machine Studies, 34*(2), 11–228.

Palfreyman, K., & Rodden, T. (1996). A protocol for user awareness on the World Wide web. In *Proceedings of CSCW'96* (pp. 130–139). New York: ACM Press.

Patterson, J. F., Day, M., & Kucan, J. (1996). Notification servers for synchronous groupware. In *Proceedings of CSCW'96* (pp. 122–129). New York: ACM Press.

Pausch, R. (1991). Virtual reality on five dollars a day. In S. P. Robertson, G. M. Olson & J. S. Olson (Eds.), *CHI'91 Conference Proceedings* (pp. 265–270). Reading, MA: Addison Wesley.

Ramduny, D., & Dix, A. (1997). Why, what, where, when: Architectures for Co-operative work on the WWW. In H. Thimbleby, B. O'Connaill, & P. Thomas (Eds.), *Proceedings of HCI'97* (pp. 283–301). Berlin, Germany: Springer.

Ramduny, D., Dix, A., & Rodden, T. (1998). Getting to know: the design space for notification servers. In *Proceedings of CSCW'98* (pp. 227–235). New York: ACM Press. http://www.hcibook.com/alan/papers/GtK98/

Resnick, P., & Varian, H. R. (Eds.). (1997). *Communications of the ACM* [Special issue on recommender systems], *40*(3), 56–89.

Rodden, T. (1996). Populating the application: a model of awareness for cooperative applications. In M. S. Ackerman (Ed.), *Proceedings of the 1996 ACM Conference on Computer-Supported Cooperative Work* (pp. 87–96). New York: ACM Press.

Rohs, M., Sheridan, J. G., & Ballagas, R. (2004). Direct manipulation techniques for large displays using camera phones. In *Proceedings of 2nd International Symposium on Ubiquitous Computing Systems*. Retrieved March 13, 2006, from http://www.equator.ac.uk/index.php/articles/113

Roussel, N. (1999). Beyond webcams and videoconferencing: Informal video communication on the web. In *Proceedings of The Active web*. Retrieved March 13, 2006, from http://www.hiraeth.com/conf/activeweb/

Roussel, N. (2001). Exploring new uses of video with videoSpace. In *Proceedings of EHCI'01, the 8th IFIP Working Conference on Engineering for Human–Computer Interaction, Lecture Notes in Computer Science* (pp.73–90). Berlin: Springer-Verlag. Retrieved March 13, 2006, from http://www-iiuf.unifr.ch/%7erousseln/publications/EHCI01.pdf

Sandor, O., Bogdan, C., & Bowers, J. (1997). Aether: An Awareness Engine for CSCW. In J. Hughes (Ed.), *Proceedings of the Fifth European Conference on Computer Supported Cooperative Work* (ECSCW'97) (pp. 221–236). Dordrecht, The Netherlands: Kluwer Academic.

Sangha, A. (2001, June). *Legal implications of location based advertising*. Interview for the WAP Group. Retrieved March 13, 2006, from http://www.thewapgroup.com/53762_1.DOC

Sarvas, R., Herrarte, E., Wilhelm, A., & Davis, M. (2004). Metadata creation system for mobile images. In *Proceedings of Second International Conference on Mobile Systems, Applications, and Services* (pp. 36–48). New York: ACM Press.

Semacode. (2006). *Semacode*. Retrieved March 13, 2006, from http://semacode.org/

Schneier, B. (1996). *Applied cryptography* (2nd ed.). New York: Wiley.

SETI@home. (2001). *SETI@home—the search for extra-terrestrial intelligence*. http://setiathome.ssl.berkeley.edu/

Shneiderman, B. (1984). Response time and display rate in human performance with computers. *ACM computing surveys, 16*(3), 265–286. Retrieved March 13, 2006, New York: ACM Press.

Siegal, D. (1999). *Futurize your enterprise*. New York: Wiley. http://www.futurizenow.com

Skype. (2006). *Skype* [On-line]. http://www.skype.com/

Stefik, M., Bobrow, D. G., Foster, G., Lanning, S., & Tatar, D. (1987). WYSIWIS revisited: early experiences with multiuser interfaces. *ACM Transactions on Office Information Systems, 5*(2), 147–167.

Stevens, W. R. (1998). *UNIX network programming, Networking APIs: Sockets and XTI* (Vol. 1). Englewood Cliffs, NJ: Prentice Hall.

Stevens, W. R. (1999). *UNIX network programming, Interprocess communications* (2nd ed., Vol. 2). NJ: Prentice Hall.

Stifelman, L., Arons, B., & Schmandt, C. (2001). The audio notebook: Paper and pen interaction with structured speech. In *Proceedings of CHI2001, CHI Letters, 3*(1), 182–189.

Sun, C., & Ellis, C. (1998). Operational transformation in real-time group editors: Issues, algorithms, and achievements. In *Proceedings of CSCW'98* (pp. 59–68). New York: ACM Press.

SUN Microsystems. (2001). *Awarenex*. Retrieved March 13, 2006, from http://www.sun.com/research/features/awarenex/

Taylor, A., Izadi, S., Swan, L., Harper, R., & Buxton, W. (2006). *Building Bowls for Miscellaneous Media. Physicality 2006 workshop*, Lancaster University, 6–7 Feb. 2006 [On-line]. http://www.physicality.org/

thePooch (2006) .:*thePooch*:. Retrieved March 13, 2006, from http://www.thepooch.com/

Tillotson, J. (2005). *Scent whisper*. Central Saint Martins College of Art & Design. Exhibited in SIGGRAPH CyberFashion 0100. Retrieved March 13, 2006), from http://psymbiote.org/cyfash/2005/

Toye, E., Madhavapeddy, A., Sharp, R., Scott, D., Blackwell, A., & Upton, E. (2004). *Using camera-phones to interact with context-aware mobile services* (Technical Report UCAM-CL-TR-609). Retrieved (date) from University of Cambridge, Computer Laboratory website. Retrieved March 13, 2006, from http://www.cl.cam.ac.uk/TechReports/UCAM-CL-TR-609.pdf

Vidot, N., Cart, M., Ferriz, J., & Suleiman, M. (2000). Copies convergence in a distributed real-time collaborative environment. In *Proceedings of CSCW'2000* (pp. 171–180). New York: ACM Press.

Wikipedia. (2006). File sharing. *Wikipedia, The Free Encyclopedia*. http://en.wikipedia.org/w/index.php?title=File_sharing&oldid=34285993.

WiMedia. (2006). *WiMedia alliance*. Retrieved March 13, 2006, from http://wimedia.org/

Wisneski, G., Ishii, H., Dahley, A., Gorbet, M., Brave, S., Ullmer, B., et al. (1998). Ambient Display: Turning Architectural Space into an Interface between People and Digital Information. In *Proceedings of the First International Workshop on Cooperative Buildings (CoBuild'98), Lecture Notes in Computer Science* (Vol. 1370; pp. 22–32). Heidelberg: Springer-Verlag.

Witten, I. H., Thimbleby, H. W., Coulouris, G., & Greenberg, S. (1991). Liveware: A new approach to sharing data in social networks. *International Journal of Man–Machine Studies, 34*, 337–348.

Xia, S., Sun, D., Sun, C., Chen, D., & Shen, H. (2004). Leveraging single-user applications for multi-user collaboration: the CoWord approach. In *Proceedings of ACM 2004 Conference on Computer Supported Cooperative Work* (pp. 162–171). Retrieved March 13, 2006, New York: ACM Press.

Zigbee. (2006). *ZigBee Alliance*. http://zigbee.org/

Zimmerman, T. G. (1996). Personal area networks: Near-field intrabody communication. *IBM Systems Journal, 35*(3/4), 609–617. Retrieved March 13, 2006, from http://isj.www.media.mit.edu/projects/isj/SectionE/609.htm

·15·

WEARABLE COMPUTERS

Dan Siewiorek and Asim Smailagic
Carnegie Mellon University

Thad Starner
Georgia Institute of Technology

INTRODUCTION

Computers have become a primary tool for office workers, allowing them to access the information they need to perform their jobs; however, accessing information is more difficult for more mobile users. With current computer interfaces, the user must focus both physically and mentally on the computing device instead of the environs. In a mobile environment, such interfaces may interfere with the user's primary task. Many mobile tasks could, however, benefit from computer support. Our focus is the design of wearable computers that augment, instead of interfere, with the user's tasks. Carnegie Mellon University's VuMan 3 project provides an example of how the introduction of wearable computing to a task can reap many rewards.

VUMAN 3 MAINTENANCE INSPECTION WEARABLE COMPUTER

Many maintenance activities begin with an inspection, during which problems are identified. Job orders and repair instructions are generated from the results of the inspection. The VuMan 3 wearable computer was designed for streamlining Limited Technical Inspections (LTI) of amphibious tractors for the U.S. Marines at Camp Pendleton, California (Smailagic, Siewiorek, Martin, & Stivoric, 1998). The LTI is a 600-element, 50-page checklist that usually takes four to six hours to complete. The inspection includes an item for each part of the vehicle (e.g., front left track, rear axle, windshield wipers, etc.). VuMan 3 created an electronic version of this checklist. The system's interface was arranged as a menu hierarchy and a physical dial and selection buttons controlled navigation. The top level consisted of a menu that gave a choice of function. Once the inspection function was chosen, the component being inspected was selected by its location on the vehicle. At each stage, the user could go up one level of the hierarchy.

The inspector selects one of four possible options about the status of the item: (a) Serviceable, (b) Unserviceable, (c) Missing, or (d) On Equipment Repair Order (ERO). Further explanatory comments about the item can be selected (e.g., the part is unserviceable due to four missing bolts).

The LTI checklist consists of a number of sections, with approximately one hundred items in each section. The user sequences through each item by using the dial to select "Next Item," or "Next Field." A "smart cursor" helps automate some of the navigation by positioning the user at the next most likely action.

As part of the design process, a field study was performed. In typical troubleshooting tasks, one Marine would read the maintenance manual to a second Marine, who performs the inspection. With the VuMan 3, only one Marine is needed for the task, as that Marine has the electronic maintenance manual. Thus, the physical manual does not have to be carried into hard-to-reach places.

The most unanticipated result was a 40% reduction in inspection time. The bottom right image of Fig. 15.1 demonstrates

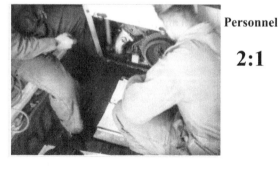

Current Practice — **SAVINGS FACTOR**

Personnel

2:1

Current Practice — **SAVINGS FACTOR** — **VuMan 3 Field Trials**

Inspection time

70% less

FIGURE 15.1. VuMan 3 Savings Factors.

the reason for this result. Here, the Marine is on his or her side looking up at the bottom of the amphibious tractor. In such places, it is hard to read or write on the clipboard typically used for inspections. The Marine constantly gets into position, crawls out to read instructions, crawls back into position for the inspection, and then crawls out again to record the results. In addition, the Marine tends to do one task at a time when the Marine might have five things to inspect in one place. This extra motion has a major impact on the time required to do a task. By making information truly portable, wearable computers can improve the efficiency of this application and many other similar ones.

The second form of time savings with the VuMan 3 occurred when the inspection is finished. The wearable computer requires a couple of minutes to upload its data to the logistics computer. The manual process, however, required a typist to enter the Marine's handwritten text into the computer. Given that the soldier may have written the notes in cold weather while wearing gloves, the writing may require some interpretation. This manual process represents another 30% of the time.

Such redundant data entry is common when users are mobile (Starner, Snoeck, Wong, & McGuire, 2004). There are numerous checklist-based applications including plant operations, preflight checkout of aircraft, inventory, and so forth that may benefit from a form-filling application run on a wearable computer. In the case of the VuMan 3 project, the results were striking. From the time the inspection was started until the data was entered into the logistics computer, 70% of the time was saved by using the wearable. There was a potential savings by reducing maintenance crews from two to one. Finally, there was also a savings in weight over paper manuals.

THE WEARABLE COMPUTING CAMP

Designing wearable computer interfaces requires attention to many different factors due to their closeness to the body and their use while performing other tasks. For the purposes of discussion, we have created the "CAMP" framework, which consists of the following factors:

Corporal: Wearables should be designed to interface physically with the user without discomfort or distraction.
Attention: Interfaces should be designed for the user's divided attention between the physical and virtual worlds.
Manipulation: When mobile, users lose some of the dexterity assumed by desktop interfaces. Controls should be quick to find and simple to manipulate.
Perception: A user's ability to perceive displays, both visual and audio, is also reduced while mobile. Displays should be simple, distinct, and quick to navigate.

Power, heat, on-body, and off-body networking, privacy, and many other factors also affect on-body computing (Starner, 2001). Many of these topics are the subjects of current research, and much work will be required to examine how these factors interrelate. Due to space, we will concentrate mainly on CAMP principles and practice in the remainder of this chapter.

CORPORAL: DESIGN GUIDES FOR WEARABILITY

The term *wearable* implies the use of the human body as a support environment for the object described. Society has historically evolved its tools and products into more portable, mobile, and wearable form factors. Clocks, radios, and telephones are examples of this trend. Computers are undergoing a similar evolution. Simply shrinking computing tools from the desktop paradigm to a more portable scale does not take advantage of a whole new context of use. While it is possible to miniaturize keyboards, human evolution has not kept pace by shrinking our fingers. There is no Moore's Law for humans. The human anatomy introduces minimal and maximal dimensions that define the shape of wearable objects, and the mobile context also defines dynamic interactions. Conventional methods of interaction, including keyboard, mouse, joystick, and monitor, have mostly assumed a fixed physical relationship between user and device. With wearable computers, the user's physical context may be constantly changing. Symbol's development of a wearable computer for shipping hubs provides an example of how computing must be adapted for the human body.

As a company, Symbol is well known for its barcode technology; however, it is also one of the first successful wearable computer companies, having sold over 100,000 units from its WSS 1000 line of wearable computers (see Fig. 15.2). The WS-1000 consists of a wrist-mounted wearable computer that features a laser barcode scanner encapsulated in a ring worn on the user's finger. This configuration allows the user to scan barcodes while keeping both hands free to manipulate the item being scanned. Because the user no longer has to fumble with a desk-tethered scanner, these devices increase the speed at which the user can manipulate packages and decrease the overall strain on the user's body. Such features are important in shipping hubs, where millions of packages are scanned by hand every year. Symbol spent over U.S. $5 million and devoted 40,000 hours of

FIGURE 15.2. Symbol's WSS 1000 series wrist-mounted wearable computer with ring scanner.

testing to develop this new class of device, and one of the major challenges was adapting the computer technology to the needs of the human body (Stein, Ferrero, Hetfield, Quinn, & Krichever, 1998).

One of the first observations made was that users may be widely varying shapes and sizes. Specifically, Symbol's scanner had to fit the fingers of both large men and small women. Similarly, the wrist unit had to be mounted on both large and small wrists. Even though the system's wires were designed to be unobtrusive, the system must be designed to break away if entangled and subjected to strain. This policy provided a safeguard for the user.

Initial testing discovered other needs that were obvious in hindsight. For example, the system was strapped to the user's forearm while the user exerted him- or herself moving boxes. Soon, the "soft-good" materials, which were designed for the comfort of the user, became sodden with sweat. After one shift, the user was expected to pass the computer to the operator on the next shift. Not only was the sweat-laden computer mount considered "gross," it also presented a possible health risk. This problem was solved by separating the computer mount from the computer itself. Each user received his or her own mount, which could be kept adjusted to the user's own needs. After each shift, the computer could be removed from the user's mount and placed in the replacement user's mount.

Another unexpected discovery is that the users tended to use the computer as body armor. When a shipping box would begin to fall on the user, the user would block the box with the computer mounted on her or his forearm, as that was the least sensitive part of the user's body. Symbol's designers were surprised to see users adapt their work practices to use the rigid forearm computer to force boxes into position. Accordingly, the computer's case was designed out of high impact materials; however, another surprise came with longer term testing of the computer.

Employees in the test company's shipping hubs constantly reached into wooden crates to remove boxes. As they reached into the crates, the computer would grind along the side. After extended use, holes would appear in the computer's casing, eventually damaging the circuitry. Changing the composition of the casing to be resistant to both abrasion and impact finally fixed the problem.

After several design cycles, Symbol presented the finished system to new employees in a shipping hub. After a couple of weeks' work, test results showed that the new employees felt the system was cumbersome, whereas established employees who had participated in the design of the project felt that the wearable computer provided a considerable improvement over the old system of package scanning. After consideration, Symbol's engineers realized that these new employees had no experiential basis for comparing the new system to the past requirements of the job. As employees in shipping hubs are often short term, a new group of employees were recruited. For two weeks, these employees were taught their job using the old system of package scanning: the employee would reach into a crate, grasp a package, transfer it to a table, grasp a handheld scanner, scan the package, replace the scanner, grasp the package, and transfer it to its appropriate conveyer belt. The employees were then introduced to the forearm-mounted WS-1000.

With the wearable computer, the employee would squeeze her or his index and middle finger together to trigger the ring-mounted scanner to scan the package while reaching for it, grasp the package, and transfer it to the appropriate convey belt in one fluid motion. These employees returned very positive scores for the wearable computer.

This lesson, that perceived value and comfort of a wearable computer is relative, was also investigated by Bodine and Gemperle (2003). In short interviews, users were fitted with a backpack or armband wearable and told that the system was either a police monitoring device (similar to those used for house arrest), a medical device for monitoring health, or a device for use during parties. The subjects were then asked to rate the devices on various scales of desirability and comfort. Not surprisingly, the police wearable was considered the least desirable; however, the police function elicited more negative *physical* comfort ratings, and the medical function elicited more positive physical comfort ratings, even though they were the same device. In other words, perceived comfort can be affected by the supposed function of the device.

Researchers have also explored wearability in more general terms. Wearability is defined as the interaction between the human body and the wearable object. Dynamic wearability includes the human body in motion. Design for wearability considers the physical shape of objects and their active relationship with the human form. Gemperle, Kasabach, Stivoric, Bauer, and Martin (1998) explored history and cultures including topics such as clothing, costumes, protective wearables, and carried devices (Siewiorek, 2002). They studied physiology, biomechanics, and the movements of modern dancers and athletes. Drawing upon the experience of CMU's wearables group, over two dozen generations of machines representing over a hundred person years of research, they codified the results into guidelines for designing wearable systems. These results are summarized in Table 15.1.

This team also developed a set of wearable forms to demonstrate how wearable computers might be mounted on the body. Each of the forms was developed by applying design guidelines and follows a simple pattern for ensuring wearability. The pods were designed to house electronic components. All of the forms are between 3/8″ and 1″ thick, and flexible circuits can fit comfortably into the 1/4″ thick flex zones. Beginning with acceptable areas and the humanistic form language, the team considered human movement in each individual area. Each area is unique, and some study of the muscle and bone structure was required along with common movements. Perception of size was studied for each individual area. For testing, minimal amounts of spandex were stretched around the body to attach the forms. The results are shown in Fig. 15.3.

These studies and guidelines provide a starting point for wearable systems designers; however, there is much work to be done in this area. Weight was not considered in these studies, nor were the long-term physiological effects such systems might have on the wearer's body. Similarly, fashion can affect the perception of comfort and desirability of a wearable component. As wearable systems become more common and are used for longer periods of time, it will be important to test these components of wearability.

TABLE 15.1. Design for Wearability Attributes.

Attribute	Comments
Placement	Identify where the computer should be placed on the body. Issues include identifying areas of similar size across a population, areas of low movement/flexibility, and areas large in surface area.
Humanistic Form Language	The form of the object should work with the dynamic human form to ensure a comfortable fit. Principles include inside surface concave to fit body, outside surface convex to deflect objects, tapering sides to stabilize form on body, and radiusing edges/corners to provide soft form.
Human Movement	Many elements make up a single human movement: mechanics of joints, shifting of flesh, and the flexing and extending of muscles and tendons beneath the skin. Allowing for freedom of movement can be accomplished in one of two ways: (a) by designing around the more active areas of the joints, or (b) by creating spaces on the wearable form into which the body can move.
Human Perception of Size	The brain perceives an aura around the body. Forms should stay within the wearer's intimate space, so that perceptually they become a part of the body. The intimate space is between zero and five inches off the body and varies with position on the body.
Size Variations	Wearables must be designed to fit many types of users. Allowing for size variations is achieved in two ways: (a) static anthropometric data, which details point to point distances on different sized bodies, and (b) consideration of human muscle and fat growth in three dimensions using solid rigid areas coupled with flexible areas.
Attachment	Comfortable attachment of forms can be created by wrapping the form around the body, rather than using single point fastening systems such as clips or shoulder straps.
Contents	The system much have sufficient volume to house electronics, batteries, and so forth that, in turn, constrains the outer form.
Weight	The weight of a wearable should not hinder the body's movement or balance. The bulk of the wearable object weight should be close to the center of gravity of the human body, minimizing the weight that spreads to the extremities.
Accessibility	Before purchasing a wearable system, walk and move with the wearable object to test its comfort and accessibility.
Interaction	Passive and active sensory interaction with the wearable should be simple and intuitive.
Thermal	The body needs to breathe and is very sensitive to products that create, focus, or trap heat.
Aesthetics	Culture and context will dictate shapes, materials, textures, and colors that perceptually fit the user and their environment.

ATTENTION

Humans have a finite and nonincreasing capacity that limits the number of concurrent activities they can perform. Herb Simon observed that human effectiveness is reduced as they try to multiplex more activities. Frequent interruptions require a refocusing of attention. After each refocus of attention, a period of time is required to reestablish the context prior to the interruption. In addition, human short-term memory can hold seven plus or minus two (e.g., five to nine) chunks of information. With this limited capacity, today's systems can overwhelm users with data, leading to information overload. The challenge to human computer interaction design is to use advances in technology to preserve human attention and to avoid information saturation.

In the mobile context, the user's attention is divided between the computing task and the activities in the physical environs. Some interfaces, like some augmented realities (Azuma, 1997) and Dual Purpose Speech (Lyons, Skeels, et al., 2004), try to integrate the computing task with the user's behavior in the physical world. The VuMan 3 interface did not tightly couple the virtual and real worlds, but the computer interface was designed specifically for the user's task and allowed the user to switch rapidly between a virtual interface and his or her hands-on vehicle inspection.

Many office productivity tasks such as e-mail or web searching, however, have little relation to the user's environment. The

mobile user must continually assess what attentional resources she or he can commit to the interface and for how long before switching attention back to the primary task. Oulasvirta, Tamminen, Roto, and Kuorelahit (2005) specifically examined such situations by fitting cameras to mobile phones and observing users attempting web search tasks while following predescribed routes. Subjects performed these tasks in a laboratory, in a subway car, riding a bus, waiting at a subway station, walking on a quiet street, riding an escalator, eating at a cafeteria and conversing, and navigating a busy street. Web pages required an average of 16.2 seconds to load and had considerable variance requiring the user to attend the interface. The subjects shifted their attention from the phone interface more often depending on the task: 35% of page loadings in the laboratory versus 80% of the page loadings while walking a quiet street. The duration of continuous attention on the mobile device also varied depending on the physical environment: 8–16 seconds for the laboratory and cafe versus below 6 seconds for the riding the escalator or navigating a busy street. Similarly, the number of attention switches depended on the demands of the environment.

The authors noted that even riding an escalator requires demands on attention (e.g., choosing a correct standing position, monitoring personal space for passers-by, and determining when the end is in order to step off). Accordingly, they are working on a "Resource Competition Framework," based on the Multiple Resource Theory of attention (Wickens, 1984), to relate mobile task demands to the user's cognitive resources. This

FIGURE 15.3. Forms studied for wearability.

framework helps predict when the mobile user will need to adopt attentional strategies to cope with the demands of a mobile task. The authors reported four such strategies observed in their study. The first, "calibrating attention," refers to the process by which the mobile user first attends to the environment and determines the amount of attention the user needs to devote to the environment versus the interface. "Brief sampling over long intervals" refers to the practice of only attending to the environment in occasional, brief bursts to monitor for changes that may require a deviation from plan, such as when reading while walking a empty street. "Task finalization" refers to subjects' preference to finish, when sufficiently close, a task or subtask before switching attention back to the physical environment. "Turntaking capture" occurs when the user is conversing with another person. Attending and responding to another person requires significant concentration, leading to minimal or no attention to the mobile interface.

The third author, who has been using his wearable computer to take notes on his everyday life since 1993, has remarked on similar strategies in his interactions. Describing these attentional strategies and designing interfaces that leverage them will be important in future mobile interfaces. Much research has been performed on aircraft and automobile cockpit design-to-design interfaces that augment, but do not interfere with, the pilot's primary task of navigating the vehicle; however, only recently has it begun to be possible to instrument mobile users and examine interface use (and misuse) "in-the-field" for the mobile computer user (Lyons & Starner, 2001). Now, theories of attention can be applied and tested to everyday life situations.

This newfound ability to monitor mobile workers may help us determine how **not** to design interfaces. In contrast to the VuMan 3 success described above, Ockerman's (2000) PhD thesis "Task Guidance and Procedure Context: Aiding Workers in Appropriate Procedure Following" warns that mobile interfaces, if not properly designed, may hinder the user's primary task.

Ockerman studied experienced pilots inspecting their small aircraft before flying. When a wearable computer was introduced as an aid to completing the aircraft's safety inspection checklist, the expert pilots touched the aircraft less (a way many pilots develop an intuition as to the aircraft's condition). In addition, the pilots relied too much on the wearable computer system, which was purposely designed to neglect certain safety steps. The pilots trusted the wearable computer checklist to be complete instead of relying on their own mental checklists. Ockerman showed how such interfaces might be improved by providing more context for each step in the procedure. Another approach would be integrating the aircraft itself into the interface (e.g., use augmented reality to overlay graphics on the aircraft indicating where the pilot physically inspects the plane).

Most recently, DARPA's Augmented Cognition project (Kollmorgen, Schmorrow, Kruse, & Patrey, 2005) aims to create mobile systems that monitor their user's attentional resources and records or delays incoming information in order to present it to the user in a more orderly and digestible time sequence. These systems exploit mobile electroencephalogram (EEG) readings or functional near infrared imaging (fNIR) to monitor the user's brain activations and relate these results to the user's current state. Such projects, if successful on a larger scale, could reveal much about the mental resources required for truly mobile computing.

The Attention Matrix, shown in Fig. 15.4 (Anhalt et al., 2001), categorizes activities by the amount of attention they require. The activities are (a) Information, (b) Communication, and (c) Creation. Individual activities are categorized by the amount of distraction they introduce in units of increasing time: (a) Snap, (b) Pause, (c) Tangent, and (d) Extended. The Snap duration is an activity that is usually completed in a few seconds, such as checking a watch for the time. The user should not have to interrupt their primary activity to perform this activity. The Pause action requires the user to stop their current activity, switch to the new but related activity, and then return to their previous task within a few minutes. Pulling over to the side of the road and checking directions is an example of a pause. A Tangent action is a medium length task that is unrelated to the action in which the user is engaged. Receiving an unrelated phone call is an example of a tangent activity. An Extended action occurs when the user deliberately switches tasks, beginning a wholly new, long-term activity. For the car driver, stopping at a motel and resting for the night is an extended activity.

As distractions on the left of the matrix take less time from the user's primary activity, our intent is to move activities of the matrix towards the left side (Snap). Our goal is to evaluate how this process extends to a larger sample of applications.

MANIPULATION

VuMan 3 Dial Pointing

VuMan 3 added a novel manipulation interface suitable for use when physical attention is occupied. The VuMan3 has a low-resolution display and, consequently, a purely textual interface. Figure 15.5 shows a sample screen from the user interface. The

Time

	Snap	Pause	Tangent	Extended
Information				
active · Receiving · Notifying · Monitoring · Serendipity · Seeking	Message arrival Information accessible Audition Stocks, Sports, Matching similar needs Free food Line length Bus arrival Locate person	Exam calendar Software/hardware help Calendaring Navigation	Looking for Class Notes Who else is doing this now? Access personal data	Audio, Walkman Transferring files from network Reading news
passive · Browsing · Finding · Verifying		Information on web or built environment Recall previous queries Double checking information	Poster, bulletin board information	Web Research Reviewing Class Notes
Communication				
artificial · Initiating · Participating	S.O.S. Emergency Instant messaging	Introductions Queries	Team building Collaborative work Event planning Assassins game Social Planning	Chatting (public or private)
informal · Broadcasting		Information exchange Scheduling	Posting information to bulletin board Advertising	
formal	· One to One communications with an individual			
	· One to Group communications with select group, team or family			
	· One to All Possible broadcast communications with unknown people			
Creation				
work · Recording · Synthesizing · Generating	Remember that Add a todo or call list	Forwarding x to y	Class note taking Meeting Filling out survey Registration New ideas Adding information to existing projects	Generating messages Summarizing lecture Mobile tool building

FIGURE 15.4.   Attention Matrix.

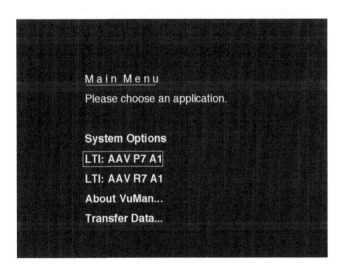

Main Menu

Please choose an application.

System Options

LTI: AAV P7 A1

LTI: AAV R7 A1

About VuMan...

Transfer Data...

FIGURE 15.5.   VuMan 3 Options Screen.

user navigates through a geographically organized hierarchy: top, bottom, front, rear; then left, right, and more detail. Eventually, at the node leafs, individual components are identified. There are over 600 of these components. Each component is indicated to be "Serviceable" or "Unserviceable." If it is serviceable,

then no further information is given. If it is unserviceable, then one of a small list of reasons is the next screen.

The user can return up the hierarchy by choosing the category name in the upper right corner, or sequence to the next selection in an ordering of the components. Once a component is marked as serviceable or unserviceable, the next selection in the sequence is automatically displayed for the user. Furthermore, each component has a probability of being serviceable associated with it, and the cursor is positioned over the most likely response for that component.

The screen contains navigational information. Sometimes there is more on a logical screen than can fit on a physical screen. Screen navigation icons are on the left-hand side of the screen. The user can go to the previous physical screen or next physical screen that are functional parts of the logical screen. The user can always go back to the main menu. In Fig. 15.6 the options include the vehicle number, number of hours on the vehicle, vehicle serial number, and so forth, which are used to distinguish this report apart from other reports. The inspection is divided into sections, and different people can be inspecting different sections in parallel. The inspector would pick a section, highlight it by rotating the dial, and then select the highlighted item by pressing a button. The inspector would then receive a detailed set of instructions on how to proceed. In Fig. 15.7, the inspector is instructed to check for damage and bare metal. The "smart cursor" anticipates that the inspector will be filling in

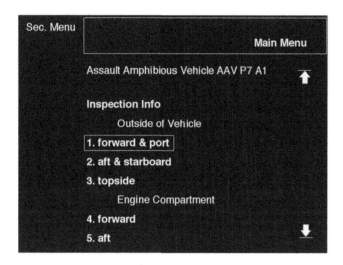

FIGURE 15.6. VuMan 3 Information Screen.

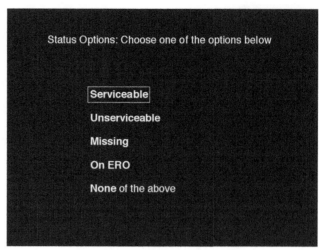

FIGURE 15.8. VuMan 3 Status Screen.

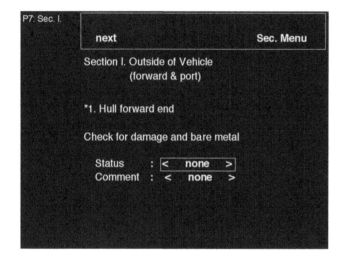

FIGURE 15.7. VuMan 3 Hull Forward Screen.

the "Status Field," whose current value is "none." By clicking, a list of options is displayed, the first of which is "Serviceable." With the Marine LTI, the item is serviceable in 80% of the cases. By ordering the most probable selection first, the interface emulates a paper checklist where most of the items will be checked as "OK." The smart cursor then assumes the most likely step. There is no need to even move the dial—the user merely clicks on the highlighted option. For example, in Fig. 15.8, "Serviceable" has been filled in and the box signifying the next activity is "Next." If all entries are "Serviceable," one would simply tap the button multiple times. If an item is "Unserviceable," the dial is turned and "Unserviceable" is selected. Next, a list of reasons why that particular device was unserviceable would appear. The dial is rotated and one or more of the options are selected. Since more than one reason may be selected as to why it's unserviceable, "Done" is selected to indicate completion. The selected items would appear in the "Comment" field. When the

checklist is completed, the data is uploaded to the logistics computer, which would then generate the job work orders.

Several lessons were derived from building the system. As part of the design cycle, a mouse (essentially, a disk with buttons) was tested; however, the physical configuration of the device could be ambiguous. Was the left button in the proper position when the mouse's tail was towards the user or away from the user? Were the buttons supposed to be at the top? The dial removed this ambiguous orientation.

Another design lesson was to minimize cables. An earlier system had a cable connecting the battery, a cable for the mouse, and a cable for the display. These wires quickly became knotted. To avoid this problem, the VuMan 3 design used internal batteries, and the dial was built into the housing. The only remaining wire connected it to the display.

A third lesson was that wearable computers have a minimum footprint that is comfortable for a user's hand. While the keyboards of palmtop computers are getting smaller, evolution has not correspondingly shrunk our fingers. The thickness of the electronics will become thinner. Eventually, it will be as thick as a sheet of plastic or incorporated into clothing; however, the interface will have a minimal footprint. Furthermore, the interface—no matter where it is located on the user's body—is operated in the same way. This is a major feature of the dial. It can be worn on a user's hip or in the small of the user's back. In airplane manufacturing, where workers navigate small spaces, the hip defines the smallest diameter through which the person can enter. Here a shoulder holster is preferred for the wearable computer. The Marines' oversized coverall pockets were an advantage for the system.

The soldiers could drop the computer into their coveralls and operate it through the cloth of the pocket. In terms of simplicity, as well as orientation independence, the dial integrated with the presentation of information on the screen. Everything on the screen could be considered to be on a circular list. In most cases, less than a dozen items on a screen are selectable. This sparse screen is an advantage on a head mounted display where the user may be reading while are moving. The font must be large enough to read while the screen is bouncing. The dial should be an intuitive interface for web browsing. Probably there are less than a half dozen items on a typical page to select, and it is rotated clockwise or counter clockwise. A button is then used to select the highlighted item. VuMan 3 had three types of buttons that all performed the same function. The buttons support left-hand and right-hand thumb-dominant as well as a central button for finger-dominant users.

Mobile Keyboards

The VuMan 3 addressed the problem of menu selection in the mobile domain and effectively used a 1D dial to create a pointing device that can be used in many different mobile domains. For tasks such as wireless messaging, however, more free-form text entry is needed. While speech technology has made great strides in the last decade, speech recognition is very difficult in the mobile environment and often suffers from high error rates. In addition, speech is often not socially acceptable (e.g., in hospitals, meetings, classrooms, etc.). Keyboard interfaces still provide one of the most accurate and reliable methods of text entry.

Since 2000, wireless messaging has been creating billions of dollars of revenue for mobile phone service providers, and over 1 trillion messages are currently being typed per year. Until recently, many of these messages were created using the Multitap or T9 input method on phone keypads. Yet studies have shown that users average a slow 10–20 words per minute (wpm) using these common typing methods (for comparison, a highly skilled secretary on a desktop averages 70–90 wpm). Given the obvious desire for mobile text input, HCI researchers have begun re-examining keyboards. While keyboard entry has been well studied in the past, mobility suggests intriguing possibilities. For example, if an adequate method of typing can be combined with a sufficient display for the mobile market, computing may move "off-the-desktop" permanently.

Traditionally, text entry studies emphasize learnability, speed, and accuracy; however, a mobile user may not be able to devote all of his or her attention to the text entry process. For example, the user may be taking notes on a conversation and wish to maintain eye contact with his or her conversational partner, or the user may be in a meeting and may hide the keyboard under the desk to avoid distracting others with the keyboard's noise and his or her finger motions. The user might also attempt to enter text while walking and need to attend to the physical environment instead of looking at the screen. These conditions all describe "blind" typing, where the user enters text with only occasional glances at the screen to ensure that the text has been entered correctly.

Lyons, Starner, and colleagues (2004) and Clawson, Lyons, Starner, and Clarkson (2005) have performed longitudinal studies on two keyboards—(a) Handykey's Twiddler (Fig. 15.9) and (b) the mini-QWERTY "thumb" keyboard (Fig. 15.10)—to determine if they might achieve desktop-level text entry in the mobile domain. As the "average" desktop entry rate was considered to be 30 wpm, including hunt-and-peck typists, this benchmark was chosen as the minimum for speed. Traditionally, very high accuracy is desired for desktop typing. As a culture of informal e-mail and SMS messaging has developed, however, less accurate

FIGURE 15.9. Handykey's Twiddler.

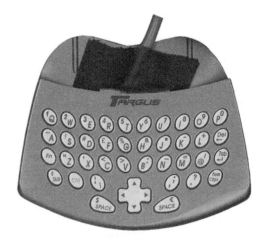

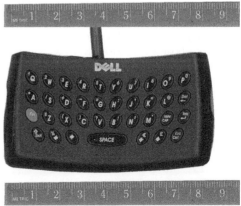

FIGURE 15.10. Mini-QWERTY Thumb Keyboard.

typing has become common. The community is debating how to reconcile speed and accuracy measures; however, error rates of approximately 5% per character are common in current mobile keyboard studies.

With the Twiddler, novices averaged 4 wpm during the first 20-minute session and averaged 47 wpm after 25 hours of practice (75 20-minute sessions). The fastest user averaged 67 wpm, which is approximately the speed of one of the authors who has been using the Twiddler for twelve years. While 25 hours of practice seems extreme, a normal high school typing class involves almost three times that training time to achieve a goal of 40 wpm.

Even so, mobile computer users may already have experience with desktop QWERTY keyboards. Due to their familiarity with the key layout, these users might more readily adopt a mini-QWERTY keyboard for mobile use. Can a mini-QWERTY keyboard achieve desktop rates? The study performed by Clawson and colleagues (2005) examined the speed and accuracy of experienced desktop typists on two different mini-QWERTY designs. These subjects averaged 30 wpm during the first twenty-minute session and increased to 60 wpm by the end of four hundred minutes of practice!

While both of these studies easily achieved desktop typing rates and had error rates comparable to past studies, can these keyboards be used while mobile? While neither study tested keyboard use while the user was walking or riding in a car, both experimented with blind text entry (in that, in at least one condition, typists could not look at the keyboard nor the output of their typing). When there was a statistically significant difference between blind and normal typing conditions, experienced Twiddler typist slightly improved their speeds and decreased their error rates. Experienced mini-QWERTY typists, however, were significantly inhibited by the blind condition, with speeds of 46 wpm and approximately three times the error rate even after 100 minutes of practice. These results might be expected in that Twiddler users are trained to type without visual feedback from the keyboard, whereas the mini-QWERTY keyboard design assumes that the user can see the keyboard to help disambiguate the horizontal rows of small keys.

The results of these studies demonstrate that there are multiple ways that desktop typing rates can be achieved on a mobile device. The question remains, however, whether the benefits of typing quickly while "blind" or moving will be sufficient to cause users to learn a new text entry method. Other benefits might also affect the adoption of keyboards in the future. For example, a 12-button device such as the Twiddler can be the size of a small mobile phone and still perform well, while 40-button mini-QWERTY keyboards may have already shrunk as much as is possible for users' hands. Another factor may be adoption of mobile computing in developing countries. According to Techweb, almost 1 billon mobile phones were shipped in 2005. Many new mobile phone users will not have learned to type on a Roman alphabet keyboard and may be more concerned with quick learning than compatibility with desktop-input skills.

Speech Interfaces

Vocollect. Mobile keyboards are not suitable for applications in which hands-free control is necessary, such as

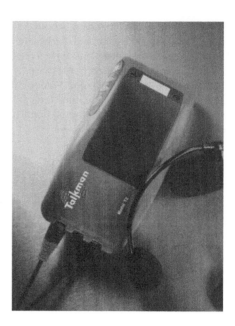

FIGURE 15.11. Vocollect's audio-based wearable computer.

warehouse applications. Pittsburgh-based Vocollect focuses on package manipulation—in particular, the warehouse-picking problem. In this scenario, a customer places an order consisting of several different items stored in a supplier's warehouse. The order transmits from the warehouse's computer to an employee's wearable computer. In turn, each item and its location are spoken to the employee through a pair of headphones. The employee can control how this list is announced through feedback via speech recognition, and can also report inventory errors as they occur. The employee accumulates the customer's order from the warehouse's shelves and ships it. This audio-only interface also frees the employee to manipulate packages with both hands, whereas a pen-based system would be considerably more awkward. As of December 2000, Vocollect had approximately 15,000 users and revenues between U.S. $10 and $25 million.

Navigator wearable computer with speech input. Boeing has been pioneering "augmented reality" using a head-mounted, see-through display. As the user looks at the aircraft, the next manufacturing step is superimposed on the appropriate portion of the aircraft. One of their first applications is fabrication of wire harnesses. Every aircraft is essentially unique. They may be from different airlines. Even if they are from the same airline, one might be configured for a long-haul route and another for a short-haul route. The airline may specify different configurations. For example, their galleys will be in different places, the wire harnesses would change, and so forth. Wire harnesses are fabricated months before they are assembled into the aircraft. The assembly worker starts with a Peg-board measuring about three feet high and six feet long. Mounted on the board is a full-sized diagram of the final wire harness. Pegs provide support for the bundles of wire as they form. The worker (a) selects a precut wire, (b) reads its identification number, (c) looks up the wire number on a paper list to find the starting coordinates of the wire, (d) searches for the wire on the diagram, and (e) threads the wire following the route on the diagram. With augmented reality, the worker selects a wire and reads the wire identification from the bar code. A head tracker provides the computer with information on where the worker is looking and superimposes the route for that particular wire on the board. Trial evaluations indicate a savings of 25% of the assembly effort primarily due to elimination of cross-referencing the wire with paper lists.

The Navigator 2, circa 1995, was designed for a voice-controlled aircraft inspection application (Siewiorek, Smailagic, & Lee, 1994). The speech recognition system, with a secondary, manually controlled cursor, offers complete control over the application in a hands-free manner, allowing the operator to perform an inspection with minimal interference from the wearable system. Entire—or portions of—aircraft manuals can be brought on-site as needed, using wireless communication. The results of inspection can then be downloaded to a maintenance logistic computer.

Consider one portion of Navigator 2's application, three-dimensional inspections. The application was developed for McClellan Air Force Base in Sacramento, California and the KC-135 aerial refueling tankers. Every five years, these aircraft are stripped down to bare metal. The inspectors use magnifying glasses and pocket knives to hunt for corrosion and cracks.

At startup (Fig. 15.12) the application prompts the user for either their choice of activating the speech recognition system or not. The user then proceeds to the Main Menu. From this location, several options are available, including online documentation, assistance, and the inspection task (Fig. 15.13). Once the user chooses to begin an inspection, information about the inspection is entered, an aircraft type to examine is selected, and the field of interest is narrowed from major features (Left Wing, Right Tail, etc.; Fig. 15.14) to more specific details (individual panes in the cockpit window glass; Fig. 15.15). A coordinate system is superimposed on the inspection region. The horizontal coordinates begin from the nose and the vertical coordinates are "water lines" derived as if the airplane was floating. The inspector records each imperfection in the skin at the corresponding location on the display. The area covered by each defect is recorded, as well as the type of defect, such as Corroded, Cracked, or Missing. To maximize usability, each item or control may be selected simply by speaking its name. Figure 15.16 shows the Navigator 2 systems in use.

The user navigates to the display corresponding to the portion of the skin currently being inspected. This navigation is partially textual based on buttons (choose aircraft type to be inspected) and partially graphical based on side perspectives of the aircraft (choose area of aircraft currently being inspected). The navigation can be performed either through a joystick input device or through speech input. The speech input is exactly the text that would be selected. The positioning of the imperfection is done solely through the joystick, since speech is not well suited for the pointing necessary to indicate the position of the imperfection. As the cursor is moved by the joystick, the coordinates and the type of material represented by the cursor is

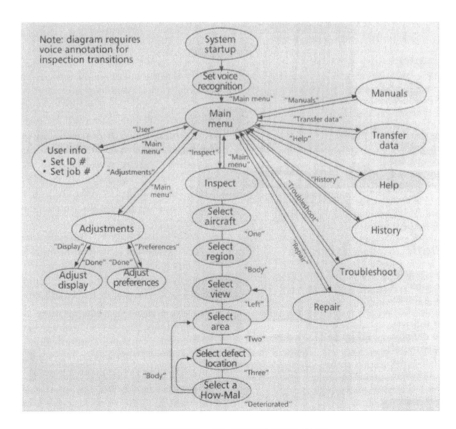

FIGURE 15.12   Navigator 2 Finite State.

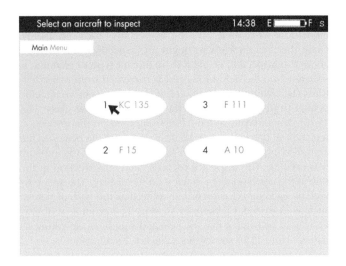

FIGURE 15.13.   Navigator 2 Main Menu.

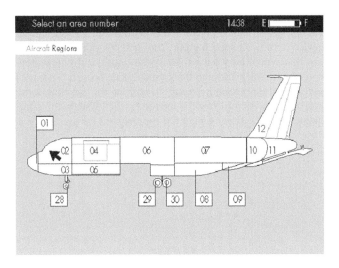

FIGURE 15.14.   Navigator 2 Region Selection.

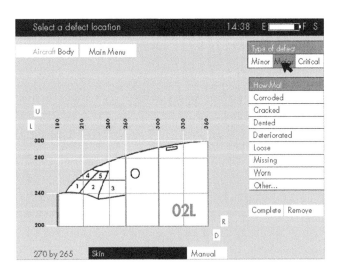

FIGURE 15.15.   Sheet metal.

FIGURE 15.16.   Navigator 2.

displayed at the bottom of the screen. If a defect is at the current position, a click produces a list of reasons why that material would be defective, such as corrosion, scratch, and so forth. The defect type can be selected by the joystick or by speaking its name, and the information would go into the database. The user can navigate to the main selection screen by selecting the "Main menu" option on all of the screens. One level up in the hierarchy can also be achieved through a single selection.

The relationship between the user interface design principles and the Navigator 2 user interface is:

• Simplicity of function. The only functions available to the user are to enter skin imperfections for one of four aircrafts, to transfer data to another computer, to enter identification information both for the vehicle and for the inspector, and to see a screen that describes the Navigator 2 project.

• No textual input. The identification information required entering numbers. A special dialogue was developed to enable the entering of numeric information using the joystick as an input device. This was cumbersome for the users, but only needed to be performed once per inspection.

• Controlled navigation. The interface was arranged as a hierarchy. The top level consisted of a menu that gave a choice of function. Once the inspection function and then the vehicle were chosen, the area of the skin inspected was navigated to via selecting an area of the aircraft to expand. Once an imperfection was indicated, the user had to select one of the allowable types of imperfections. At each stage, the user could go up one level of the hierarchy or return to the main menu.

One of the lessons learned with Navigator 2 is the power of forcing the use of a common vocabulary. Since the average age of the aircraft is 35 years, the types of defects encountered is a very stable set. Previously, one inspector would call a defect "gouged," while another inspector would call the same defect a "scratch." What is the difference between a gouge and a scratch? How much material does it take? How much time does it take to repair? What skill of labor is needed? The logistics problem is much more difficult without a standardized vocabulary. Thus there is a serendipitous advantage in injecting more technology.

A second lesson is that in some cases, the speech recognition front end mistakenly produces the wrong output. Speech recognition systems typically have an error rate of 2% to 10%. The unexpected output may cause the application to produce the wrong result. In one of Navigator 2's early demonstrations, the user was attempting to exit the application; however, the speech recognition system thought a number was spoken. At that point, the application was expecting a second number, but there was no match since the user was saying "Exit," "Quit," "Bye," and so forth. The system appeared to be frozen, when in actuality, there was a mismatch between what the application software was expecting and what the user thought the application state was. The solution was to give the user more feedback on the state of the application by additional on-screen clues. Also developed was a novel application input test generator that took a description of the interface screens and created a list of all possible legal exits from each screen.

A third lesson learned was the criticality of response time. When speech recognition was done in software on Navigator (circa 1995) it was 12 times real time, which became very frustrating. People are less patient when they are on the move than they are when they are at a desktop. People at a desk are willing to wait three minutes for the operating system to boot up, but when a user is on the move, expectations are for instant response like that of portable tools such as a flashlight. For example, some airplanes have a digital computer to control the passengers' overhead lights. It is disconcerting that when the button is pushed, two or three seconds may pass before the light turns on. Even a couple-second delay in a handheld device is disruptive. Users typically continue to push buttons until they receive a response. The extra inputs cause a disconnect between the software and the user. The software receives a stream of inputs, but the user sees outputs that are related to inputs given a long time before the screen actually appears. The situation is similar to listening to yourself talk when there is a second or two delay in the sound played back. Such delays can easily confuse users.

The field evaluation indicated that inspection is composed of three phases. The inspectors would spend the same amount of time maneuvering their cherry picker to access a region of the airplane, visually inspecting and feeling the airplane's skin, and recording the defect's type and location. Navigator 2 reduced the paperwork time by half resulting in an overall time savings of about 18%. Training time to familiarize inspectors with the use of Navigator 2 was about five minutes, after which they would proceed with actual inspections. A major goal of field evaluations is that users perform productive work. They do not want to redo something that was already done once.

The typical inspection, which discovers approximately 100 defects, requires about 36 hours. Today, the inspector takes notes on a clipboard. Upon completion, the inspector fills out forms on a computer. The inspector spends two to three minutes entering each defect. The data entry is thus an additional three-to-four-hour task. Navigator 2 transmits the results of the inspection by radio in less than two minutes.

In summary, evaluations of inspectors before and after the introduction of Navigator 2 indicated a 50% reduction in the time to record inspection information (for an overall reduction of 18% in inspection time) and almost two orders of magnitude reduction in time to enter inspection information into the logistics computer (from over three hours to two minutes). In addition, Navigator 2 weighs two pounds, compared to the cart the inspectors currently use, which carries 25 pounds of manuals.

Speech translation. The SR/LT application (Speech Recognition/Language Translation) consists of three phases: (a) speech-to-text language recognition, (b) text-to-text language translation, and (c) text-to-speech synthesis. The application running on TIA-P (Tactical Information Assistant-Prototype, circa 1996) is the Dragon Multilingual Interview System (MIS), jointly developed by Dragon Systems and the Naval Aerospace and Operational Medical Institute (NAOMI). It is a keyword-triggered, multilingual playback system, which listens to a spoken phrase in English, proceeds through a speech recognition front-end, plays back the recognized phrase in English, and after some delay (~8–10 secs), synthesizes the phrase in a foreign language (Croatian). The other, local person can answer with "Yes," "No," and some pointing gestures. The Dragon MIS has about 45,000 active phrases, in the following domains: medical examination, mine fields, road checkpoints, and interrogation. Therefore, a key characteristic of this application is that it deals with a fixed set of phrases, and includes one-way communication. A similar system is used in Iraq as a briefing aid to interrogate former Iraqi intelligence officials and to speak with civilians about information relevant to locating individuals (Chisholm, 2004). This shows the viability of the approach.

TIA-P is a commercially available system, developed by CMU, incorporating a 133 MHz 586 processor, 32MB DRAM, 2 GB IDE Disk, full-duplex sound chip, and spread spectrum radio (2Mbps, 2.4 GHz) in a ruggedized, handheld, pen-based system designed to support speech translation applications. TIA-P is shown in Fig. 15.17.

Dragon loads into memory and stays memory resident. The translation uses uncompressed ~20 KB of .WAV files per phrase. There are two channels of output: (a) the first plays in English, and (b) the second plays in Croatian. A stereo signal can be split, with one channel directed to an earphone, and the second to a speaker. This is done in hardware attached to the external speaker. An Andrea noise-canceling microphone is used with an on-off switch.

Speech translation for one language (Croatian) requires a total of 60 MB disk space. The speech recognition requires an additional 20–30 MB of disk space.

FIGURE 15.17.   TIA-P Wearable Computer.

FIGURE 15.18.   U.S. Soldier in Balkans Using TIA-P.

TIA-P has been tested with the Dragon speech translation system in several foreign countries: Bosnia (Fig. 15.18), Korea, and Guantanamo Bay, Cuba. TIA-P has also been used in human intelligence data collection and experimentation with the use of electronic maintenance manuals for F-16 maintenance.

The following lessons were learned during the TIA-P field tests:

• Wires should be kept to a minimum.
• Handheld display was convenient for checking the translated text.
• Standard, external electrical power should be available for use internationally.
• Battery lifetime should be extended.
• Ruggedness is important.

The smart modules (circa 1997) are a family of wearable computers dedicated to the speech processing application (Smailagic, Siewiorek, & Reilly, 2001). A smart module provides a service almost instantaneously and is configurable for different applications. The design goals also included (a) reduce latency, (b) remove context swaps, and (c) minimize weight, volume, and power consumption (Reilly, 1998; Martin, 1999). The functional prototype consists of two functionally specialized modules: (a) performing language translation and (b) speech recognition. The first module incorporates speech to text language recognition and text to speech synthesis. The second module performs text-to-text language translation. The LT module runs

the PANLITE language translation software (Frederking & Brown, 1996), and the SR module runs CMU's Sphinx II continuous, speaker-independent speech recognition software (Ravishankar, 1996; Li, Hon, Hwang, & Reddy, 1989) and Phonebox Speech Synthesis software.

Figure 15.19 depicts the structure of the speech translator, from English to a foreign language, and vice versa. The speech is input into the system through the Speech Recognition subsystem. A user wears a microphone as an input device, and background noise is eliminated using filtering procedures. A language model, generated from a variety of audio recordings and data, provides guidance for the speech recognition system by acting as a knowledge source about the language properties. The Language Translation engine uses an Example-Based Machine Translation (EBMT) system, which takes individual sentence phrases and compares them to a corpus of examples it has in memory to find phases it knows how to translate. A lexical MT (glossary) translates any unknown word that may be left. The EBMT engine translates individual "chunks" of the sentence using the source language model and then combines them with a model of the target language to ensure correct syntax. When reading from the EBMT corpus, the system makes several random-access reads while searching for the appropriate phrase. Since random reads are done multiple times, instead of loading large, continuous chunks of the corpus into memory, the disk latency times will be far more important than the disk bandwidth. The Speech Generation subsystem performs text-to-speech conversion at the output stage. To make sure that misrecognized words are corrected, a Clarification Dialog takes place on-screen. It includes the option to speak the word again, or to write it in. As indicated in Fig. 15.19, an alternative input modality could be the text from the Optical Character Recognition subsystem (such as scanned documents in a foreign language), which is fed into the Language Translation subsystem.

User interface design went through several iterations based on feedback during field tests. The emphasis was on getting completely correct two-way speech translation, and having an easy to use, straightforward interface for the clarification dialogue.

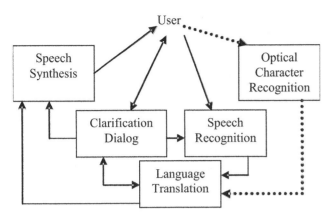

FIGURE 15.19. Speech Translator System Structure.

Speech to Speech

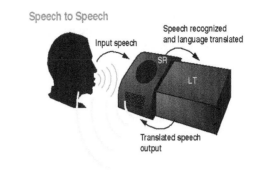

FIGURE 15.20. Speech Recognizer (SR) and Language Translator (LT) Smart Module.

The speech recognition code was profiled and tuned. Profiling was performed to identify "hot spots" for hardware and software acceleration, as well as to reduce the required computational and storage resources. A six-times speedup was achieved over the original desktop PC system implementation of language translation, and five times smaller memory requirements (Christakos, 1998). Reducing OS swapping and code optimization made a major impact. Input to the module is audio and output is ASCII text. The speech recognition module is augmented with speech synthesis. Figure 15.20 illustrates a combination of the language translation module (LT), and speech recognizer (SR) module, forming a complete stand-alone, audio-based, interactive dialogue system for speech translation.

Target languages included Serbo-Croatian, Korean, Creole French, and Arabic. Average language translation performance was one second per sentence.

The key factors that determine how many processes can be run on a module are memory, storage space, and available CPU cycles. To minimize latency, the entirety of an application's working dataset should be able to stay memory resident.

Figure 15.21 depicts the functional prototype of the Speech Translator Smart Module, with one module performing language translation, and another one speech recognition and synthesis.

FIGURE 15.21. Speech Translator SM Functional Prototype.

PERFORMANCE EVALUATION

Figure 15.22 illustrates the response time for speech recognition applications running on TIA-P and SR Smart Module. As SR is using a lightweight operating system (Linux) versus Windows 95 on TIA-P and the speech recognition code is more customized, it has a shorter response time. An efficient mapping of the speech recognition application onto the SR Smart Module architecture provided a response time very close to real-time. To ensure system responsiveness, it was important to provide feedback to the person in near real time.

The lessons learned from tests and demonstrations include: manual intervention process to correct misrecognized words incurs some delay; swapping can diminish the performance of the language translation module; the size of display can be as small as a deck of cards.

Performance Comparison

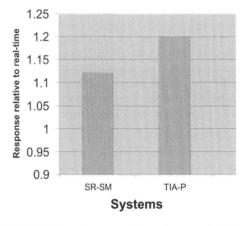

FIGURE 15.22. Response Times (lower is better).

The required system resources for speech translator software are several times smaller than for the laptop/workstation version, as shown in Table 15.2.

Dual Purpose Speech

In industry, most speech recognition on mobile computers concentrates on the tasks of form filling or simple interface commands and navigation. One reason is that speech interfaces are often socially interruptive when other people are nearby. Speech translation, as with the TIA system above, is a different class of interface. The computer is an essential enabler of the conversation. Lyons, Skeels, and colleagues (2004) introduced a different type of conversation enabler in their Dual Purpose Speech work.

Dual Purpose Speech is easiest to discuss using a scenario. Tracy, a wearable user equipped with a head-up display (HUD) and a Twidder keyboard, is in conversation with a recently introduced colleague. Pressing a button on the keyboard, the wearable user enables speech recognition and says, "Bob, what is your phone number so that I have it for later?"

The wearable recognizes that its user wants to record a phone number and starts the user's contact list application. It attempts to recognize the name spoken and enters that into the application; however, it also saves the speech so that the user can correct the text later if there is an error.

> Bob responds, "Area code 404"
> "404," repeats Tracy.
> "555-1212," completes Bob.
> "555-1212," continues Tracy, who presses another button on her keyboard indicating the interaction is over. "Okay, I have it!"

On Tracy's head-up display a new contact has been made for "Bob (404) 555-1212." When Tracy finishes her conversation, she clicks an "Accept" button on the application because she has recognized the information correctly. Tracy could also edit the information or play back the audio recorded during the interaction with Bob. Note that Tracy verbally repeated the information that Bob provided—a good conversational practice. Tracy both confirmed that she understood the information and provided Bob with an opportunity to correct her if necessary; however, this practice is also good from a privacy standpoint. Tracy wears a noise-canceling microphone, which is thresholded to record only her own voice and not that of her conversational partners. In this way, Tracy respects the privacy of her colleagues.

Lyons, Skeels, and colleagues (2004) have designed Dual Purpose Speech applications for scheduling appointments, providing reminders for the user, and communicating important information to close colleague. The key point of this research

from the perspective of this section, however, is that these applications allow the user to manipulate information on their wearables as part of the process of communicating (thus, the "dual purpose" name). The users may actively format their speech so the system can better understand them, and they may have to correct the system afterwards; however, the interface is manipulated and the information is entered as part of a social process.

This style of interface provides a contrast to the traditional desktop computer, where the user's attention is assumed to be dedicated to the interface. Other wearable-computing-related fields also attempt to create interfaces that are driven by the user's interactions with the environment. For example, Feiner, MacIntyre, and Seligmann (1993) early augmented reality systems attempted to display appropriate repair instructions based on the user's actions during the repair process. Such awareness of the user's context and goals may allow wearable computers to be utilized where a user's lack of attentional or physical resources would normally preclude traditional desktop applications.

Perception

Just as dexterity is impaired when a user is on the go, the user's ability to perceive a wearable's interface is also lessened. The vibration and visual interference from a moving background interferes with visual tasks. Background noise and the noise from the body itself affect hearing. The moving of clothes over the body and the coupling of mechanical shock through the body can lessen the user's ability to perceive tactile displays. Sears, Lin, Jacko, and Xiao (2003) described these detriments to mobile interaction caused by environmental and situational factors as "Situationally-Induced Impairments and Disabilities". These researchers and others are developing procedures to test human performance in mobile computing tasks in context (in this case, walking a path; Barnard, 2005, in press). Such research is sorely needed, as not enough is known about how to adequately simulate mobile computing scenarios in testing. For example, in Barnard and colleagues' (in press) work on performing reading comprehension tasks on PDAs while walking, lighting levels affected workload measures more when walking a path than when walking on a treadmill. The community needs to develop understanding about the interactions between mobility, attention, and perception in common mobile computing scenarios in order to adequately develop testing environments for mobile interfaces.

In the past, such work focused on cockpits, both for aviation and automobiles (Wickens & Hollands, 1999; Melzer, & Moffit, 1997; Velger, 1998); however, the U.S. military's Land Warrior project has highlighted the need for such research for dismounted users who are on the go (Blackwood, 1997). Some researchers have begun exploring mobile output devices for very specific tasks. For example, Krum (2004) described experiments with a head-up display that focused on determining how to render overhead views of an area to encourage learning of the layout of the surrounding environment while the user is navigating to a goal on foot. As mobile-augmented reality is becoming practical from a technical standpoint, researchers have begun to address perceptual issues. While not a mobile experiment, Laramee and Ware (2002) have investigated

TABLE 15.2. Comparison of Required System Resources

	Laptop/ Workstation	Functional Module SR/LT	Optimized Module SR/LT
Memory Size	195 MB	53 MB	41 MB
Disk Space	1 GB	350 MB	200 MB

head-mounted displays to determine the relative effects of rivalry and visual interference between binocular and monocular displays with varying levels of transparency. As the market determines which mobile contexts are most important for users, experiments such as these will help determine how to design interfaces to least interfere with the user's primary tasks while providing the most value in terms of augmentation.

CONCLUSION AND FUTURE CHALLENGES

Wearable computers are an attractive way to deliver a ubiquitous computing system's interface to a user, especially in non-office-building environments. The biggest challenges in this area deal with fitting the computer to the human in terms of interface, cognitive model, contextual awareness, and adaptation to tasks being performed. These challenges include

- User interface models. What is the appropriate set of metaphors for providing mobile access to information (e.g., what is the next "desktop" or "spreadsheet")? These metaphors typically take over a decade to develop (e.g., the desktop metaphor started in early 1970s at Xerox PARC and required over a decade before it was widely available to consumers). Extensive experimentation working with end-user applications will be required. Furthermore, a set of metaphors may each be tailored to a specific application or a specific information type.

- Input/output modalities. While several modalities mimicking the input/output capabilities of the human brain have been the subject of computer science research for decades, the accuracy and ease of use (e.g., many current modalities require extensive training periods) are not yet acceptable. Inaccuracies produce user frustrations. In addition, most of these modalities require extensive computing resources that will not be available in low-weight, low-energy, wearable computers. There is room for new, easy-to-use input devices such as the dial developed at Carnegie Mellon University for list-oriented applications.

- Quick interface evaluation methodology. Current approaches to evaluate a human computer interface requires elaborate procedures with scores of subjects. Such an evaluation may take months and is not appropriate for use during interface design. These evaluation techniques should especially focus on decreasing human errors and frustration.

- Matched capability with applications. The current thought is that technology should provide the highest performance capability; however, this capability is often unnecessary to complete an application and enhancements such as full-color graphics require substantial resources and may actually decrease ease of use by generating information overload for the user. Interface design and evaluation should focus on the most effective means for information access and resist the temptation to provide extra capabilities simply because they are available.

- Context-aware applications. How do we develop social and cognitive models of applications? How do we integrate input from multiple sensors and map them into user social and cognitive states? How do we anticipate user needs? How do we interact with the user? These, plus many other questions, have to be addressed before context aware computing becomes possible. Some initial results have been reported in Krause, Smailagic, and Siewiorek (2006).

References

Anhalt, J., Smailagic, A., Siewiorek, D., et al. (2001). Towards context aware computing. *IEEE Intelligent Systems, 6*(3), 38–46.

Azuma, R. (1997). A survey of augmented reality. *Presence, 6*(4), 355–386.

Barnard, L., Yi, J. S., Jacko, J. A., & Sears, A. (2005). An Empirical Comparison of Use-in Motion Evaluation Scenarios for Mobile Computing Devices. *International Journal of Human Computer Studies, 62*, 487–520.

Barnard, L., Yi, J., Jacko, J. A., & Sears, A. (in press). Capturing the Effects of Context on Human Performance in Mobile Computing Systems. *Personal and Ubiquitous Computing.*

Blackwood, W. (1997). *Tactical display for soldiers.* Washington, DC: National Academy of Sciences. pp. 57–61

Chisholm, M. (2004). Technology that speaks in tongues. *Military Information Technology, 8*(2). Retrieved from http://www.military-information-technology.com/article.cfm?DocID=424

Christakos, C. K. (1998). *Optimizing a language translation application for mobile use.* Unpublished master's thesis, Carnegie Mellon University, Department of Electrical and Computer Engineering, Pittsburgh, PA.

Clawson, J., Lyons, K., Starner, T., & Clarkson, E. (2005). The impacts of limited visual feedback on mobile text entry using the mini-QWERTY and Twiddler keyboards. *Proceedings of the IEEE International Symposium on Wearable Computers.* Los Alamitos, CA: IEEE Computer Society Press.

Feiner, S., MacIntyre B., & Seligmann, D. (1993). Knowledge-based augmented reality. *Communications of the ACM, 36*(7), 52–62.

Frederking, R. E., & Brown, R. (1996). The Pangloss-lite machine translation system: Expanding MT horizons. *Proceedings of the Second Conference of the Association for Machine Translation in the Americas* (pp. 268–272). Montreal, Canada.

Gemperle, F., Kasabach, C., Stivoric, J., Bauer, B., & Martin, R. (1998). Design for wearability. *Second International Symposium on Wearable Computers* (pp. 116–122). Los Alamitos, CA: IEEE Computer Society Press.

Iachello, G., Smith, I., Consolvo, S., Abowd, G., Hughes, J., Howard, J., et al., (2005). Control, deception, and communication: evaluating the deployment of a location-enhanced messaging service. *Proceedings of the Seventh International Conference on Ubiquitous Computing* (pp. 213–231). Los Alamitos, CA: IEEE Computer Society Press.

Kollmorgen, G. S., Schmorrow. D., Kruse, A., & Patrey, J. (2005). The cognitive cockpit-state of the art human-system integration. *Proceedings of the November 2005 Interservice/Interindustry Training Simulation and Education Conference*, DARPA, Arlington, VA.

Krause, A., Smailagic, A., & Siewiorek, D. P. (2006). Context-aware mobile computing: learning context-dependent personal preferences from a wearable sensor array. *IEEE Transactions on Mobile Computing, 5*(2), 113–127.

Krum, D. (2004). *Wearable computers and spatial cognition*. Unpublished doctoral dissertation, Georgia Institute of Technology, College of Computing, Atlanta.

Lee, K.-F., Hon, H.-W., Hwang, M.-Y., Mahajan, S., Reddy, R. (1989). *International Conference Acoustics, Speech, and Signal Processing* (ICASSP'89) May 23–26, (vol. 1, pp. 445–448). IEEE Computer Society Press, Los Alamitos, CA.

Laramee, R., & Ware, C. (2002). Rivalry and interference with a head mounted display. *ACM Transactions on Computer-Human Interface, 9*(3), 238–251. New York: ACM Press

Lyons, K., Skeels, C., Starner T., Snoeck, B., Wong, B., & Ashbrook, D. (2004). Augmenting conversations using dual-purpose speech. *Proceedings, User Interface and Software Technology* (pp. 237–246). Los Alamitos, CA:

Lyons, K., Starner, T., & Plaisted, D. (2004) Expert typing using the Twiddler one handed chord keyboard. *Proceedings of the IEEE International Symposium on Wearable Computers* (pp. 94–101).

Lyons, K., & Starner, T. (2001). Mobile capture for wearable computer usability testing. *Proceedings of the International Symposium on Wearable Computers.*

Martin, T. (1999). *Balancing batteries, power, and performance: System issues in CPU speed-setting for mobile computing*. Unpublished doctoral dissertation, Carnegie Mellon University, Department of Electrical and Computer Engineering, Pittsburgh, PA.

Melzer, J., & Moffitt, K. (1997). *Head mounted displays: Designing for the user. New York:* McGraw-Hill.

Ockerman, J. (2000). Task guidance and procedure context: Aiding workers in appropriate procedure following. Unpublished doctoral dissertation, Georgia Institute of Technology, School of Industrial System Engineering, Atlanta. New York: ACM Press

Oulasvirta, A., Tamminen, S., Roto, V., & Kuorelahit, J. (2005). Interaction in 4-second bursts: the fragmented nature of attentional resources in mobile HCI. *Proceedings of the SIGCHI Conference on Human Factors in Computing Systems.* (pp. 919–928).

Ravishankar, M. (1996). *Efficient algorithms for speech recognition*. Unpublished doctoral dissertation, Carnegie Mellon University, School of Computer Science, Pittsburgh, PA.

Reilly, D. (1998). *Power consumption and performance of a wearable computing system*. Unpublished master's thesis, Carnegie Mellon University, Electrical and Computer Engineering Department, Pittsburgh, PA. New York: Springer

Sears, A., Lin, M., Jacko, J., & Xiao, Y. (2003). When Computers Fade: Pervasive Computing and Situationally-Induced Impairments and Disabilities. *Proceedings of HCII 2003* (pp. 1298–1302).

Siewiorek, D. P. (2002). Issues and challenges in ubiquitous computing: new frontiers of application design. *Communications of the ACM, 45*(12), 79–82.

Siewiorek, D. P., Smailagic, A., & Lee, J. C. (1994). An interdisciplinary concurrent design methodology as applied to the Navigator wearable computer system. *Journal of Computer and Software Engineering, 2*(2), 259–292.

Smailagic, A. (1997). ISAAC: A voice activated speech response system for wearable computers. *Proceedings of the IEEE International Conference on Wearable Computers.* IEEE Computer Society Press, Los Alamitos, CA.

Smailagic A., Siewiorek D. P., Martin, R., & Stivoric, J. (1998). Very rapid prototyping of wearable computers: A case study of VuMan 3 custom versus off-the-shelf design methodologies. *Journal of Design Automation for Embedded Systems, 3*(2–3), 219–232. IEEE Computer Society Press, Los Alamitos, CA.

Smailagic, A., & Siewiorek, D. P. (1994). The CMU mobile computers: A new generation of computer systems. *Proceedings of the IEEE COMPCON 94* (pp. 467–473). Los Alamitos, CA: IEEE Computer Society Press.

Smailagic, A., & Siewiorek, D. (1996). Modalities of interaction with CMU wearable computers. *IEEE Personal Communications, 3*(1), 14–25. IEEE Computer Society Press, Los Alamitos, CA.

Smailagic, A., Siewiorek, D. P., & Reilly, D. (2001). CMU wearable computers for real-time speech translation. *IEEE Personal Communications, 8*(2), 6–12. New York: ACM Press

Smailagic, A., & Siewiorek, D. (1993). A case study in embedded system design: the VuMan 2 wearable computer. *IEEE Design and Test of Computers, 10*(3), 56–67.

Starner, T., Snoeck, C., Wong, B., & McGuire, R. (2004) Use of mobile appointment scheduling devices. *Proceedings of the ACM Conference Human Factors in Computing Systems* (pp. 1501–1504). Los Alamitos, CA: IEEE Computer Society Press.

Starner, T. (2001). The challenges of wearable computing: Part 1 + 2. *IEEE Micro, 21*(4), 44–67.

Stein, R., Ferrero, S., Hetfield, M., Quinn, A., & Krichever, M. (1998). Development of a commercially successful wearable data collection system. *Proceedings of the IEEE International Symposium on Wearable Computers* (pp. 18–24).

Velger, M. (1998). *Helmet-mounted displays and sights*. Norwood, MA: Artech House, Inc.

Wickens, C. D. (1984). Processing resources in attention. In R. Parasuraman & R. Davies (Eds), *Varieties of attention*. New York: Academic Press.

Wickens, C. D., & Hollands, J. (1999). *Engineering psychology and human performance* (3rd edition). Prentice Hall.

·16·

DESIGN OF COMPUTER WORKSTATIONS

Michael J. Smith and Pascale Carayon
University of Wisconsin–Madison

William J. Cohen
Exponent, Inc.

In our last version of this chapter, we focused on fixed location computing workstations and situations, but acknowledged the burgeoning use of portable information devices (PIDs) and provided a few ideas to consider about their use. We indicated that fixed computer environments, while still a significant part of workstation design considerations, have been surpassed by highly mobile information technology that does not use fixed workstations. These portable technologies are now in use in almost every venue and human activity, and the nature of their characteristics and activities of use do not lend them to traditional fixed workstation considerations. This introduces a host of potential ergonomic concerns related to the design of work areas (and activities) in which PIDs and other forms of computing are used. Decades of research and applications have defined important considerations in the ergonomic design of fixed computer work areas (Grandjean, 1979, 1984; Stammerjohn, Smith, & Cohen, 1981; Cakir, Hart, & Stewart, 1979; ANSI/HFES-100, 1988; Smith & Cohen, 1997; Smith, Carayon, & Cohen, 2003; BSR/HFES-100, 2005). However, very little has been done to define the design of work areas for PIDs and mobile computing. In this chapter, we will propose some ideas and considerations for dealing with ergonomic concerns for these mobile technologies in addition to updating information on fixed computer workstation applications. However, we admit that the challenge to define good ergonomic practice for the use of PIDs is large, and our advice is still very limited.

Ergonomics is the science of fitting the environment and activities to the capabilities, dimensions, and needs of people. Ergonomic knowledge and principles are applied to adapt working conditions to the physical, psychological, and social nature of the person. The goal of ergonomics is to improve performance while enhancing comfort, health, and safety. Computer workstation design is more than just making the computer interfaces easier to use, or making furniture adjustable in various dimensions. It also involves integrating design considerations with the work environment, the task requirements, the social aspects of work, and job design. Critical considerations for good ergonomics practice are to limit biomechanical loading on the back and joints to low forces of short duration; to keep the back, neck, and joints in good postures while working; to limit the amount of time you are in static postures; to limit the duration and extent of repetitive motions of any part of your body including your voice; to work in environments where you can easily see and hear; and to take frequent rest breaks from working.

A fundamental perspective in this chapter is that work area (workstation) design influences employee comfort, health, motivation, and performance. We will examine basic ergonomic considerations (principles, practices, concerns) that can be used to develop guidance for the design of work areas (workstations) for the use of computing and related IT such as PIDs. Some of today's technologies have the capability to interact directly with each other, sometimes without human intervention, while other times requiring human action. The wide myriad of environments of use, interaction schemes, and activities of use make specific guidance for PIDs complex and difficult. Thus, we will provide general guidance to reduce musculoskeletal stress and to enhance physical comfort.

ERGONOMIC PERSPECTIVE

Forty years ago, a person would interact with several different types of information sources and technologies when engaging in activities, such as work. The person might look at hardcopy documents, take notes with a pen and a paper tablet, use a fixed location telephone, talk face-to-face with colleagues, type on a typewriter, and perform many other tasks during the course of the working day. The diversity of activities led people to move actively around during the day. Then, thirty years ago, many people started to spend most of their workdays in sedentary work sitting in front of a computer terminal. This type of human-machine system led to restricted physical movement and employee attention directed toward the computer monitor. Today there are mixed exposures in a multidimensional environment where some people are still at fixed sedentary workstations, while others are interacting with multiple technologies that have different physical characteristics. In some instances, people are again on the move and not in fixed, sedentary situations. Movement is good for the muscles while sedentary sitting is not if it lasts for too long. However, the interaction with technology while moving may create dynamic postural loading and the awkward application of perceptual-motor skills when making inputs to technology or when reading and listening to displays. For example, you might be walking through the airport while checking your email on your cell phone.

In Sweden, since the initial work of Hultgren and Knave (1973), Ostberg (1975), and Gunnarsson and Ostberg (1977), thousands of research studies from every corner of the globe have examined the working conditions of computer users and their associated health complaints. There have been several international conferences devoted to these issues starting in 1980, and conferences on these issues are already programmed through 2010. The findings from this research and the meetings have generally indicated that poor ergonomic and task design conditions are associated with large numbers of computer users complaining about visual discomfort, musculoskeletal discomfort and pain, and psychological distress. In fact, if these adverse ergonomic conditions are repeated daily over a long period, more or less chronic aches and pains of the upper extremities and back can occur, and may involve not only muscles but also other soft tissues, such as tendons and nerves. Long lasting, adverse ergonomic conditions may lead to a deterioration of joints, ligaments, and tendons. Reviews of field studies for fixed computer workstations (Bergquist, 1984; Bergquist et al., 1992; Smith, 1984, 1987, 1997; Grandjean, 1979, 1987; Hagberg et al., 1995; Bernard, 1997; Carayon, Smith, & Haims, 1999), as well as general experience, have shown that these conditions may be associated with a higher risk of

1. Inflammation of the joints
2. Inflammation of the tendon-sheaths
3. Inflammation of the attachment-points of tendons
4. Symptoms of degeneration of the joints in the form of chronic arthroses
5. Painful muscles
6. Disc troubles
7. Peripheral nerve disorders

The relationship between the user and technology is reciprocal. It is a system where the constraints of one element affect the performance of the other. Various aspects of the system, such as the task requirements, the work demands, the environment, and the workstation, all influence how effectively and comfortably the technology can be used by the person (Smith & Sainfort, 1989; Smith & Carayon, 1995; Carayon & Smith, 2000). One of the consequences of this is that the design of the workspace limits the nature and effectiveness of the interaction between the person and the technology. For instance, inadequate space for carrying out physical activities can lead to constrained postures, which together with long lasting static loading produces discomfort in the back and shoulders. This leads to symptoms of tiredness, muscle cramps, and pain. In addition, heavy workload, chronic repetition, and other biomechanical strains can cause similar problems. These adverse postural, repetition, and workload exposures lead to reduced performance and productivity, and in the long run, they may also affect employee well being and health.

Today, people carry their computing and communications with them and engage in activities in any available work area. The potential for serious ergonomic risks is high for these new modes of using technology. We will first examine problems and design considerations for fixed workstations where PCs and laptops are used. Then we will look at PIDs.

STUDIES ON THE DESIGN OF FIXED WORKSTATIONS

Workstation design is a major element in ergonomic strategies for improving user comfort and particularly for reducing musculoskeletal problems. Often, the task requirements will have a role in defining the layout and dimensional characteristics of the workstation. The relative importance of the screen, input devices, and hardcopy (e.g., source documents) depends primarily on the task, and this can influence the design considerations necessary to improve operator performance, comfort, and health. Many studies have examined fixed computer workstation design issues and generally have found that the quality of the fit between the user and the interactive devices being used has a substantial role in user musculoskeletal comfort and musculoskeletal pain (Smith, Carayon, & Cohen, 2003). For instance, Cohen, James, Taveira, Karsh, Scholz, and Smith (1995) identified the following working conditions that led to awkward posture and undue loads on the musculoskeletal system:

1. Static postures of the trunk neck and arms
2. Awkward twisting and reaching motions
3. Poor lighting and glare
4. The placement of the keyboard on uneven working surfaces
5. Insufficient work surface space
6. Insufficient knee and toe space
7. The inability for the chair armrests to fit under the working surfaces

8. Chairs with poor back and shoulder support, inadequate padding in the backrest and seat pan, arm rests that did not fit under working surfaces, and a lack of appropriate seat pan height adjustment

Based on this and several other studies, ergonomics experts have proposed guidance for designing fixed computing workstations. Grandjean (1984) proposed the consideration of the following features of fixed computer workstation design:

1. The furniture should be as flexible as possible with adjustment ranges to accommodate the anthropometric diversity of the users.
2. Controls for workstation adjustment should be easy to use.
3. There should be sufficient knee space for seated operators.
4. The chair should have an elongated backrest with an adjustable inclination and a lumbar support.
5. The keyboard should be moveable on the desk surface. (Today we might say the input devices should be able to be positioned at several locations on the working surfaces.)

Derjani-Bayeh and Smith (1999; Smith & Derjani-Bayeh, 2003) conducted a prospective intervention study to examine the benefits of ergonomic redesign for computer users at fixed workstations. The study took place in a consumer products call center where shoppers could order products from a catalog using a telephone or an Internet site. There were three ergonomic interventions studied. In the first condition, ergonomics experts provided modifications to current workstation configurations to maximize their fit with the incumbent employee. In the second condition, new workstation accessories (keyboard tray, monitor holder, document holder, wrist rest, footrest, and task lighting) were added as needed to improve the employees' fit with their workstations and general environment. In the third condition, the same factors as the second condition were added but in addition, a new chair with multiple adjustments was also added.

Eighty volunteer subjects participated. They were drawn from a larger pool of volunteers. The participants for the third condition were randomly selected from the larger pool, and then subjects for conditions one and two were matched to these selections based on type of job, age, gender, and length of experience with the company. Baseline measurements of self-reported health status were collected using a questionnaire survey. Follow-up measurements were taken directly after implementation of the ergonomic improvements and then 12 months later. In addition, productivity measurements were obtained for each participant and a control group of approximately 375 employees in the same departments. The results indicated that subjects working under conditions two and three showed reductions in the extent and intensity of musculoskeletal health complaints, but not the subjects in condition one. However, the subjects in condition one showed greater average improvement in productivity than in conditions two and three and to the control group receiving no treatment. Of importance to designers was the finding that not all subjects showed improved productivity with ergonomic improvements. In fact about one-half of the subjects showed reduced productivity with the ergonomic improvements, even though the overall average for the ergonomic improvements showed a positive effect.

GUIDELINES FOR FIXED COMPUTER WORKSTATION DESIGN FEATURES

Based on the findings from several field studies and standards dealing with computer workstation design (ANSI/HFES-100, 1988; CSA, 1989; BSR/HFES-100, 2005) the following general guidance is proposed for the design of fixed computer workstations.

The recommended size of the work surface is dependent upon the tasks, the documents, and the technologies being used. The primary working surface (e.g., those supporting the keyboard, the mouse, the displays, and documents) should be sufficient to:

1. Allow the display screens to be moved forward or backward for a comfortable viewing distance for different employees with differing visual capabilities.
2. Allow a detachable keyboard and a detachable mouse (or other pointing device) to be placed in several locations on the working surface.
3. Allow source documents to be positioned for easy viewing and proper musculoskeletal alignment of the upper extremities and the back.
4. Additional working surfaces (e.g., secondary working surfaces) may be necessary to store, layout, read, write on, or manipulate documents, materials, or technologies (input devices, displays, computers, PIDs).
5. Provide adequate knee and legroom for repositioning movements while working.
6. The tabletop should be as thin as possible for better thigh and knee clearance.
7. Establish a comfortable table height that provides the necessary thigh and knee clearance, and which allows input devices (keyboard, mouse) to be at comfortable heights. Adjustable tables provide the opportunity to better fit the users.

Sometimes computer workstations are configured so that multiple pieces of equipment and source materials can be equally accessible to the user. In this case, additional working surfaces are necessary to support these additional tools and should be arranged to allow for easy movement from one surface to another. Proper clearances under each working surface should be maintained, as well as a comfortable height.

It is important to provide unobstructed room under the working surface for the feet and legs so that operators can easily shift their posture. Knee space height and width and toe depth are the three key factors for the design of clearance space under the working surfaces. The recommended minimum width for leg clearance is 51 cm, while the preferred minimum width is 61 cm (ANSI/HFS-100, 1988). The BSR/HFES-100 (2005) guidance calls for minimum width of leg clearance of 50 cm. The minimum depth under the work surface from the operator edge of the work surface should be 38 cm for clearance at the knee level and 59 cm at the toe level (ANSI/HFS-100, 1988). The BSR/HFES-100 (2005) recommends a minimum depth at the extended leg toe level of 60 cm. A good workstation design accounts for individual body sizes and often exceeds minimum clearances to allow for free postural movement.

Table height has been shown to be an important contributor to computer user musculoskeletal problems (Hünting, Läubli, & Grandjean, 1981; Grandjean, Nishiyama, Hünting & Pidermann, 1982; Grandjean, Hünting & Nishiyama, 1982, 1984; Grandjean, Hünting, & Pidermann, 1983). Normal desk height of 30 inches (76 cm) is often too high for keyboard and mouse use by most people. It is desirable for table heights to vary with the height of the user, particularly if the chair is not height adjustable. Height adjustable working surfaces are effective for this. Adjustable multisurface tables encourage good posture by allowing the keyboard, mouse, and displays to be independently adjusted to appropriate keying, pointing, and viewing heights for each user and each task. Tables that cannot be adjusted easily can be a problem when used by multiple users of differing sizes, especially if the chair is not height adjustable. When adjustable tables are used, the ease of making the adjustments is essential. Adjustments should be easy to make, and operators should be instructed on how to adjust the workstation to be comfortable.

Specifications for seated working surfaces' heights vary with whether the table is adjustable or at one fixed height, and with a single working surface or multiple working surfaces. The proper height for a nonadjustable working surface is about 70 cm (27.5 inches) from the floor to top of the working surface. However, it must be emphasized that fixed height table surfaces fit only a portion of the user population. In the absence of a height adjustable chair, such fixed height table surfaces will produce a misfit in a portion of the users that could influence musculoskeletal comfort and pain.

The BSR/HFES-100 (2005) guidance requires that all surfaces supporting input devices must be height adjustable (or a combination of height and tilt adjustable). In addition, the input-device support surfaces should provide for adjustment fore and aft and side-to-side if possible.

Adjustable tables allow vertical adjustments of the keyboard (and mouse) and displays. Some allow for the independent adjustment of the keyboard and display. For a single adjustable seated working surface, the working-surface height adjustment range should be between 60 cm to 80 cm (23.5–31.5 inches). The BSR/HFES-100 (2005) guidance is for an adjustment range of 56 cm to 72 cm (22.0–28.5 inches). In the case of independently adjustable working surfaces for the keyboard and the screen, the appropriate height ranges are 59 cm to 71 cm (23.0–28.0 inches) for the keyboard surface (ANSI/HFS-100, 1988), and 90 cm to 115 cm (35.5–45 inches) for the screen surface (Grandjean, 1987).

The BSR/HFES-100 (2005) guidance indicates that the worktable surfaces should provide for tilt from −15 degrees to +20 degrees if possible.

For standing only workstations, the BSR/HFES-100 (2005) recommends a table surface height adjustment range of 95 cm to 118 cm (37.5–46.5 inches). However, if both height and tilt surface adjustments are included, then the range of height adjustment recommended is from 78 cm to 118 cm (30.5–46.5 inches) with tilt adjustment ranges from −45 degrees to +20 degree. For sit and stand workstations, the BSR/HFES-100 recommendation for height adjustment of the input device surfaces is between 56 cm to 118 cm (22.0–46.5 inches).

THE DESIGN OF WORKSTATIONS FOR USE WITH LAPTOP COMPUTERS

We will start with the laptop as a prime example of the influence of portability and efficiency, and then we will move on to other portable IT devices. Recently the Human–Computer Interaction Committee of the International Ergonomics Association (IEA) produced a guideline for the use of laptop computers to improve ergonomic conditions (Saito et al., 2000). This was prompted by the ever-increasing sales of laptops and the replacement of fixed PCs by portable laptops on fixed working surfaces.

The primary advantage of the laptop is easy portability so the user can take the computer anywhere to do work. He or she can use it at the office, take it home to finish work, or take it with him or her to make a presentation at a meeting. The convenience and effectiveness of easy, lightweight portability are very high. In addition, all of the files are with the laptop, so nothing is mistakenly left behind at the office (or home). However, the comfort and health factors can be very low if the person uses the laptop in all manner of environments, workstations, and tasks that diminish the consistent application of good ergonomic principles. An important feature of the IEA laptop guideline (Saito, 2000) is to encourage situations of use that mirror the best practices of ergonomic conditions for fixed computer workstations.

Work Environment and Workstation Layout, "Create an environment that fits your work"

1. Use your laptop in a proper environment (lighting, temperature, noise, and so on). In particular, make sure the work area is neither too bright nor too dark.
2. Allocate enough space on your desk when placing a laptop.

Chair and Desk, "Adjust chair height to match your physique"

1. Adjust your chair height based on the height of the keyboard, such that your forearm is parallel to the surface of the keyboard.
2. If your feet do not lie flat on the floor, provide a footrest.
3. Provide enough space underneath the desk.

Keyboard, "Set the keyboard to a desirable angle, and use a palm rest if necessary"

1. Adjust the angle of the keyboard based on your posture and preferences.
2. Make sure there is space in front of the keyboard for you to rest your wrists comfortably (this space can be on the desktop surface itself if the keyboard is thin).
3. If the keyboard seems difficult to use, use an external keyboard.

Working Posture, "Avoid unnatural postures, and change your posture occasionally"

1. Avoid staying in postures where you are bent too far forward or backward, or twisted, for an extended duration
2. Laptop users tend to view the display from too close, so make sure you maintain a distance of at least 40–50 cm between the display and your eyes.
3. Alternate near vision with far vision (i.e., observe object located at least 6 m far) as frequent as possible.
4. Make sure your wrists are not at an unnatural angle.

Nonkeyboard Input Devices, "Use a mouse as your pointing device if at all possible"

1. If a mouse can be connected to your laptop, then do so as often as possible. Use a mouse pad whenever you use a mouse.
2. When you cannot connect a mouse, make sure you understand the built-in pointing device, and use the pointing device appropriately (Saito et al., 2000, pp. 421–434).

These laptop guidelines provided by Saito et al. (2000) are useful when the laptop is used as a fixed PC at a docking station or at a desk (worktable). However, they do not provide as much help in the situations where there is no fixed workstation. We will describe some situations below where this can occur.

Imagine sitting at the airport, and your flight has been delayed for two hours. Your have your laptop with you, so you decide the get some work done while you wait. You could rent a cubicle or kiosk at the airport that would provide you with a high-speed Internet connection, a stationary telephone, a working surface (desk or table), a height adjustable chair, and some privacy (noise control, personal space). The characteristics of these work areas do not often conform to the best principles of ergonomic design. It is likely that the cubicle will provide some improvement over sitting with the laptop on your lap, but the characteristics may not meet the recommendations presented in this chapter. Such situations are acceptable for short exposures of up to 60 minutes, but longer exposures may lead to musculoskeletal discomfort, pain, and injury (if chronic). Now imagine you have been told to stay in the boarding area because the departure may be sooner than two hours. You get out your laptop, connect it to your cell phone, and place them on your lap. (That is why they are called laptops.) You are sitting in a nonadjustable chair with poor back support.

This scenario is all too common. You can walk through O'Hare International Airport on any given day and see hundreds of people sitting at the boarding gate working on laptops that are sitting on their laps. Now imagine a palm held device that allows you to access your e-mail or to connect to the Internet. This device can be operated while you are standing in line at the airport to check in, or sitting at the boarding gate like the laptop users. You can stand or sit punching at miniature buttons (sometimes with a stylus because they are so small) and interact with the interconnected world. Again, this scene is all too familiar in almost any venue (e.g., airport, restaurant, street, office).

In situations where there is not a fixed workstation, the device is typically positioned wherever it is convenient. Very often such positioning creates bad postures for the legs, back, shoulders, arms, wrists/hands, or neck. In addition, the smaller dimensions of the manual input devices (touch pad, buttons, keyboard, joystick, roller ball) make motions much more difficult, and these often produce constrained postures and/or the use of too much force to operate the device. If the devices are used continuously for a prolonged period (such as one hour or more), muscle tension builds up and discomfort in joints, muscles, ligaments, tendons, and nerves can occur. Some devices use voice and audio interfaces and can be used when you are walking (or even running). These might be headsets with earplugs and a microphone. These devices put additional load on the neck, and the voice and audio interfaces can strain the voice or the ears.

We know very little about the discomfort, health effects, and psychosocial effects of these new strains. However, it is intuitive that frequent use of these interfaces will lead to strain and potential adverse consequences.

To reduce the undesirable effects of the poor workstation characteristics that lead to the discomfort the following recommendations are given:

1. If you are using a laptop on your lap, find a work area where you can put the laptop on a table (rather than on your lap). Then arrange the work area as closely as possible with the recommendations presented above in the IEA laptop guidelines (Saito, 2000).
2. If you are using a handheld PID, you should position yourself so that your back is supported. It is preferable to use the device sitting down. Of course, if you are using the PID as you are walking, then this is not possible. If the PID has a voice interface, then use an earpiece and a microphone so that you do not have to hold it with your hand. Be sure not to overuse the device such that your voice and ears are strained. Take frequent breaks from use to provide recovery for your voice and ears.
3. Never work in poor postural conditions for more than 30 minutes continuously. Take at least a 5-minute break (preferably 10 minutes) away from the laptop/PID use, put the device down (away), get up and stretch for 1 minute or more, and then walk for 2–3 minutes (unless you are already walking in which case you should sit down for 5 minutes). If you are using a handheld PID in a standing position, then during your break put it away, do 1 minute of stretching, and then sit down for 4 minutes. That may mean sitting on the floor, but preferably, you will sit where you can support your back (against a wall, or a seat back).
4. Buy equipment that provides the best possible input interfaces and displays (screens, headphones, typing pads). Since these devices are small, the perceptual motor requirements for their use are much more difficult (sensory requirements, motion patterns, skill requirements, postural demands, and force demands). Therefore, screens should provide easily readable characters (large, understandable), and input buttons should be easy to operate (large, properly spaced, easily accessible, low force).
5. Only use these devices when you do not have access to fixed workstations that have better ergonomic characteristics. Do not use these devices continuously for more than 30 minutes.

PIDS, CARRY ALONG, AND WEARABLE IT

We have already discussed some ergonomic issues of using PIDs. We can foresee an even greater potential for ergonomic concerns with the use of PIDs than with laptop computers. PIDs are made to be as small and light as possible for portability and convenience of carrying. This is good as the lower weight produces smaller loads of the body. Small sized devices are more difficult to manipulate with the hands, and to observe displays with the eyes. This has led many designers to emphasize verbal, auditory, and haptic interfaces that do not require substan-

tial manipulation by the hands or good vision (Hirose & Hirota, 2005; HCII2005, 2005). In fact, several new PIDs have the capability to communicate with other IT devices without human intervention (HCII2205, 2005).

Now, we will explore some current applications of PIDs and some possible future applications to see where ergonomic and workstation issues might emerge. Millions and millions of cell phones are in use worldwide, and cell phones are a good representative of the PID. There is virtually nowhere in the world where cell phones cannot be used (with a few exceptions), and there is virtually no one (with adequate finances) who cannot access a cell phone and connect with the world. Cell phones can have many capabilities including telephoning, e-mailing, texting, auditory streaming, internet surfing, walkie-talkie communicating, photographing, video camera picturing and recording, and television/cable/satellite broadcast receiving. Even with all of these built in features, the size of cell phones is shrinking. As cell phones shrink, the manual, hand interfaces are getting smaller, as are the visual displays. With their small size cell phone can be carried easily and can be used when a person is walking, sitting, running, lying down, or hanging upside down. What are the ergonomic concerns with their use, and in particular the workstation design issues?

Small manual interfaces and displays make the accurate and comfortable application of perceptual-motor skills difficult (Albers & Kim, 2002; Hinckley, 2003; Haggerty & Tarasewich, 2005; Myers & Wobbrock, 2005). A cell phone can be held in one hand (this hand becomes the workstation) and then be manipulated with the other hand. In some instances, the hand holding the cell phone is also used to manipulate the manual interface. Either of these situations leads to workstation conditions where the users cannot apply their highest level of perceptual-motor skills. For example, they cannot use both hands for inputting into the interface, the device is held at an awkward angle for manual inputting, or the posture of the trunk is unstable which limits the capacity to use the hands effectively.

Due to this, a preferred form of input to control the action of the PID is speech (Gamm & Haeb-Umback, 1995; Karat, Vergo, & Nabamoo, 2003), while the displays are typically a combination of visual and auditory information. These interfaces may lead to ergonomic problems of overuse of the voice, increased duration of mental concentration, increased eyestrain, visual discomfort, and increased error rates of the communication with the devices. No body of research data tells us if these problems are or will become prevalent among PID users or if there will be long-term comfort, health, and performance effects. For now, we can only conjecture about the possibilities for problems. It is clear that many people are using PIDs and other computer-based technologies many hours per day, and this increased extent of use will likely lead to discomfort and some health effects for the voice, ears, upper extremities, and mental stress (Eklof, Gustafsson, Hagberg, & Thomee, 2005).

Now we will discuss workstation issues with PIDs, using the cell phone as an example. We will go back to the airport and I will use my cell phone to communicate with my office. At this time, I am standing in a long line at the airline check in counter. My cell phone has voice activation and control capability. I ask my cell phone for my e-mail service provider, and my e-mails come up on the display screen. I am using my thumb to scroll through the e-mails one at a time. I hold the cell phone close to my face to enhance my ability to read the display screen. I shuffle forward as the line moves toward the airline counter (automated check in station). I am standing, and I have no postural support for my back, buttocks, and legs. It is true that I would be standing in line feeling the postural strain even if I were not using my cell phone. However, my cell phone use adds extra postural loading, since I am manually (vocally) manipulating the interface to operate the e-mail processing. The small visual display may create eyestrain, and if I talk on the cell phone for too long, it may create voice strain.

One way to reduce the load would be if I had a workstation where I could sit and support my back, buttocks, legs, and even my arms and elbows as I manually manipulate the interface. I could use a vocal and auditory interface with a head mounted earplug and microphone (headset), and thus eliminate the manual input and visual display viewing. The headset is a workstation improvement that reduces the manual manipulation and some of the postural loading on the upper extremities and back. The headset adds some weight to the head, and this increases loading on the neck, shoulders, and back. One difficulty with using an auditory interface is the high ambient noise in the airport lounge that interferes with and masks the cell phone's auditory signals. To reduce this auditory interference, a helmet with acoustical privacy could serve as a workstation improvement. However, a helmet adds substantial weight to the head, which puts increased loading on the neck, shoulders, and back. A major workstation improvement for users waiting in line, that would reduce the postural loading when using a cell phone, is to provide chairs so a user can sit down. A chair provides postural support for the back, buttocks, and legs. A chair on casters or wheels that can be scooted along as the line moves forward could be beneficial. However, this could increase the congestion in the lounge area as people are moving through the line. This example illustrates the new loads that may be added to technology users' lives by the use of PIDs. We will reiterate some of the advice from the laptop section as this is also applicable to the use of PIDs and other portable devices.

To reduce the undesirable effects of a lack of a workstation that can lead to the users' discomfort the following recommendations are given:

1. If you are using a laptop on your lap, find a work area where you can put the laptop on a table (rather than on your lap). Then arrange the work area as closely as possible with the recommendations presented above in the IEA laptop guidelines (Saito, 2000).
2. Never work in poor postural conditions for more than 30 minutes continuously. Take at least a 5-minute break (preferably 10 minutes) away from the laptop/PID use, put the device down (away), get up and stretch for 1 minute or more, and then walk for 2–3 minutes (unless you are already walking in which case you should sit down for 5 minutes). If you are using a handheld PID in a standing position, then during your break put it away, do 1 minute of stretching, and then sit down for 4 minutes. That may mean sitting on the floor, but preferably, you will sit where you can support your back (against a wall, or a seat back).

3. Buy equipment that provides the best possible input interfaces and displays (screens, headphones, typing pads). Since these devices are small, the perceptual motor requirements for their use are much more difficult (sensory requirements, motion patterns, skill requirements, postural demands, and force demands). Therefore, screens should provide easily readable characters (large, understandable), and input buttons should be easy to operate (large, properly spaced, easily accessible, low force).
4. Do not use these devices continuously for more than 30 minutes.

THE CHAIR AS A CRITICAL ELEMENT OF THE WORKSTATION

It was not until the last 40 years that sitting posture and chairs (seats) became topics for scientific research, especially for ergonomics and orthopedics. Studies have revealed that the sitting position, as compared to the standing position, reduces static muscular efforts in legs and hips, but increases the physical load on the intervertebral discs in the lumbar region of the spine.

The debate over what constitutes proper-seated posture is not yet fully resolved. Is an upright-seated posture most healthy, or is a relaxed posture with a backward-leaning trunk healthier? Interesting experiments by the Swedish surgeons Nachemson and Elfstrom (1970) and Andersson and Ortengreen (1974) offered some guidance about this. These authors measured the pressure inside the intervertebral discs as well as the electrical activity of the back muscles in relation to different sitting postures. When the backrest angle of the seat was increased from 90 to 120 degrees, subjects exhibited an important decrease of the intervertebral disc pressure and of the electromyographic activity of the back. Since heightened pressure inside intervertebral discs means that they have more stress, it was concluded that a sitting posture with reduced disc pressure is more healthy and desirable.

The results of the Swedish studies indicated that leaning the back against an inclined backrest transfers some of the weight of the upper part of the body to the backrest. This reduces considerably the physical load on the intervertebral discs and the static strain of the back and shoulder muscles. Thus, some computer users seem to instinctively get into the proper posture when they leaning backwards.

Most ergonomic standards for computer workstations are based on a more traditional view about a healthy sitting posture. Mandal (1982) reported that the "correct seated position" goes back to 1884 when the German surgeon Staffel recommended the well-known upright position. Mandal stated:

But no normal person has ever been able to sit in this peculiar position (upright trunk, inward curve of the spine in the lumbar region and thighs in a right angle to the trunk) for more than 1–2 minutes, and one can hardly do any work as the axis of vision is horizontal. Staffel never gave any real explanation why this particular posture should be better than any other posture. Nevertheless, this posture has been accepted ever since quite uncritically by all experts all over the world as the only correct one.

The sitting posture of students in the lecture hall or of any other audience is very seldom a "correct upright position of the trunk." On the contrary, most people lean backward (even with unsuitable chairs) or, in some cases, lean forward with elbows resting on the desk. These two preferred trunk positions are probably associated with a substantial decrease of intervertebral disc pressure, as well as lessened tension of muscles and other tissues in the lumbar and thoracic spine (Nachemson & Elfstrom, 1970; Andersson & Ortengreen, 1974). Thus, sitting in general and particularly when using a computer is probably most comfortable and healthy when leaning back, or when the arms are supported while leaning forward.

Poorly designed chairs can contribute to computer user discomfort. Chair adjustability in terms of height, seat angle, and lumbar support helps to provide trunk, shoulder, neck, and leg postures that reduce strain on the muscles, tendons, and discs. The postural support and action of the chair help maintain proper-seated posture and encourage good movement patterns. A chair that provides swivel action encourages movement while backward tilting increases the number of postures that can be assumed.

The chair height should be adjustable so that the computer operator's feet can rest firmly on the floor with minimal pressure beneath the thighs. The minimum range of adjustment for seat pan height should be between 38 and 56 cm (15–22 inches) to accommodate a wide range of stature (BSR/HFES, 2005).

To enable short users to sit with their feet on the floor without compressing their thighs, it may be necessary to add a footrest. A well-designed footrest has the following features:

1. It is inclined upwards slightly (about 5–15 degrees).
2. It has a nonskid surface.
3. It is heavy enough that it does not slide easily across the floor.
4. It is large enough for the feet to be firmly planted.
5. It accommodates persons of different stature.

The seat pan is where the person sits on the chair. This part of the chair directly supports the weight of the buttocks. The seat pan should be wide enough to permit operators to make slight shifts in posture from side to side. This not only helps to avoid static postures, but also accommodates a large range of individual buttock sizes. The seat pan should not be overly U-shaped because this can lead to static sitting postures. The minimum seat pan width should be 46 cm (18 inches), and the depth between 38 and 43 cm (15–17 inches) (BSR/HFES, 2005; ANSI/HFES-100, 1988). The front edge of the seat pan should be well rounded downward to reduce pressure on the underside of the thighs that can affect blood flow to the legs and feet. This feature is often referred to as a "waterfall" design. The seat needs to be padded to the proper firmness that ensures an even distribution of pressure on the thighs and buttocks. A properly padded seat should compress about one-half inch to one inch when a person sits on it.

Some experts feel that the seat front should be elevated slightly (up to seven degrees), while others feel it should be lowered slightly (about five degrees). There is some disagreement among the experts about the correct answer. Due to this disagreement, many chairs allow for both front and backward angling of the front edge of the seat pan. The operator can then angle the chair's front edge to a comfortable position. The BSR/HFES (2005) guidelines recommend a user adjustable range of at least six degrees of forward and backward adjustment with at reclined position of three degrees. The seat pan height and angle adjustments should be accessible and easy to use from a seated position.

The tension and tilt angle of the chair's backrest should be adjustable. Inclination of chair backrest is important for operators to be able to lean forward or back in a comfortable manner while maintaining a correct relationship between the seat pan angle and the backrest inclination. A backrest inclination of about 110 degrees is considered an appropriate posture by many experts. However, studies have shown that operators may incline backwards as much as 125 degrees, which also is an appropriate posture. Backrests that tilt to allow an inclination of up to 125 degrees are therefore a good idea. The backrest tilt adjustments should be accessible and easy to use. An advantage of having an independent tilt angle adjustment is that the backrest tilt will then have little or no effect on the front seat height or angle. This also allows operators to shift postures readily. BSR/HFES (2005) recommends backrest adjustment of up to 105 degrees must be achievable with up to 120 degrees preferred.

Chairs with high backrests are preferred since they provide support to both lower back and the upper back (shoulder). This allows employees to lean backward or forward, adopting a relaxed posture and resting the back and shoulder muscles. A full backrest with a height around 45–51 cm (18–20 inches) is recommended. BSR/HFES (2005) recommends a backrest be at least 45 cm (18 inches) high. To prevent back strain it is also recommended that chairs have lumbar (midback) support, since the lumbar region is one of the most highly strained parts of the spine when sitting. BSR/HFES (2005) recommends that the width of the backrest should be at least 36 cm (14 inches).

For most computer workstations, chairs with rolling castors or wheels are desirable: They are easy to move and facilitate postural adjustment, particularly when the operator has to reach for equipment or materials that are on the secondary working surfaces. Chairs should have five supporting legs.

Another important chair feature is armrests. Both pros and cons to the use of armrests at computer workstations have been advanced. On the one hand, some chair armrests can present problems of restricted arm movement, interference with the operation of input devices, pinching of fingers between the armrest and table, restriction of chair movement, such as under the worktable, irritation of the arm or elbows due to tissue compression when resting on the armrest, and adoption of awkward postures. Properly designed armrests can overcome the problems mentioned above. Armrests can provide support for resting the arms to prevent or reduce arm, shoulder, and neck fatigue. Removable armrests are an advantage because they provide greater flexibility for individual operator preference. For specific tasks such as using a numeric keypad, a full armrest can be beneficial in supporting the arms. Many chairs have height adjustable armrests that are helpful for operator comfort, and some allow for adjusting the angle of the armrests as well.

ADDITIONAL WORKSTATION CONSIDERATIONS

Providing the capability for the screen to swivel and tilt up and down gives the user the ability to better position the screen for easier viewing. Reorientation of the screen around its vertical and horizontal axes can help to position a screen to reduce screen reflections and glare. Reflections can be reduced by simply tilting the display slightly back or down, or to the left or right away from the source of glare. The perception of screen reflections depends not only upon screen tilt, but also upon the operator's line of sight.

An important component of the workstation that can help reduce musculoskeletal loading is a document holder. When properly designed, proportioned, and placed, document holders reduce awkward inclinations of the head and neck and frequent movements of the head up and down and back and forth. They permit source documents to be placed in a central location at the same viewing distance as the computer screen. This eliminates needless head and neck movements and reduces eyestrain. In practice, some flexibility about the location, adjustment, and position of the document holder should be maintained to accommodate both task requirements and operator preferences. Dainoff (1982) showed the effectiveness of an inline document holder. The document holder should have a matte finish so that it does not reflect light.

Privacy requirements include both visual and acoustical control of the workplace. Visual control prevents physical intrusions, contributes to confidential/private conversations, and prevents the individual from feeling constantly watched. Acoustical control prevents distracting and unwanted noise (from machine or conversation) and permits speech privacy. While certain acoustical methods and materials, such as freestanding panels, are used to control general office noise level, they can also be used for privacy. Planning for privacy should not be made at the expense of visual interest or spatial clarity. For instance, providing wide visual views can prevent the individual from feeling isolated. Thus, a balance between privacy and openness enhances user comfort, work effectiveness, and office communications. Involving the employee in decisions of privacy can help in deciding the compromises between privacy and openness.

The use of a wrist rest when keying can help to minimize extension (backward bending) of the hand/wrist, but the use of a wrist rest for operator comfort and health has generated some debate because there are trade-offs for comfort and health. When the hand or wrist is resting on the wrist rest, there is compression of the tissue, which may create increased carpal canal pressure or local tissue ischemia. On the other hand, the wrist rest allows the hands and shoulders to be supported with less muscular tension, which is beneficial to computer operator comfort. At this time, there is no scientific evidence that the use of a wrist rest either causes or prevents serious musculoskeletal disorders of the hands, wrists, or shoulders. Thus, the choice to use a wrist rest should be based on employee comfort and performance considerations until scientific evidence suggests otherwise.

If used, the wrist rest should have a broad surface (five cm minimum) with a rounded front edge to prevent cutting pressure on the wrist and hand. Padding further minimizes skin compression and irritation. Height adjustability is important so that the wrist rest can be set to a preferred level in concert with the keyboard height and slope.

Arm holders are also available to provide support for the hands, wrists, and arms while keyboarding and have shown to be useful for shoulder comfort. The placement of the arm holder should not induce awkward postures in its use. The device should be placed within easy reach of the operator, especially when it will be used frequently during work.

When keyboard trays are used, they should allow for the placement of other input devices directly on the tray instead of on other working surfaces.

THE VISUAL ENVIRONMENT

Visual displays tend to be of two types, the cathode ray tube (CRT) and the diode matrix (LED/LCD) flat panels. Fixed workstations with personal computers tend to use CRTs because they are cheap, while laptop computers and hand-held devices are almost exclusively flat panel due to size and weight limitations. Both types of displays have characteristics that lead to problems from environmental influences. For instance, luminance sources in the environment that fall on the screen wash out characters on the screen, and the accumulation of dust particles on the screen may distort images. These conditions affect the ability to read the screen and can lead to visual fatigue and dysfunction. Specific characteristics of the environment, such as illumination and glare, have been related to computer-operator vision strain problems. The main visual functions involved in computer work are accommodation, convergence, and adaptation.

The alignment of lighting in relation to the computer workstation, as well as levels of illumination in the area surrounding a computer workstation, have been shown to influence the ability of the computer operator to read hardcopy and the computer screen (Cakir et al., 1979; Stammerjohn et al., 1981; Dainoff, 1983; Grandjean, 1987). Readability is also affected by the differences in luminance contrast in the work area. The level of illumination affects the extent of reflections from working surfaces and from the screen surface. Mismatches in these characteristics and the nature of the job tasks have been postulated to cause the visual system to overwork and lead to visual fatigue and discomfort (Cakir et al., 1979; NAS, 1983).

Generally, it has been shown that excessive illumination leads to increased screen and environmental glare, and poorer luminance contrast (Gunnarsson & Ostberg, 1977; Läubli, Hünting, & Grandjean, 1981; Ghiringhelli, 1980). Several studies have shown that screen and working surface glare are problematic for visual disturbances (Gunnarsson & Ostberg, 1977; Läubli et al., 1981; Cakir et al., 1979; Stammerjohn et al., 1981). Research by van der Heiden, Braeuninger, and Grandjean (1984) showed that computer users spend a considerable amount of their viewing time looking at objects other than the screen. Bright luminance sources in the environment can produce reflections and excessive luminance contrasts that may create excessive pupillary response, which leads to visual fatigue.

All surfaces within the visual field of an operator should be of a similar order of brightness as much as practical to achieve. The temporal uniformity of the surface luminance is as important as the static spatial uniformity. Rhythmically fluctuating surface luminance in the visual field is distracting and reduces visual performance. Such unfavorable conditions prevail if the work requires the operator to glance alternately at a bright and then a dark surface, or if the light source generates an oscillating light. Sometimes fluorescent lights can have a noticeable strobo- scopic effect on moving reflective objects, and sometimes when reflected from a computer CRT screen. When fluorescent tubes wear out or are defective, they develop a slow, easily perceptible flicker, especially at the visual periphery. Flickering light is ex- tremely annoying and causes visual discomfort.

Lighting is an important aspect of the visual environment that influences computer screen and hardcopy readability, glare on the screen, and viewing in the general environment.

The intensity of illumination, or the illuminance being mea- sured, is the amount of light falling on a surface. In practice, this level depends on both the direction of flow of the light and the spatial position of the surface being illuminated in relation to the light flow. Illuminance is measured in both the horizontal and vertical planes. At computer workplaces, both the horizontal and vertical illuminances are important. A document lying on a desk is illuminated by the horizontal illuminance, whereas the com- puter screen is illuminated by the vertical illuminance. In an of- fice that is illuminated from overhead luminaries, the ratio be- tween the horizontal and vertical illuminances is usually between 0.3 and 0.5. If the illuminance in a room is said to be 500 lux, this implies that the horizontal illuminance is 500 lux while the vertical illuminance is between 150 and 250 lux (0.3 to 0.5 of the horizontal illuminance).

The illumination required for a particular task is determined by the visual requirements of the task and the visual ability of the employees concerned. The illuminance in workplaces that use computer screens should not be as high as in workplaces that exclusively use hardcopy. Lower levels of illumination will provide better computer-screen image quality and reduced screen glare. Illuminance in the range of 300 to 700 lux mea- sured on the horizontal working surface (not the computer screen) is normally preferable. The lighting level should be set up according to the visual demands of the tasks performed. For instance, higher illumination levels are necessary to read hard- copy and lower illumination levels are better for work that just uses the computer screen. Thus, a job in which both a hardcopy and a computer screen are used should have a general work- area illumination level of about 500 to 700 lux; a job that re- quires reading only the computer screen would have a general work area illumination of 300 to 500 lux.

Conflicts can arise when both hardcopy and computer screens are used by different employees who having differing job task requirements or differing visual capabilities and are working in the same room. As a compromise, room lighting can be set at the lower level (300 lux) or intermediate level (500 lux) and additional task lighting for hardcopy tasks can be provided at each workstation as needed. Such additional lighting must be carefully shielded and properly placed to avoid glare and reflec- tions on the computer screens and adjacent working surfaces of other employees. Furthermore, task lighting should not be too bright in comparison to the general work area lighting, since the

contrast between these two different light levels may produce eyestrain.

The surface of the computer screen reflects light and images. The luminance of the reflections decreases character contrast and disturbs legibility; it can be so strong that it produces a glare. Image reflections are annoying, especially since they also interfere with focusing mechanisms; the eye is induced into fo- cusing between the text and the reflected image. Thus, reflec- tions are also a source of distraction. Stammerjohn et al. (1981), as well as Elias and Cail (1983), observed that bright reflections on the screen are often the principal complaint of operators.

Luminance is a measure of the brightness of a surface, the amount of light leaving the surface of an object, either reflected by the surface (as from a wall or ceiling), emitted by the sur- face (as from the CRT characters), or transmitted (as light from the sun that passes through translucent curtains). High inten- sity luminance sources (such as windows) in the peripheral field of view should be avoided. In addition, a balance among lumi- nance levels within the computer user's field of view should be maintained. To reduce environmental glare, the luminance ratio within the user's near field of vision should be approximately 1:3 and within the far field of vision should be approximately 1:10. For luminance on the screen itself, the character-to-screen background-luminance contrast ratio should be at least 7:1. To give the best readability for each operator, it is important to pro- vide screens with adjustments for character contrast and bright- ness. These adjustments should have controls that are obvious and easily accessible from the normal working position (e.g., located at the front of the screen).

Experts have traditionally recommended a viewing distance between the screen and the operator's eye of 45–50 cm, but no more than 70 cm. However, experience in field studies has shown that users may adopt a viewing distance greater than 70 cm and still be able to work efficiently and comfortably. Thus, viewing distance should be determined in context with other considerations. It will vary depending upon the task require- ments, computer screen characteristics, and individual visual ca- pabilities. For instance, with poor screen or hardcopy quality, it may be necessary to reduce viewing distance for easier charac- ter recognition. Typically, the viewing distance should be 50 cm or less because of the small size of the characters on the com- puter or IT device screen.

THE AUDITORY ENVIRONMENT

As the use of PIDs in a wide variety of environments increases, the need for a quiet auditory environment that allows the user to easily hear and speak at a normal loudness level becomes more important. Crowded, noisy environments detract from users' abilities to hear. Many PIDs have auditory interfaces, such as headphones, which can concentrate the primary auditory signal and block out some of the environmental noise. One concern is that users may increase the loudness beyond levels that are safe for the auditory sensory system (CDC, 2003). High ambient noise in the environment leads to the need for greater mental concentration (attention) to the auditory signals, increased intensity of the primary auditory signal (up to levels that may cause temporary auditory threshold shifts), and mood disturbances (irritation, anger, and discouragement; Crocker, 1997). These effects can result in increased mental fatigue and psychosocial stress, and long-term exposures may lead to discomfort and health consequences.

While headphones may provide benefits to concentrate the primary auditory signal, they also have the drawback of providing too much auditory energy to the ears. Prolonged exposure to load primary auditory signals can cause the adverse effects described above. To protect the auditory sensory system when using PIDs, we proposed the following:

1. Find environments to use your PID where the ambient auditory levels are 50 dba or less.
2. Do not use your PID for more than 30 minutes without taking a break where you can rest your eyes and ears. You should rest for at least 10 minutes before you start using your PID again. As your total use of the PID increases over the course of a day, the rest breaks should become longer. While the literature suggests that taking a sufficient break after 30 minutes of PID use should be sufficient for your eyes to recover (given a proper visual environment as defined above), the literature does not provide guidance on the maximum amount of time of PID use before a break is necessary to provide recovery for your ears and mental mood. Nor does the literature indicate the minimum amount of break time necessary to achieve auditory or mental recovery.
3. Keep the volume of the headphones of the PID at the lowest level necessary to hear the primary auditory signal properly. In no instance exceed 85 dba, and do not exceed 30 minutes of continuous exposure (see 2 above). It is best to not exceed 70 dba, and if you adhere to 1 above, you will not have to exceed 60–70 dba if you have normal hearing.
4. Be courteous when using your microphone or cell phone. Do not talk directly at other people such that your conversation may produce masking noise that interferes with their conversations. Keep your voice as low as possible that allows your interface or listener to understand your signal.

ERGONOMIC IMPLEMENTATION ISSUES

Implementing a workplace change such as improving workstation design or work methods is a complex process because it impacts many elements of the work system (Derjani & Smith, 1999; Smith & Carayon, 1995; Hagberg et al., 1995; Smith & Sainfort, 1989; Carayon & Smith, 2000; Smith & Derjani-Bayeh, 2003). Managers, designers, and engineers often like to believe that technological enhancements are easy to make, and that performance and health improvements will be immediate and substantial. Proper implementation involves changes in more than the workstation; for instance, there needs to be consideration of the work organization, job content, task improvements, job demands, and socialization issues. Planning for change can help the success of implementation and reduce the stress generated by the change. The success of implementing change depends heavily on the involvement and commitment of the concerned parties, in particular management, technical staff, support staff, first line supervision, and the employees.

There is universal agreement among change management experts that the most successful strategies for workplace improvements involve all elements (subsystems) of the work system that will be affected by the change (Hendrick, 1986; Lawler, 1986; Smith & Carayon, 1995; Carayon & Smith, 2000). Involvement assumes that there is an active role in the change process, not just providing strategic information. Active participation generates greater motivation and better acceptance of solutions than passively providing information and taking orders. Active participation is achieved by soliciting opinions and sharing authority to make decisions about solutions. However, one drawback of active participation is the need to develop consensus among participants who have differing opinions and motives. This usually takes more time than traditional decision making, and can bring about conflict among subsystems. Another drawback is that line employees often do not have the technical expertise necessary to form effective solutions.

Participative ergonomics can take various forms, such as design decision groups, quality circles and worker-management committees. Some of the common characteristics of these various programs are employee involvement in developing and implementing ergonomic solutions, dissemination and exchange of information, pushing ergonomics expertise down to lower levels, co-operation between experts and nonexperts. One of the characteristics of participatory ergonomics is the dissemination of information (Noro, 1991). Participative ergonomics can be beneficial to reduce or prevent resistance to change because of the information provided to the various members of the organization concerned with the new technology. Uncertainty and lack of information are two major causes of resistance to change and have been linked to increased employee stress. If employees are informed about potential ergonomics changes in advance, they are less likely to actively resist the change.

Training computer users about how the new workstation functions and operations are important especially if the adjustment controls are neither obvious nor intuitive. Hagberg et al. (1995) indicated that employee training is a necessary component to any ergonomic program for reducing work-related musculoskeletal disorders. Green and Briggs (1989) found that adjustable workstations are not always effective without appropriate information about benefits of adjustments and training in how to use the equipment. Hagberg et al. (1995) suggested the following considerations for ergonomics training programs:

1. Have employees involved in the development and process of training. Using employee work experiences can be helpful in illustrating principles to be learned during training. In addition, using employees as instructors can be motivational for the instructors and learners.
2. Use active learning processes where learners participate in the process and apply hands-on methods of knowledge and skill acquisition. This approach to learning enhances acquisition of inputs and motivation to participate.
3. Apply technology, such as audio-visual equipment and computers, to illustrate principles. Much like active processes,

technology provides opportunities for learners to visualize the course materials and to test their knowledge dynamically and immediately.

4. Use of on-the-job training is preferred over classroom training. Either can be effective when used together.

SUMMARY OF RECOMMENDATIONS

A computer workstation (work area) is comprised of the computer, input and output interfaces, the furniture where the computer is used, and the physical environment in which the computer is used. The design of these elements and how they fit together play a crucial role in user performance and in minimizing potential adverse discomfort and health consequences. The recommendations presented in this chapter address the physical environment and implementation issues. Important consideration should also be given to organizational factors and task related factors as they affect and depend on the individual. Unique situational factors need to be considered when an ergonomic intervention is being implemented. Generalizing these recommendations to all situations in which computers or PIDs are used is a mistake. The approach presented in this chapter emphasizes the adaptation of the work area to a user's needs so that sensory, musculoskeletal, and mental loads are minimized.

Positioning of computers in the workplace is important in order to provide a more productive work environment. Computers workstations should be placed at right angles to the windows, and windows should not be behind or in front of the operator. This will reduce the possibility of reflections on the screen, which can otherwise reduce legibility. In addition, bright reflections coming from light sources can be reduced by placing these light sources on either side of or parallel to the line of vision of the operator.

Moreover, illumination should be adapted to the quality of the source documents and the task required. This is done by having a high enough illumination to enhance legibility, yet low enough to avoid excessive luminance contrasts. Recommended levels range between 300 and 700 lux.

The keyboard and other input devices should be movable on the work surface, but stable when in use. A wrist rest or support surface of 15 cm in depth is recommended to rest the wrist and forearms. Computer workstations should allow for adjustments that promote good postures. A computer workstation without adjustable keyboard (input device) height and without adjustable height and distance of the screen is not reasonable for continuous work of more than 30 minutes. The controls for adjusting the dimensions of a workstation and chair should be easy to use. Such adjustability is particularly important at workstations and chairs used by more than one employee. Furthermore, sufficient space for the user's thighs, knees, and legs should be provided to allow for comfort and to avoid unnatural or constrained postures.

With regard to chair design, a backward-leaning posture allows for relaxation of the back muscles and decreases the load on the intervertebral discs. Chair seat pan height should be easily adjustable and fit a wide range of statures. The chair should have a full backrest that can incline backward of up to 130 degrees. In addition, the backrest should have a lumbar support and a slightly concave form at the thoracic level.

PIDs pose unique problems and ergonomic sense indicates that users need to take actions that will minimize the loads on their musculoskeletal system, sensory systems, and mental processes. PIDs have inherent ergonomic problems due to their ubiquitous applications, small size, and potential for users' over exposures to prolonged use. Using PIDs in the proper environment, limiting the extent of use by taking frequent rest breaks, and having the best possible interfaces is good ergonomic sense.

These recommendations are not exhaustively inclusive of all aspects of computer (and PID) workstation (environment) design. The chapter described other important workstation criteria in various sections.

References

Albers, M., & Kim, L. (2002). Information design for small-screen interface: An overview of web design issues for personal digital assistants. *Technical Communication, 49*(1), 45–60.

Andersson, B. J. G., & Ortengreen, R. (1974). Lumbar disc pressure and myoeletric back muscle activity. *Scandinavian Journal of Rehabilitation Medicine, 3*, 115–121.

ANSI/HFES. (1988). *American national standard for human factors engineering of visual display terminal workstations* (ANSI/HFS Standard No. 100-1988). Santa Monica, CA: The Human Factors Society.

Bergqvist, U., Knave, B., Voss, M., & Wibom, R. (1992). A longitudinal study of VDT work and health. *International Journal of Human–Computer Interaction, 4*(2), 197–219.

Bergqvist, U. O. (1984). Video display terminals and health: A technical and medical appraisal of the state of the art. *Scandinavian Journal of Work, Environment and Health, 10*(Suppl. 2), 87.

Bernard, P. B. (1997). *Musculoskeletal disorders and workplace factors DHHS.* (NIOSH) Washington, DC: National Technical Information Service. (NTIS No. 97-141)

BSR/HFES. (2005). *Human factors engineering of computer workstations* (BSR/HFES-100), Santa Monica, CA: The Human Factors and Ergonomics Society.

Cakir, A., Hart, D. J., & Stewart, T. F. M. (1979). *The VDT manual.* Darmstadt: Inca-Fiej Research Association.

Carayon, P., Smith, M. J., & Haims, M. C. (1999). Work organization, job stress, and work-related musculoskeletal disorders. *Human Factors, 41*, 644–663.

Carayon, P., & Smith, M. J. (2000). Work organization and ergonomics. *Applied Ergonomics, 31*, 649–662.

Centers for Disease Control. (2003). Third national health and nutrition examination survey (NHANES lll). Atlanta: Author.

Cohen, W. J., James, C. A., Taveira, A. D., Karsh, B., Scholz, J., & Smith, M. J. (1995). Analysis and design recommendations for workstations: A case study in an insurance company. *Proceedings of the Human Factors and Ergonomics Society 39th Annual Meeting, 1, San Diego,* 412–416.

Crocker, M. J. (1997). Noise. In G. Salvendy (Ed.), *Handbook of human factors and ergonomics* (2nd ed., pp. 790–827). New York: John Wiley & Sons, Inc.

Canadian Standards Association. (1989). *A guideline on office ergonomics: A national standard of Canada* (Document No. CAN/CSA-Z412-m89). Rexdale, Ontario: Author.

Dainoff, M. J. (1982). Occupational stress factors in visual display terminal (VDT) operation: A review of empirical research. *Behaviour and Information Technology, 1*(2), 141–176.

Dainoff, M. J. (1983, December). Video display terminals: The relationship between ergonomic design, health complaints and operator performance. *Occupational Health Nursing, 12*, 29–33.

Derjani-Bayeh, A., & Smith, M. J. (1999). Effect of physical ergonomics on VDT workers' health: A longitudinal intervention field study in a service organization. *International Journal of Human Computer Interaction, 11*(2), 109–135.

Eklof, M., Gustafsson, E., Habberg, M., & Thomee, S. (2005). ICT use and the prevalence of upper extremity symptoms and psychological symptoms among young adults in a prospective cohort study. In *Proceedings of the 11th International Conference on Human–Computer Interaction: Vol. I. Las Vegas, July 22–27, 2005* [Electronic publication]. Mahwah, NJ: Lawrence Erlbaum Associates.

Elias, R., & Cail, F. (1983). Constraints et astreints devant les terminaux a ecran cathodique. (1109). Securite, Paris. Institut National de Recherche et de

Gamm, S., & Haeb-Umback (1995). User interface design of voice controlled consumer electronics. *Philips Journal of Research, 49*(4).

Ghiringhelli, L. (1980). Collection of subjective opinions on use of VDUs. In E. Grandjean (Ed.), *Ergonomical and medical aspects of cathode ray tube displays*. Zurich, Switzerland: Federal Institute of Technology.

Grandjean, E. (1979). *Ergonomical and medical aspects of cathode ray tube displays*. Zurich, Switzerland: Federal Institute of Technology.

Grandjean, E. (1984). Postural problems at office machine work stations. In E. Grandjean (Ed.), *Ergonomics and health in modern offices* (pp. 445–455). London: Taylor & Francis, Ltd.

Grandjean, E. (1987). Design of VDT workstations. In G. Salvendy (Ed.), *Handbook of human factors* (pp. 1359–1397). New York: John Wiley & Sons.

Grandjean, E., Hünting, W., & Nishiyama, K. (1982). Preferred VDT workstation settings, body posture and physical impairments. *Journal of Human Ergology, 11*(1), 45–53.

Grandjean, E., Hünting, W., & Nishiyama, K. (1984). Preferred VDT workstation settings, body posture and physical impairments. *Applied Ergonomics, 15*(2), 99–104.

Grandjean, E., Hünting, W., & Pidermann, M. (1983). VDT workstation design: Preferred settings and their effects. *Human Factors, 25*(2), 161–175.

Grandjean, E., Nishiyama, K., Hünting, W., & Pidermann, M. (1982). A laboratory study on preferred and imposed settings of a VDT workstation. *Behaviour and Information Technology, 1*, 289–304.

Green, R. A., & Briggs, C. A. (1989). Effect of overuse injury and the importance of training on the use of adjustable workstations by keyboard operators. *Journal of Occupational Medicine, 31*, 557–562.

Gunnarsson, E., & Ostberg, O. (1977). *Physical and emotional job environment in a terminal-based data system* (1977:35): Department of Occupational Safety, Occupational Medical Division, Section for Physical Occupational Hygiene.

Hagberg, M., Silverstein, B., Wells, R., Smith, M. J., Hendrick, H., Carayon, P., et al. (1995). *Work related musculoskeletal disorders (WRMSDs): A reference book for prevention*. London: Taylor & Francis, Ltd.

Haggerty, B., & Tarasewich, P. (2005). A new stylus-based method for text entry on small devices. In *Proceedings of the 11th International Conference on Human–Computer Interaction: Vol. 4. Las Vegas, July 22–27, 2005* [Electronic publication]. Mahwah, NJ: Lawrence Erlbaum Associates.

HCII2005. (2005). *Proceedings of the 11th International Conference on Human–Computer Interaction, Las Vegas, July 2–27,*

2005 [Electronic publication]. Mahwah, NJ: Lawrence Erlbaum Associates.

Heiden, G. H. van der, Braeuninger, U., & Grandjean, E., (1984). Ergonomic studies on computer aided design. In E. Grandjean (Ed.), *Ergonomic and health aspects in modern offices*. London: Taylor & Francis, Ltd.

Hendrick, H. (1986). Macroergonomics: A conceptual model for integrating human factors with organizational design. In O. Brown & H. Hendrick (Eds.), *Human factors in organizational design and management* (pp. 467–477). Amsterdam: Elsevier Science Publishers.

Hinckley, K. (2003). Input technologies and techniques. In J. A. Jacko & A. Sears (Eds.), *The human–computer interaction handbook* (pp. 151–168). Mahwah, NJ: Lawrence Erlbaum Associates.

Hirose, M., & Hirota, K. (2005). PUI (Perceptual User Interface) In *Proceedings of the 11th International Conference on Human–Compter Interaction: Vol. 5, Las Vegas, July 22–27, 2005* [Electronic publication]. Mahwah, NJ: Lawrence Erlbaum Associates.

Hultgren, G., & Knave, B. (1973). Contrast blinding and reflection disturbances in the office environment with display terminals. *Arbete Och Halsa*.

Hünting, W., Läubli, T., & Grandjean, E. (1981). Postural and visual loads at VDT workplaces: I. Constrained postures. *Ergonomics, 24*(12), 917–931.

Karat, C.-M., Vergo, J., & D. Nahamoo. (2003). Conversational interface technologies. In J. A. Jacko & A. Sears (Eds.), *The human–computer interaction handbook* (pp. 169–186). Mahwah, NJ: Lawrence Erlbaum Associates.

Läubli, T., Hünting, W., & Grandjean, E. (1981). Postural and visual loads at VDT workplaces: II. Lighting conditions and visual impairments. *Ergonomics, 24*(12), 933–944.

Lawler, E. E. (1986). *High-involvement management*. San Francisco, CA: Jossey-Bass Publishers.

Mandal, A. C. (1982). The correct height of school furniture. *Human Factors, 24*(3), 257–269.

Myers, B. A., & Wobbrock, J. O. (2005). Text input to handheld devices for people with physical disabilities. In *Proceedings of the 11th International Conference on Human–Computer Interaction: Vol. 4. Las Vegas, July 22–27, 2005* [Electronic publication]. Mahwah, NJ: Lawrence Erlbaum Associates.

Nachemson, A., & Elfstrom, G. (1970). Intravital dynamic pressure measurements in lumbar discs. *Scandinavian Journal of Rehabilitation Medicine*, (Suppl. 1).

NAS. (1983). *Video terminals, work and vision*. Washington, DC: National Academy Press.

Noro, K. (1991). Concepts, methods and people. In K. Noro & A. Imada (Eds.), *Participatory ergonomics* (pp. 3–29). London: Taylor & Francis, Ltd.

Ostberg, O. (1975, November/December). Health problems for operators working with CRT displays. *International Journal of Occupational Health and Safety, 6*, 24–52.

Saito, S., Piccoli, B., Smith, M. J., Sotoyama, M., Sweitzer, G., Villanuela, M. B. G., & Yoshitake, R. (2000). Ergonomics guidelines for using notebook personal computers. *Industrial Health, 38*, 421–434.

Smith, M. J. (1984). Health issues in VDT work. In J. Bennet, D. Case, J. Sandlin, & M. J. Smith (Eds.), *Visual display terminals* (pp. 193–228). Englewood Cliffs, NJ: Prentice Hall.

Smith, M. J. (1987). Mental and physical strain at VDT workstations. *Behaviour and Information Technology, 6*(3), 243–255.

Smith, M. J. (1997). Psychosocial aspects of working with video display terminals (VDT's) and employee physical and mental health. *Ergonomics, 40*(10), 1002–1015.

Smith, M. J., & Carayon, P. C. (1995). New technology, automation and work organization: Stress problems and improved technology implementation strategies. *International Journal of Human Factors in Manufacturing, 5*, 99–116.

Smith, M. J., & Sainfort, P. C. (1989). A balance theory of job design for stress reduction. *International Journal of Industrial Ergonomics, 4*, 67–79.

Smith, M. J., & Cohen, W. J. (1997). Design of computer terminal workstations. In G. Salvendy (Ed.), *Handbook of human factors and ergonomics* (2nd ed., pp. 1637–1688). New York: John Wiley & Sons, Inc.

Smith, M. J., & Derjani-Bayeh, A. (2003). Do ergonomic improvements increase computer workers' productivity?: An intervention study in a call center. *Ergonomics, 46*(1), 3–18.

Smith, M. J., Carayon, P., & Cohen, W. (2003). Design of computer workstations. In J. Jacko & A. Sears (Eds.), *The human–computer interaction handbook* (pp. 384–395). Mahwah, NJ: Lawrence Erlbaum Associates.

Stammerjohn, L. W., Smith, M. J., & Cohen, B. G. F. (1981). Evaluation of workstation design factors in VDT operations. *Human Factors, 23*(4), 401–412.

AUTHOR INDEX

Page references followed by *f* indicate figures.
Page numbers followed by *t* indicate tables.
Page numbers followed by *n* indicate footnotes.
Italicized page numbers indicate references.

SUBJECT INDEX

Printed and bound by CPI Group (UK) Ltd, Croydon, CR0 4YY

21/10/2024

01777040-0020